# Methods in Molecular Biology

*Series Editor*
**John M. Walker**
**School of Life and Medical Sciences**
**University of Hertfordshire**
**Hatfield, Hertfordshire, UK**

For further volumes:
http://www.springer.com/series/7651

For over 35 years, biological scientists have come to rely on the research protocols and methodologies in the critically acclaimed *Methods in Molecular Biology* series. The series was the first to introduce the step-by-step protocols approach that has become the standard in all biomedical protocol publishing. Each protocol is provided in readily-reproducible step-by-step fashion, opening with an introductory overview, a list of the materials and reagents needed to complete the experiment, and followed by a detailed procedure that is supported with a helpful notes section offering tips and tricks of the trade as well as troubleshooting advice. These hallmark features were introduced by series editor Dr. John Walker and constitute the key ingredient in each and every volume of the *Methods in Molecular Biology* series. Tested and trusted, comprehensive and reliable, all protocols from the series are indexed in PubMed.

# Mass Spectrometry of Glycoproteins

## Methods and Protocols

Edited by

**Arnaud Delobel**

*Quality Assistance S.A., Donstiennes, Belgium*

Humana Press

*Editor*
Arnaud Delobel
Quality Assistance S.A.
Donstiennes, Belgium

ISSN 1064-3745 ISSN 1940-6029 (electronic)
Methods in Molecular Biology
ISBN 978-1-0716-1240-8 ISBN 978-1-0716-1241-5 (eBook)
https://doi.org/10.1007/978-1-0716-1241-5



This Humana imprint is published by the registered company Springer Science+Business Media, LLC, part of Springer Nature.
The registered company address is: 1 New York Plaza, New York, NY 10004, U.S.A.

# Preface

Glycosylation is one of the most important post-translational modifications of proteins. It is estimated that half of all proteins are glycosylated. Glycosylation has a critical biological role, as it is involved in many functions such as cell signaling, cell–cell contact, innate immune response, protein stability, or host-pathogen recognition. For therapeutic proteins, glycosylation is also considered a critical quality attribute, as it can affect the safety and efficacy of the drug.

This post-translational modification adds complexity to proteins that are usually already quite heterogeneous. To achieve a complete characterization of glycoproteins, state-of-the-art analytical techniques are therefore required. Over the past decade, mass spectrometry has become the method of choice for the analysis of glycoproteins, as it allows a multi-level characterization, from the intact protein to the building blocks of the glycans: the monosaccharides.

The aim of this book is to present methods that can be used for the analysis of glycoproteins at different levels (intact, subunit, glycopeptide, glycan, monosaccharide), to solve most analytical challenges that a scientist working on glycoproteins may face.

The book is divided into 26 chapters. The first chapter is a short review dedicated to therapeutic glycoproteins, which aims to highlight the role of glycosylation on their properties. In chapters 2–20, the reader will find different analytical methods to characterize glycosylation, from the intact protein to the glycan level, for both N-linked and O-linked glycoproteins. Several of these methods were designed for the analysis of therapeutic glycoproteins but can be easily adapted to any glycoprotein. Chapter 21 describes a mass spectrometry imaging methodology for glycosylation analysis in tissues. Chapters 22 and 23 propose two approaches to characterizing glycosylation in cultured cells. As scientists, it is crucial to remain open-minded and to recognize that mass spectrometry can have limitations. Chapters 24 and 25 focus on two alternative methods that can be extremely useful in some cases, especially for highly glycosylated proteins: SEC/MALS and FT-IR. Finally, Chapter 26 is dedicated to a new aspect of modern analytical chemistry: the use of cloud computing to deploy mass spectrometry data analysis.

These chapters differ from conventional articles, because primary emphasis is placed on providing reliable procedures for users. Special attention is given to important experimental data, and practical hints in the "Notes" section enable the reader to adapt these procedures to one's specific problems.

I sincerely hope that the readers enjoy the information provided in this book and find its content interesting and stimulating. Even if it is never possible to be exhaustive on the subject, I also hope that I have prepared a successful compilation of chapters within the exciting field of MS of glycoproteins.

Eight-seven authors from 30 research laboratories all over the world have contributed to *Mass Spectrometry of Glycoproteins: Methods and Protocols*. I want to express my thanks to all the authors and coauthors for making their expertise and knowledge available to those who are not already versed in this area.

I would especially like to thank Dr. John Walker, the series editor, for his invitation to edit this volume of *Methods in Molecular Biology* and for his enthusiasm and his support.

I dedicate the book to my wife Aurélie and to my children Lauric, Eden, and Illan.

***Donstiennes, Belgium*** ***Arnaud Delobel***

# Contents

# Contributors


JESÚS S. AGUILAR DÍAZ DE LEÓN • *School of Molecular Sciences, The Biodesign Institute—Center for Personalized Diagnostics, Arizona State University, Tempe, AZ, USA*

PEGGY M. ANGEL • *Department of Cell and Molecular Pharmacology & Experimental Therapeutics, Medical University of South Carolina, Charleston, SC, USA*

HYUN JOO AN • *Asia-Pacific Glycomics Reference Site, Chungnam National University, Daejeon, South Korea; Graduate School of Analytical Science and Technology, Chungnam National University, Daejeon, South Korea*

PARASTOO AZADI • *Complex Carbohydrate Research Center, University of Georgia, Athens, GA, USA*

ALIREZA BANAZADEH • *Department of Chemistry and Biochemistry, Texas Tech University, Lubbock, TX, USA*

ALAIN BECK • *Pierre Fabre Laboratories, IRPF—Centre d'Immunologie Pierre-Fabre (CIPF), Saint-Julien-en-Genevois, France*

FERNANDO BENAVENTE • *Department of Chemical Engineering and Analytical Chemistry, Institute for Research on Nutrition and Food Safety (INSA·UB), University of Barcelona, Barcelona, Spain*

ABEL BERMUDEZ • *Canary Center at Stanford for Cancer Early Detection, Department of Radiology, Stanford University School of Medicine, Palo Alto, CA, USA*

MICHAEL BIACCHI • *Laboratoire de Spectrométrie de Masse des Interactions et des Systèmes (LSMIS), UMR 7140 (Unistra-CNRS), Université de Strasbourg, Strasbourg, France*

IAN BLACK • *Complex Carbohydrate Research Center, University of Georgia, Athens, GA, USA*

JEAN-FRANÇOIS BOE • *SGS M-Scan SA, Plan-les-Ouates, Switzerland*

CHAD R. BORGES • *School of Molecular Sciences, The Biodesign Institute—Center for Personalized Diagnostics, Arizona State University, Tempe, AZ, USA*

CLAIRE I. BUTRÉ • *Quality Assistance sa, Donstiennes, Belgium*

KEVIN CANIS • *SGS M-Scan SA, Plan-les-Ouates, Switzerland*

FABRICE CANTAIS • *Quality Assistance sa, Donstiennes, Belgium*

RUSSELL W. CARLSON • *Complex Carbohydrate Research Center, University of Georgia, Athens, GA, USA*

STÉPHANE CHENU • *Pierre Fabre Laboratories, IRPF—Centre d'Immunologie Pierre-Fabre (CIPF), Saint-Julien-en-Genevois, France*

BYEONG GWAN CHO • *Department of Chemistry and Biochemistry, Texas Tech University, Lubbock, TX, USA*

SARAH CIANFERANI • *Laboratoire de Spectrométrie de Masse BioOrganique, IPHC UMR 7178, Université de Strasbourg, CNRS, Strasbourg, France*

SAM CLAWSON • *Covance Laboratories Ltd, Harrogate, UK*

OLIVIER COLAS • *Pierre Fabre Laboratories, IRPF—Centre d'Immunologie Pierre-Fabre (CIPF), Saint-Julien-en-Genevois, France*

BEN COWPER • *National Institute for Biological standards and Control, Hertfordshire, UK*

VALENTINA D'ATRI • *School of Pharmaceutical Sciences, University of Geneva, CMU, Geneva, Switzerland; Institute of Pharmaceutical Sciences of Western Switzerland (ISPSO), University of Geneva, CMU, Geneva, Switzerland*

ARNAUD DELOBEL • *Quality Assistance S.A., Donstiennes, Belgium*
CÉDRIC DELPORTE • *Pharmacognosy, Bioanalysis and Drug Discovery and Analytical Platform, RD3 Department, Faculty of Pharmacy, Université libre de Bruxelles, Brussels, Belgium*
ALLISON DERENNE • *Center for Structural Biology and Bioinformatics, Laboratory for the Structure and Function of Biological Membranes, Université Libre de Bruxelles, Brussels, Belgium*
KHEIRO-MOUNA DERFOUFI • *Center for Structural Biology and Bioinformatics, Laboratory for the Structure and Function of Biological Membranes, Université Libre de Bruxelles, Brussels, Belgium*
RICHARD R. DRAKE • *Department of Cell and Molecular Pharmacology & Experimental Therapeutics, Medical University of South Carolina, Charleston, SC, USA*
JAMES DUFFY • *Covance Laboratories Ltd, Harrogate, UK*
DAVID FIRTH • *Covance Laboratories Ltd, Harrogate, UK*
YANNIS-NICOLAS FRANÇOIS • *Laboratoire de Spectrométrie de Masse des Interactions et des Systèmes (LSMIS), UMR 7140 (Unistra-CNRS), Université de Strasbourg, Strasbourg, France*
ESTELLE GARÉNAUX • *SGS M-Scan SA, Plan-les-Ouates, Switzerland*
SAKSHI GAUTAM • *Department of Chemistry and Biochemistry, Texas Tech University, Lubbock, TX, USA*
ESTELA GIMÉNEZ • *Department of Chemical Engineering and Analytical Chemistry, Institute for Research on Nutrition and Food Safety (INSA·UB), University of Barcelona, Barcelona, Spain*
JÉRÉMIE GIORGETTI • *Laboratoire de Spectrométrie de Masse des Interactions et des Systèmes (LSMIS), UMR 7140 (Unistra-CNRS), Université de Strasbourg, Strasbourg, France*
MONA GOLI • *Department of Chemistry and Biochemistry, Texas Tech University, Lubbock, TX, USA*
ERIK GOORMAGHTIGH • *Center for Structural Biology and Bioinformatics, Laboratory for the Structure and Function of Biological Membranes, Université Libre de Bruxelles, Brussels, Belgium*
GRACE GRIMSLEY • *Department of Cell and Molecular Pharmacology, Medical University of South Carolina, Charleston, SC, USA*
DAVY GUILLARME • *School of Pharmaceutical Sciences, University of Geneva, CMU, Geneva, Switzerland; Institute of Pharmaceutical Sciences of Western Switzerland (ISPSO), University of Geneva, CMU, Geneva, Switzerland*
NORITAKA HASHII • *Division of Biological Chemistry and Biologicals, National Institute of Health Sciences, Kanagawa, Japan*
KATHRYN HASTIE • *La Jolla Institute for Immunology, La Jolla, CA, USA*
CHRISTIAN HEISS • *Complex Carbohydrate Research Center, University of Georgia, Athens, GA, USA*
YOSRA HELALI • *Pharmacognosy, Bioanalysis, and Drug Discovery and Analytical Platform, RD3 Department, Faculty of Pharmacy, Université libre de Bruxelles, Brussels, Belgium*
AHMED HUSSEIN • *Department of Biotechnology, Institute of Graduate Studies and Research, University of Alexandria, Alexandria, Egypt*
PERNILLE FOGED JENSEN • *Symphogen A/S, Ballerup, Denmark*
JONATHAN E. KATZ • *Lawrence J. Ellison Institute for Transformative Medicine of USC, Los Angeles, CA, USA; Department of Medicine/Oncology, Keck School of Medicine of USC, Los Angeles, CA, USA*

UNYONG KIM • *Biocomplet Inc., Seoul, South Korea*
SONJA KRIEGER • *Agilent Technologies, Waldbronn, Germany*
DAN BACH KRISTENSEN • *Symphogen A/S, Ballerup, Denmark*
ERIC LARGY • *University of Bordeaux, INSERM and CNRS, Laboratoires Acides Nucléiques: Régulations Naturelle et Artificielle (ARNA, U1212, UMR5320), IECB, Pessac, France*
ANTONY LECHNER • *Laboratoire de Spectrométrie de Masse des Interactions et des Systèmes (LSMIS), UMR 7140 (Unistra-CNRS), Université de Strasbourg, Strasbourg, France*
EMMANUELLE LEIZE-WAGNER • *Laboratoire de Spectrométrie de Masse des Interactions et des Systèmes (LSMIS), UMR 7140 (Unistra-CNRS), Université de Strasbourg, Strasbourg, France*
XIAOWEI LU • *Department of Cell and Molecular Pharmacology, Medical University of South Carolina, Charleston, SC, USA*
PAULA MAGNELLI • *New England Biolabs, Ipswich, MA, USA*
MONTSERRAT MANCERA-ARTEU • *Department of Chemical Engineering and Analytical Chemistry, Institute for Research on Nutrition and Food Safety (INSA·UB), University of Barcelona, Barcelona, Spain*
ARNAUD MARCHANT • *Institute for Medical Immunology, Faculty of Medicine, Université libre de Bruxelles, Brussels, Belgium*
ELIZABETH MCLEOD • *New England Biolabs, Ipswich, MA, USA*
YEHIA MECHREF • *Department of Chemistry and Biochemistry, Texas Tech University, Lubbock, TX, USA*
ANAND S. MEHTA • *Department of Cell and Molecular Pharmacology & Experimental Therapeutics, Medical University of South Carolina, Charleston, SC, USA*
EDGAR NAEGELE • *Agilent Technologies, Waldbronn, Germany*
ANDREAS NÄGELI • *Genovis AB, Lund, Sweden*
MARIA NORDGREN • *Genovis AB, Lund, Sweden*
KIM NORRIS-CANEDA • *Department of Cell and Molecular Pharmacology, Medical University of South Carolina, Charleston, SC, USA*
HELÉN NYHLÉN • *Genovis AB, Lund, Sweden*
MYUNG JIN OH • *Asia-Pacific Glycomics Reference Site, Chungnam National University, Daejeon, South Korea; Graduate School of Analytical Science and Technology, Chungnam National University, Daejeon, South Korea*
MARTIN ØRGAARD • *Symphogen A/S, Ballerup, Denmark*
MITUL PATEL • *GlaxoSmithKline, Stevenage, UK*
WEIJING PENG • *Department of Chemistry and Biochemistry, Texas Tech University, Lubbock, TX, USA*
SHARON J. PITTERI • *Canary Center at Stanford for Cancer Early Detection, Department of Radiology, Stanford University School of Medicine, Palo Alto, CA, USA*
VAMSEEDHAR RAYAPROLU • *La Jolla Institute for Immunology, La Jolla, CA, USA*
JOSIANE SAADÉ • *Laboratoire de Spectrométrie de Masse des Interactions et des Systèmes (LSMIS), UMR 7140 (Unistra-CNRS), Université de Strasbourg, Strasbourg, France*
VICTORIA SANZ-NEBOT • *Department of Chemical Engineering and Analytical Chemistry, Institute for Research on Nutrition and Food Safety (INSA·UB), University of Barcelona, Barcelona, Spain*
ERICA OLLMANN SAPHIRE • *La Jolla Institute for Immunology, La Jolla, CA, USA*
SONJA SCHNEIDER • *Agilent Technologies, Waldbronn, Germany*

YOUNGSUK SEO • *Asia-Pacific Glycomics Reference Site, Chungnam National University, Daejeon, South Korea; Graduate School of Analytical Science and Technology, Chungnam National University, Daejeon, South Korea*
SHILPEE SHARMA • *Institute for Medical Immunology, Faculty of Medicine, Université libre de Bruxelles, Brussels, Belgium*
XIAOFENG SHI • *New England Biolabs, Ipswich, MA, USA*
JONATHAN SJÖGREN • *Genovis AB, Lund, Sweden*
TRINE MEIBORG SLOTH • *Symphogen A/S, Ballerup, Denmark*
JUNYA SUZUKI • *Division of Biological Chemistry and Biologicals, National Institute of Health Sciences, Kanagawa, Japan*
YURY O. TSYBIN • *Spectroswiss Sarl, Lausanne, Switzerland*
ROSIE UPTON • *Covance Laboratories Ltd, Harrogate, UK*
PIERRE VAN ANTWERPEN • *Pharmacognosy, Bioanalysis and Drug Discovery and Analytical Platform, RD3 Department, Faculty of Pharmacy, Université libre de Bruxelles, Brussels, Belgium*
MARIE VANDEPUT • *Pharmacognosy, Bioanalysis, and Drug Discovery and Analytical Platform, RD3 Department, Faculty of Pharmacy, Université libre de Bruxelles, Brussels, Belgium*
VITHIYA VIMALRAJ • *GlaxoSmithKline, Stevenage, UK*
ELSA WAGNER-ROUSSET • *Pierre Fabre Laboratories, IRPF—Centre d'Immunologie Pierre-Fabre (CIPF), Saint-Julien-en-Genevois, France*
DANSALA WELBA • *Institute for Medical Immunology, Faculty of Medicine, Université libre de Bruxelles, Brussels, Belgium*
CONNOR A. WEST • *Department of Cell and Molecular Pharmacology, Medical University of South Carolina, Charleston, SC, USA*
XIMO ZHANG • *Waters Corporation, Milford, MA, USA*
JINGFU ZHAO • *Department of Chemistry and Biochemistry, Texas Tech University, Lubbock, TX, USA*

# Chapter 1

# Glycosylation of Therapeutic Proteins: A Critical Quality Attribute

## Arnaud Delobel

## Abstract

Glycosylation is a common posttranslational modification of therapeutic proteins. The glycosylation pattern is dependent on many parameters such as the host cell line or the culture conditions. N- and O-linked glycans usually play a great role on the stability, safety, and efficacy of the drug. For this reason, glycosylation is considered as a critical quality attribute of therapeutic glycoproteins, and a thorough characterization should be performed, as well as a systematic control for each batch produced. This chapter gives a short presentation of the structure of glycans commonly found on recombinant therapeutic proteins, and their role on the properties of the drug, in terms of stability, pharmacokinetics, safety, and efficacy. Lastly, the use of mass spectrometry for the analysis of glycoproteins is briefly described.

**Key words** Glycoprotein, Monoclonal antibody, Glycosylation, Pharmacokinetics, Immunogenicity, Critical quality attribute, Mass spectrometry

## 1 Introduction

### *1.1 Glycoproteins as Therapeutics*

Therapeutic proteins are among the most promising drugs currently under development. They are used for many different indications, such as cancer, autoimmunity and inflammation, genetic disorders, or infections. In 2018, the market share of biopharmaceuticals was estimated to 28% of the total market, and to 53% when only Top 100 pharma companies are considered [1]. This figure keeps increasing year after year. The global demand for biopharmaceuticals by product is led by monoclonal antibodies (mAbs), estimated at US$ 82.3 billion (33.2% share) in 2018, and is also expected to grow with a 10.8% CAGR (compound annual growth rate) between 2018 and 2025. The global market for biopharmaceuticals in 2019 is expected to reach US$ 269.3 billion [2].

Glycosylation is one of the most common posttranslational modification of proteins in nature. It is estimated that more than half of all proteins found in nature are glycoproteins [3]. The share of glycoproteins among therapeutic proteins is even higher. Among

Arnaud Delobel (ed.), *Mass Spectrometry of Glycoproteins: Methods and Protocols*, Methods in Molecular Biology, vol. 2271, https://doi.org/10.1007/978-1-0716-1241-5_1, 

these are monoclonal antibodies, that usually contain (at least) two sites of *N*-glycosylation [4]. Monoclonal antibodies (and related compounds such as bispecific antibodies, antibody–drug conjugates, or Fc-fusion proteins) represent today most of the therapeutic proteins under development. Among the 10 top-selling drugs in 2018, seven were glycosylated proteins [5], all being monoclonal antibodies or related products (such as Fc-fusion proteins).

In most cases, the glycosylation pattern of the drug can have a significant impact on the efficacy and safety of the drug. Therefore, glycosylation is commonly considered as a critical quality attribute (CQA) of glycoproteins, and more specifically of monoclonal antibodies [6–8]. Regulatory authorities have issued guidelines that require the characterization of the glycosylation using state-of-the-art techniques, and the control of this quality attribute for each batch produced [9–11].

## 1.2 Structure of Glycosylated Proteins

### 1.2.1 Monosaccharides, the Building Blocks of Glycans

Glycans found on therapeutic glycoproteins are oligomers of monosaccharides [12], linked together via glycosidic bonds. The different monosaccharides commonly found on therapeutic glycoproteins are presented in Fig. 1. Different types of monosaccharides are present: neutral sugars such as glucose, mannose, galactose, fucose, and xylose (found only in proteins expressed in plant cells); amino sugars such as *N*-acetylglucosamine and *N*-acetylgalactosamine; and sialic acids such as *N*-acetylneuraminic acid and *N*-glycolylneuraminic acid (not found in human glycoproteins).

The nomenclature of the linkage is presented in Fig. 2. The two numbers correspond to the carbons involved in the linkage, and the terms α and β correspond to the stereochemistry on the anomeric carbon: if the oxygen is axial/down, the carbon is α, if the oxygen is equatorial/up, then the carbon is β.

### 1.2.2 *N*-Glycans

*N*-glycosylation is probably the most important type of glycosylation as regards therapeutic proteins, as most of them contain only N-glycosylation. *N*-glycans are linked to asparagine residues via a glycosidic bond involving the nitrogen atom of the side chain (Fig. 3).

Not all asparagine can be *N*-glycosylated: they must be part of a so-called "consensus sequence", consisting of an asparagine residue followed by any amino acid (except proline) and a serine or threonine (N-X-S/T). But even in the presence of this sequon, not all asparagine residues may be glycosylated. They can either be glycosylated, nonglycosylated, or partially glycosylated, which adds heterogeneity to the protein.

*N*-glycans of therapeutic proteins share a common core sequence: Manα1-3(Manα1-6)Manβ1-4GlcNAcβ1–4GlcNAcβ1–Asn-X-Ser/Thr.

They can be classified into three types: (1) oligomannose (also known as "high mannose"), in which only mannose residues are added to the core structure; (2) complex types, for which core

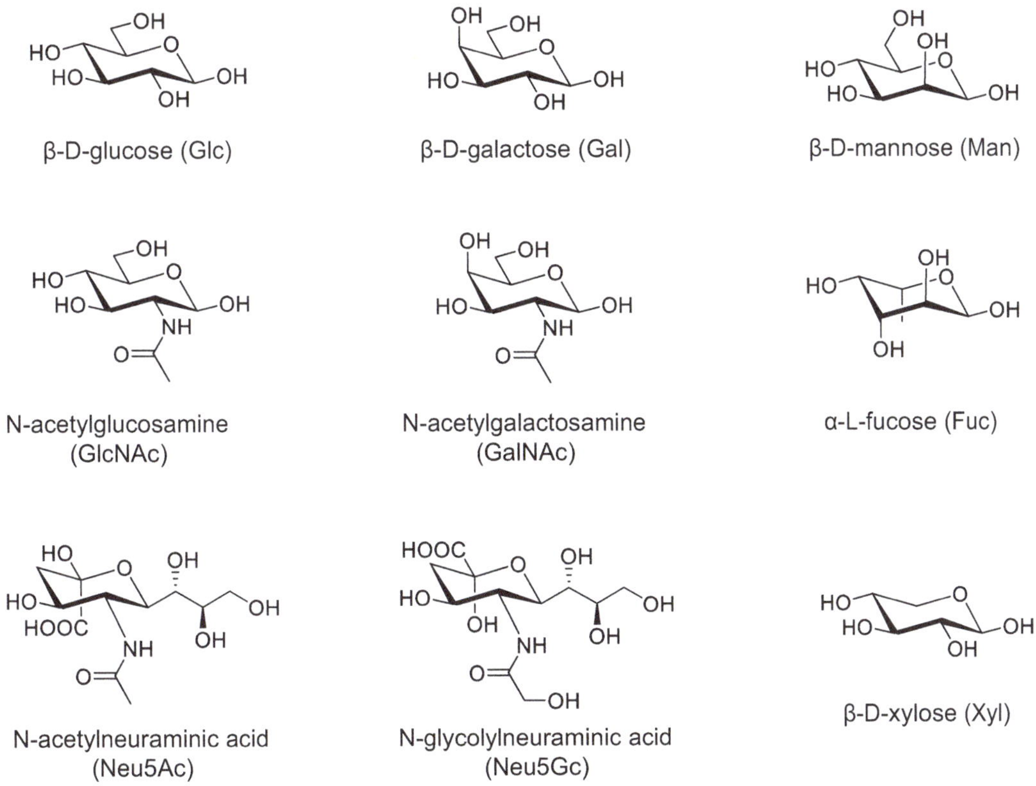


**Fig. 1** Structure of monosaccharides commonly found in recombinant glycoproteins

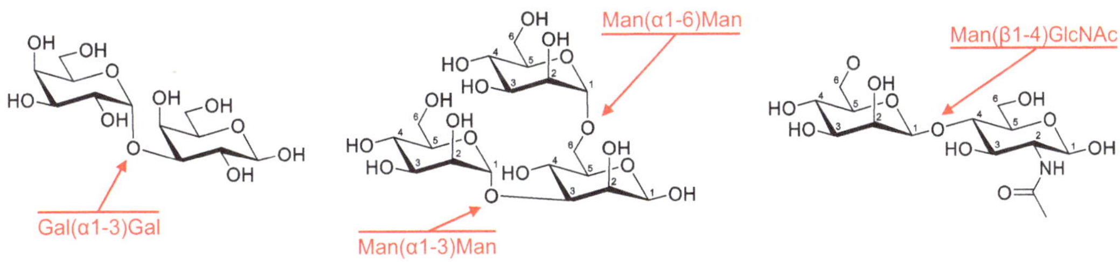


**Fig. 2** Nomenclature of glycosidic linkages

GlcNAc residues will be extended by antennae; and (3) hybrid types, in which mannose residues extend the Manα1-6 arm of the core and one or two GlcNAcs extend the Manα1-3 arm (*see* Fig. 4).

As it will be discussed in Subheading 2, the structure of *N*-glycans will have an impact on the properties of the drug, including its safety and efficacy. A precise control of the *N*-glycosylation will therefore be mandatory for the therapeutic use of the glycoproteins. The analysis of *N*-glycosylation can be performed by using specific enzymes such as PNGases [13], that allow the cleavage of the oligosaccharide from the peptidic backbone. But many other endo- or exoglycosidases with various specificities can also be used for precise characterization of *N*-glycosylated proteins [14].

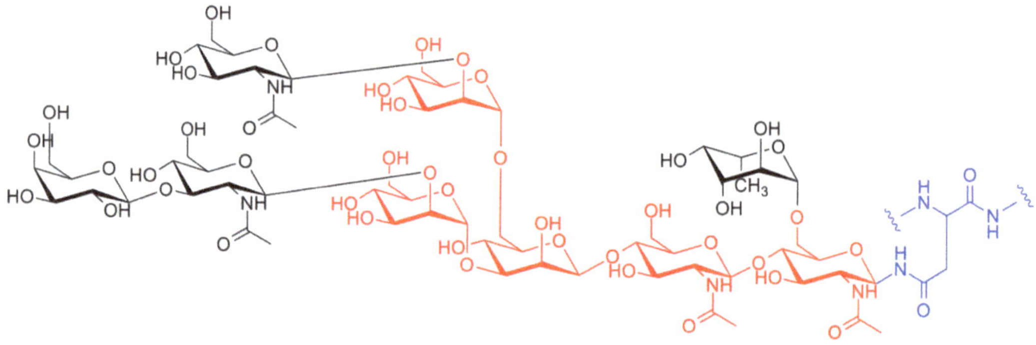

**Fig. 3** Chemical structure of N-linked glycans (asparagine residue in blue; core structure of the N-glycan in red)

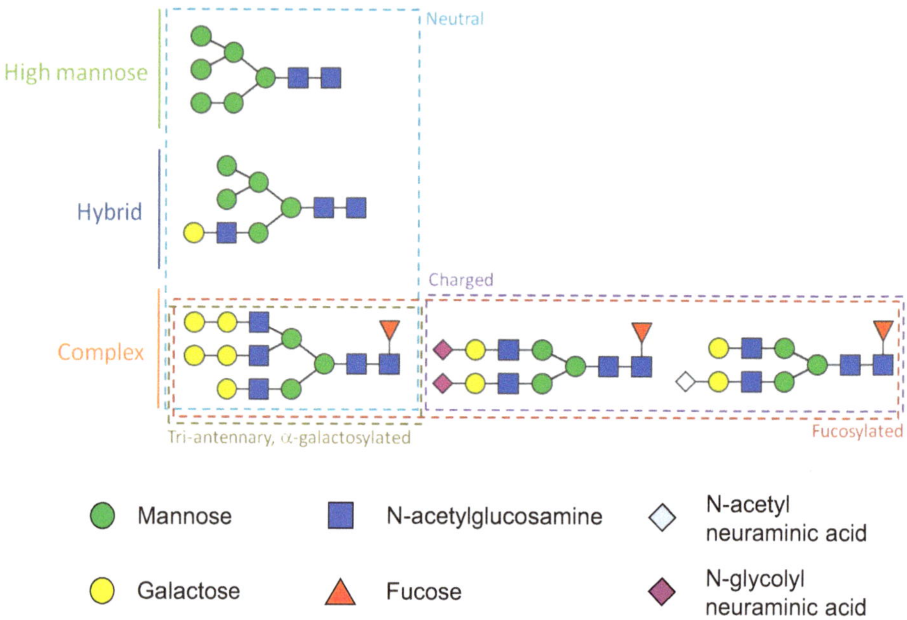


**Fig. 4** Different types of *N*-glycans commonly found in therapeutic glycoproteins

### 1.2.3 O-*Glycans*

*O*-glycosylation is less common in therapeutic glycoproteins, but it can be found in some of them such as Enbrel® (etanercept), a fusion protein composed of the Fc portion of an IgG1 monoclonal antibody fused to TNFα receptor [15]. The linker between the Fc portion and the TNFα receptor is heavily *O*-glycosylated.

*O*-glycans are linked to the protein via the oxygen atom of a serine or threonine residue, as shown in Fig. 5, which shows an example of mucin-type Core 1 glycan containing two *N*-acetyl-neuraminic acids. Unlike *N*-glycosylation, there is no consensus sequence for *O*-glycosylation, and virtually any serine or threonine residue could be *O*-glycosylated. Heavily *O*-glycosylated proteins usually contain a cluster of serine and/or threonine residues that can be fully or partially modified.

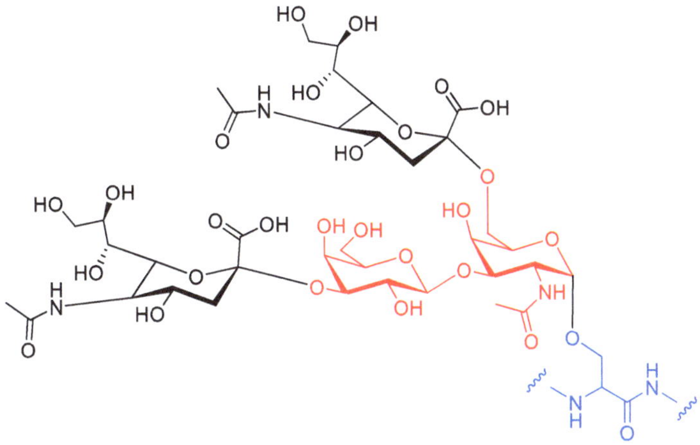

**Fig. 5** Chemical structure of a Core 1 *O*-glycan (in red) linked to a serine residue (in blue) and decorated with two *N*-acetylneuraminic acids (in black)

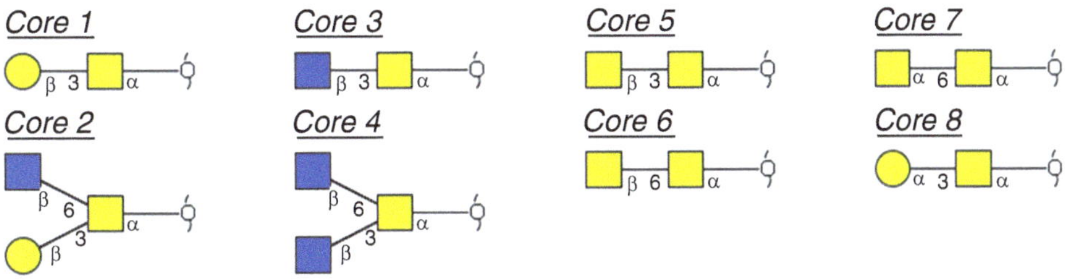


**Fig. 6** Core structures of mucin-type *O*-glycans

*O*-glycosylation is more complex than *N*-glycosylation, and different types of core structures are found. The most common modification is mucin-type *O*-glycosylation, in which a serine or a threonine is linked to *N*-acetylgalactosamine, giving rise to different core structures as presented in Fig. 6, the most common being Core 1 to Core 4. Sialic acids can then be linked to these core structures, leading to complex glycans, as shown in Fig. 7. *O*-fucosylation, *O*-glucosylation, and *O*-mannosylation can also be found [16].

Unlike *N*-glycosylation, there is no enzyme that can specifically cleave *O*-glycans, and chemical release is commonly used, which makes the characterization of *O*-glycosylation quite complex. However, new enzymes were recently developed for the characterization of *O*-glycosylation [17], that may facilitate this type of work in the future.

When choosing an expression system to produce a recombinant protein for further therapeutic use, its impact on glycosylation should be carefully studied. One will want to choose a system that is able to produce proteins with glycans that will positively affect the

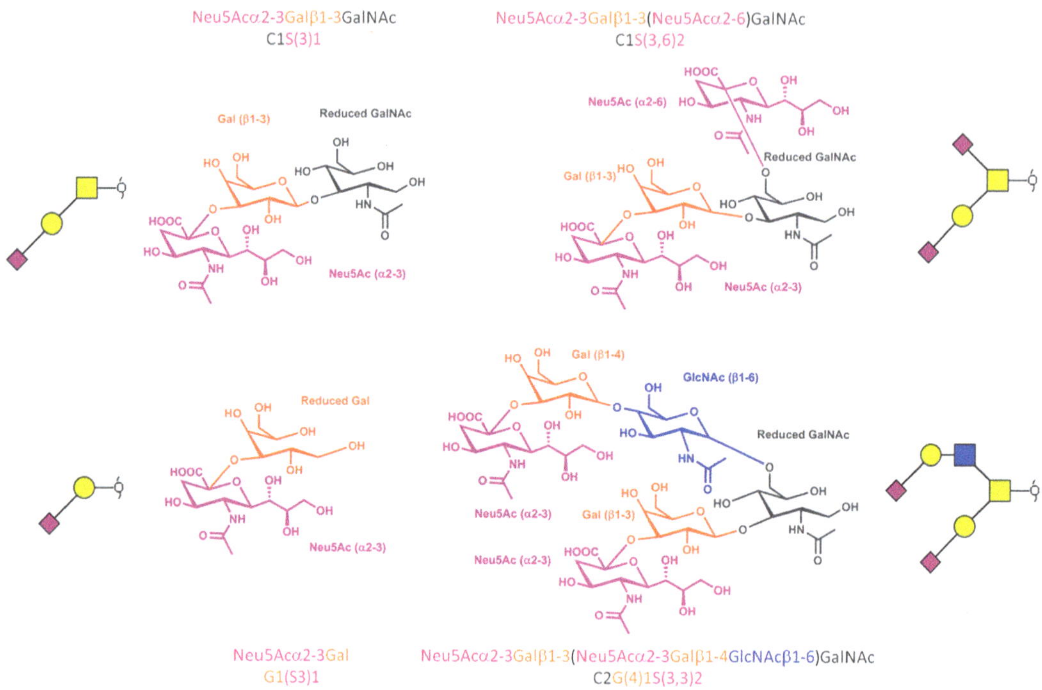


**Fig. 7** Examples of mucin-type *O*-glycans

### *1.3 Impact of the Expression System on the Glycosylation Profile*

potency of the drug, while avoiding safety issues to the patient. Therefore, glycosylation patterns as similar as possible to the ones of human proteins are preferred. Indeed, nonhuman cells may produce glycans that are not present in human proteins, or decorate proteins with glycans that contain chemical groups, monosaccharides or linkages between monosaccharides that are not present in human. Nonhuman glycosylation is one of the major causes of immunogenicity [18]. These glycoproteins may also be rapidly cleared from circulation, leading to a loss in efficacy (*see* Subheading 2).

The choice of the expression system will have a great impact on the glycosylation of the protein produced [19–21]. The repertoire of glycans that will be produced will mainly depend on the glycosyltransferase and glycosidase enzymes available, but also on the cell culture conditions. For example, yeast cells will mainly produce high-mannose structures [22], while insect [23] and plant [24] cell lines will produce unique glycans, different from those of mammalian cell lines. But even when mammalian cell lines are used, significant differences can be observed from one host to another [25].

In the recent years, glycoengineering of cell lines has been an important matter of research [26–29] with the aim of producing glycoproteins with optimal glycosylation, both in terms of safety and efficacy, with high production yields [30, 31]. Using such approaches, the first therapeutic protein produced in plant (taliglucerase alfa, Elelyso®) was approved by the US FDA in 2012 [32].

### 1.4 Impact of the Manufacturing Process on Protein Glycosylation

As shown in the previous section, the cell line used for the production of a recombinant protein will have a great impact on the glycosylation pattern. Even for a given cell line, using different cell culture conditions could make a difference. Therefore, optimizing these conditions may allow for fine tuning of the glycan structure, and then modify the physicochemical and biological properties of the drug. For example, high concentrations of ammonia have an impact on terminal glycosylation [33, 34], while pH can have an impact on galactosylation and sialylation microheterogeneity [35]. Reduced culture temperature [36] and the production method [37] (fed batch vs. perfusion) are also parameters that affect sialylation. Other parameters such as dissolved oxygen concentration should also be considered, but their effect is more variable and cell-specific [38].

### 1.5 Biosimilar Drugs

With the loss of patent protection for the first generation of biotherapeutics, biosimilars have been of major interest for biopharmaceutical companies over the last few years [39, 40]. At the end of 2019, the EMA had approved 56 biosimilar products via the European Union–wide authorization procedure, and the USA had only begun to catch up with the EU with 19 products approved since the first biosimilar approval in 2015.

The analytical similarity assessment is a critical step in the biosimilarity assessment, as it can potentially avoid the need for costly clinical studies, or at least limit the extent of these studies [41, 42]. Unsurprisingly, as glycosylation is a critical quality attribute of biotherapeutics, the characterization of glycans is of great importance in the demonstration of similarity. Therefore, the regulatory guidelines related to biosimilars highlight the importance of this characterization [43–45]. In order to detect minor differences in glycosylation patterns that may significantly impact the safety and/or efficacy of biosimilar drugs, state-of-the-art analytical methods are used, and the extent of the glycosylation characterization studies for biosimilars can be even higher than those of originators [46, 47]. However, recent studies showed a significant variability in originators products, without significant impact on biological activity [48, 49]. This may broaden the range of glycosylation pattern for biosimilars.

## 2 Impact of the Glycosylation Profile on the Properties of Therapeutic Proteins

Since the advent of biotherapeutics, the influence of glycosylation on the physical and biological properties of glycoproteins has been studied in detail. *N*-glycosylation was studied extensively [50], while the role of *O*-glycosylation has historically received less attention [51]. However, both types of glycosylation can have a great impact on properties such as stability, half-life in blood circulation, efficacy, propensity to aggregation, solubility, and immunogenicity. All these parameters are inextricably linked, and glycosylation should therefore be clearly considered as a critical quality attribute of therapeutic glycoproteins. In Subheadings 2.1–2.4, glycoproteins in general will be considered; the specific subject of monoclonal antibodies will be covered in Subheading 2.5.

### 2.1 Influence of Glycosylation on Solubility and Stability of Glycoproteins

The first impact of glycosylation is related to the physicochemical properties and the stability of the protein [52]. Glycans can interact with the peptidic backbone, and then modify the intrinsic properties of the protein. By masking hydrophobic amino acids, or regions that are prone to aggregation, glycosylation can improve the solubility of the protein and limit aggregation–oligomerization phenomena. For example, charged glycans such as sialic acids, will impact the charge of the protein, and then its stability. The stabilization will be dependent on the number of glycans, their length, their branching, and their charge. The negative charge due to the presence of sialic acids also confers resistance toward proteolytic cleavage, and consequently improve the in vivo stability. For example, it was demonstrated that sialylation had a positive impact on resistance to proteolysis for von Willebrand factor [53].

The stabilization of the native structure also has a positive influence on the protection against factors such as temperature and pH, as shown in studies on granulocyte-colony stimulating factor (G-CSF) and erythropoietin (EPO) [54, 55], but also chemical denaturation, for example induced by agents such as guanidine hydrochloride (GdnHCl) or sodium dodecyl sulfate (SDS) [55]. A study also demonstrated that EPO, whose activity is affected by oxidation, can be protected against oxidation when it is glycosylated [56]. A study on G-CSF showed that glycosylation protected the protein against intermolecular crosslinking via disulfide bridges [57]. Finally, glycosylation was shown to facilitate the folding of some proteins, such as EPO [58].

The improved stability conferred by glycosylation is very important for therapeutic proteins. As aggregation can induce immunogenicity reactions, limiting the aggregation phenomena can improve the safety of the drug. The stabilization of the native structure and the resistance against proteolytic cleavage will help maintaining the biological activity of the drug, and then its efficacy in the patient.

### 2.2 Influence of Glycosylation on Pharmacokinetics

The impact of glycosylation on pharmacokinetics (PK) is due notably to two types of endocytic receptors highly expressed on liver cells: mannose receptors (ManR) and asialoglycoprotein receptors (ASGPR) [59]. Mannose receptors will mainly bind terminal mannose, *N*-acetylglucosamine, and fucose residues, while asialoglycoprotein receptors will bind terminal galactose and *N*-acetylglucosamine residues, especially on highly branched glycans (such as tri- and tetraantennary glycans). The proteins that bind these receptors will be adsorbed from serum by receptor-mediated endocytosis to the lysosomes, where they will be degraded by proteases and glycosidases.

The presence of sialic acids is thus able to improve the serum half-life of glycoproteins, by masking galactose residues and avoiding the binding to ASGPR in hepatocytes. This was demonstrated for example for recombinant erythropoietin and PEGylated erythropoietin [60–62], for von Willebrand factor [53] and for recombinant Factor VIII [63]. For EPO, a linear relationship was demonstrated between the sialic acid content (especially on *N*-glycans) and the in vivo activity [64]. If serum half-life is increased, systemic exposure is also improved, which can improve the efficacy of the drug and avoid too frequent dosing for the patient and increase the quality of life. For this reason, the influence of glycosylation on the pharmacokinetics should be carefully studied during the design of a therapeutic glycoprotein.

Proteins containing nonhuman glycans can also be cleared more rapidly due to their binding to preexisting antibodies targeting these glycosylation patterns. The choice of the cell lines used to produce recombinant glycoproteins is critical to avoid these nonhuman glycans.

Finally, glycosylation and its impact on PK can be used to improve the stability of some drugs. In the recent years, small antibody fragments with two specificities ("diabodies") were developed for oncology indications [65]. Their structure that is simpler than the one of full antibodies has some advantages. They are indeed easier to produce and penetrate tumors more efficiently thanks to their small size. This small size is also a problem, as diabodies are cleared quite rapidly from serum. Stork et al. have shown that glycosylation of diabodies was a way to improve the pharmacokinetics of those constructs [66]. FSH (follicle-stimulating hormone) is also a nice example of how glycosylation can improve the properties of a protein. The clinical interest of FSH is limited due to its very rapid clearance. Many attempts to improve the PK properties were reported, the most interesting one being the development of corifollitropin alfa, a fusion product of human follicle-stimulating hormone and the C-terminal peptide of the β-subunit of human chorionic gonadotropin (hCG) [67]. In this fusion product, additional glycosylation sites are introduced, which improves greatly the pharmacokinetics of the product. Corifollitropin alfa is considered as a "long-acting FSH."

### 2.3 Influence of Glycosylation on Receptor Binding

As it will be discussed in Subheading 2.5, the influence of glycosylation on receptor binding is more critical for monoclonal antibodies. However, for some other recombinant proteins, glycosylation can also play a role. It was shown that EPO binding to its receptor was inversely proportional to its sialylation content: the more sialic acids are present on the protein, the lower the association constant [68]. The same phenomenon was observed on darpoetin alfa, a glycoengineered analog of human recombinant EPO with two additional *N*-glycans [69]. The glycoengineering allowed to improve half-life by a factor of three and improve the in vivo activity, but the binding to receptor was decreased. It was proved that this was due to sialic acids, and not to steric hindrance related to additional *N*-glycans. However, the loss in receptor binding was overcome by the improved pharmacokinetics properties [70].

### 2.4 Influence of Glycosylation on Immunogenicity

Glycosylation can, directly or indirectly, induce an immune response in the patient. When used as a drug, the glycoproteins should be carefully designed to avoid these adverse effects.

Nonhuman glycans could indeed be recognized by preexisting antibodies in the patient. For example, human serum contains antibodies against Gal-α-1,3-Gal motifs, that are not present in human glycoproteins due to the lack of the enzyme α1,3 galactosyltransferase in human. The presence of such glycoforms in therapeutic drugs could induce a fast clearance of the drug, and then a loss of efficacy, or an immune response that can in some cases be a life-threatening safety issue [71].

*N*-glycolylneuraminic acid (Neu5Gc) is a sialic acid found in nonhuman mammals. Host cells commonly used to produce therapeutic glycoproteins, such as CHO, NS0, or Sp2/0 cells, may produce glycans containing Neu5Gc. As most humans have preexisting antibodies against Neu5Gc, sometimes at high levels, therapeutic drugs with Neu5Gc-containing glycoforms may represent a safety issue. To overcome this problem, the use of specific cell culture conditions can limit the presence of this sialic acid [72, 73].

Plant and insect cells also produce glycans that are not found on human glycoproteins. Plant *N*-glycans can be oligomannose, complex, hybrid and paucimannose (short mannose structures) with the *N*-glycans often modified by a xylose residue [24]. The GlcNAc core of the *N*-glycan can also be modified by an α1,3 fucose. This core modification is known to be allergenic in humans. Generally, insect *N*-glycans are either high-mannose or paucimannose glycans and the *N*-glycans can also be modified at the GlcNAc core by an α1,3 fucose [23]. Unless genetically engineered, these cells are not used for therapeutic glycoproteins production, due to the risk of immune reactions in patients.

However, such immune reactions should be put in perspective: a recent study was conducted to evaluate the impact of plant glycans on immunogenicity in patients. It was concluded that there was a low prevalence of preexisting antibodies against plant glycans, and that the exposure to a therapeutic drug containing these glycans had no significant impact on the antibodies [74]. Moreover, the mechanism for immunogenicity can be quite complex, and for monoclonal antibodies, the presence of an immunogenic glycan will not have the same impact if it is present on a Fab or on a Fc glycan [75, 76].

In other cases, glycosylation can have a positive impact on immunogenicity. This is the case of recombinant human granulocyte-macrophage colony-stimulating factor (GM-CSF), that can be expressed in yeast, bacteria, or mammalian cells. GM-CSF expressed in *E. coli* was found to be immunogenic [77]. This immunogenicity comes from antibodies that react with an epitope of the protein that is masked by glycans in proteins expressed in other hosts that are capable of producing glycosylated proteins.

### 2.5 Glycosylation of Monoclonal Antibodies (mAbs) and Related Constructs

Although some aglycosylated antibodies are in clinical development [78], most mAbs are glycoproteins and the glycosylation is critical for their biological activity. Most of these products currently in clinical development have one glycosylation site on each heavy chain in the Fc region, but some of them, such as cetuximab [79] (approved for oncology indications), also contain a glycosylation site in the Fab region, with usually more complex glycan structures. It was estimated that about 20% of polyclonal IgG contain a second glycosylation site in the Fab region [80].

Even more significantly than for other therapeutic proteins, the glycosylation of therapeutic antibodies will have a strong influence on their properties. The *N*-glycan in the Fc region will indeed modulate the binding to Fc receptors, and then the activity and half-life of the protein in the circulation. Nonhuman glycans may also trigger immunogenicity responses. Examples of glycans that are found on monoclonal antibodies are presented in Fig. 8.

The most commonly found glycans in human antibodies are G0F, G1F, and G2F, complex-type glycans containing a core fucosylation and 0–2 terminal galactose residues. A terminal sialic acid, *N*-acetyl neuraminic acid, is sometimes also present. CHO cells produce very similar glycosylated antibodies, except for the presence of glycans containing a bisecting *N*-acetylglucosamine, and low levels of terminal sialylation [82].

Murine cells (NS0 and SP2/0) also mainly produce similar glycans, but also low levels of Gal-α-1,3-Gal and terminal *N*-glycolyl neuraminic acid, which are both immunogenic in human. However, the amount of such glycoforms remains relatively low, and no serious adverse effects were reported for approved antibodies produced in these murine cells [83]. The case of cetuximab is somewhat different, as higher amounts of immunogenic glycoforms are present: on the second *N*-glycosylation site in the Fab portion of the antibody, it was shown that about 30% of the glycans contain a Gal-α-1,3-Gal motif, and more than 10% contain *N*-glycolyl neuraminic acid [84]. Consequently, more than 20% of patients treated with cetuximab had severe hypersensitivity reactions [85].

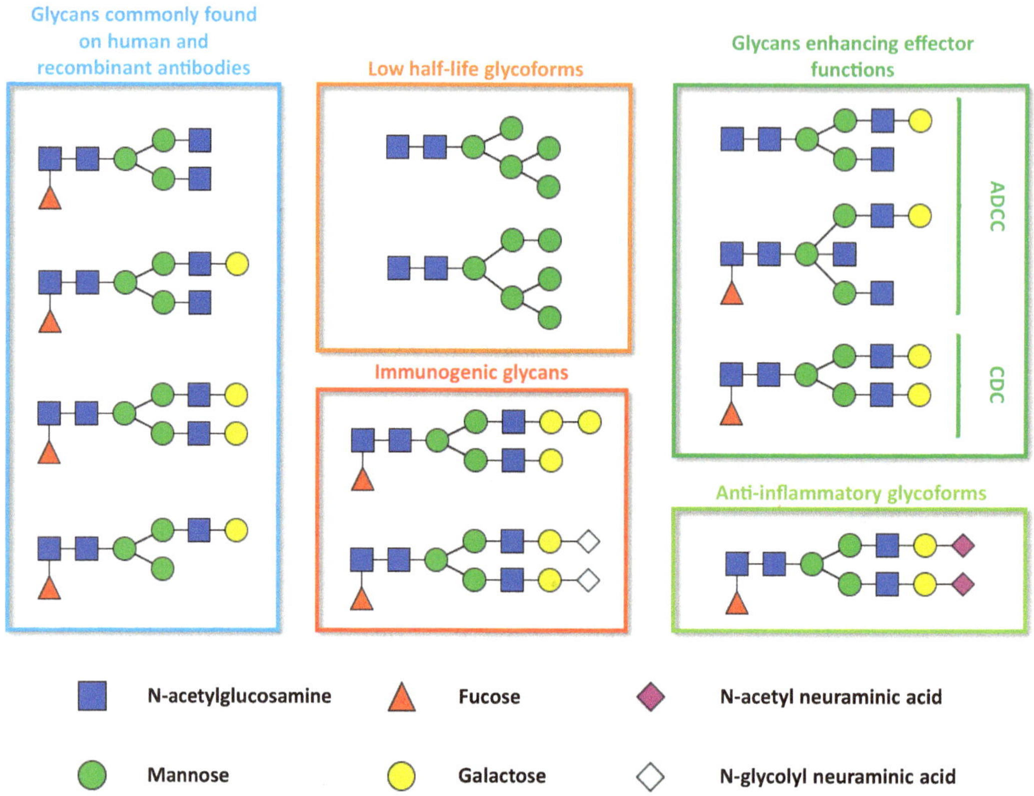


**Fig. 8** Examples of glycan structures that can be found in recombinant monoclonal antibodies, with their impact on pharmacokinetics, safety and efficacy (adapted from [81])

Oligomannose species (also known as "high mannose," glycans that contain between 5 and 9 mannose residues) are commonly found in proteins expressed in mammalian, yeast, plant or insect cells, but are present at very low levels in normal human antibodies [86]. These glycoforms are rapidly cleared from serum, and should therefore be considered as a critical quality attribute as they may affect pharmacokinetic properties of the drug [87].

The biological activity of monoclonal antibodies and other therapeutic proteins containing a Fc domain is linked to their interaction with Fcγ (fragment crystallizable γ) and FcRn receptors. Fcγ receptors are cell surface receptors that can be found on innate immune effector cells. The binding of monoclonal antibodies to those receptors is one of the modes of action, especially for oncology indications [88, 89]. Fc glycosylation of antibodies and Fc-fusion proteins is critical for the binding to Fc receptors, and consequently for the biological activities of these proteins used as therapeutics. Many groups have studied the influence of the glycosylation on Fc binding, as for example a recent study by Wada et al. [90]. Using a combination of surface plasmon resonance (SPR) assays, affinity chromatography and ADCC (antibody-dependent cell-mediated cytotoxicity) reporter bioassay, they showed that defucosylation, and to a lesser extent galactosylation, had a positive effect on FcγRIIIa binding and on ADCC activity, whereas sialylation decreased the activity. They also showed that terminal galactosylation had a small positive impact on binding to C1q complex, which is involved in the CDC (complement-dependent cytotoxicity) activity of monoclonal antibodies.

Binding to FcRn (neonatal receptor) was also studied, as it has an impact on the half-life of antibodies [91]. FcRn extends the half-life of antibodies by reducing lysosomal degradation in endothelial cells. IgGs are continuously internalized through pinocytosis. While most serum proteins are generally transported from the endosomes to the lysosome, where they are degraded, IgGs bind to FcRn at slightly acidic pH, and recycled to the cell surface where they are released in blood at a neutral pH. In this way they avoid lysosomal degradation and they can have a long serum circulation half-life. SPR assays showed no significant influence of glycosylation on FcRn binding, whereas affinity chromatography showed that galactosylated and sialylated species increased the binding [90]. FcRn binding is the most critical parameter for the serum half-life of antibodies, and the nature of the glycans is less important than for other receptors. FcRn binding and its influence on serum clearance is used in fusion proteins, in which the Fc domain of an antibody is fused to another protein. This fusion is a way to improve the half-life of proteins that would otherwise be cleared rapidly from serum. One well-known example is etanercept (Enbrel®), a fusion protein containing a Fc part fused to TNFα receptor via an *O*-glycosylated linker [15].

Studies were also performed to study the influence of glycosylation on the stability of antibodies. It was shown that glycosylation protected these proteins against proteolysis and thermal denaturation [52, 92]: deglycosylated antibodies exhibit less thermal stability, especially in the $CH_2$ domain, and are less resistant to unfolding induced by guanidine hydrochloride. They are also more susceptible to papain digestion.

The influence of glycosylation on antibody properties is now a great area of research, and glycoengineering is widely used to improve the pharmacokinetics, affinity and stability of monoclonal antibodies. A summary of the influence of glycosylation on the properties of a monoclonal antibody are summarized in Fig. 9. As cell lines are not able to produce naturally the optimal glycoforms, they are genetically modified to have them produce antibodies with an optimized glycosylation based on the intended mode of action [21, 29, 31, 81, 94–96].

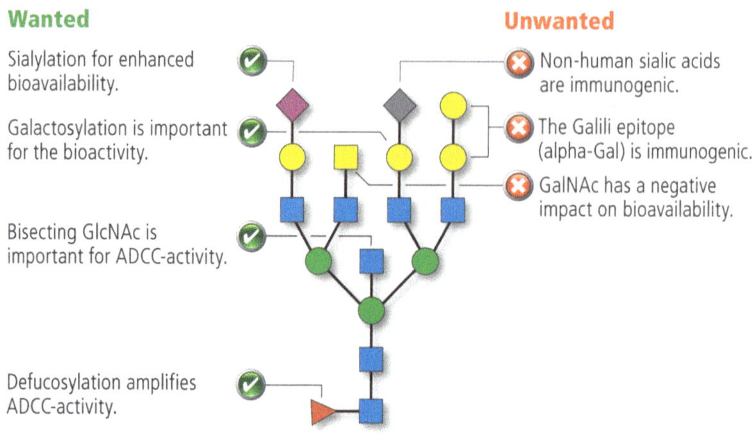


**Fig. 9** Influence of the glycosylation pattern on the properties of monoclonal antibodies (reproduced with permission of Glycotope GmbH [93])

## 3 The Role of Mass Spectrometry on the Analysis of Glycosylation

Glycosylation makes the analytical characterization of proteins very complex. Macroheterogeneity (heterogeneity related to the presence or absence of glycans on a protein) as well as microheterogeneity (heterogeneity of glycans linked to a specific glycosylation site) should both be characterized, and state-of-the-art analytical techniques are required. Advances in analytical sciences over the years have allowed a much more in-depth characterization of glycosylated proteins. The techniques that are commonly used are (ultra)high-performance liquid chromatography and capillary electrophoresis, with optical (UV or fluorescence) or mass spectrometric detection. Mass spectrometry alone is not commonly used for this kind of studies, and most of the time it will be used after a chromatographic separation. The aim of this section is not to provide an exhaustive view on the use of mass spectrometry for the characterization of glycoproteins, but rather to discuss briefly how it can be used and what information this technique can provide. The focus will be made on purified therapeutic glycoproteins.

The easiest way to analyze glycoproteins is to work at the intact level, without sample preparation, in order to have a view of the macroheterogeneity of the protein. In this case, chromatography can be used to desalt the sample before MS analysis and avoid offline desalting. Excellent mass accuracies can be obtained using QTOF or Orbitrap analyzers, down to 10 ppm or less, depending on the equipment and the mass of the protein. However, this approach is limited when glycoproteins are highly heterogeneous (this is the case for proteins containing many glycosylation sites, and/or both *N*- and *O*-glycans). In that case, MALDI-TOF mass spectrometry can be better suited to obtain the average molecular mass.

Although intact level analysis is useful to get a quick view of the glycosylation profile, limitations are quickly encountered. The next step is to work at the subunit level. This approach is commonly used for monoclonal antibodies and Fc fusion proteins, that can be either reduced or digested with enzymes such as papain or IdeS for example. In that case, shorter fragments are obtained, which allows to improve mass accuracy and chromatographic resolution. Reversed-phase chromatography is not the best option to separate the subunits, and HILIC chromatography was successfully used for this purpose, as it is able not only to separate glycosylated and nonglycosylated species but also to separate quite efficiently the different glycoforms of a protein [97, 98].

The next step is to analyze glycopeptides. Using an enzyme such as trypsin, the glycoprotein is digested into peptides, some of which are glycosylated. After chromatographic separation, mass spectrometry can identify the peptides based on their molecular mass and determine which glycans are linked to the peptides. MALDI-TOF mass spectrometry, without separation, can also be used but the analysis of complex peptide mixtures will be more challenging. Using this approach, a determination of site-specific glycosylation is possible. Reversed-phase chromatography is widely used for the analysis of tryptic digests, but HILIC can also be valuable as with this separation mode glycosylated and nonglycosylated peptides are well separated [99].

If site-specific glycosylation is not the focus of the analysis, released glycans analysis is the method of choice. This is usually the method chosen for the control of glycosylation during batch release of monoclonal antibodies. For *N*-glycans, enzymes such as PNGases can be used, whereas a chemical release will be required for *O*-glycans. The released glycans can be analyzed by MS without derivatization, for example after separation on a porous graphitic carbon (PGC) column, or, after derivatization, by MALDI-TOF, LC/MS (using HILIC columns), or CE/MS. Many derivatizing agents have been developed for LC/MS analysis of glycans that now allow the detection of minor glycoforms below 0.05% [100]. Ultrahigh-resolution separations also allow the separation between isomers of some glycans [101]. Using more complex analyses such as GC/MS, data can also be obtained on the linkages between the sugar units [102].

All these basic principles that were briefly described can be incorporated in more complex workflows, combining different separation and/or detection modes and involving sample preparation and prefractionation. Due to the complexity of some therapeutic glycoproteins, a combination of orthogonal techniques, most of which involving mass spectrometry, is necessary to get a fine characterization of the glycosylation [103]. An example of workflow describing the techniques that can be used for the characterization of *N*- and *O*-glycosylated therapeutic proteins is presented in Fig. 10.

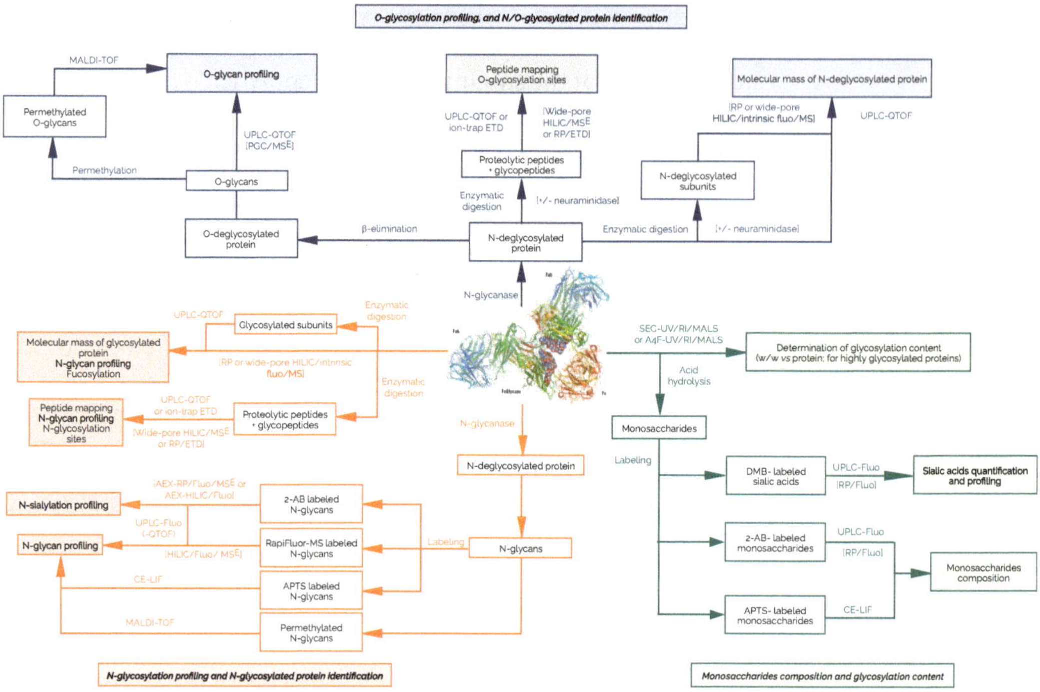


**Fig. 10** Analytical workflows that can be used for the characterization of therapeutic glycoproteins (reproduced with permission of Quality Assistance sa [104])

## References


1. Evaluate Pharma (2019) World preview 2019, Outlook 2024
2. Industry Experts (2019) Biopharmaceuticals—A Global Market Overview
3. Apweiler R, Hermjakob H, Sharon N (1999) On the frequency of protein glycosylation, as deduced from analysis of the SWISS-PROT database. Biochim Biophys Acta 1473:4–8. https://doi.org/10.1016/s0304-4165(99)00165-8
4. Cymer F, Beck H, Rohde A, Reusch D (2017) Therapeutic monoclonal antibody N-glycosylation – structure, function and therapeutic potential. Biologicals 52:1–11. https://doi.org/10.1016/j.biologicals.2017.11.001
5. Urquhart L (2019) Top drugs and companies by sales in 2018. Nat Rev Drug Discov 18:245. https://doi.org/10.1038/d41573-019-00049-0
6. Alt N, Zhang TY, Motchnik P et al (2016) Determination of critical quality attributes for monoclonal antibodies using quality by design principles. Biologicals 44:291–305. https://doi.org/10.1016/j.biologicals.2016.06.005
7. Xu Y, Wang D, Mason B et al (2019) Structure, heterogeneity and developability assessment of therapeutic antibodies. MAbs 11:239–264. https://doi.org/10.1080/19420862.2018.1553476
8. Beck A, Liu H (2019) Macro- and micro-heterogeneity of natural and recombinant IgG antibodies. Antibodies 8:18. https://doi.org/10.3390/antib8010018
9. Committee for Medicinal Products for Human Use (CHMP) - European Medicines Agency (2016) Guideline on development, production, Characterisation and specification for monoclonal antibodies and related products
10. International Council for Harmonisation of Technical Requirements for Pharmaceuticals for Human Use (1999) Guideline Q6B - Specifications: Test procedures and acceptance criteria for biotechnological/biological products
11. U. S. Department of Health and Human Services, – Food and Drug Administration (1997) Points to Consider in the Manufacture and Testing of Monoclonal Antibody Products for Human Use

12. Stick R, Williams SJ (2009) Carbohydrates: the essential molecules of life. Elsevier, Amsterdam
13. Wang T, Voglmeir J (2014) PNGases as valuable tools in glycoprotein analysis. Protein Pept Lett 21:976–985. https://doi.org/10.2174/09298665216661406261l1237
14. Kobata A (2013) Exo- and endoglycosidases revisited. Proc Japan Acad Ser B Phys Biol Sci 89:97–117. https://doi.org/10.2183/pjab.89.97
15. Goldenberg MM (1999) Etanercept, a novel drug for the treatment of patients with severe, active rheumatoid arthritis. Clin Ther 21:75–87. https://doi.org/10.1016/S0149-2918(00)88269-7
16. Zhang P, Wang T, Bardor M, Song Z (2013) Deciphering O-glycomics for the development and production of biopharmaceuticals. Pharm Bioprocess 1:89–104. https://doi.org/10.4155/pbp.13.7
17. Yang W, Ao M, Hu Y et al (2018) Mapping the O-glycoproteome using site-specific extraction of O-linked glycopeptides (EXoO). Mol Syst Biol 14:e8486. https://doi.org/10.15252/msb.20188486
18. Kuriakose A, Chirmule N, Nair P (2016) Immunogenicity of biotherapeutics: causes and association with posttranslational modifications. J Immunol Res 2016:1298473. https://doi.org/10.1155/2016/1298473
19. Brooks SA (2004) Appropriate glycosylation of recombinant proteins for human use: implications of choice of expression system. Mol Biotechnol 28:241–256. https://doi.org/10.1385/MB:28:3:241
20. Raju TS, Briggs JB, Borge SM, Jones AJS (2000) Species-specific variation in glycosylation of Igc: evidence for the species-specific sialylation and branch-specific galactosylation and importance for engineering recombinant glycoprotein therapeutics. Glycobiology 10:477–486. https://doi.org/10.1093/glycob/10.5.477
21. Gupta SK, Shukla P (2018) Glycosylation control technologies for recombinant therapeutic proteins. Appl Microbiol Biotechnol 102:10457–10468. https://doi.org/10.1007/s00253-018-9430-6
22. Lehle L (1992) Protein glycosylation in yeast. Antonie Van Leeuwenhoek 61:133–134. https://doi.org/10.1007/BF00580620
23. Walski T, De Schutter K, Van Damme EJM, Smagghe G (2017) Diversity and functions of protein glycosylation in insects. Insect Biochem Mol Biol 83:21–34. https://doi.org/10.1016/j.ibmb.2017.02.005
24. Strasser R (2016) Plant protein glycosylation. Glycobiology 26:926–939. https://doi.org/10.1093/glycob/cww023
25. Lalonde M-E, Durocher Y (2017) Therapeutic glycoprotein production in mammalian cells. J Biotechnol 251:128–140. https://doi.org/10.1016/j.jbiotec.2017.04.028
26. Elliott S, Lorenzini T, Asher S et al (2003) Enhancement of therapeutic protein in vivo activities through glycoengineering. Nat Biotechnol 21:414–421. https://doi.org/10.1038/nbt799
27. Niwa R, Satoh M (2015) The current status and prospects of antibody engineering for therapeutic use: focus on Glycoengineering technology. J Pharm Sci 104:930–941. https://doi.org/10.1002/jps.24316
28. Werner RG, Kopp K, Schlueter M (2007) Glycosylation of therapeutic proteins in different production systems. Acta Paediatr Int J Paediatr 96:17–22. https://doi.org/10.1111/j.1651-2227.2007.00199.x
29. Beck A, Wagner-Rousset E, Bussat M et al (2008) Trends in glycosylation, glycoanalysis and glycoengineering of therapeutic antibodies and fc-fusion proteins. Curr Pharm Biotechnol 9:482–501. https://doi.org/10.2174/138920108786786411
30. Du T, Buenbrazo N, Kell L et al (2019) A bacterial expression platform for production of therapeutic proteins containing human-like O-linked Glycans. Cell Chem Biol 26:203–212.e5. https://doi.org/10.1016/j.chembiol.2018.10.017
31. Jacobs P, Callewaert N (2009) N-glycosylation engineering of biopharmaceutical expression systems. Curr Mol Med 9:774–800. https://doi.org/10.2174/156652409789105552
32. Fox JL (2012) First plant-made biologic approved. Nat Biotechnol 30:472. https://doi.org/10.1038/nbt0612-472
33. Yang M, Butler M (2000) Effects of ammonia on CHO cell growth, erythropoietin production, and glycosylation. Biotechnol Bioeng 68:370–380. https://doi.org/10.1002/(SICI)1097-0290(20000520)68:4<370::AID-BIT2>3.0.CO;2-K
34. Yang M, Butler M (2002) Effects of ammonia and glucosamine on the heterogeneity of erythropoietin glycoforms. Biotechnol Prog 18:129–138. https://doi.org/10.1021/bp0101334
35. Müthing J, Kemminer SE, Conradt HS et al (2003) Effects of buffering conditions and culture pH on production rates and glycosylation of clinical phase I anti-melanoma mouse

IgG3 monoclonal antibody R24. Biotechnol Bioeng 83:321–334. https://doi.org/10.1002/bit.10673

36. Trummer E, Fauland K, Seidinger S et al (2006) Process parameter shifting: part II. Biphasic cultivation - a tool for enhancing the volumetric productivity of batch processes using Epo-fc expressing CHO cells. Biotechnol Bioeng 94:1045–1052. https://doi.org/10.1002/bit.20958
37. Lipscomb ML, Palomares LA, Hernández V et al (2005) Effect of production method and gene amplification on the glycosylation pattern of a secreted reporter protein in CHO cells. Biotechnol Prog 21:40–49. https://doi.org/10.1021/bp049761m
38. Restelli V, Wang MD, Huzel N et al (2006) The effect of dissolved oxygen on the production and the glycosylation profile of recombinant human erythropoietin produced from CHO cells. Biotechnol Bioeng 94:481–494. https://doi.org/10.1002/bit.20875
39. Harvey RD (2017) Science of biosimilars. J Oncol Pract 13:17s–23s. https://doi.org/10.1200/jop.2017.026062
40. Kabir ER, Moreino SS, Siam MKS (2019) The breakthrough of biosimilars: a twist in the narrative of biological therapy. Biomol Ther 9:1–34. https://doi.org/10.3390/biom9090410
41. Chow S-C, Song F, Bai H (2016) Analytical similarity assessment in biosimilar studies. AAPS J 18:670–677. https://doi.org/10.1208/s12248-016-9882-5
42. Chung Chow S (2014) On assessment of analytical similarity in biosimilar studies. Drug Des Open Access 03. https://doi.org/10.4172/2169-0138.1000e124
43. Committee for Medicinal Products for Human Use (CHMP) - European Medicines Agency (2014) Similar biological medicinal products containing biotechnology-derived proteins as active substance: quality issue
44. U.S. Department of Health and Human Services Food and drug administration (2019) Development of Therapeutic Protein Biosimilars: Comparative Analytical Assessment and Other Quality-Related Considerations (draft)
45. World Health Organization (2009) Guidelines on evaluation of similar biotherapeutic products (SBPs)
46. Duivelshof BL, Jiskoot W, Beck A et al (2019) Glycosylation of biosimilars: recent advances in analytical characterization and clinical implications. Anal Chim Acta 1089:1–18. https://doi.org/10.1016/j.aca.2019.08.044
47. Hajba L, Szekrényes Á, Borza B, Guttman A (2018) On the glycosylation aspects of biosimilarity. Drug Discov Today 23:616–625. https://doi.org/10.1016/j.drudis.2018.01.009
48. Kim S, Song J, Park S et al (2017) Drifts in ADCC-related quality attributes of Herceptin®: impact on development of a trastuzumab biosimilar. MAbs 9:704–714. https://doi.org/10.1080/19420862.2017.1305530
49. Planinc A, Dejaegher B, Vander HY et al (2017) Batch-to-batch N-glycosylation study of infliximab, trastuzumab and bevacizumab, and stability study of bevacizumab. Eur J Hosp Pharm 24:286–292. https://doi.org/10.1136/ejhpharm-2016-001022
50. Zhou Q, Qiu H (2018) The mechanistic impact of N-glycosylation on stability, pharmacokinetics and immunogenicity of therapeutic proteins. J Pharm Sci 108:1366–1377. https://doi.org/10.1016/j.xphs.2018.11.029
51. Sjögren J, Lood R, Nägeli A (2019) On enzymatic remodeling of IgG glycosylation; unique tools with broad applications. Glycobiology 30(4):254–267. https://doi.org/10.1093/glycob/cwz085
52. Solá RJ, Griebenow K (2009) Effects of glycosylation on the stability of protein pharmaceuticals. J Pharm Sci 98:1223–1245. https://doi.org/10.1002/jps.21504
53. Ward S, O'Sullivan JM, O'Donnell JS (2019) von Willebrand factor sialylation—a critical regulator of biological function. J Thromb Haemost 17:1018–1029. https://doi.org/10.1111/jth.14471
54. Ono M (1994) Physicochemical and biochemical characteristics of glycosylated recombinant human granulocyte colony stimulating factor (lenograstim). Eur J Cancer 30:S7–S11
55. Narhi LO, Arakawa T, Aoki KH et al (1991) The effect of carbohydrate on the structure and stability of erythropoietin. J Biol Chem 266:23022–23026
56. Uchida E, Morimoto K, Kawasaki N et al (1997) Effect of active oxygen radicals on protein and carbohydrate moieties of recombinant human erythropoietin. Free Radic Res 27:311–323. https://doi.org/10.3109/10715769709065769
57. Raso SW, Abel J, Barnes JM et al (2005) Aggregation of granulocyte-colony stimulating factor in vitro involves a conformationally altered monomeric state. Protein Sci 14:2246–2257. https://doi.org/10.1110/ps.051489405
58. Banks DD (2011) The effect of glycosylation on the folding kinetics of erythropoietin. J

Mol Biol 412:536–550. https://doi.org/10.1016/j.jmb.2011.07.061
59. Raju TS (2003) Glycosylation variation in expression systems and their impact on biological activity of therapeutic immunoglobulins. Bio Process Int 1:44–53
60. Jelkmann W (2002) Pharmacology, pharmacokinetics and safety of recombinant human erythropoietin (rhEPO). Recomb Hum Erythrop Clin Oncol:203–221. https://doi.org/10.1007/978-3-7091-7658-0_11
61. Liu L, Li H, Hamilton SR et al (2012) The impact of sialic acids on the pharmacokinetics of a PEGylated erythropoietin. J Pharm Sci 101:4414–4418. https://doi.org/10.1002/jps.23320
62. Kwak CY, Park SY, Lee CG et al (2017) Enhancing the sialylation of recombinant EPO produced in CHO cells via the inhibition of glycosphingolipid biosynthesis. Sci Rep 7:6–10. https://doi.org/10.1038/s41598-017-13609-4
63. Teare JM, Kates DS, Shah A, Garger S (2019) Increased branching and sialylation of n-linked glycans correlate with an improved pharmacokinetic profile for BAY 81–8973 compared with other full-length rFVIII products. Drug Des Devel Ther 13:941–948. https://doi.org/10.2147/DDDT.S188171
64. Morimoto K, Tsuda E, Said AA et al (1996) Biological and physicochemical characterization of recombinant human erythropoietins fractionated by mono Q column chromatography and their modification with sialyltransferase. Glycoconj J 13:1013–1020. https://doi.org/10.1007/BF01053197
65. Holliger P, Prospero T, Winter G (1993) "Diabodies": small bivalent and bispecific antibody fragments. Proc Natl Acad Sci U S A 90:6444–6448. https://doi.org/10.1073/pnas.90.14.6444
66. Stork R, Zettlitz KA, Müller D et al (2008) N-glycosylation as novel strategy to improve pharmacokinetic properties of bispecific single-chain Diabodies. J Biol Chem 283:7804–7812. https://doi.org/10.1074/jbc.M709179200
67. Loutradis D, Vlismas A, Drakakis P (2010) Corifollitropin alfa: a novel long-acting recombinant follicle-stimulating hormone agonist for controlled ovarian stimulation. Women's Heal 6:655–664. https://doi.org/10.2217/WHE.10.56
68. Darling RJ, Kuchibhotla U, Glaesner W et al (2002) Glycosylation of erythropoietin affects receptor binding kinetics: role of electrostatic interactions. Biochemistry 41:14524–14531. https://doi.org/10.1021/bi0265022
69. Egrie JC, Dwyer E, Browne JK et al (2003) Darbepoetin alfa has a longer circulating half-life and greater in vivo potency than recombinant human erythropoietin. Exp Hematol 31:290–299. https://doi.org/10.1016/S0301-472X(03)00006-7
70. Elliott S, Egrie J, Browne J et al (2004) Control of rHuEPO biological activity: the role of carbohydrate. Exp Hematol 32:1146–1155. https://doi.org/10.1016/j.exphem.2004.08.004
71. Steinke JW, Platts-Mills TAE, Commins SP (2015) The alpha-gal story: lessons learned from connecting the dots. J Allergy Clin Immunol 135:589–596
72. Ghaderi D, Taylor RE, Padler-Karavani V et al (2010) Implications of the presence of N-glycolylneuraminic acid in recombinant therapeutic glycoproteins. Nat Biotechnol 28:863–867. https://doi.org/10.1038/nbt.1651
73. Ghaderi D, Zhang M, Hurtado-Ziola N, Varki A (2012) Production platforms for biotherapeutic glycoproteins. Occurrence, impact, and challenges of non-human sialylation. Biotechnol Genet Eng Rev 28:147–176. https://doi.org/10.5661/bger-28-147
74. Rup B, Alon S, Amit-Cohen B-C et al (2017) Immunogenicity of glycans on biotherapeutic drugs produced in plant expression systems—the taliglucerase alfa story. PLoS One 12:e0186211. https://doi.org/10.1371/journal.pone.0186211
75. Yu C, Gao K, Zhu L et al (2016) At least two fc Neu5Gc residues of monoclonal antibodies are required for binding to anti-Neu5Gc antibody. Sci Rep 6:20029. https://doi.org/10.1038/srep20029
76. van Bueren JJL, Rispens T, Verploegen S et al (2011) Anti-galactose-α-1,3-galactose IgE from allergic patients does not bind α-galactosylated glycans on intact therapeutic antibody fc domains. Nat Biotechnol 29:574–576. https://doi.org/10.1038/nbt.1912
77. Gribben JG, Devereux S, Thomas NSB et al (1990) Development of antibodies to unprotected glycosylation sites on recombinant human GM-CSF. Lancet 335:434–437. https://doi.org/10.1016/0140-6736(90)90665-R
78. Ju M-S, Jung ST (2014) Aglycosylated full-length IgG antibodies: steps toward next-generation immunotherapeutics. Curr Opin

Biotechnol 30:128–139. https://doi.org/10.1016/j.copbio.2014.06.013
79. Goldberg RM, Kirkpatrick P (2005) Cetuximab. Nat Rev Drug Discov 4:10–12. https://doi.org/10.1038/nrd1728
80. Holland M, Yagi H, Takahashi N et al (2006) Differential glycosylation of polyclonal IgG, IgG-fc and IgG-fab isolated from the sera of patients with ANCA-associated systemic vasculitis. Biochim Biophys Acta 1760:669–677. https://doi.org/10.1016/j.bbagen.2005.11.021
81. Beck A, Reichert JM (2012) Marketing approval of mogamulizumab - a triumph for glycoengineering. MAbs 4:419–425. https://doi.org/10.4161/mabs.20996
82. Jefferis R (2012) Isotype and glycoform selection for antibody therapeutics. Arch Biochem Biophys 526:159–166. https://doi.org/10.1016/j.abb.2012.03.021
83. Raju TS, Jordan RE (2012) Galactosylation variations in marketed therapeutic antibodies. MAbs 4:385–391. https://doi.org/10.4161/mabs.19868
84. Sundaram S, Matathia A, Qian J et al (2011) An innovative approach for the characterization of the isoforms of a monoclonal antibody product. MAbs 3:505–512. https://doi.org/10.4161/mabs.3.6.18090
85. Chung CH, Mirakhur B, Chan E et al (2008) Cetuximab-induced anaphylaxis and IgE specific for galactose-α-1,3-galactose. N Engl J Med 358:1109–1117. https://doi.org/10.1056/NEJMoa074943
86. Beck A, Cochet O, Wurch T (2009) GlycoFi's technology to control the glycosylation of recombinant therapeutic proteins. Expert Opin Drug Discov 5:95–111. https://doi.org/10.1517/17460440903413504
87. Goetze AM, Liu YD, Zhang Z et al (2011) High-mannose glycans on the fc region of therapeutic IgG antibodies increase serum clearance in humans. Glycobiology 21:949–959. https://doi.org/10.1093/glycob/cwr027
88. Jiang X-R, Song A, Bergelson S et al (2011) Advances in the assessment and control of the effector functions of therapeutic antibodies. Nat Rev Drug Discov 10:101. https://doi.org/10.1038/nrd3365
89. Hogarth PM, Pietersz GA (2012) Fc receptor-targeted therapies for the treatment of inflammation, cancer and beyond. Nat Rev Drug Discov 11:311–331. https://doi.org/10.1038/nrd2909
90. Wada R, Matsui M, Kawasaki N (2019) Influence of N-glycosylation on effector functions and thermal stability of glycoengineered IgG1 monoclonal antibody with homogeneous glycoforms. MAbs 11:350–372. https://doi.org/10.1080/19420862.2018.1551044
91. Mackness BC, Jaworski JA, Boudanova E et al (2019) Antibody fc engineering for enhanced neonatal fc receptor binding and prolonged circulation half-life. MAbs 11:1276–1288. https://doi.org/10.1080/19420862.2019.1633883
92. Zheng K, Bantog C, Bayer R (2011) The impact of glycosylation on monoclonal antibody conformation and stability. MAbs 3:568–576. https://doi.org/10.4161/mabs.3.6.17922
93. Glycotope GmbH. https://www.glycotope.com/biopharmaceuticals-services/cell-lines/
94. Liu R, Giddens J, McClung CM et al (2015) Evaluation of a glycoengineered monoclonal antibody via LC-MS analysis in combination with multiple enzymatic digestion. MAbs 8:340–346. https://doi.org/10.1080/19420862.2015.1113361
95. Buettner MJ, Shah SR, Saeui CT et al (2018) Improving immunotherapy through Glycodesign. Front Immunol 9:2485. https://doi.org/10.3389/fimmu.2018.02485
96. Mimura Y, Katoh T, Saldova R et al (2018) Glycosylation engineering of therapeutic IgG antibodies: challenges for the safety, functionality and efficacy. Protein Cell 9:47–62. https://doi.org/10.1007/s13238-017-0433-3
97. Periat A, Fekete S, Cusumano A et al (2016) Potential of hydrophilic interaction chromatography for the analytical characterization of protein biopharmaceuticals. J Chromatogr A 1448:81–92. https://doi.org/10.1016/j.chroma.2016.04.056
98. Domínguez-Vega E, Tengattini S, Peintner C et al (2018) High-resolution glycoform profiling of intact therapeutic proteins by hydrophilic interaction chromatography-mass spectrometry. Talanta 184:375–381. https://doi.org/10.1016/j.talanta.2018.03.015
99. Gilar M, Yu Y-Q, Ahn J et al (2011) Characterization of glycoprotein digests with hydrophilic interaction chromatography and mass spectrometry. Anal Biochem 417:80–88. https://doi.org/10.1016/j.ab.2011.05.028
100. Zhou S, Veillon L, Dong X et al (2017) Direct comparison of derivatization strategies for LC-MS/MS analysis of N -glycans. Analyst 142:4446–4455. https://doi.org/10.1039/c7an01262d

101. Hilliard M, Alley WR, McManus CA et al (2017) Glycan characterization of the NIST RM monoclonal antibody using a total analytical solution: from sample preparation to data analysis. MAbs 9:1349–1359. https://doi.org/10.1080/19420862.2017.1377381
102. Hounsell EF, Davies MJ, Smith KD (2003) Determination of monosaccharide linkage and substitution patterns by GC-MS methylation analysis. In: Protein protocols handbook. Humana Press, New Jersey, pp 811–814
103. Largy E, Cantais F, Van Vyncht G et al (2017) Orthogonal liquid chromatography–mass spectrometry methods for the comprehensive characterization of therapeutic glycoproteins, from released glycans to intact protein level. J Chromatogr A 1498:128–146. https://doi.org/10.1016/j.chroma.2017.02.072
104. Quality Assistance sa. https://www.quality-assistance.com/resources/study-and-technical-sheets

# Chapter 2

# Characterization of Protein Glycoforms at Intact Level by Orbitrap Mass Spectrometry


## Dan Bach Kristensen, Trine Meiborg Sloth, Martin Ørgaard, and Pernille Foged Jensen



## Abstract

Intact mass analysis of proteins is simple, fast, and specific, and it effectively provides structural insight into the proteoforms or variants of the analyzed protein. For instance, the multiple glycoforms of recombinant monoclonal antibodies can be effectively analyzed by intact mass spectrometry (MS). A recent development in the Orbitrap technology has made this platform particularly well suited for analysis of large intact biomolecules, and here we describe procedures for performing intact mass analysis of intact glycoproteins using the Orbitrap platform, with the aim of identifying and quantitating the glycoforms. Emphasis is placed on the analysis of biopharmaceutical immunoglobulins (IgGs), but the procedures can be extended to other glycoproteins as needed.



**Key words** Antibody, Biopharmaceutical, Glycoprotein, IgG, Native intact mass spectrometry, Orbitrap, Glycoform


## 1 Introduction

*N*-linked glycosylation plays a critical role in the structure and function of glycoproteins. For instance, biopharmaceutical IgGs are glycosylated in a conserved Asn in the constant region of the heavy chain, and this glycosylation plays an important role for effector functions, including antibody-dependent cell-mediated cytotoxicity (ADCC) and complement-dependent cytotoxicity (CDC) [1]. Furthermore, the glycosylation pattern of biopharmaceuticals may impact drug stability [2], in vivo pharmacokinetics [3], as well as immunogenicity and safety in humans, as the IgGs are often product by nonhuman cell lines [4]. Consequently, it is critical to control and monitor glycosylation of biopharmaceuticals. A broad range analytical techniques have been established for glycan analysis of glycoproteins, ranging from analysis of (a) intact glycoproteins, (b) glycopeptides (following glycoprotein digestion), (c) release glycans (following enzymatic or chemical release


Arnaud Delobel (ed.), *Mass Spectrometry of Glycoproteins: Methods and Protocols*, Methods in Molecular Biology, vol. 2271, https://doi.org/10.1007/978-1-0716-1241-5_2, © Springer Science+Business Media, LLC, part of Springer Nature 2021

of glycans) and (d) monosaccharides (following acid hydrolysis of glycans) [5]. Analysis of intact glycoproteins requires the ability to resolve or partially resolve glycoforms, and analytical methodologies applied for this purpose include reversed-phase (RP) liquid chromatography (LC), ion exchange (IEX) LC, capillary electrophoresis (CE) and lectin arrays [5]. Common for these technologies are that they rely on physical separation of glycoforms before detection and quantitation. In contract, mass spectrometry (MS) analysis of intact glycoproteins primarily relies on the spectral resolution of the MS instrument, that is, separation of glycoforms is not per se required before MS analysis. The recent advancement of MS instruments purpose built for analysis of large biomolecules, has made intact MS one of the most simple, fast, specific and robust tools available for characterization of glycoproteins.

In recent years the Orbitrap technology has become a powerful platform for characterization of large intact biomolecules. While the Orbitrap MS was originally optimized for small molecules, recent developments has focused on compatibility with large biomolecules, and the latest instrument generations, such as the Q Exactive HF and Orbitrap Exploris 480, have been designed for measuring large biomolecules [6]. Several Orbitrap instruments now come with a so-called BioPharma Option, which includes a High Mass Range Mode (HMR) with an extended mass range up to 8000 $m/z$. The HMR Mode is designed for the analysis of large intact biomolecules. Analysis can be performed under protein denaturing conditions (e.g. RP LC-MS), or more recently using LC conditions, under which the protein remain folded or in its "native" state. This intact MS approach is referred to as native MS [7]. Native MS can be performed under gentle conditions, that is, neutral or near neutral pH, room temperature, and without organic solvents, thereby minimizing the risk of introducing artefacts during analysis. Furthermore, by preserving protein folding during native MS, the resulting average charge state and the total number of charge states are significantly reduced compared to denaturing LC-MS approaches, such as RP LC-MS. Consequently, for native MS the protein signal is distributed over a lower number of lower charge states, which results in simpler, cleaner spectral data with high spatial separation between the individual charge states (Fig. 1). Finally, current Orbitraps have a very high effective (i.e., measured) spectral resolution. Indeed, at optimal settings the measured spectral peak width of a 150 kDa IgG is very close to the theoretical peak width (natural isotope distribution), as illustrated in Fig. 2. This illustrates that a near ideal spectral peak width can be obtained for large biomolecules on Orbitrap instruments equipped with the BioPharma Option (in this case data was acquired at resolution 35.000 on a Q Exactive Plus with Biopharma Option).

A range of chromatographic techniques can be hyphenated to the Orbitrap MS for denatured and native MS analysis, including

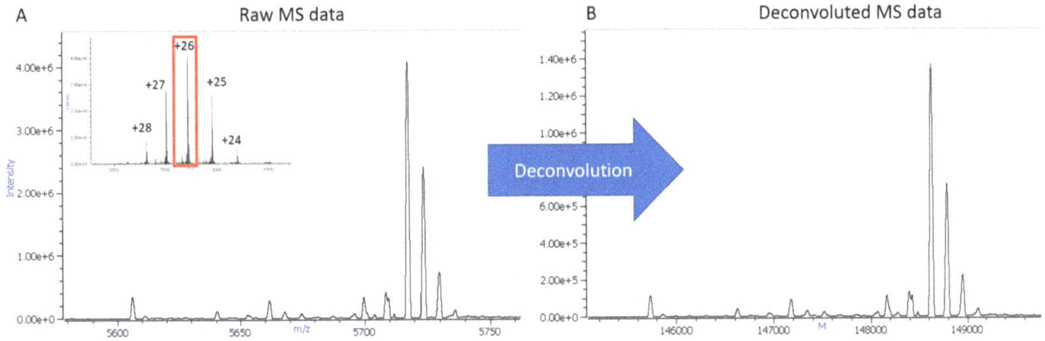


**Fig. 1** Raw and deconvoluted spectra of a Symphogen reference IgG (IgG1). (**a**) The main charge states are shown in the small spectrum and a zoom of the most intense charge state (+26) is shown in the large spectrum. (**b**) The deconvoluted spectrum. Deconvolution was performed in Protein Metrics Intact Mass software (*see* Fig. 10). The data quality of the Orbitrap spectral raw data is very high and has minimal background noise, and consequently the deconvolution algorithm has minimal impact spectral data quality

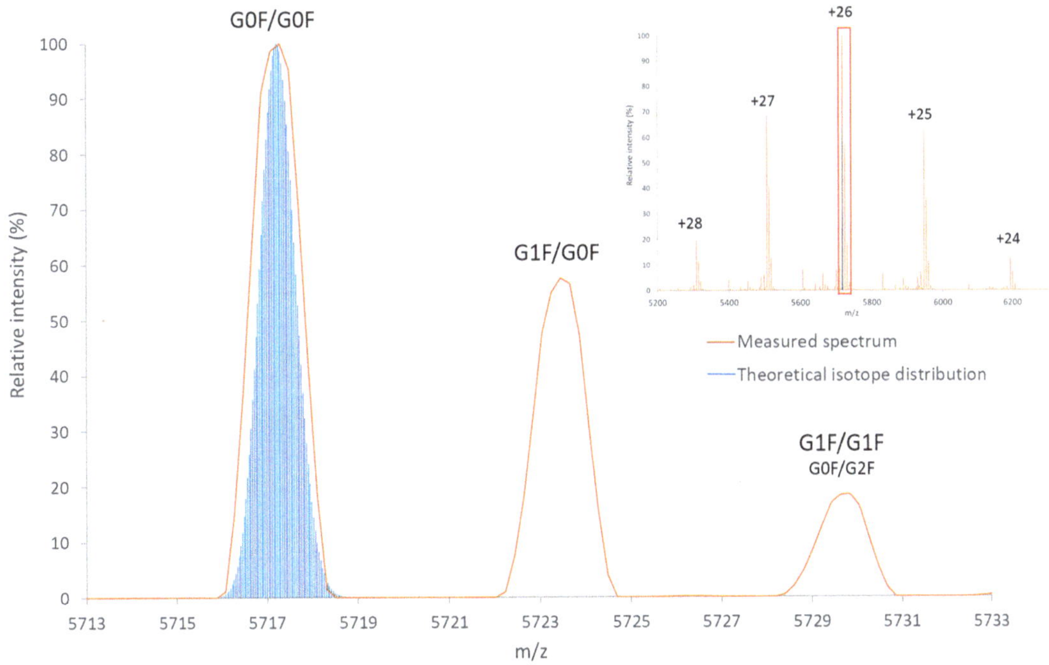


**Fig. 2** Raw spectra showing the charge states (upper right corner) and zoomed view of the +26 charge state of a reference IgG. The reference IgG was measured by native SEC MS on a Q Exactive Plus with BioPharma Option. The theoretical spectral peak of the main form of the reference IgG (glycoform G0F/G0F) is shown in blue. The theoretical spectral peak was determined based on the elemental composition (C6588 H10236 N1730 O2092 S46) obtained from GPMAW (http:/www.gpmaw.com/). The theoretical peak was determined using IDCalc (https:/proteome.gs.washington.edu/software/IDCalc/). The measured spectral peak of the G0F/G0F peak matches the theoretical peak well, illustrating the high effective (or measured) resolution of the Orbitrap MS. In other words, the measured spectral peak width matches the natural peak width (isotope distribution) well for large biomolecules on the Orbitrap MS

RP, size exclusion chromatography (SEC), cation exchange chromatography (CIEX), and Protein A for IgGs, and in recent years a multitude of native MS applications have been developed with success [8–12]. At Symphogen, native MS is established as the principal intact MS platform for biopharmaceutical development, and native MS is extensively used during e.g. lead selection studies, where up to 400 IgGs are analyzed consecutively [13]. Here three intact MS approaches are presented: (1) native SEC MS, (2) native CIEX MS, and (3) denaturing RP LC-MS. Native SEC MS method is the default method for intact mass analysis, but for complex samples (e.g., antibody mixtures or antibody–drug conjugates) requiring chromatographic separation the native CIEX MS or RP LC-MS approaches may be applied (*see* **Note 1**).

The presented methods were developed for intact MS analysis of IgG molecules, but the methods are to some extent generic in nature and can be applied to other glycoproteins with little or no modification to the presented methods. No sample preparation is required for intact MS analysis, the main challenge lies in selecting the appropriate solvents, LC columns, MS instrument settings, and how the raw MS data is processed. Consequently, the focus of this chapter is on the preparation of solvents, choice of LC columns, optimal MS settings, and considerations relating to processing of raw MS data with the aim of detecting and quantifying protein glycoforms.

## 2 Materials

All solutions should be prepared using MS grade reagents, unless otherwise indicated. Prepare and store all solvents at room temperature. LC solvents are prepared by adding reagents directly to the purchased solvent bottles (*see* **Note 2**). Native MS solvents should be discarded after 1 week. Choice of column is critical for optimal performance during native MS, particularly for native SEC MS, and recommendations should be followed carefully (*see* **Note 3**).

### 2.1 LC-MS System

A Thermo Scientific Vanquish Horizon LC was used but other biocompatible LC systems can be used. An Orbitrap MS equipped with a BioPharma Option (HMR Mode) is required to run native MS. The methods described here are based on the Q Exactive Plus with BioPharma Option. Other suitable instruments include Q Exactive HF-X, Q Exactive HF, Orbitrap Exploris 480, and Orbitrap Eclipse Tribrid equipped with BioPharma Option. The LC-MS instrument configuration used in the current chapter is summarized in Table 1.

### 2.2 MS Data Processing

Data processing and reporting was done using Protein Metrics Intact Mass™ software [14], but other software solution can be

**Table 1**
**Instrument Configuration. Configuration of LC components and MS instrument used in the current chapter**

| Category | Description | Producer | Catalogue no. |
|---|---|---|---|
| Vanquish horizon system, Equipped with: VH-C10-A column compartment | UHPLC system | Thermo Scientific | VH-C10-A |
| Pump duo | VH-P10-A high pressure binary pump | Thermo Scientific | VH-P10-A |
| Mixer | 10 μL + 25 μL | Thermo Scientific | – |
| Autosampler | VH-A10-A autosampler | Thermo Scientific | VH-A10-A |
| Sample loop | 100 μL ($V$ = 130 μL) | Thermo Scientific | – |
| Column compartment | Column compartment H | Thermo Scientific | – |
| Detector | VF-D40-A VWD detector | Thermo Scientific | VF-D40-A |
| Flow cell | Bio 2.5 μL—7 mm | Thermo Scientific | – |
| Q Exactive plus with biopharma option | MS—Orbitrap | Thermo Scientific | – |

used as needed [15, 16]. Data processing requirements for intact raw MS data include (1) deconvolution of raw spectra, (2) assignment of glycoforms to peaks in the deconvoluted spectra, and (3) relative quantitation of peak intensities for assigned glycoforms. It is not the intention of the current chapter to describe the use of these software tools. Instead, reference is made to the software user manuals.

### 2.3 Columns

The choice of column, including dimensions, is critical for optimal chromatographic performance during LC-MS (*see* **Note 3**). The recommended columns for the different LC-MS methods are summarized in Table 2.

### 2.4 Solvents and Solutions

#### 2.4.1 Native MS Solvents

1. Native CIEX solvent A (also the solvent used for native SEC MS): 25 mM ammonium acetate pH 5.4. Add 1.93 g ammonium acetate and 220 μL 100% acetic acid directly to the 1 L MS grade $H_2O$ bottle. Mix by gently vortexing and inverting the bottle. Aspirate solvent and measure the pH (should be between 5.3 and 5.5). Do not measure pH directly in the bottle. Store at room temperature. Discard after 1 week.

**Table 2**
**Recommended Columns. Choice of column is critical for optimal chromatographic performance during native LC-MS. This includes column dimension, particularly for native CIEX MS**

| LC-MS method | Column | Producer | Catalogue no. |
|---|---|---|---|
| Native SEC MS | ACQUITY UPLC BEH200 SEC, 4.6 × 150 mm, 1.7 μm | Waters | VH-C10-A |
| Native CIEX MS | MabPac SCX-10 RS, 2.1 × 50 mm, 5 μm | Thermo Scientific | 082675 |
| RP LC-MS | MabPac RP, 2.1 × 150 mm, 4 μm | Thermo Scientific | 303270 |

2. Native CIEX solvent B: 10 mM ammonium hydroxide pH 10.7. Remove 10.0 mL water from a new 1 L bottle of MS grade $H_2O$. Add 10.0 mL of 1 N ammonium hydroxide directly to the 1 L MS grade $H_2O$ bottle. Mix by gently vortexing and inverting the bottle. Aspirate 1 mL and measure the pH (should be between 10.6 and 10.8). Do not measure pH directly in the bottle. Store at room temperature. Discard after 1 week.
3. Column storage solution: 20 mM MES, 0.1% (w/v) sodium azide, pH 6.5. Add 19.5 g MES, 50 mL 10% (w/v) sodium azide and 4500 mL ultrapure water. Dissolve by mixing using magnetic stirrer. Adjust pH to 6.5 with 1 N NaOH. Remove magnet and add ultrapure water to 5000 mL. Store at room temperature for up to 6 months.

*2.4.2 RP LC-MS Solvents*

1. RP LC-MS solvent A: 0.1% difluoroacetic acid (DFA) in $H_2O$ (*see* **Note 4**). Add 1 mL DFA directly to the 1 L MS grade $H_2O$ bottle. Mix by swirling in hand.
2. RP LC-MS solvent B: 0.1% DFA in acetonitrile (*see* **Note 4**). Add 1 mL DFA directly to purchased 1 L acetonitrile bottle (*see* **Note 4**). Mix by swirling in hand.
3. Column storage solution: 80% acetonitrile in water. Add pure acetonitrile and $H_2O$ to independent solvent lines on the LC, mix to 80% acetonitrile and flush the column with a minimum of two column volumes.

## 3 Methods

Native SEC MS is preferred method for analyzing most glycoproteins due to ease of data processing (typically only one chromatographic peak to process) and high quality of the MS data (*see*

Subheading 3.1). However, native CIEX MS or RP LC-MS may be applied in cases where sample complexity (e.g., IgG mixtures or complex antibody–drug conjugates) requires chromatographic reduction of sample complexity prior to MS analysis (*see* Subheadings 3.2 and 3.3) (*see* **Note 5**). Please note that, although focus is on analysis of glycoforms of glycoproteins in the current chapter, all methods presented here have the potential analyze quality attributes not related to glycosylation, including aggregation, charge variants, and oxidized forms (*see* **Note 6**).

### 3.1 Native SEC MS

#### 3.1.1 Instrument Settings and Methods

MS instrument scan and source settings play a critical role in native MS (*see* **Note 7**). The LC and MS settings used for native SEC MS are shown in Tables 3 and 4, respectively. The MS settings from the Chromeleon™ interface is furthermore shown in Fig. 3. The chromatographic run is performed according to Table 3 and a chromatogram of a Symphogen IgG reference sample in shown in Fig. 4 (*see* **Note 8**).

**Table 3**
**Native SEC MS—LC settings**

| Vanquish Horizon | |
|---|---|
| **Parameter** | **Value** |
| Flow | 0.3 mL/min |
| Run time | 8 min |
| Column temperature | Setpoint: 20.0 °C, acceptable range: 18.0–22.0 °C |
| Thermostatting mode: | Still air |
| Preinject wash | 100 s |
| Postinject wash | 100 s |
| Max. column pressure | 483 bar |
| Autosampler temperature | Setpoint: 5.0 °C |
| Detection type:<br>· Primary wavelength (reporting)<br>· Secondary wavelength (characterization) | UV detection<br>· 280 nm<br>· 214 nm |
| Data collection rate | 4.0 Hz |
| Response time | 1.00 s |
| Narrowest peak width | 0.100 min |
| Sample injection amount | Target: 10 μg (minimum 2 μg, maximum 20 μg) |
| Sample injection volume | Maximum 20 μL |

**Table 4**
**Intact MS settings for native MS and RP LC-MS. Same settings apply to native SEC and CIEX MS**

| Q exactive plus with biopharma option | | |
|---|---|---|
| **Scan parameter** | **Native SEC and native CIEX—values** | **RP LC-MS—values** |
| Scan type | HMR—Full MS | HMR—Full MS |
| Scan range | 2500 to 8000 $m/z$ | 1500 to 6000 $m/z$ |
| Fragmentation | In-source CID 130.0 eV | In-source CID 80.0 eV |
| Resolution | 35,000 | 35,000 |
| Polarity | Positive | Positive |
| Microscans | 10 | 10 |
| Lock masses | Off | Off |
| AGC target | $3 \times 10^6$ | $3 \times 10^6$ |
| Maximum inject time | 200 | 200 |
| *HESI source* | *Value* | |
| Sheath gas flow rate | 25 | 25 |
| Aux gas flow rate | 5 | 5 |
| Sweep gas flow rate | 0 | 0 |
| Spray voltage (kv) | 4.20 | 3.5 |
| Spray current (μA) | – | – |
| Capillary temp. (°C) | 275 | 300 |
| S-lens RF level | 200.0 | 100 |
| Aux gas heater temp (°C) | 175 | 150 |

#### 3.1.2 Start-Up (New Column)

1. Equilibrate with mobile phase A by ramping up flow slowly from 0.1 mL to 0.3 mL/min.
2. Equilibration must be with at least 10 column volumes mobile phase A (140 mL).
3. Obtain stable baseline.
4. Obtain stable chromatography with a suitable reference sample (minimum 2 injections of 10 μg, up to 10 injections of 10 μg may be required).
5. It is recommended to define acceptance criteria for the reference material, e.g. retention time window for main peak, relative area limits for monomer peak and high molecular weight (HMW) peak(s). The last two injections should be pass acceptance criteria.

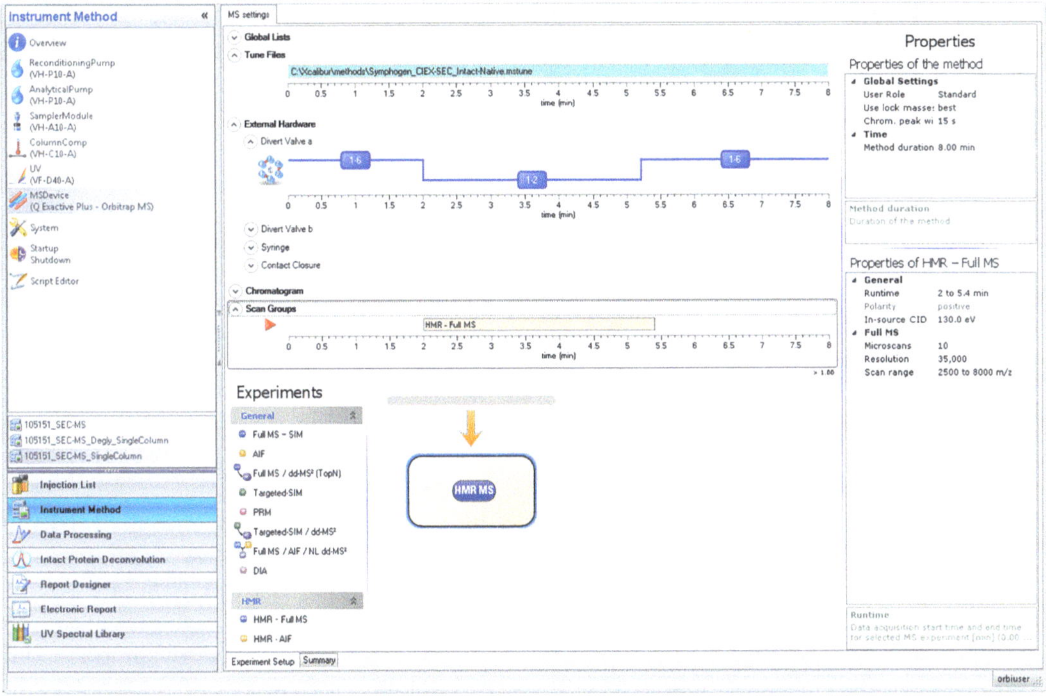


**Fig. 3** Chromeleon™ MS Settings for native SEC MS. Total run time is 8 min. Solvent path is directed to the MS between 2 and 5.2 min using the divert valve, and MS data acquisition is active between 2 and 5.4 min

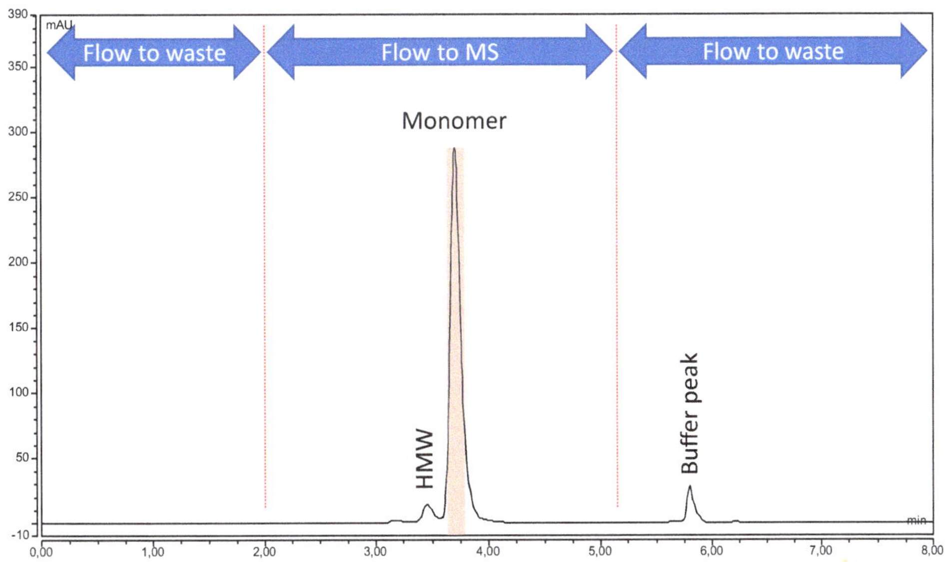


**Fig. 4** SEC UV trace of Symphogen's IgG reference sample. The flow is directed to the MS between 2 and 5.5 min. It is important to direct the flow to waste before the buffer peak elutes, to avoid contamination of the MS with the potentially harmful components (e.g., salts and buffer components) that are part of the sample formulation. HMW: high molecular weight form (e.g., IgG dimer). The summed average spectrum time window is highlighted in red (final spectrum is an average of all mass spectra in the time window). The time window is defined tightly around the monomer peak to ensure good MS signal intensity

*3.1.3 Start-Up (Used Column)*

If the column has been stored at long term storage conditions, perform the following with the flow path going to waste:

1. Equilibrate with Mobile phase A.
2. Continue, until a stable baseline is obtained.

*3.1.4 Sample Analysis*

1. Build a sample sequence in Chromeleon and run samples using the LC-MS conditions described in Subheading 3.1.1.
2. Run reference sample as system suitability control at regular intervals (*see* **Note 9**).

*3.1.5 Column Storage and Cleaning*

1. Short-term storage (<24 h): Mobile phase A at 0.1 mL/min.
2. Long-term storage (>24 h): flush column with at least 2 column volumes of storage solution (Subheading 2.4.1) (*see* **Note 10**).

## 3.2 Native CIEX MS

*3.2.1 Instrument Settings and Methods*

If sample complexity is high (i.e., there is a significant overlap in isotope pattern of different glycoproteins) native CIEX MS can reduce complexity prior to MS analysis. Native CIEX MS may be relevant for antibody mixtures and antibody–drug conjugates (*see* **Note 11**). Two native CIEX MS methods, named low gradient and high gradient, have been developed for analysis of IgGs with a pI below and above approximately 8.5, respectively (*see* **Note 12**). The LC gradient and MS settings for low and high gradient are shown in Figs. 5 and 6, respectively. Details on the MS scan and source

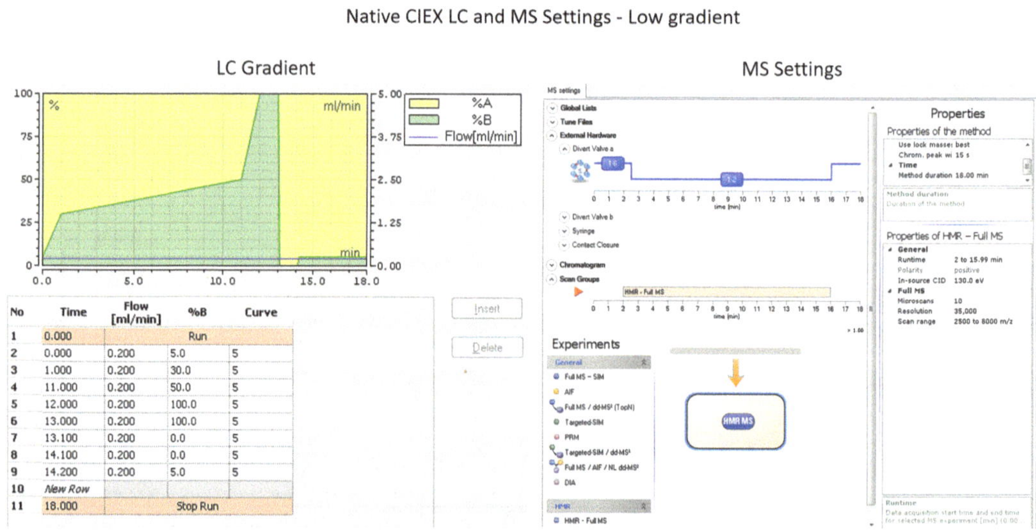


**Fig. 5** Native CIEX LC and MS Settings for Low Gradient Method. The low gradient is generally applicable to IgGs with a pI below 8.5. The optimal gradient will however need to be determined experimentally. Theoretical pIs calculated in GPMAW software (http:/www.gpmaw.com/) according to [19]. The MS data acquisition windows and divert valve settings are shown for each method. For the Low Gradient Method, the solvent path is directed to the MS between 2.5 and 16 min using the divert valve, and MS data acquisition is active between 2 and 16 min

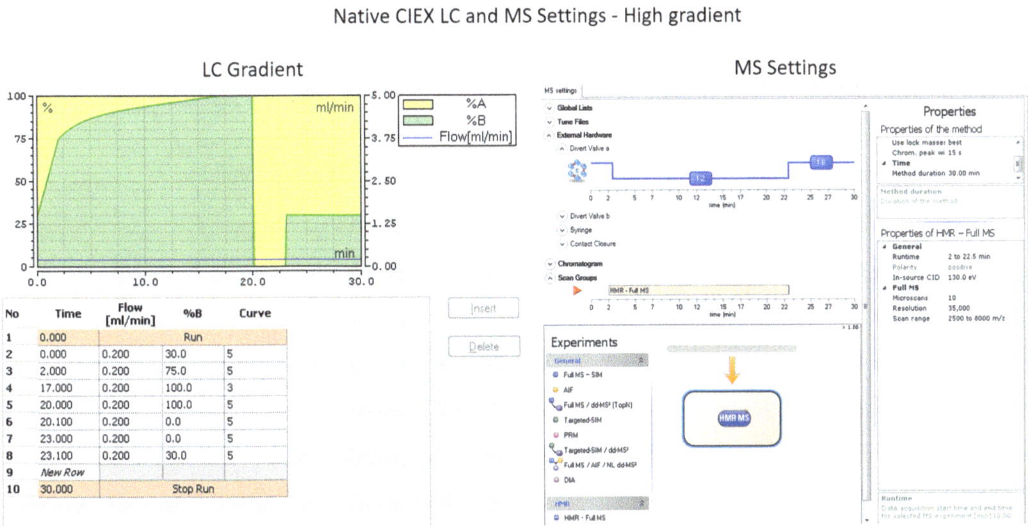


**Fig. 6** Native CIEX LC and MS Settings for High Gradient Method. The High Gradient Method is generally applicable to IgGs with a pI above 8.5. The optimal gradient will however need to be determined experimentally. Theoretical pIs calculated in GPMAW software (http:/www.gpmaw.com/) according to [19]. The MS data acquisition windows and divert valve settings are shown for each method. For the High Gradient Method, the solvent path is directed to the MS between 2.5 and 22.5 min using the divert valve, and MS data acquisition is active between 2 and 22.5 min

settings are shown in Table 4. Details of the LC settings are summarized in Table 5. A native CIEX chromatogram of a mixture of six IgGs is shown in Fig. 7. For such a complex, sample chromatographic separation prior to MS detection is required and this can be achieved by native CIEX MS.

### *3.2.2 Start-Up (New Column)*

1. Ramp flow up slowly from 0.1 mL to 0.2 mL/min using solvent according to gradient start conditions (*see* Figs. 5 or 6).
2. Equilibration must be with at least 10 column volumes mobile phase.
3. Obtain stable baseline.
4. Obtain stable chromatography with a suitable reference material (minimum 2 injections of 25 μg, more injections may be required).
5. It is recommended to define acceptance criteria for the reference material, e.g. retention time window for major charge variant peaks, relative area limits for major charge variant peak(s). The last two injections should be pass acceptance criteria.

**Table 5**
**LC Methods. For gradient conditions of the gradient methods refer to Figs. 5, 6 and 8**

| Vanquish Horizon | | |
|---|---|---|
| **Parameter** | **Native CIEX MS—Values** | **RP LC-MS—Values** |
| Flow | 0.2 mL/min | 0.3 mL/min |
| Run time | Low-gradient: 18 min<br>High-gradient: 30 min | 35 min |
| Column temperature | Setpoint: 25.0 °C, acceptable range: 23.0–27.0 °C | Setpoint: 70.0 °C, acceptable range: 69.0–71.0 °C |
| Thermostatting mode: | Still air | Still air |
| Preinject wash | 60 s | 60 s |
| Postinject wash | 60 s | 60 s |
| Max. Column pressure | 483 Bar | 275 bar |
| Autosampler temperature | Setpoint: 5.0 °C | Setpoint: 5.0 °C |
| Detection type:<br>· Primary wavelength<br>· Secondary wavelength | UV detection<br>1. 280 nm<br>2. 214 nm | UV detection<br>1. 280 nm<br>2. 214 nm |
| Data collection rate | 4.0 Hz | 4 Hz |
| Response time | 1.00 s | 1.00 s |
| Narrowest peak width | 0.100 min | 0.100 min |
| Sample injection amount | Target: 25 μg (per IgG for mixtures) | Target: 10 μg (per IgG for mixtures) |
| Sample injection volume | Maximum 130 μL | Maximum 20 μL |

#### *3.2.3 Start-Up (Used Column)*

1. Equilibrate column with solvent using gradient start conditions (*see* Figs. 5 or 6).
2. Continue, until a stable baseline is obtained.

#### *3.2.4 Sample Analysis*

1. Build a sample sequence in Chromeleon and run samples using the LC-MS conditions described in Subheading 3.2.1.
2. Run reference material as system suitability control at regular intervals (*see* **Note 9**).

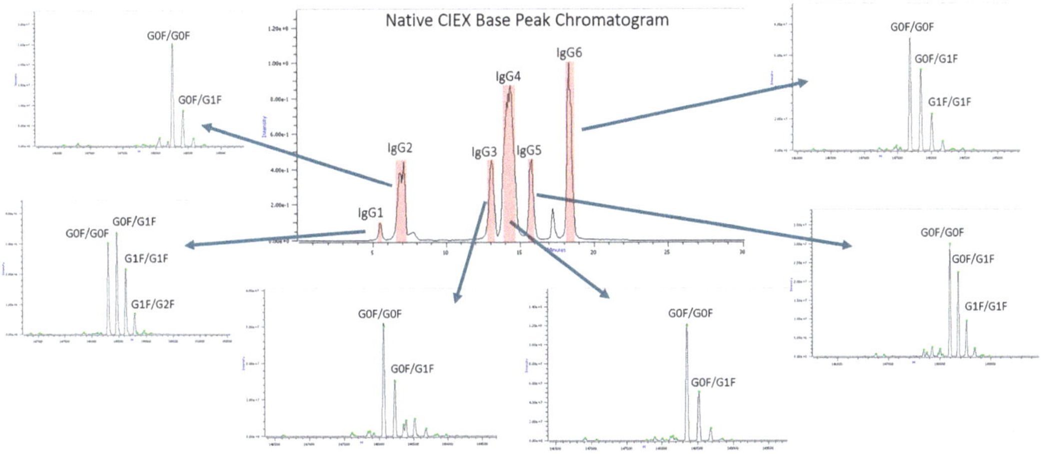


**Fig. 7** Native CIEX base peak chromatogram of a mixture of six IgGs, including deconvoluted spectra. For complex samples native CIEX MS can be applied to reduced sample complexity prior to MS detection. When glycoproteins are chromatographically resolved as shown in this example, a deconvoluted spectrum, and thus glycoprofile, can be generated for each glycoprotein in the sample. The summed average spectrum time windows are highlighted in red. These are defined tightly around the main peak for each IgG to avoid overlapping IgG signals and to ensure good MS signal intensity. The peak eluting between 17 and 18 min represents an acidic (glycated) form of IgG6

#### 3.2.5 Column Storage (See Subheading 3.1.5)

### 3.3 Intact RP LC-MS

#### 3.3.1 Instrument Settings and Methods

If sample complexity is high (e.g., antibody mixtures and complex antibody–drug conjugates) RP LC-MS can reduce complexity prior to MS analysis. However, it is recommended to run native SEC MS whenever possible, due to simple data processing and the high quality of the MS data (*see* **Note 13**). The RP LC gradient and MS settings are shown in Fig. 8, and details on the MS scan and source settings are shown in Table 4. The LC settings are summarized in Table 5. A RP LC chromatogram of a mixture of two IgGs is shown in Fig. 9.

#### 3.3.2 Start-Up (New Column)

1. Equilibrate column with solvent using gradient start conditions (*see* Fig. 8).
2. Inject a suitable number of reference material samples (10 μg) until acceptable and reproducible chromatographic performance is obtained with respect to retention time, peak resolution, and selectivity.

#### 3.3.3 Start-Up (Used Column)

1. Equilibrate column with solvent using gradient start conditions (*see* Fig. 8).
2. Continue, until a stable baseline is obtained.

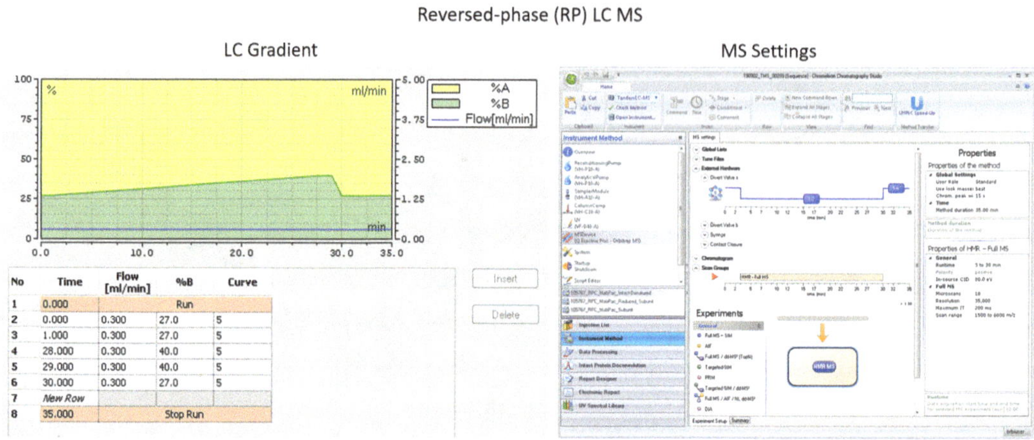


**Fig. 8** RP LC-MS—LC and MS Settings. The RP gradient was optimized for IgG1 molecules. Other gradients may be required for other IgG molecules and especially for non-IgG glycoproteins. In those cases, the optimal gradient will have to be determined experimentally. The solvent path is directed to the MS between 3 and 30 min using the divert valve, and MS data acquisition is active between 3 and 30 min. Notice that the scan range and in-source CID differ from the native MS setting. Refer to Table 4 for further details on the MS settings

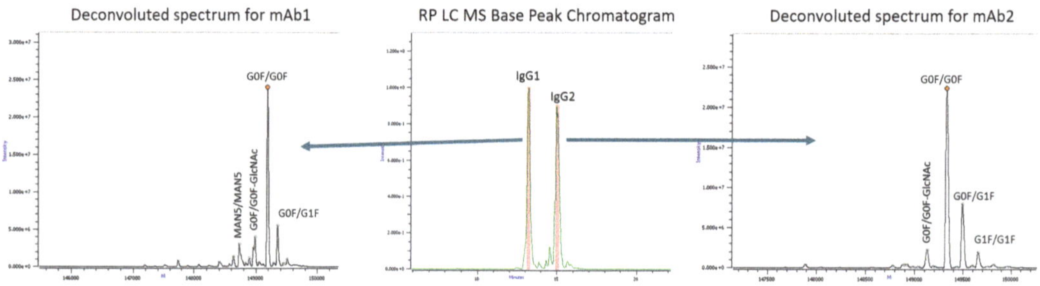


**Fig. 9** RP LC-MS base peak chromatogram of a mixture of two IgGs, including deconvoluted spectra. For complex samples RP LC-MS can be applied to reduced sample complexity prior to MS detection. However, when reduction of sample complexity is not required it is recommended to use native SEC MS, due to simpler data processing and higher spectral data quality (*see* Subheading 3.1). The RP LC gradient may need to be optimized for individual glycoproteins, but the presented gradient is a good starting point for IgG1 samples (*see* Fig. 8). The summed average spectrum time windows are highlighted in red. These are defined tightly around the main peak for each IgG to avoid overlapping IgG signals and to ensure good MS signal intensity

#### 3.3.4 Sample Analysis

1. Build a sample sequence in Chromeleon and run samples using the LC-MS conditions described in Subheading 3.3.1.
2. Run reference material as system suitability control at regular intervals (*see* **Note 9**).

#### 3.3.5 Column Storage

1. Short term storage (<24 h): run 27% mobile phase A and 73% mobile phase B at 0.05 mL/min.
2. Long term storage (>24 h): flush column with at least 2 column volumes of 80% acetonitrile and 20% water. Store at room temperature.

### 3.4 Data Processing

Protein deconvolution, peak assignment and relative quantitation of glycoforms were performed using Protein Metrics Intact Mass™ software [14]. Refer to user manual for details. Other software solution may be used, such as Genedata Expressionist [16] and Thermo BioPharma Finder [15], but it is beyond the scope of this chapter to provide instruction on the use of the different software applications. Instead, general considerations and critical settings for processing of data from the current native SEC, native CIEX MS and RP LC-MS methods (in Protein Metrics Intact Mass™) are presented here.

The intact MS data processing involves (1) deconvolution of raw mass spectra, (2) assignment of glycoforms to the peaks in the deconvoluted spectrum, and (3) relative quantitation of the intensities of the assigned glycoforms in the deconvoluted spectrum. Assignment of glycoforms to the peaks in the deconvoluted spectrum is based on the theoretical molecular mass of the analyzed and a delta mass list, containing a list of mass changes associated with the relevant glycoforms. A list of biantennary glycans typically associated with recombinant IgGs expressed in Chinese hamster ovary (CHO) expression systems is shown in Table 6. The list includes both IgGs carrying a single glycan on one heavy chain or IgG carrying two glycans, one on each heavy chain. Figures 10 and 11 summarizes the settings used for processing native SEC/CIEX MS and RP LC-MS data in Protein Metrics Intact Mass™, respectively, including the deconvolution settings (*see* **Note 14**). Figure 12 shows the delta mass list used at Symphogen, including mass tolerance (6 Da). For a correctly calibrated Orbitrap MS the measured, deconvoluted mass should not deviate more than 2 Da (~13 ppm) form the theoretical mass for an IgG. The list contains all the typical glycoforms derived from a recombinant CHO-based expression system, including the non-glycosylated form, single glycan forms and glycoforms carrying a glycan on each heavy chain, including the G0F/G0F form which is typically the dominant form in CHO-based expression systems (*see* **Note 15**). Figure 13 shows the deconvoluted spectrum of a Symphogen reference sample, including assignment of glycoforms, and Fig. 14 shows relative intensities of assigned glycoforms, both graphically and numerically. Figure 14 illustrates a key advantage of the intact MS approach, which is that information is obtained at the intact glycoprotein level, that is, structural insight is obtained on the level of nonglycosylated IgG forms, IgG forms glycosylated on only one of the heavy chains (monoglycosylated) and IgG forms glycosylated on both heavy chains (diglycosylated). This information is lost when employing strategies based on analysis of released glycans or glycopeptides. Native MS on the Orbitrap platform thus allows confident assignment and quantitation of approximately 20 glycoforms of Symphogen's reference material, which represents

**Table 6**
**Delta Mass List—IgG Glycoforms. The list contains glycoforms that are typically observed for IgGs in a CHO-based expression system**

| Glycoform | Description (single glycans) | Delta Mass (average, Dalton) |
|---|---|---|
| Nonglycosylated | Nonglycosylated | 0.0 |
| Man3 | Hex(3)HexNAc(2) | 892.8 |
| Man4 | Hex(4)HexNAc(2) | 1054.9 |
| Man5 | Hex(5)HexNAc(2) | 1217.2 |
| Man6 | Hex(6)HexNAc(2) | 1379.2 |
| G0F-GlcNAc | dHex(1)Hex(3)HexNAc(3) | 1242.2 |
| G0 | Hex(3)HexNAc(4) | 1299.3 |
| Man5F | dHex(1)Hex(5)HexNAc(2) | 1363.3 |
| G0F | dHex(1)Hex(3)HexNAc(4) | 1445.4 |
| G1 | Hex(4)HexNAc(4) | 1461.3 |
| G1F | dHex(1)Hex(4)HexNAc(4) | 1607.6 |
| G2 | Hex(5)HexNAc(4) | 1623.5 |
| G2F | dHex(1)Hex(5)HexNAc(4) | 1769.7 |
| Man3/Man3 | | 1785.6 |
| G1F + NeuAc | dHex(1)Hex(4)HexNAc(4)NeuAc(1) | 1898.8 |
| G2F + NeuAc | dHex(1)Hex(5)HexNAc(4)NeuAc(1) | 2061.0 |
| Man4/Man4 | | 2109.8 |
| G2F + 2NeuAc | dHex(1)Hex(5)HexNAc(4)NeuAc(2) | 2352.2 |
| Man5/Man5 | | 2434.4 |
| Man5/G0 | | 2662.6 |
| G0F/G0F-GlcNAc | | 2687.6 |
| G0F/G0F-Fuc | | 2744.6 |
| Man6/Man6 | | 2758.4 |
| G0F/G0F | | 2890.8 |
| G0F/G1F | | 3053.0 |
| G1F/G1F | | 3215.1 |
| G0F/G1F + NeuAc | | 3344.2 |
| G1F/G2F | | 3377.3 |
| G1F/G1F + NeuAc | | 3506.4 |
| G2F/G2F | | 3539.4 |
| G1F/G2F + NeuAc | | 3668.5 |
| G2F/G2F + NeuAc | | 3830.7 |
| G1F/G2F + 2NeuAc | | 3959.8 |
| G2F/G2F + 2NeuAc | | 4121.9 |

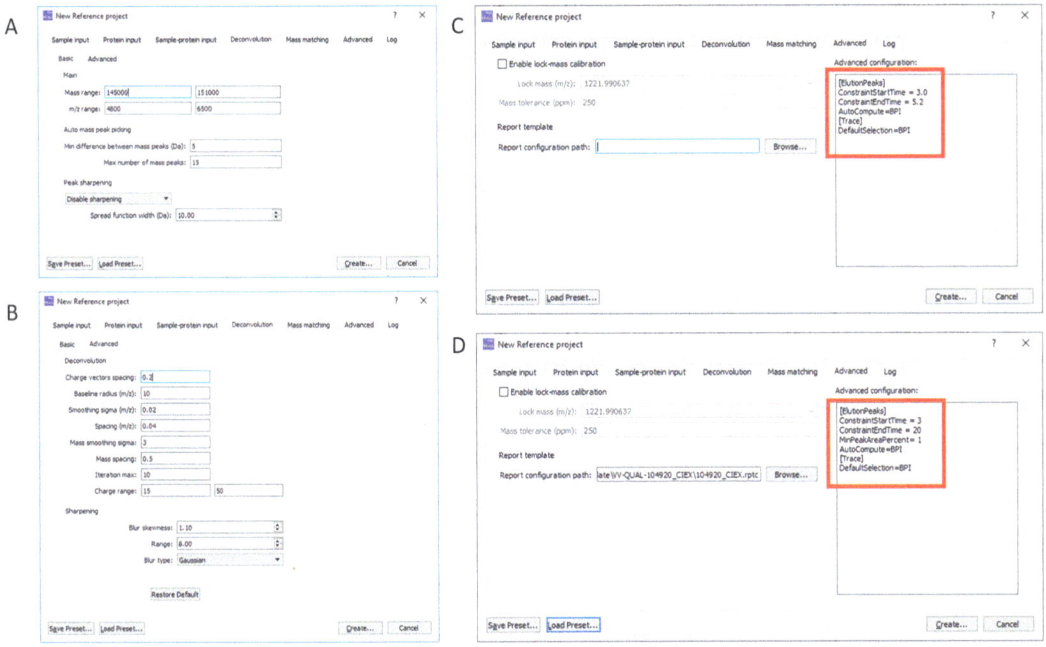


**Fig. 10** Critical settings for processing of native SEC and CIEX MS data in Protein Metrics Intact Mass software. Settings are based on the analysis of recombinant IgGs and will have to be adjusted (e.g., mass range for deconvolution) if different glycoproteins are being processed. (**a**) Deconvolution basic settings. These are the same for all native SEC MS and native CIEX MS methods. (**b**) Deconvolution advanced settings. These are the same for all native SEC MS and native CIEX MS methods. The parameters have been adjusted to have minimal impact on the spectral data quality (*see* **Note 14**). (**c**) Advanced settings for native SEC MS method only. The settings return the base peak intensity (BPI) trace. Integration window adjusted to elute time of monomer peak. (**d**) Advanced settings for native CIEX MS method only. The settings return the base peak intensity (BPI) trace in the result windows. Integration window adjusted to native CIEX gradient

a typical recombinant IgG1. Please be aware that IgG glycation (chemical addition of hexose to primary amines [17]) may impact the glycosylation distribution (*see* **Note 16**).

## 4 Notes

1. For glycoproteins with multiple glycosylation sites and multiple glycan structures at each site intact MS may not be the optimal choice. Such glycoproteins have an inherently complex heterogeneity (i.e., a very large number of glycoforms), which may result in MS signal dilution/suppression and challenges with overlapping isotopic patterns which cannot be resolved spectrally. For such glycoproteins alternative methods are recommended, such as peptide mapping, which can provide site specific and quantitative information about glycan structures.

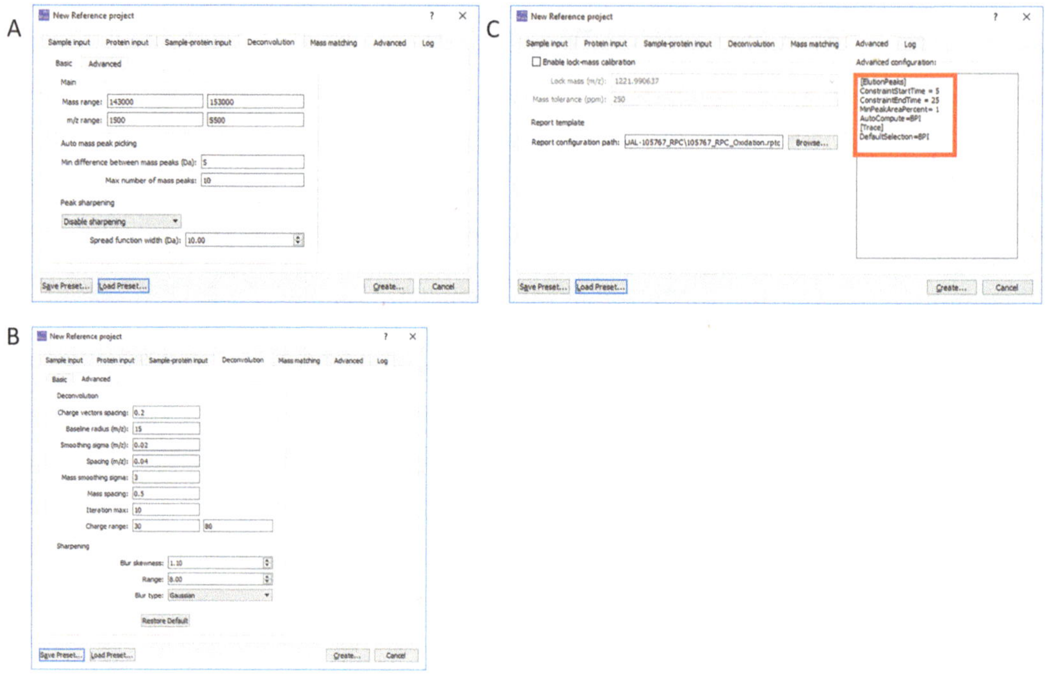


**Fig. 11** Critical settings for processing of intact RP LC-MS data in Protein Metrics Intact Mass software. Settings are based on the analysis of recombinant IgGs and will have to be adjusted (e.g., mass range for deconvolution) if different glycoproteins are being processed. (**a**) Deconvolution Basic settings. *m*/*z* range should be lower than the values for native MS (**b**) Deconvolution advanced settings. Charge range should be higher than the values for native MS. The parameters have been adjusted to have minimal impact on the spectral data quality (*see* **Note 13**). (**c**) Advanced settings. The settings return the base peak intensity (BPI) trace in the results windows. Integration window adjusted to RP LC gradient

2. Do not filter or expose solvents to reusable glass ware (e.g., measuring cylinder). Simply dissolve reagents directly in the purchased glass bottle, assuming the purchased bottle contains 1 L of MS grade $H_2O$.
3. Native MS employs solvents with low buffering capacity and ionic strength. Consequently, choice of column is critical for effective chromatographic performance, such as the separation of glycoproteins from, for example, buffer and salt components of the sample which may harm the MS instrument. Column reconditioning may be challenging with native solvents, and consequently column dimensions are critical, that is, when possible smaller column volume is preferable to larger column volume (particularly for native CIEX MS).
4. It is essential to use difluoroacetic acid (DFA) to obtain the best combination of chromatographic performance and MS signal intensity. Trifluoroacetic acid (TFA) provides good chromatographic performance but is not compatible with MS of

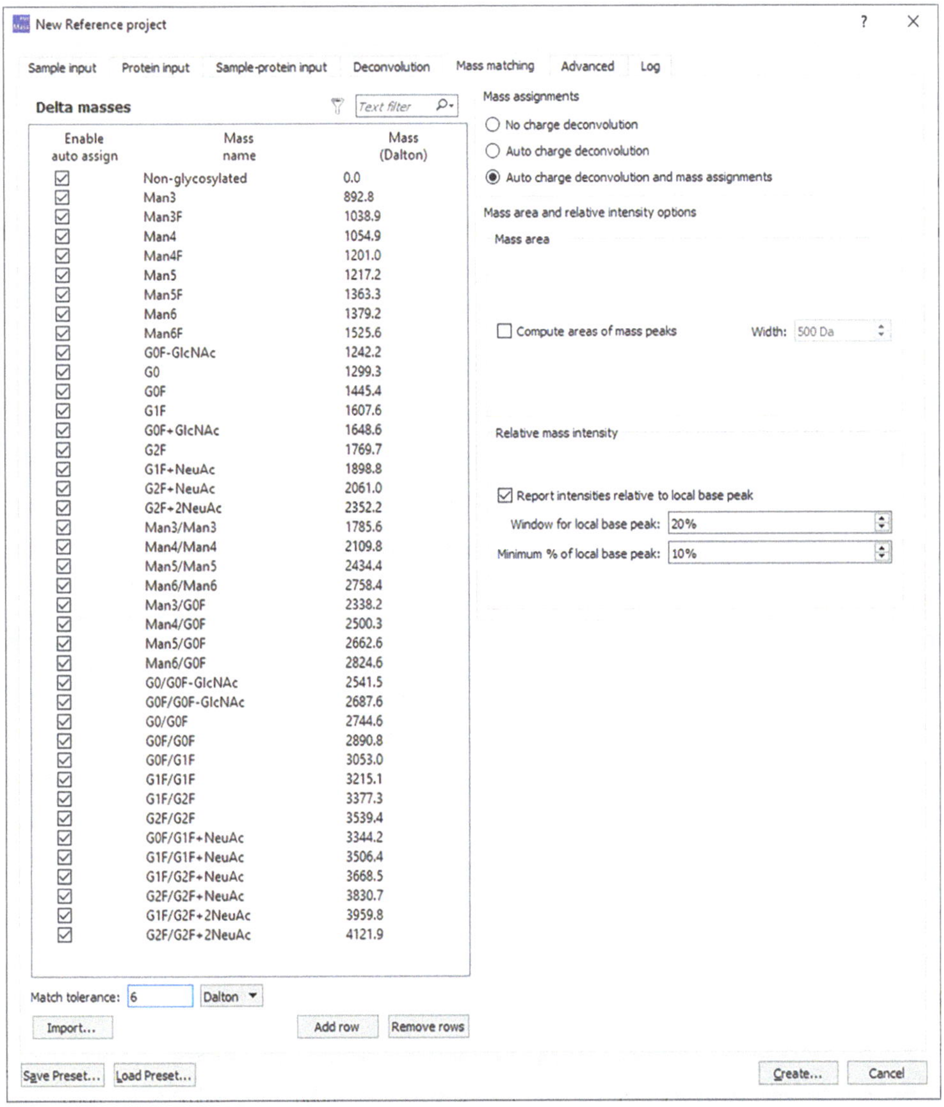


**Fig. 12** Delta mass list of typical CHO-derived glycoforms for recombinant IgGs. The non-glycosylated (a-glycosylated) mass of the analyzed IgG is defined as the reference mass and the Protein Metrics Intact Mass™ associates glycoforms with the masses on the deconvoluted spectrum based on the delta mass list. An example of the assigned glycoforms for Symphogen reference material is shown in Fig. 13

large biomolecules. Formic acid results in good MS signal intensity, but chromatographic performance is poor for large, intact biomolecules.

5. Use native SEC MS whenever this is adequate, that is, when sample complexity does not require separation of glycoproteins prior to MS detection. As a rule of thumb the Orbitrap MS detector can adequately resolve and quantitate glycoforms that differ by more than 25 Da in average molecular mass for IgGs

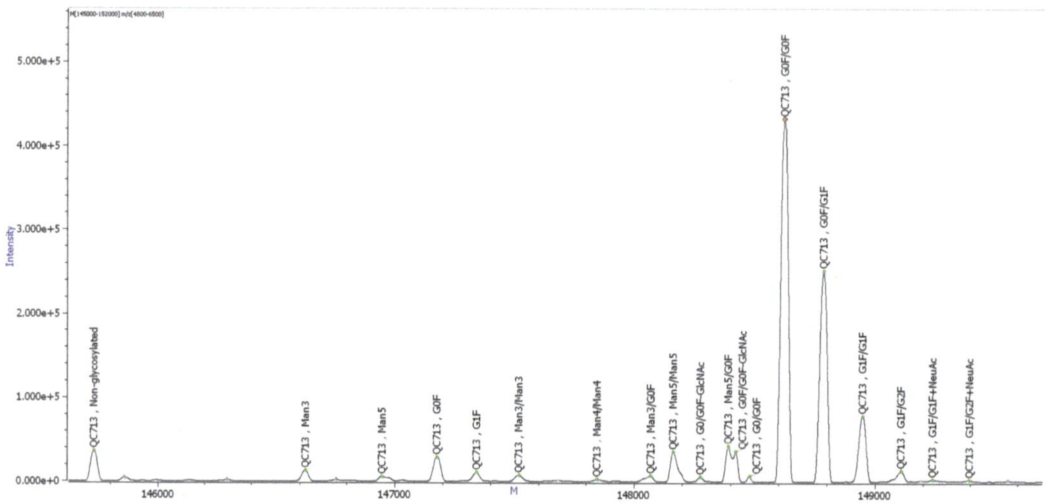


**Fig. 13** Deconvoluted mass spectrum of Symphogen reference material with glycoforms assigned by Protein Metrics Intact Mass™ software. A key strength of intact glycoprotein analysis is that it provides information about the glycoforms of the assembled glycoprotein, including the presence of non-glycosylated form and forms carrying a single glycan on one of the heavy chains of IgGs. The high spectral resolution and accuracy of the of the Orbitrap MS furthermore allow distinction between glycoforms with a small mass differences and overlapping isotope distributions, such as Man5/G0F and G0F/G0F-GlcNAc (delta mass 25 Da)

A

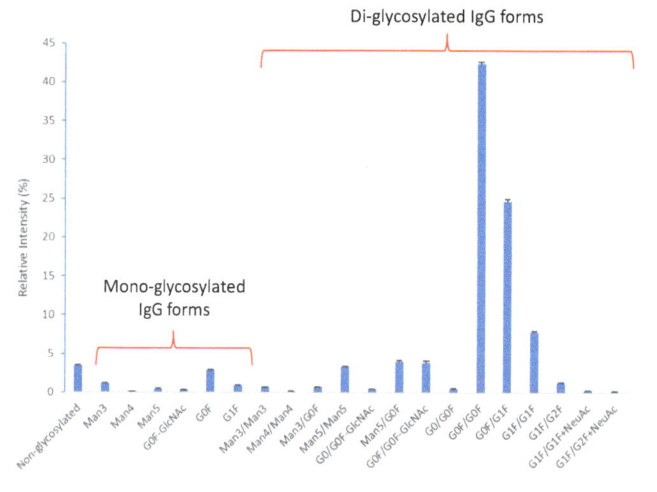


B

| Glycoform | Average Relative Intensity | RSD |
|---|---|---|
| G0F/G0F | 42.3 % | 0.7% |
| G0F/G1F | 24.6 % | 1.4% |
| G1F/G1F | 7.8 % | 2.6% |
| Man5/G0F | 4.1 % | 3.0% |
| G0F/G0F-GlcNAc | 3.8 % | 6.4% |
| Non-glycosylated | 3.5 % | 2.4% |
| Man5/Man5 | 3.3 % | 3.3% |
| G0F | 2.9 % | 3.1% |
| G1F/G2F | 1.2 % | 7.9% |
| Man3 | 1.2 % | 4.5% |
| G1F | 0.9 % | 6.6% |
| Man3/G0F | 0.7 % | 5.2% |
| Man3/Man3 | 0.7 % | 6.5% |
| Man5 | 0.5 % | 6.2% |
| G0/G0F | 0.5 % | 22.6% |
| G0/G0F-GlcNAc | 0.5 % | 11.5% |
| G0F-GlcNAc | 0.4 % | 11.6% |
| G1F/G1F+NeuAc | 0.3 % | 10.0% |
| Man4/Man4 | 0.2 % | 13.1% |
| G1F/G2F+NeuAc | 0.2 % | 14.3% |
| Man4 | 0.1 % | 19.6% |

**Fig. 14** Relative peak intensities for identified glycoforms of Symphogen reference material, which represents a typical IgG1. The values were generated using Protein Metrics Intact Mass™ software [14]. Results are based on a lead selection study in which 384 IgG leads were analyzed by native SEC MS together with Symphogen reference material, which was analyzed after every 25th sample. A total of 18 reference standard runs were performed. (**a**) Bar chart of the average ($n = 18$) relative intensity of identified glycoforms, including error bars (standard deviation). Glycoforms are sorted after increasing mass from left to right. (**b**) Average relative intensity ($n = 18$) for identified glycoforms, including relative standard deviation (RSD). The glycoforms are sorted based relative intensity, going from high to low

(less for smaller glycoproteins, more for larger glycoproteins). For IgG mixtures, in which significant isotopic overlap of different of glycoforms may be present, it may be preferable to use native CIEX MS or RP LC-MS. The aim of native CIEX MS and RP LC-MS is to chromatographically separate glycoproteins prior to MS analysis when this is required.

6. SEC can be used for the analysis of aggregation and to some extent fragmentation. CIEX can be used for the analysis of charge variants derived from e.g. deamidation, glycation and C-terminal lysine (on IgGs). RP LC can be used for the analysis variants with altered hydrophobicity, for example variants derived from oxidation, isomerization, and fragmentation.
7. MS instrument source and scan settings are critical for native MS performance. The in-source CID is particularly critical, and a good starting point for native MS is 130 eV. For a general recommendation of source and scan settings for intact MS refer to [18].
8. Make sure to switch the solvent flow to waste using the divert valve before the buffer peak elutes as this may contain components that are not compatible with the MS instrument (*see* Fig. 4).
9. Run the system suitability test at start and end of sequence and at regular intervals for large sample sets (e.g., for every 25th sample). Define relevant system suitability criteria, such as minimum main peak intensity, retention time window for main peak, and relative peak area(s) for the UV trace, and mass accuracy and relative peak intensity limits for major glycoforms in the deconvoluted spectrum.
10. Make sure the flow goes to waste when flushing the column with storage solution, as the storage solution is damaging to the MS instrument. Alternatively, flush the column with storage solution on a different LC system.
11. Be aware that some glycoforms (such sialic acid containing forms) may separate chromatographically from other glycoforms of the same glycoprotein by CIEX. In such cases, the data processing software needs to deconvolute, assign, and quantitate chromatographically resolved glycoforms (*see* Subheading 3.4).
12. The optimal pH gradient will have to be determined experimentally; this is just a rough guideline. Be aware that the native CIEX methods are only suitable for glycoproteins with a pI above 7. For glycoproteins with a pI below 7 anion exchange chromatography (AIEX) may be a better option.
13. RP LC-MS is only recommended if native MS is not an option (i.e. if the Orbitrap is not equipped with a Biopharma option or sample complexity requires the use of RP separation prior to

MS), since higher quality MS data is obtained by native MS. Furthermore, RP LC-MS is more prone to introduce sample artefacts (e.g., cleavage of acid labile sites) due to the harsher analysis condition (high column temperature, acidic pH, organic solvents).

14. The Orbitrap platform with BioPharma Option produces excellent spectral raw data in native MS mode. Consequently, the need for spectral data processing (e.g. spectral smoothing) is minimal, and the deconvolution settings should be set to have minimal impact on spectral data quality. That is, the appearance of spectral features should highly similar between the raw and the deconvoluted data, in order to preserve quantitative information and minor spectral features, as shown in Fig. 1.
15. For non-CHO expression systems and non-IgG glycoproteins other glycoform list may be appropriate [5].
16. The molecular mass of G0F/G1F equals that of G0F/G0F + 1 glycation, the mass of G1F/G1F equals that of G0F/G1F + 1 glycation, and so on. To get an estimate of the glycation level (using the methods presented in the current chapter) the IgG samples can be deglycosylated using PNGase F according to manufacturer's instruction prior to intact MS analysis. This will result in a deconvoluted mass spectrum in which the main peak represent the deglycosylated IgG. IgGs carrying a single glycation are seen as the main peak +162 Da, IgGs carrying two glycations are seen as the main peak $+2 \times 162$ Da, and so on.

## References


1. Beck A, Wagner-Rousset E, Ayoub D et al (2013) Characterization of therapeutic antibodies and related products. Anal Chem 85:715–736. https://doi.org/10.1021/ac3032355
2. Kayser V, Chennamsetty N, Voynov V et al (2011) Glycosylation influences on the aggregation propensity of therapeutic monoclonal antibodies. Biotechnol J 6:38–44. https://doi.org/10.1002/biot.201000091
3. Liu L (2015) Antibody glycosylation and its impact on the pharmacokinetics and pharmacodynamics of monoclonal antibodies and Fc-fusion proteins. J Pharm Sci 104:1866–1884. https://doi.org/10.1002/jps.24444
4. CMC Biotech Working group (2009) A-Mab: a case study in bioprocess development. CASSS 2.1:
5. Zhang L, Luo S, Zhang B (2016) Glycan analysis of therapeutic glycoproteins. MAbs 8:205–215. https://doi.org/10.1080/19420862.2015.1117719
6. Makarov A, Denisov E (2009) Dynamics of ions of intact proteins in the Orbitrap mass analyzer. J Am Soc Mass Spectrom 20:1486–1495. https://doi.org/10.1016/j.jasms.2009.03.024
7. Bailey AO (2018) The quest for highly effective resolution: therapeutic protein characterization via separations-coupled native MS. Am Chem Soc:1–15
8. van de Waterbeemd M, Fort KL, Boll D et al (2017) High-fidelity mass analysis unveils heterogeneity in intact ribosomal particles. Nat Methods 14:283–286. https://doi.org/10.1038/nmeth.4147
9. Gault J, Donlan JAC, Liko I et al (2016) High-resolution mass spectrometry of small molecules bound to membrane proteins. Nat Methods 13:333–336. https://doi.org/10.1038/nmeth.3771

10. Belov ME, Damoc E, Denisov E et al (2013) From protein complexes to subunit backbone fragments: a multi-stage approach to native mass spectrometry. Anal Chem 85:11163–11173. https://doi.org/10.1021/ac4029328
11. Füssl F, Trappe A, Cook K et al (2018) Comprehensive characterisation of the heterogeneity of adalimumab via charge variant analysis hyphenated on-line to native high resolution Orbitrap mass spectrometry. MAbs 11 (1):116–128. https://doi.org/10.1080/19420862.2018.1531664
12. Füssl F, Cook K, Scheffler K et al (2018) Charge variant analysis of monoclonal antibodies using direct coupled pH gradient cation exchange chromatography to high-resolution native mass spectrometry. Anal Chem 90:4669–4676. https://doi.org/10.1021/acs.analchem.7b05241
13. Kristensen DB, Jensen PF, Slot TM, et al (2019) Expanding applications of chromatographic techniques coupled to native mass spectrometry for routine analysis of large and complex biopharmaceutical molecules. In: HPLC 2019. Milan
14. Protein Metrics Intact Mass software. https://www.proteinmetrics.com/products/intact-mass/
15. Thermo BioPharma Finder software. https://www.thermofisher.com/dk/en/home/industrial/mass-spectrometry/liquid-chromatography-mass-spectrometry-lc-ms/lc-ms-software/multi-omics-data-analysis/biopharma-finder-software.html
16. Genedata Expressionist. https://www.genedata.com/products/expressionist/biotherapeutics/intact-proteins/
17. Fischer S, Hoernschemeyer J, Mahler H-C (2008) Glycation during storage and administration of monoclonal antibody formulations. Eur J Pharm Biopharm 70:42–50. https://doi.org/10.1016/j.ejpb.2008.04.021
18. Scheffler K, Damoc E (2017) Full characterization of heterogeneous antibody samples under denaturing and native conditions on the Q Exactive BioPharma mass spectrometer. Appl Note 72348
19. Skoog B, Wichman A (1986) Calculation of the isoelectric points of polypeptides from the amino acid composition. TrAC Trends Anal Chem 5:82–83. https://doi.org/10.1016/0165-9936(86)80045-0

# Chapter 3

# Analysis of Intact Glycoproteins by Matrix-Assisted Laser Desorption/Ionization Time-of-Flight Mass Spectrometry

**Estela Giménez, Montserrat Mancera-Arteu, Fernando Benavente, and Victoria Sanz-Nebot**

## Abstract

Matrix-assisted laser desorption/ionization time-of-flight mass spectrometry (MALDI-TOF MS) can be regarded as a key tool to rapidly obtain molecular mass information of intact glycoproteins in glycoproteomic studies and quality control of recombinant biopharmaceuticals. However, MALDI-TOF MS of these glycosylated compounds is a tricky task due to its low ionization efficiency and fragmentation of labile groups such as sialic acids.

Here, we offer the reader a practical overview of the available methodologies for the confident analysis of intact glycoproteins with different glycosylation degree by MALDI-TOF MS. The three proposed methods fulfil the requirements of reproducibility and low extent of glycan fragmentation required to successfully analyze intact glycoproteins.

**Key words** MALDI, Intact glycoprotein, Matrix, Ionic liquid, Glycosylation degree

## 1 Introduction

Glycosylation is an important posttranslational protein modification which modulates a wide variety of biological functions and plays a major role in disease initiation and progresion [1–3]. In this context, investigations in glycoproteomics, which demand sensitive and selective analytical tools for glycoprotein analysis, have gained importance in biomedical and biotech research. Mass spectrometry has proven itself to be a leading technology in the analysis of glycoproteins due to its reliability and sensitivity [4–8]. Identification and comprehensive characterization of glycoproteins usually requires multiple complementary mass spectrometry approaches, including the analysis of the intact protein (*top-down* approach) and the glycopeptides or glycans generated by enzymatic digestion (*bottom-up* approach).

Arnaud Delobel (ed.), *Mass Spectrometry of Glycoproteins: Methods and Protocols*, Methods in Molecular Biology, vol. 2271, https://doi.org/10.1007/978-1-0716-1241-5_3, 

Matrix-assisted laser desorption/ionization time-of-flight mass spectrometry (MALDI-TOF MS) is widely used in glycoproteomics because its specific advantages such as tolerance toward different buffers, uncomplicated spectra due to the almost exclusive generation of singly charged molecular ions, appropriate sensitivity, simplicity of operation, and possibility of rapid and high-throughput analyses with low sample and reagents consumption [8–10]. Thus, this technique is an excellent tool to rapidly obtain molecular mass information of intact glycoproteins in glycoproteomic studies or quality control of recombinant biopharmaceuticals. However, MALDI-TOF MS of glycosylated compounds such as intact glycoproteins is a difficult task due to its poor ionization efficiency and the presence of labile groups such as sialic acids or *N*-acetylglucosamine (GlcNAc) units, which are prone to fragmentation in the ion source or during acceleration [11, 12]. Moreover, the quality of the mass spectra is strongly dependent on the choice of the matrix compound and its capability to absorb laser energy and assist the ionization process. Many authors have described the properties that an ideal matrix should have, and a wide variety of substances have been tested and applied as MALDI matrices for the analysis of different compounds [13–16]. For glycoprotein analysis, benzoic acid derivatives like 2,5-dihydroxibenzoic acid (DHB) have been described as "cool" or soft matrices, since they enable minimized glycan fragmentation [9, 12, 17]. These MALDI matrices seem to be more suitable than other typical MALDI matrices derived from cinnamic acid (e.g., sinapinic acid [SA], ferulic acid [FA], or α-cyano-4-hydroxycinnamic acid [CCA]).

Besides the choice of the matrix, results obtained in MALDI experiments are strongly influenced by the sample-matrix preparation procedures. Among the different procedures available in the literature [12, 16, 18], the dried-droplet method is the most widely used in MALDI-TOF MS. However, this method is not suitable enough for some matrices and analytes as it may promote the generation of heterogeneous spots. This issue affects the reproducibility of the results and increases the time required to make the measurements because of the need to find a "sweet" spot. Therefore, alternative crystallization procedures as fast evaporation or vacuum drying methods have been investigated in the analysis of intact glycoproteins [12].

To overcome many of the shortcomings of solid matrices, Armstrong et al. [19] successfully introduced ionic liquids as MALDI matrices in 2001. Ionic liquid matrices (ILMs) are organic salts, with a melting point at or below 100 °C, formed by equimolar mixtures of the typical acidic MALDI matrix compounds (e.g., SA or DHB) with organic bases (e.g., tributylamine or pyridine) [14–16, 20]. They are easily prepared and require no cocrystallization with the analyte, providing better spot-to-spot and shot-to-shot reproducibility. This fact can make the automatic acquisition in

MALDI-TOF MS instruments easier and is crucial for quantitative analysis. In addition, ILMs have been reported to show a reduction of fragmentation of labile groups in glycoprotein analysis compared to solid matrices [21–23].

In this chapter, three methodologies are described in order to offer the reader a practical overview of the available strategies for the reliable analysis of intact glycoproteins by MALDI-TOF MS. The proposed methods fulfil the necessary requirements of reproducibility and low extent of glycan fragmentation. We have selected three different combinations of matrix compounds and sample-matrix preparation procedures depending on the percentage of glycosylation of the glycoprotein. For glycoproteins with glycosylation percentages lower than 10% (m/m), like human transferrin (hTf, ~6% (m/m)) or mouse transferrin (mTf, ~3% (m/m)) (Fig. 1), the methodology using SA as matrix and the

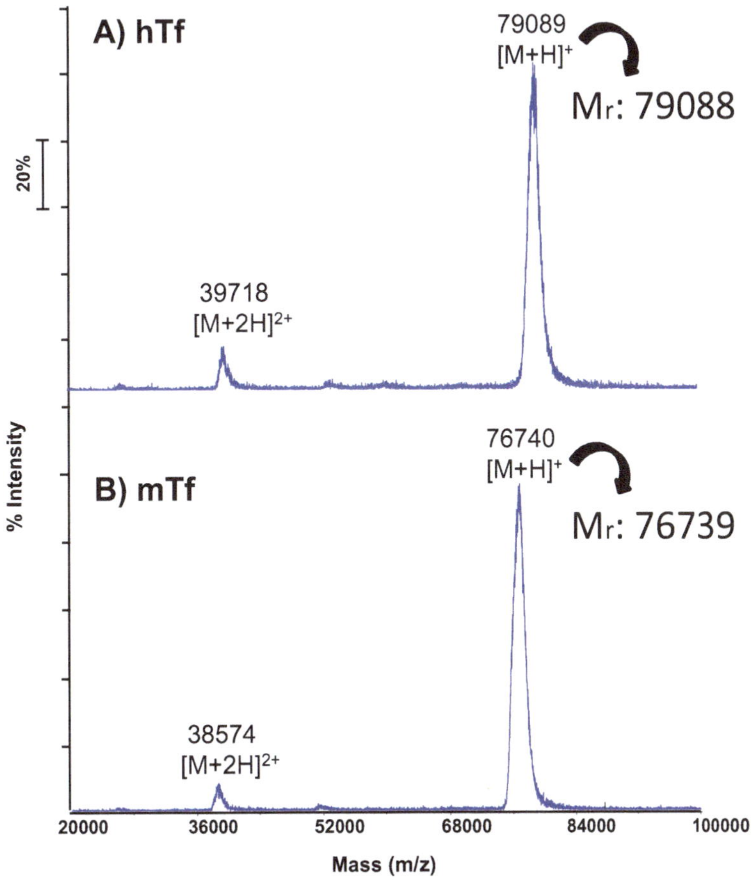


**Fig. 1** Mass spectra and experimental molecular mass obtained by MALDI-TOF MS of (**a**) human transferrin (hTf) and (**b**) mouse transferrin (mTf) with sinapinic acid (SA) as MALDI matrix and the fast evaporation method as sample-matrix preparation procedure (laser intensity close to the threshold)

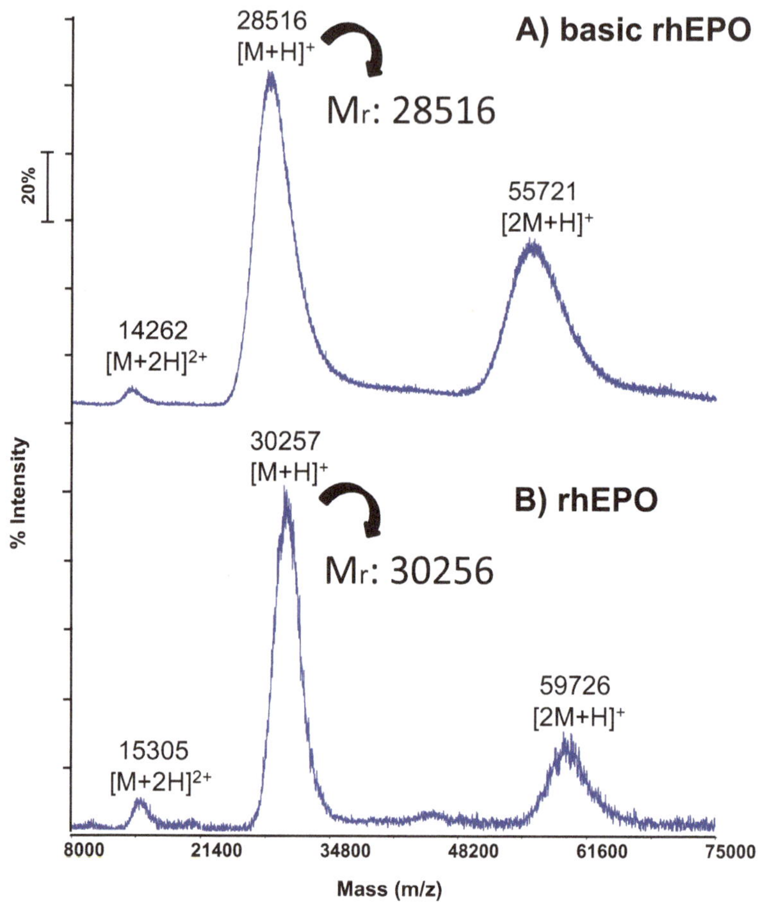


**Fig. 2** Mass spectra and experimental molecular mass obtained by MALDI-TOF MS of two recombinant human erythropoietins (**a**) basic rhEPO and (**b**) rhEPO with 2,5-dihydroxybenzoic acid (DHB) as MALDI matrix and the vacuum drying method as sample-matrix preparation procedure (laser intensity close to the threshold)

fast evaporation method as sample-matrix preparation procedure, provides homogeneous spots with a superior number of smaller crystals, which resulted in higher reproducibility of the obtained mass spectra. Moreover, working with this matrix at laser intensities close to the threshold enables the analysis of glycoproteins with low-carbohydrate content with adequate sensitivity and reduced fragmentation of the labile groups.

The combination of DHB and the vacuum drying method shows better performance for glycoproteins with higher glycosylation degree like recombinant human erythropoietins (i.e., rhEPO and a basic rhEPO with lower sialic acid content, 40% (m/m) and 35% (m/m) of glycosylation degree, respectively) (Fig. 2). Reproducibility improves owing to the more homogeneous spots and greater number of glycoprotein–matrix cocrystals compared to

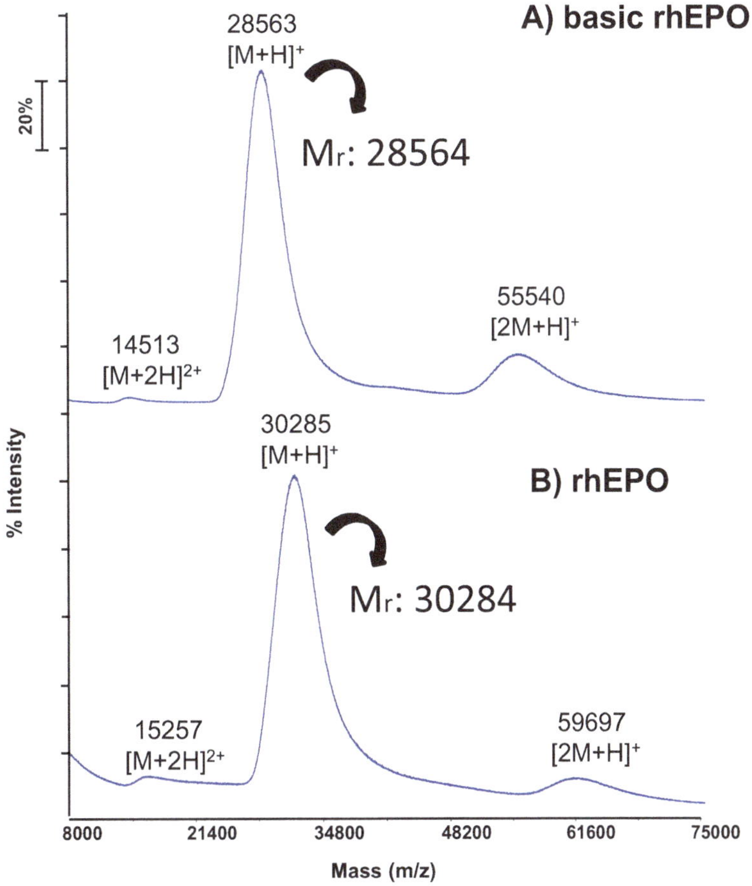


**Fig. 3** Mass spectra and experimental molecular mass obtained by MALDI-TOF MS of two recombinant human erythropoietins (**a**) basic rhEPO and (**b**) rhEPO with sinapinic acid and the ILM method (SA-ILM) (laser intensity close to the threshold)

the dried droplet method. Using DHB-vacuum drying, the loss of labile groups was less significant than with other solid matrices like SA even working at higher laser intensities.

Alternatively, for these highly glycosylated proteins, the use of a SA-butylamine ILM (SA-ILM) matrix provides better spot-to-spot and shot-to-shot reproducibility (Fig. 3), but it normally shows a slight decrease in signal-to-noise ratio values compared to DHB solid matrix. Nevertheless, this issue needs to be evaluated in depth for each studied glycoprotein. Moreover, as other authors have indicated with other ILMs [15, 23], the ionization efficiency with SA-ILM is slightly worse for glycoproteins with lower sugar content (i.e., below ~20%). Hence, SA-ILM may be regarded as a reproducible and robust alternative to DHB-vacuum drying to obtain a reliable average molecular mass value for intact glycoproteins with high glycosylation degree [21].

## 2 Materials

### 2.1 Instrumentation

1. 4800 Plus MALDI TOF/TOF mass spectrometer (AB Sciex, Framingham, Massachusetts) equipped with a nitrogen laser (355 nm) and a microchannel plate detector (MCP). Data acquisition and data processing are performed using the 4800 Series Explorer™ and Data Explorer® softwares (AB Sciex) (*see* **Note 1**).
2. For vacuum drying of the spots, the tank of a solid-phase extraction vacuum manifold from Supelco (Bellefonte, Pennsylvania, USA) connected to a vacuum pump is used.
3. Mikro 220R centrifuge (Hettich Zentrifugen, Tuttlingen, Germany).

### 2.2 Chemicals and Solutions

Prepare all solutions using ultrapure water (conductivity value lower than 0.05 μS/cm at 25 °C) and analytical grade reagents. Store all solutions at 4 °C and allow to stand at room temperature before use.

1. Protein solutions: Prepare 1000 μg/mL stock solution of each glycoprotein in water. Mouse transferrin (mTf, $M_r$ ~ 76,000) and human transferrin (hTf, $M_r$ ~ 79,000). Recombinant human erythropoietins (rhEPO and basic rhEPO, $M_r$ ~ 28,000–30,000) (*see* **Notes 2** and **3**). Prepare a 1000 μg/mL solution of bovine serum albumin (BSA) in water as calibrant.
2. Fast evaporation method solutions: 27 mg/mL of sinapinic acid (SA) in acetone–water (99:1 v/v). Weight 27 mg of SA and add 990 μL of acetone and 10 μL of water. Vortex for 15 s. 10 mg/mL of SA in acetonitrile–water (50:50 v/v) with 0.1% v/v of trifluoroacetic acid (TFA). Weight 10 mg of SA and add 500 μL of acetonitrile (ACN), 499 μL of water and 1 μL of trifluoroacetic acid (TFA). Vortex for 15 s.
3. Vacuum drying method solution: 10 mg/mL of 2,5-dihydroxybenzoic acid (DHB) in ethanol–water (50:50 v/v). Weight 10 mg of DHB and add 500 μL of ACN, 499 μL of water and 1 μL of TFA. Vortex for 15 s.
4. Ionic liquid matrix solution (SA-ILM): Weight 100 mg of SA and add 3242 μL of methanol and 64.1 μL of butylamine. Cap, vortex for 15 s, and evaporate to approximately 100 μL with air (*see* **Note 4**). Finally reconstitute the mixture with 100 μL of ACN.

## 3 Methods

### 3.1 Sample-Matrix Preparation Procedure

All sample-matrix preparation procedures are performed in a thermostatted room at 25 °C (*see* **Note 5**).

#### 3.1.1 Fast Evaporation Method

1. Deposit 1 μL of 27 mg/mL of SA in acetone–water (99:1, v/v) onto the stainless steel MALDI plate and let dry.
2. Sample-matrix solution: Mix 1 μL of glycoprotein solution with 1 μL of 10 mg/mL of SA in ACN–water (50:50 v/v, 0.1% v/v TFA). Vortex for 15 s in order to obtain an homogeneous sample (*see* **Note 6**).
3. Deposit 1 μL of sample-matrix solution over this first layer and let dry.

#### 3.1.2 Vacuum Drying Method

1. Sample-matrix solution: Mix 1 μL of glycoprotein solution with 1 μL of 10 mg/mL of DHB in ethanol–water (50:50 v/v, 0.1% v/v TFA). Vortex for 15 s in order to obtain an homogeneous sample (*see* **Note 6**).
2. Deposit 1 μL of sample-matrix solution onto the MALDI plate and place it immediately inside the vacuum tank in order to dry the spots under a vacuum of approximately 2 kPa (*see* **Note 7**).

#### 3.1.3 ILM Method

1. Sample-matrix solution: Mix 1 μL of glycoprotein solution with 1 μL of SA-ILM solution. Vortex for 15 s in order to obtain an homogeneous sample (*see* **Note 6**).
2. Deposit 1 μL of sample-matrix solution onto the MALDI plate and let dry (*see* **Note 8**).

### 3.2 MALDI-TOF MS Analysis

1. Before introducing the MALDI plate in the instrument, clean the plate and remove the moisture passing a slow stream of nitrogen over the surface of the plate.
2. Mass spectra are acquired over a range of 8000–75,000 and 20,000–100,000 *m/z* using the mid mass and high mass positive linear mode, respectively (*see* **Note 9**). The final mass spectrum is typically obtained as a summation of 100 consecutive laser shots (i.e., measured mass spectra).
3. Prior to starting acquisition, the instrument is externally calibrated using the singly and doubly charged molecular ions from bovine serum albumin (BSA), prepared following the fast evaporation method described in subheading 3.1.1 (*see* **Note 10**).
4. Measurements must be performed close to the laser intensity threshold value required for molecular ion generation to avoid carbohydrate fragmentation, specially when analyzing highly glycosylated proteins (*see* **Note 11**).

5. Acquire the MALDI mass spectrum of the glycoprotein in the selected *m/z* ratio scanning range (*see* **Note 12**). Figures 1, 2, and 3 show the spectra of the studied glycoproteins (*see* **Note 13**).
6. Determine the average molecular mass of the glycoprotein with the singly charged molecular ion (*see* **Note 14**). Relative molecular masses ($M_r$) obtained by MALDI-TOF MS are shown in Figs. 1, 2, and 3 for the studied glycoproteins.

## 4 Notes

1. Other MALDI-TOF mass spectrometers have been also reported to allow this type of MS application.
2. Interfering contaminants such as excipients or salts must be removed by ultrafiltration with centrifugal filters of 10,000 $M_r$ cutoff. Wash the filter with 50 μL of water and centrifuge at 13,680 × *g*. Add the glycoprotein sample and again centrifuge for 10 min. Wash the residue three times with 100 μL of water. Discard the filtrates from each of the five previous steps. Recover the sample from filter by centrifugation upside down in a new vial (3 min at 590 × *g*). Add water to adjust the glycoprotein concentration to 1000 μg/mL.
3. Here, we describe the analysis of intact glycoproteins at high concentration (~500 μg/mL). The analysis at lower concentration should be evaluated for each studied glycoprotein and also depends on the complexity of the sample and the instrumental set-up.
4. ILM-SA solution should exhibit a glycerol-like viscosity. Discard the ILM if solidifaction is observed during the preparation procedure as redissolution is not possible.
5. Temperature is an important parameter to control to obtain a reproducible and optimum crystallization of the spots on the MALDI target plate.
6. When working with such small volumes, centrifuge to recover all the liquid at the bottom of the vial.
7. If vacuum drying is not avaliable, the dried droplet method can be used, although poorer reproducibility will be obtained due to the lower spot homogeneity.
8. Spots with the ILM-SA take more time to dry compared to spots prepared with the solid matrix methods. Complete drying is not achieved and spots finally remain as dense and homogeneous glycerin-like drops.
9. Only select the "high mass mode" when strictly necessary as the instrument parameters of these acquisition methods can

promote glycan fragmentation in intact glycoproteins, especially those with high glycosylation degree.

10. In order to improve calibration accuracy, take special precaution in spotting the calibrant protein solution next to the sample spots being analyzed.
11. Start the measurements with a very low laser intensity and increase gradually until the detection threshold is reached. Finally, select a laser intensity value, that provides an adequate signal-to-noise ratio, but remaining always near the threshold to prevent fragmentation.
12. MALDI mass spectrum of an intact glycoprotein usually shows the singly and doubly charged molecular ions and, occasionally, the singly charged ion of a dimeric form generated during the desorption process (as can be observed in Figs. 2 and 3 when analyzing the recombinant erythropietins).
13. The percentage of glycosylation of the protein normally determines the selection of the MALDI matrix and the sample-matrix preparation procedure. For glycoproteins with a low glycosylation degree (up to ~10%), such as human and mouse transferrin (Fig. 1), SA and the fast evaporation method is preferred. For glycoporoteins with higher degrees of glycosylation, DHB with vacuum drying is an adequate method to obtain confident molecular masses of intact glycoproteins as it minimizes glycan fragmentation. However, to achieve higher spot-to-spot and shot-to-shot reproducibility, the ILM-SA is a better choice for highly glycosylated glycoproteins (above ~20% of glycosylation), although in some cases it can be less sensitive.
14. The experimental molecular mass is usually calculated with the singly charged molecular ion as, generally, it is the most intense, and thus, less mass error is achieved.

## Acknowledgments

This work was supported by a grant from the Spanish Ministry of Economy and Competitiveness (RTI2018-097411-B-I00) and the Cathedra UB Rector Francisco Buscarons Ubeda (Forensic Chemistry and Chemical Engineering). Montserrat Mancera-Arteu acknowledges the University of Barcelona for an ADR fellowship.

### References

1. Krištić J, Lauc G (2017) In: Lauc G, Wuhrer M (eds) Ubiquitous importance of protein glycosylation BT—high-throughput glycomics and glycoproteomics: methods and protocols. Springer New York, New York, NY, pp 1–12
2. Almeida A, Kolarich D (2016) The promise of protein glycosylation for personalised medicine. Biochim Biophys Acta 1860:1583–1595. https://doi.org/10.1016/j.bbagen.2016.03.012

3. Miura Y, Endo T (2016) Glycomics and glycoproteomics focused on aging and age-related diseases—Glycans as a potential biomarker for physiological alterations. Biochim Biophys Acta 1860:1608–1614. https://doi.org/10.1016/j.bbagen.2016.01.013
4. Chen Z, Huang J, Li L (2018) Recent advances in mass spectrometry (MS)-based glycoproteomics in complex biological samples. TrAC—Trends Analyt Chem 118:880–892. https://doi.org/10.1016/j.trac.2018.10.009
5. Banazadeh A, Veillon L, Wooding KM, Zabet-moghaddam M, Mechref Y (2017) Recent advances in mass spectrometric analysis of glycoproteins. Electrophoresis 38:162–189. https://doi.org/10.1002/elps.201600357
6. Ruhaak LR, Xu G, Li Q, Goonatilleke E, Lebrilla CB (2018) Mass spectrometry approaches to glycomic and glycoproteomic analyses. Chem Rev 118:7886–7930. https://doi.org/10.1021/acs.chemrev.7b00732
7. Gaunitz S, Nagy G, Pohl NLB, Novotny MV (2017) Recent advances in the analysis of complex glycoproteins. Anal Chem 89:389–413. https://doi.org/10.1021/acs.analchem.6b04343
8. Zhang Q, Li Z, Wang Y, Zheng Q, Li J (2018) Mass spectrometry for protein sialoglycosylation. Mass Spectrom Rev 37:652–680. https://doi.org/10.1002/mas.21555
9. Harvey DJ (2018) Analysis of carbohydrates and glycoconjugates by matrix-assisted laser desorption/ionization mass spectrometry: an update for 2013–2014. Mass Spectrom Rev 37:353–491. https://doi.org/10.1002/mas.21530
10. Perreault H, Lattová E, Šagi D, Peter-Katalinic J (2013) MALDI-MS of glycans and glycoconjugates. In: Hillenkamp F, Peter-Katalinic J (eds) MALDI MS. A practical guide to instrumentation, methods and applications, 2nd edn, pp 239–271
11. Nie H, Li Y, Sun XL (2012) Recent advances in sialic acid-focused glycomics. J Proteome 75:3098–3112. https://doi.org/10.1016/j.jprot.2012.03.050
12. Giménez E, Benavente F, Barbosa J, Sanz-Nebot V (2010) Towards a reliable molecular mass determination of intact glycoproteins by matrix-assisted laser desorption/ionization time-of-flight mass spectrometry. Rapid Commun Mass Spectrom 24:3567–3577. https://doi.org/10.1002/rcm
13. Fukuyama Y (2015) MALDI matrix research for biopolymers. Mass Spectrom 4:A0037. https://doi.org/10.5702/massspectrometry.a0037
14. Tholey A, Heinzle E (2006) Ionic (liquid) matrices for matrix-assisted laser desorption/ionization mass spectrometry-applications and perspectives. Anal Bioanal Chem 386:24–37. https://doi.org/10.1007/s00216-006-0600-5
15. Crank JA, Armstrong DW (2009) Towards a second generation of ionic liquid matrices (ILMs) for MALDI-MS of peptides, proteins, and carbohydrates. J Am Soc Mass Spectrom 20:1790–1800. https://doi.org/10.1016/j.jasms.2009.05.020
16. Abdelhamid HN (2017) Organic matrices, ionic liquids, and organic matrices@nanoparticles assisted laser desorption/ionization mass spectrometry. TrAC—Trends Analyt Chem 89:68–98. https://doi.org/10.1016/j.trac.2017.01.012
17. Lee C, Ni CK (2019) Soft matrix–assisted laser desorption/ionization for labile Glycoconjugates. J Am Soc Mass Spectrom 30:1455–1463. https://doi.org/10.1007/s13361-019-02208-4
18. Hillenkamp F, Jaskolla TW, Karas M (2013) The MALDI process and method. In: Hillenkamp F, Peter-Katalinic J (eds) MALDI MS. A practical guide to instrumentation, methods and applications, 2nd edn, pp 1–40
19. Armstrong DW, Zhang LK, He L, Gross ML (2001) Ionic liquids as matrixes for matrix-assisted laser desorption/ionization mass spectrometry. Anal Chem 73:3679–3686. https://doi.org/10.1021/ac010259f
20. Abdelhamid HN (2018) Ionic liquid-assisted laser desorption/ionization–mass spectrometry: matrices, microextraction, and separation. Methods Protoc 1:23. https://doi.org/10.3390/mps1020023
21. Giménez E, Benavente F, Barbosa J, Sanz-Nebot V (2010) Ionic liquid matrices for MALDI-TOF-MS analysis of intact glycoproteins. Anal Bioanal Chem 398:357–365. https://doi.org/10.1007/s00216-010-3898-y
22. Fukuyama Y, Nakaya S, Yamazaki Y, Tanaka K (2008) Ionic liquid matrixes optimized for MALDI-MS of sulfated/sialylated/neutral oligosaccharides and glycopeptides. Anal Chem 80:2171–2179. https://doi.org/10.1021/ac7021986
23. Mank M, Stahl B, Boehm G (2004) 2,5-Dihydroxybenzoic acid butylamine and other ionic liquid matrixes for enhanced MALDI-MS analysis of biomolecules. Anal Chem 76:2938–2950. https://doi.org/10.1021/ac030354j

# Chapter 4

# Fc Glycosylation Characterization of Human Immunoglobulins G Using Immunocapture and LC-MS

Yosra Helali, Shilpee Sharma, Marie Vandeput, Dansala Welba, Pierre Van Antwerpen, Arnaud Marchant, and Cédric Delporte

## Abstract

Immunoglobulins G (IgG) are proteins produced by the immune system of higher life forms that play a central role in the defense against microbial pathogens. IgG bind pathogens with the hypervariable Fab component and mediate a diversity of effector functions by binding to immune effector cells via their crystallizable (Fc) component. All IgG Fc carry a polymorphic *N*-glycan that regulates its binding properties and thereby its effector functions. The glycosylation profile of IgG Fc is modulated by physiological and pathological conditions, including infectious diseases and inflammatory disorders. Characterization of IgG Fc glycosylation profiles is a promising approach to understand the pathogenesis of diseases involving the immune system and to develop novel biomarkers of disease activity. Measuring the proportion of the different IgG Fc glycoforms remains an analytical challenge, that requires a sensitive and reproducible analytical approach.

This chapter describes an optimized approach for the preparation and the analysis of Fc *N*-glycans from total serum or plasma IgG using magnetic beads, RapiFluor MS label©, and LC-MS.

**Key words** *N*-glycosylation, LC-MS, RapiFluor-MS protocol, Immunoglobulins IgG, Fragment crystallizable region (Fc), Characterization, Fluorescence, Beads

## 1 Introduction

Immunoglobulins (Ig) are glycoproteins produced and secreted by a specialized immune cell population, called B lymphocytes. B lymphocytes can produce five different classes of Ig: IgA, IgD, IgE, IgG, and IgM. IgG is the most abundant Ig in the blood, with concentrations ranging from 7 to 18 g/L in healthy adults [1]. IgG play a central role in the defense against infectious pathogens. They bind and neutralize pathogens with their hypervariable Fab component and they mediate a diversity of effector functions via the binding of their crystallizable (Fc) component to complement proteins and to Fcγ receptors expressed by immune effector cells, including macrophages and natural killer cells [1, 2]. Although

Arnaud Delobel (ed.), *Mass Spectrometry of Glycoproteins: Methods and Protocols*, Methods in Molecular Biology, vol. 2271, https://doi.org/10.1007/978-1-0716-1241-5_4, 

human IgG are divided in four subclasses, that is, IgG1, IgG2, IgG3, and IgG4, which are named in order of decreasing abundance in serum, the four subclasses share more than 95% of their amino acid sequence. Yet they have very different affinities for complement proteins and Fcγ receptors and therefore stimulate qualitatively different immune responses [3, 4]. A second and essential layer of structural diversity of IgG is provided by posttranslational modifications, primary *N*-glycosylation, of the Fc component.

Glycosylation is one of the most prominent posttranslational modification, and it plays an important role in the maintenance of the structure and function of glycoproteins. Over the past decades, research from many laboratories has defined the role of glycans in the etiology of major diseases, and a key challenge is to understand the phenotypic consequences of changes in protein glycosylation [5]. The composition of the IgG Fc *N*-glycan has a profound impact on the quaternary structure and on the stability of the protein. Variations in glycosylation profile influence different biological processes such as signal transduction, protein-folding and immune responses [6]. The most characterized groups of glycosylation are *N*- and *O*-linked glycans. Glycans represent around 15% of IgG weight [7]. *O*-glycosylation is only present at the hinge region of 10% of IgG3, whereas all IgG subclasses carry an *N*-glycan at the Asn 297 of the $CH_2$ domain of each the two heavy chains (Fig. 1). This *N*-glycan has a constant biantennary heptasaccharide that can carry variable levels of fucose, *N*-acetylglucosamine, galactose, and sialic acid. Glycosylation of IgG Fc can be asymmetrical, that is, two different glycan structures expressed by each of the two heavy chains. On the other hand, about 15–25%

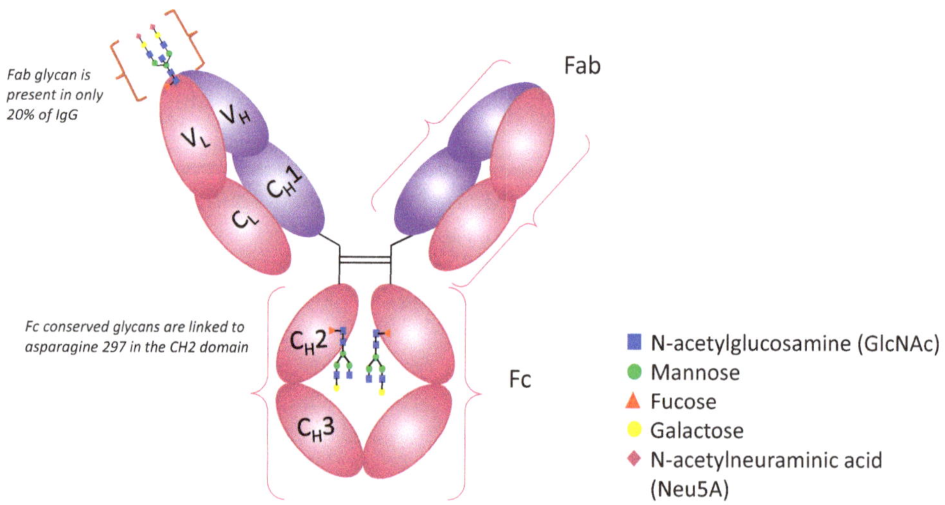


**Fig. 1** IgG glycosylation, Fc glycans are linked to asparagine 297 in the $CH_2$ domain

of the Fab component of IgG are *N*-glycosylated but the role of this glycan remains incompletely elucidated [8, 9]. IgG Fc glycosylation is a highly regulated process, influenced by age, sex, hormones, and inflammatory mediators [2]. As a result, IgG carrying different IgG Fc glycans have different affinity for complement proteins, lectins and Fcγ receptors and thereby promote different immune effector functions [2]. Characterizing IgG Fc glycosylation profiles is therefore an important approach to understand the pathogenesis of diseases involving the immune system and to develop novel biomarkers of disease activity and response to therapies [6, 10, 11].

In the present chapter we describe an optimized method to purify IgG Fc *N*-glycan by immunocapture using beads coated with antibodies against the two light chains of IgG (Lambda and Kappa) and to characterize the *N*-glycan using RapiFluor-MS label and LC coupled to fluorescence (FLD) and mass spectrometry (MS) detection. The aim of this analytical method is to study the variation of IgG Fc *N*-glycosylation between healthy adults and patients with infectious or inflammatory diseases. The purification of the Fc fragment from the total IgG is based on the use of magnetic beads: Lambda and Kappa (which are catch respectively lambda and kappa Fab fragments) and enzymes. Firstly, the total IgG is fixed on the beads (incubated for 1 h at the room temperature), then the IDEZ (IgG Digestion Enzymes) is added. These enzymes cleave below the hinge region (between two glycines for IgG 1, 3, 4 or between a glycine and an alanine for IgG 2) to give 2*Fc/2 and F(ab)′2 [12]. In the present approach, this digestion enables to release Fc in the solution and then analyze only the Fc *N*-glycosylation profile (removing the variation from Fab part) (Fig. 2).

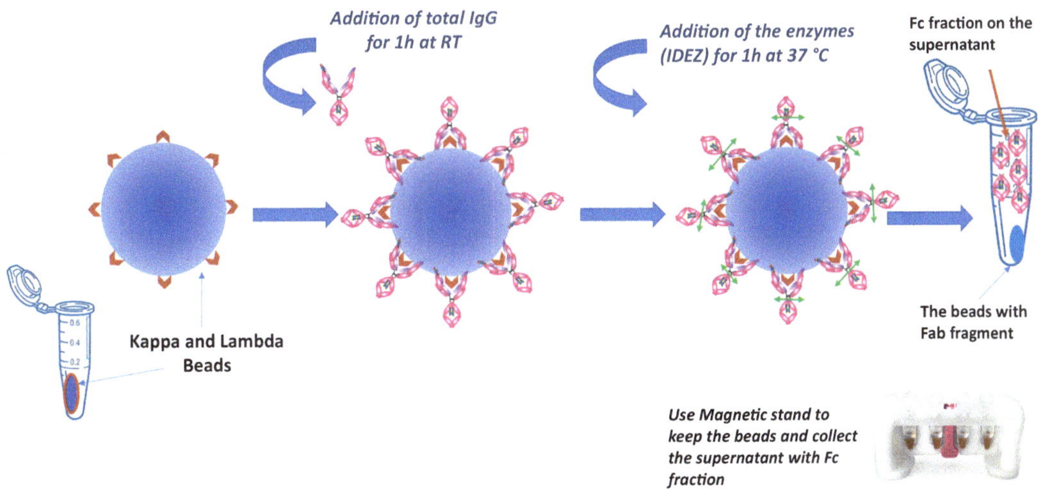


**Fig. 2** Purification of the Fc fragment using magnetic beads

## 2 Materials

### 2.1 Chemicals

1. Lambda and Kappa magnetic beads (Merck, Ref LSKMAGLM02/LSKMAGKP02).
2. Washing/binding buffer: PBS (Lonza, Ref. 17-516).
3. IdeZ enzyme (New England BioLabs, Ref P0770S).
4. Samples: Purified IgG from Serum or plasma of patients or healthy volunteers (For the purification of the total IgG a purification kit: Thermo Scientific Melon Gel IgG Spin Purification Kit is used).
5. Intact mAb check standard (intact mouse IgG1 monoclonal antibody used as a control) (Waters, from Glycoworks *Rapi*-Fluor-MS kit).
6. Rapid PNGase F (0.035 mL of PNGase F, enzyme used for rapid deglycosylation) (Waters, from Glycoworks *Rapi*Fluor-MS kit).
7. GlycoWorks Rapid Buffer (Buffer used to dissolve the RapiGest SF (0.25 mL of 250 mM HEPES pH 7.9) (Waters, from Glycoworks *Rapi*Fluor-MS kit).
8. RapiGest SF (enzyme-friendly surfactant used to denature the glycoproteins and to facilitate the deglycosylation reaction) (Waters).
9. *Rapi*Fluor-MS Label powder of the reagent *RapiFluor*-MS from Waters.
10. Anhydrous dimethylformamide (DMF) (anhydrous DMF) used to solubilize the *RapiFluor*-MS Label powder.
11. 18.2 MΩ water.
12. Formic acid (FA), LC/MS-grade.
13. Acetonitrile (ACN), LC/MS-grade.

### 2.2 Equipment

1. A 6520 series mass spectrometer: electrospray ionization (ESI)-Quadrupole-Time-of-Flight (Q-TOF) high-resolution MS (HRMS) from Agilent Technologies.
2. A 1200 series Rapid Resolution Liquid Chromatography (RRLC) from Agilent Technologies with a binary pump, a degasser, a thermostated autosampler, a column oven, and a fluorescence detector.
3. A Magnetic Bead System (Ref: LSKMAGA02 from Merk).
4. XBridge BEH Amide XP Column 130 Å, 2.5 μm, 2.1 mm × 150 mm (Waters, ref.: 186006724).
5. Heating block: Esco Provocell shaking micro incubator (Speed: 800 rpm).

6. 96-well plate extraction Vaccum Manifold, used for the μElution SPE.
7. Mobile phase solvent A: 50 mM formate solution, pH = 4.4.
8. Mobile phase solvent B: pure acetonitrile.

## 3 Methods

### 3.1 Sample Preparation

#### 3.1.1 IgG Binding

1. Calculate the required volume of bead slurry based on the number of reactions.

   30 μL of Lambda and Kappa beads mixed at the ratio of 1:1 (15 μL of Lambda +15 μL of Kappa) is required per sample.
2. Mix the bead slurry so that all the beads are uniformly resuspended.
3. Pipet the required volume of resuspended bead slurry into a microcentrifuge tube. Place the tube into the Magnetic Bead System and allow the beads to migrate to the magnet. Remove the storage buffer with a pipette and discard it.
4. Wash the beads twice using 500 μL of PBS for each wash add PBS to the tube, vortex 10 s, put the tube back on the magnet stand and discard the supernatant (*see* **Note 1**).
5. Resuspend the beads by adding the required volume of PBS (to get the same volume calculated in the **step 1**).
6. Add 30 μL of the resuspended beads solution to the fresh tubes, put the fresh tubes on the magnet for 1 min and discard the supernatant.
7. Take the tubes out of the magnet and add the required volume of purified IgG samples (Ensure that the IgG volume is at least five times the settled bead) volume mix well using the pipette. This protocol is designed for 25 μg of total IgG (*see* **Note 3**)).
8. Incubate for 1 h at the room temperature, with continuous mixing or end-over-end rotation (=spinning rotor).

#### 3.1.2 IDEZ Digestion

1. Take the tubes from the spinning rotor, briefly spin down, and put it on the magnet for 1 min, and discard the liquid.
2. Take the tubes out of the magnet, wash the beads twice by adding 200 μL of PBS, then take off the supernatant (always using the magnet sand).
3. Reconstitute the sample (=IgG attached on beads) in 20.0 μL of PBS, then add 1.0 μL of IDEZ and incubate for 1 h (spinning/shaking) at 37 °C.

*3.1.3 Collecting the Fc Fraction*

1. Use the magnet to collect the supernatant (Fc fraction) and transfer it to the new tubes (*see* **Note 4**).

*3.1.4 IgG Denaturation*

1. Prepare the buffer solution: dissolve the content of *RapiGest SF* (1 vial of 3 mg) in 60 μL of Glycoworks RapidBuffer (*see* **Note 6**).
2. Put 15.3 μL of 18.2 MΩ water into a 1 mL tube (*see* **Note 11**).
3. Add 7.5 μL of the collected Fc fraction solution into the 1 mL tube (*see* **Note 5**).
4. Add 6 μL of the buffer solution to the sample and aspirate and dispense to mix.
5. Put the sample for 3 min in a heat block set to 90 °C (*see* **Note 8**).

*3.1.5 Digestion*

1. Take out the sample from the heat block, and let it cool down for 3 min.
2. Add 1.2 μL of PNGase F, to get a sample concentration of 0.5 mg/mL of IgG, and aspirate and dispense to mix (*see* **Note 9**).
3. Incubate the sample for 5 min in the heat block set to 50 °C (*see* **Note 8**).
4. Remove the samples from the heat block, and let it cool down at the room temperature for 3 min.

*3.1.6 Labeling*

1. Prepare the labeling: dissolve the *Rapi*Fluor-MS Label powder (1 vial of 9 mg) in 131 μL of anhydrous DMF (be sure the all the powder is dissolved by aspirating and dispensing gently 5–10 times (*see* **Note 10**)).
2. Put 12 μL of the labeling solution to the sample.
3. Mix the reagent solution.
4. Let the sample a few minutes at the room temperature to allow the labeling reaction.
5. After 5 min, add 358 μL of ACN to dilute the sample before the HILIC SPE (hydrophilic interaction chromatography based solid phase extraction).

*3.1.7 Cleaning up with HILIC-SPE*

1. Connect the μElution plate to the vacuum manifold outfitted (set to 2.5–4 Hg), and put the waste tray below.
2. Condition the μElution plate with 200 μL of 18.2 MΩ water.
3. Equilibrate wells with 200 μL of 15:85 water–acetonitrile.
4. Load the diluted samples to wells.
5. Wash the well two times with 600 μL of 1:9:90 (v/v/v) FA–water–CAN.

6. Remove the waste tray and replace it with a 96-well collection plate, fitted with 600 μL tapered bottom inserts.
7. Elute glycans three times with 30 μL of GlycoWorks SPE Buffer (200 mM ammonium acetate in 5% acetonitrile). This three elutes have to be collected in the same tube.

#### 3.1.8 Prepare the Samples for the Injection

1. Add 310 μL of GlycoWorks Sample Diluent—DMF/ACN—to the 90 μL of the eluate.
2. Aspirate and dispense five times to ensure the mixing.
3. Cap the vial and store it at 4 °C if the injection will take place in the next 3 days, or put the samples at −80 °C for longer-term storage.

## 3.2 Analysis Conditions

#### 3.2.1 LC Conditions

*N*-glycans are eluted using a gradient method starting with 68% solvent B at 0.5 mL/min. From 0 to 25 min, solvent B is decreased to 57% with the same flow. During this gradient, *N*-glycans are eluted and separated. Depending on the dimension of the column (particle size, porosity, length, and diameter) these gradients might be changed. For example, starting and final % of ACN might be optimized as well as the time length of the gradient. From 26 to 29 min, solvent B is decreased to 20% with a flow rate of 0.25 mL/min. The flow is decreased, as pressure will be very high. This segment enables a complete wash of the phase. Then initial conditions are reloaded (68% solvent B from 31 min to 35 min at 0.25 mL/min). Finally, flow rate is increased at 0.5 mL/min at 40 min. Total run time is 50 min. The samples are maintained at 4 °C in the injector for the best stability the separation is performed with a column temperature sets to 35 °C.

#### 3.2.2 ESI-Q-TOF and FLD Detector Parameters

1. MS analysis is performed with an Agilent Q-TOF 6520 equipped with electrospray ionization (ESI) used in positive mode.
2. Source temperature is set at 300 °C, with 7 L/min (420 L/H) for the drying gas, 45 psi for the nebulizer and 4.5 kV for the capillary voltage. Fragmentor is at 174 V.
3. Data Storage: Centroid mode, mass spectra are recorded from 100 to 3200 m/z, and the acquisition rate is 1 spectrum/s.
4. MS data acquisition and processing respectively by Agilent Mass Hunter LC/MS Acquisition B.08.00 and Agilent Mass Hunter Qualitative Analysis B10.0.
5. Calibration is done using the ESI-L Low Concentration Mix (Ref: G1969-85000) from Agilent Technologies.
6. A reference solution was nebulized for continuous calibration in the positive mode using the reference masses of *m/z* 121.0508 and *m/z* 922.0097.

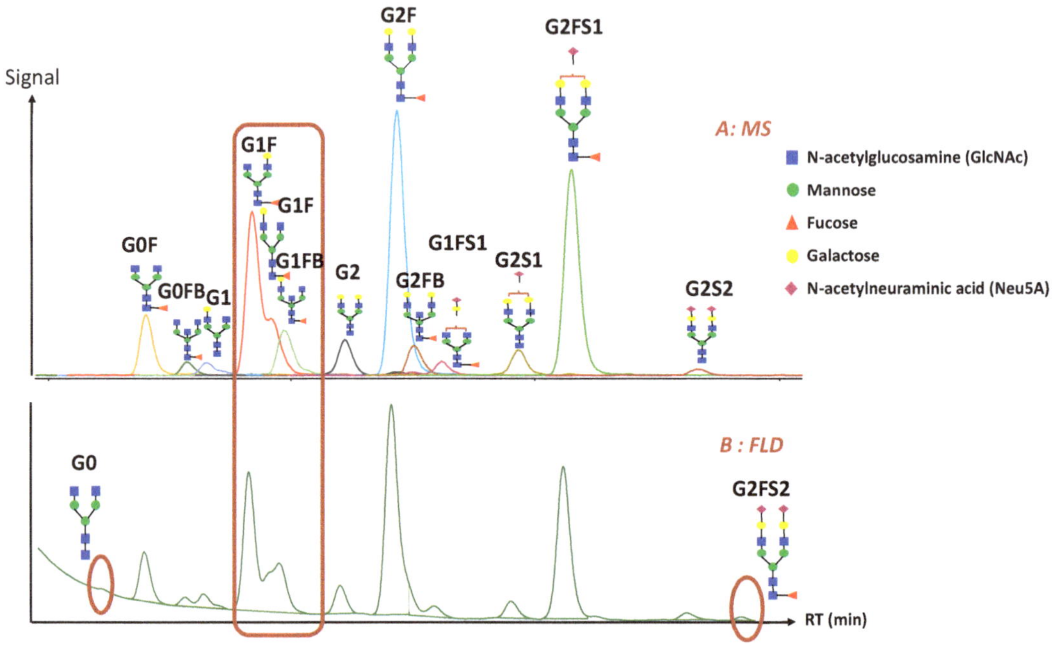


**Fig. 3** (**a**) MS-Total compound chromatogram (MS) and (**b**) Fluorescence (FLR) chromatograms obtained by HILIC-FLR-MS analysis of IgG Fc *N*-glycans. The combination of the two detectors allows the identification of the fifteen most abundant IgG glycans

7. The fluorescence detector is set to 265 nm for the excitation and 425 nm for the emission wavelengths.

*3.2.3 Data Analysis*

*N*-glycans profile characterization remains a complex task but is fundamental to investigate the impact of glycosylation of IgG in several immunological diseases. Here, we take advantages to collect data with two different detectors FLD and MS, to characterize the *N*-glycan profile of the Fc domain of IgGs. Usually, the IgG *N*-glycosylation profile includes fifteen different *N*-glycans, which are illustrated on Fig. 3.

FLD usually offers a high sensitivity. However, coeluting analytes might not be properly detected which is a common issue HILIC-FLD profiling of *N*-glycans. Higher resolution power of LC-HILIC systems has improved the resolution. However, the combination of FLD detection with MS detection enables to identify each peaks and even to refine the presence of different *N*-glycan in one FLD peak. Figure 4 is a zoom on one part of the chromatograms of Fig. 3 (see red box on Fig. 3). On the FLD chromatogram we observe only one peak unresolved triple peak (dashed green curve on Fig. 4), but using the MS we were able to differentiate three *N-glycan* (both G1F glycan and G1FB glycan). These is confirmed as those two glycans have two different $m/z$ values. Indeed, on the left side of Fig. 4, we observe the different extracted

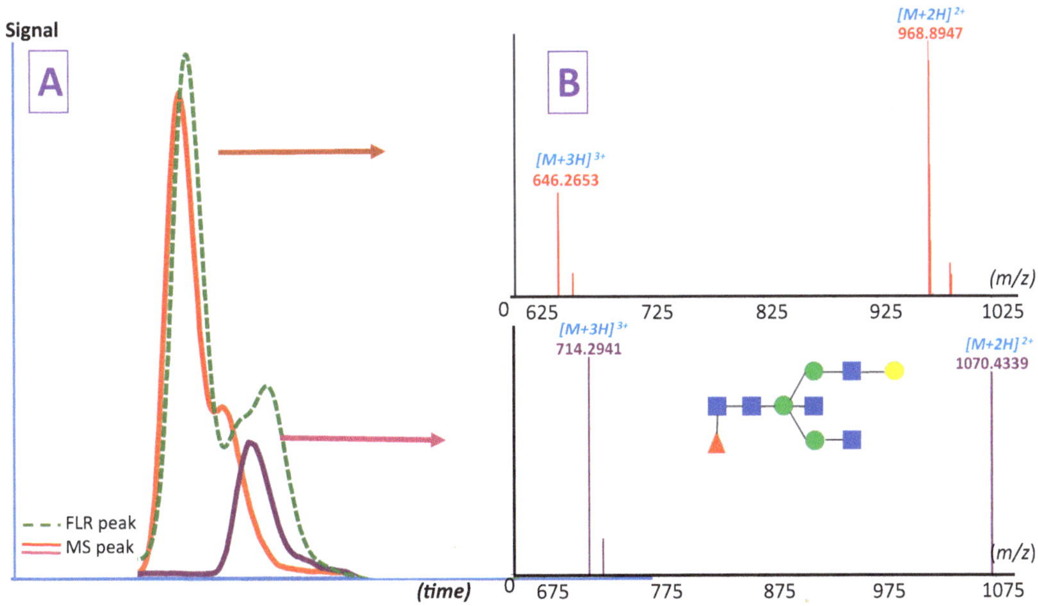


**Fig. 4** (**a**) Zoom on one peak of Fig. 3, the MS chromatogram and FLR chromatogram of the selected peak and (**b**) The two different spectra of the selected peak corresponding to G1F and G1FB

spectrum of each compound. Furthermore, we may observe that the most abundant observed charge states are $[M + 2H]^{2+}$ and $[M + 3H]^{3+}$, this may be explained by the high proton affinity of the RapiFluor-MS label.

For quantification approach, *N*-glycans profiles are usually expressed in relative percentages. However, some differences are usually observed in the glycosylation profiles using the two different detectors (namely FLD and MS). It is well known that MS responses vary between *N*-glycan as the ionization efficiency might change, even if the RapiFluor MS tag brings a highly ionizable function. Indeed, retention time change might itself involved ionization change as the mobile phase composition is different. Furthermore, inherent charge of the glycan in LC condition can explain these variations of ionization observed between glycans; one example is the charge of sialylated glycans versus neutral glycans. However, fluorescence response is stable between *N-glycan* and in a quantitative approach might be more relevant. The only disadvantage of FLD remains the coeluting peaks where resulting total AUC of the peak is the addition of both glycan AUC. In this case, the ratio of MS data remain the best way to estimate the relative composition of the FLD peak AUC between two coeluting glycans. This method of calculation of FLD coeluting peak is so used to determined AUC of each glycan.

Figure 5 shows the variation of the abundance percentage of total fucosylation in a group of eight subjects ($n = 8$) using both FLD and MS. This percentage is the sum of percentage of all

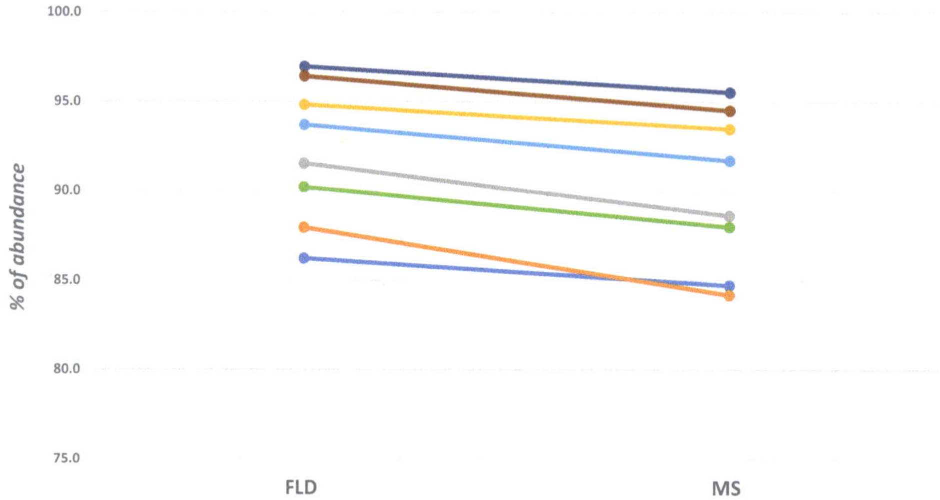


**Fig. 5** Comparison of fucosylation abundance using both detectors (FLR and MS) ($n = 8$)

detected fucosylated glycans (G0F, G0FB, G1F, G1FB, G2F, G2FB, G1FS1, G2FS1, G2FS2). This figure demonstrates the light decrease of the fucosylation percentage using the MS detector, comparing to the FLR, which confirm the variability of responses of glycans to the ionization process and the stability of the FLR response.

Nevertheless, both methods may be used for *N*-glycans profiling as the trend is conserved either in MS or in FLD detection and so enables a comparison of two groups or of several samples as soon as the same detection is used. These hyphenated approach (using HILIC-FLD and MS) was used for characterizing *N*-glycosylation profiles in two medical conditions. For the purpose of the present chapter, the number of subjects was restricted to $n = 4$.

This method is so used to monitor different glycosylation in clinical studies.

1. Galactosylation.

   In general, IgG- G0 and G1 (monogalactosylated structures) represent about 35% of total IgG Fc glycan, while G2 (digalactosylated structures) represents around 15% [13]. This quantity may varied with several diseases [7]; it increases for some cancers (Urothelial carcinoma [14], thyroid cancer and multiple myeloma) or decreases with inflammatory, autoimmune diseases such as rheumatoid arthritis and infectious diseases (leprosy, infective endocarditis, and HIV) [15]. This alteration is partially associated to the modification of the estrogen level. Indeed, it increases for the pregnant women and decreases for the postmenopausal women [7].

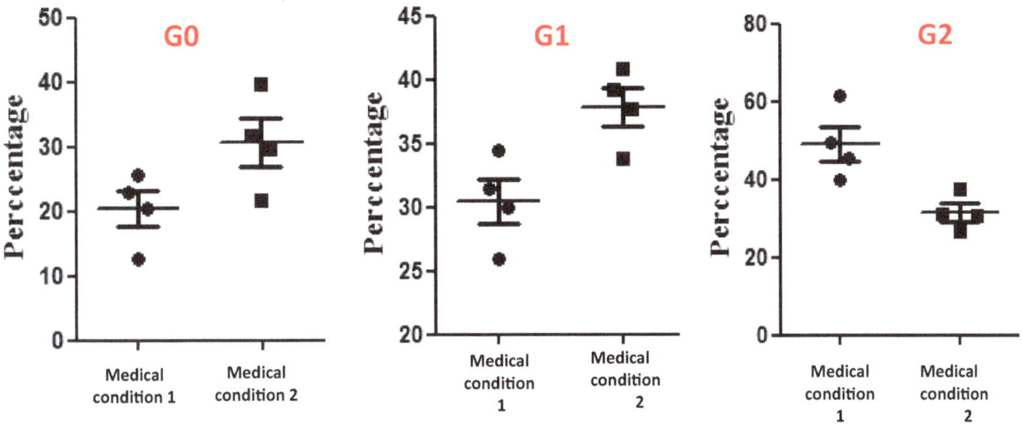


**Fig. 6** Variation of galactosylation between medical condition 1 and medical condition 2

Thanks to the HILIC-FLD-MS approach and the sample treatment, we could monitor galactosylation of Fc from circulant IgG in patients under two medical conditions.

Figure 6 shows these variations of galactosylation: a significant decrease of G0 and G1, and a rise of G2 is observed for the medical condition 1. This makes sense as G2 is produced from G1 which is produced from G0. So an increase of G2 might involve a decrease of G1/G0.

2. Core Fucosylation and Bisecting *N*-acetylglucosamine:

The fucosylated glycans represent the major part of *N*-glycan. A lack of the core fucose can improve the ADCP and ADCC (Antibody-dependent cellular cytotoxicity) activity by macrophages, monocytes, naturel killer (NK) cells, and granulocytes [16, 17].

On the other hand, an increase of fucosylation can significantly reduce the affinity to FcγR classes, and particularly FcγIII (50–100 fold), due to a sterically disturbed interaction of the IgG (Asn 297) and receptor glycan (Asn 162) [18, 19].

*N-glycan* with bisecting *N*-acetylglucosamine represent a small part of IgG Fc glycans (10–15%). This modification is partially the opposite of the addition of core fucose during the synthesis of *N-glycan* which making it difficult to discriminate the effect of the bisecting *N*-acetylglucosamine from the core fucosylation.

Using the protocol described above, a small decrease of bisecting-glycans was observed in medical condition 1. However, there was no significant variation of the fucosylation observed in this case (Fig. 7). This may be due to the reciprocal manner with proximal bisecting GlcAc blocking the core fucose. However, the number of subjects in each group is only four ($n = 4$) and might be not enough for a significant change observation.

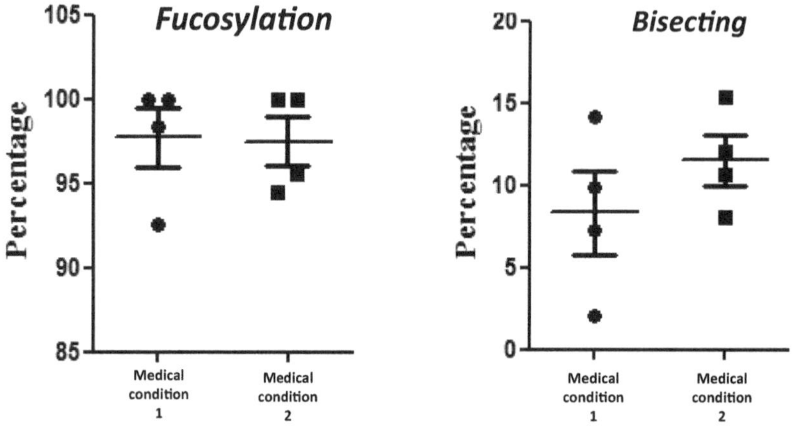


**Fig. 7** Variation of fucosylation and bisecting between medical condition 1 and medical condition 2

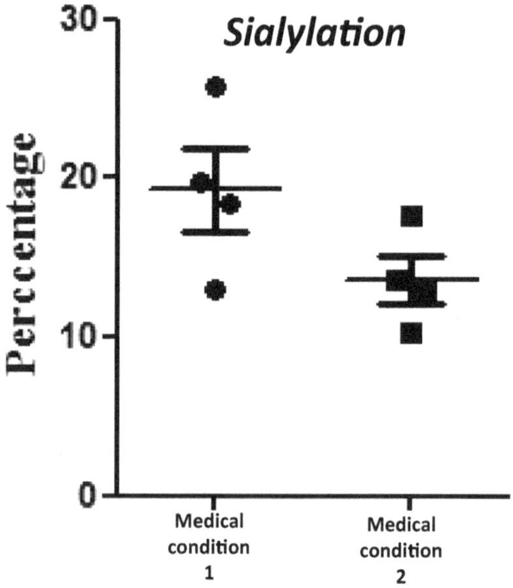


**Fig. 8** Variation of sialylation between medical condition 1 and medical condition 2

3. Sialylation:

For the healthy adults, sialylated *N-glycan* represent 10–15% of total glycans Fc [7].

The sialic acid is reported to have the most effect on the structure of the Fc domain of the antibody. The sialylated Fc glycans (G2FSX) associated to the protein of the Cγ2 domain cause the closed conformation, resulting in a cryptic binding site for place for the DC-SIGN in the $CH_2$–$CH_3$ interface which raise the immunomodulatory function [20].

Furthermore, the presence or absence of sialic acid can negatively impact IgG binding to FcγR different classes, reducing ADCC and CDC. [16]

Different changes of sialylation were reported, example, an increase of the quantity of terminally sialylated glycans with some cancer (stage-dependent) [7], or with pregnancy [15].

Figure 8 shows the increase of the quantity of sialylated glycans between the two medical conditions 1 and 2. This variation is generally correlated with an increase of galactosylation, this may be explained by the fact that galactosylated IgG is the substrate for sialytransferases. This leads to different hypothesis as for example the rapid processing of proteins in Golgi apparatus than might be not sufficient for a sialylation.

### 3.3 General Discussion on Data Interpretation in Disease

IgG glycosylation is a complex posttranslational modification. It may be altered with the availability of glycosidase and glycosyltransferase. For this reason, the interpretation of the different variation should not be studied separately for each glycan, without the rest of the other glycoforms. Instead, the different derived traits (galactosylation, fucosylation, bisecting, and sialylation) has to be expressed using a normalization in percentage over the total signal (=100%).

For example, the decrease of the galactosylation is generally associated to a simultaneous decrease, of the sialylation and an increase for the agalactosylation, which is the case when the majority of IgG glycans persist at the level of G0. This IgG glycoprofile can be typical for highly inflammatory cases. On the other hand, the same variation of galactosylation correlated with an increase of sialylation and a decrease of agalactosylation may be explained by the use of the terminally galactosylated IgG as a substrate for subsequent sialylation. This change may be an indicator for a highly anti-inflammatory IgG.

So in conclusion, the approach using a Fc characterization after IgG purification and digestion by IDEZ on beads enables the rapid and easy characterization of Fc glycosylation profiles by HILIC-FLD-MS. As illustrated above, FLD and MS detections are complementary as MS enables the correct identifications of *N*-glycans observed in FLD profile and even enables the correct attribution of AUC to each coeluting glycans. Furthermore, FLD profiling enables a better quantitative approach than MS because FLD is not sensitive to response variation with glycan nature and retention time. Nevertheless, both detection methods might be used to compare different conditions of glycosylation as illustrated here with two medical conditions.

## 4 Notes

1. For the steps of washing of the beads using the magnet, it is recommended to withdraw the washing solution slowly, otherwise, you lose a part of the beads.
2. Make sure that the volume of the added IgG is at least five times the settled bead volume.
3. To purify the IgG from plasma or serum, we used a Thermo Scientific™ Melon Gel IgG Spin Purification Kit, which provides recovery greater than 90% and a purity higher than 80%. The yield was measured using a Nano drop (spectrophotometer) upon the purification.
4. For the Fc collecting step, make sure to collect all supernatant (all the Fc fraction).
5. The purification protocol using the beads is optimized for a glycoprotein quantity of 25 μg, while, the RFMS label protocol is designed for 15 μg.
6. When you add the GlycoWorks RapidBuffer to the RapiGest SF powder, do not mix the solution, otherwise, bubbles will be observed. In this case you have to wait or centrifuge.
7. Just before using the Rapid PNGase F, get it out from the fridge and centrifuge it.
8. For the incubation steps (at 50 °C and 90 °C), the heat block has to be settled a bit higher in temperature to be sure that the sample are incubated at the recommended temperature (check the temperature before use for each instrument).
9. Here the deglycosylation step with PNGase F is short (5 min). It is so mandatory to use Rapid PNGase F, to assume a rapid and complete deglycosylation.
10. Different forms of the RapiFluor-MS kit are available for 96 samples or 24 samples; therefore, different forms of reagents are available, so pay attention to the indicated reactive quantity/concentration on the vials.
11. To obtain optimal reaction of the protein without any lost, use Protein Low binding Eppendorf tube.
12. To ensure the stability, the prepared samples (up to labeling and purification steps) are stored at −80 °C.
13. Columns for glycans analysis are also available from many suppliers. However, each column has its own chemistry and dimensions, so modification in the gradient might be performed to get the best resolution of the *N*-glycans. Here we also illustrated that MS might help in the identification and

quantification of coeluting *N*-glycans. So when both technics of detection are combined, full resolution of coeluting glycans is less critical.

14. For low sensitivity mass spectrometers, dry the eluted 90 μL (end of purification step) and resuspended it with 4.5 μL of water, 5 μL of DMF, and finally 10.5 μL of ACN.

## References


1. Yau LF, Liu J, Jiang M, Bai G, Wang JR, Jiang ZH (2019) An integrated approach for comprehensive profiling and quantitation of IgG-fc glycopeptides with application to rheumatoid arthritis. J Chromatogr B Anal Technol Biomed Life Sci 1122–1123:64–72
2. Jennewein MF, Alter G (2017) The Immunoregulatory roles of antibody glycosylation. Trends Immunol 38(5):358–372
3. Vidarsson G, Dekkers G, Rispens T (2014) IgG subclasses and allotypes: from structure to effector functions. Front Immunol 5:1–17
4. Kronimus Y, Dodel R, Galuska SP, Neumann S (2019) IgG Fc *N*-glycosylation: alterations in neurologic diseases and potential therapeutic target? J Autoimmun 96:14–23
5. Campbell MP, Packer NH (2016) UniCarbKB: new database features for integrating glycan structure abundance, compositional glycoproteomics data, and disease associations. Biochim Biophys Acta 1860(8):1669–1675
6. Verhelst X et al (2020) Protein glycosylation as a diagnostic and prognostic marker of chronic inflammatory gastrointestinal and liver diseases. Gastroenterology 158(1):95–110
7. Gudelj I, Lauc G, Pezer M (2018) Immunoglobulin G glycosylation in aging and diseases. Cell Immunol 333:65–79
8. van de Bovenkamp FS, Hafkenscheid L, Rispens T, Rombouts Y (2016) The emerging importance of IgG fab glycosylation in immunity. J Immunol 196(4):1435–1441
9. Bondt A et al (2014) Immunoglobulin G (IgG) fab glycosylation analysis using a new mass spectrometric high-throughput profiling method reveals pregnancy-associated changes. Mol Cell Proteomics 13(11):3029–3039
10. Xiang T et al (2017) Alteration of N-glycan expression profile and glycan pattern of glycoproteins in human hepatoma cells after HCV infection. Biochim Biophys Acta Gen Subj 1861(5):1036–1045
11. Hu M, Lan Y, Lu A, Ma X, Zhang L (2019) Glycan-based biomarkers for diagnosis of cancers and other diseases: past, pre, vol 162, 1st edn. Elsevier Inc., Amsterdam
12. Tonini L, Huillet C, Colas O, Klinguer-hamour C, Corvaïa N, Beck A (2013) Glycosylation engineering of biopharmaceuticals, vol 988. Humana Press, Totowa, New Jersey
13. Arnold JN, Wormald MR, Sim RB, Rudd PM, Dwek RA (2007) The impact of glycosylation on the biological function and structure of human immunoglobulins. Annu Rev Immunol 25(1):21–50
14. Oikawa M et al (2018) Significance of serum N-glycan profiling as a diagnostic biomarker in urothelial carcinoma. Eur Urol Focus 4 (3):405–411
15. Reiding KR et al (2017) Serum protein *N*-glycosylation changes with rheumatoid arthritis disease activity during and after pregnancy. Front Med 4:241
16. Kronimus Y, Dodel R, Galuska SP, Neumann S (2019) IgG fc N-glycosylation: alterations in neurologic diseases and potential therapeutic target? J Autoimmun 96:14–23
17. Chen G et al (2012) Human IgG fc-glycosylation profiling reveals associations with age, sex, female sex hormones and thyroid cancer. J Proteome 75(10):2824–2834
18. Shields RL et al (2002) Lack of fucose on human IgG1 N-linked oligosaccharide improves binding to human FcγRIII and antibody-dependent cellular toxicity. J Biol Chem 277(30):26733–26740
19. Chung S et al (2012) Quantitative evaluation of fucose reducing effects in a humanized antibody on Fcγ receptor binding and antibody-dependent cell-mediated cytotoxicity activities. MAbs 4(3):326–340
20. Sondermann P, Pincetic A, Maamary J, Lammens K, Ravetch JV (2013) General mechanism for modulating immunoglobulin effector function. Proc Natl Acad Sci U S A 110 (24):9868–9872

# Chapter 5

# Fast Afucosylation Profiling of Glycoengineered Antibody Subunits by Middle-Up Mass Spectrometry

**Elsa Wagner-Rousset, Olivier Colas, Stéphane Chenu, Yannis-Nicolas François, Davy Guillarme, Sarah Cianferani, Yury O. Tsybin, Jonathan Sjögren, Arnaud Delobel, and Alain Beck**

## Abstract

Middle-up LC-MS antibody characterization workflows using reduction or IdeS digestion for a focused assessment of *N*-glycan profiling of three representative glycoengineered monoclonal antibodies (mAbs), namely, obinutuzumab (GlycomAb technology, Glycart/Roche), benralizumab (Potelligent Technology, BioWa, Kyowa Kirin) and mAb B (kifunensine) and compared to mAb A, produced in a common CHO cell line. In addition, EndoS or EndoS2 enzyme are used for quantitative determination of Fc-glycan core afucosylation and high mannose for these antibodies, as requested by health authorities for Fc-competent therapeutics mAbs critical quality attributes (CQAs).

**Key words** IdeS, EndoS, EndoS2, Kifunensine, Obinutuzumab, Benralizumab, Low fucose, High mannose, ADCC, Mass spectrometry

## 1 Introduction

Chinese hamster ovary cells (CHO) and mouse myeloma cells (NS0, SP2/0) have become the gold-standard mammalian host cells to produce therapeutic antibodies and Fc-fusion proteins [1]. The glycoforms identified on recombinant IgGs produced from CHO cells are close to human ones except for the third GlcNac bisecting arm, which represents ~10% of human IgGs glycoforms, and very low amounts of terminal *N*-acetylneuraminic acid (NANA) [2]. NS0 produced IgGs such as NISTmAb reference material show additional complex glycoforms [3].

Glycoengineering technologies are developed to control the composition of carbohydrates and to enhance the pharmacological properties of monoclonal antibodies (mAbs) and other proteins. The approval, in 2012 in Japan, of mogamulizumab (POTELIGEO, Kyowa Kirin), the first glycoengineered antibody

Arnaud Delobel (ed.), *Mass Spectrometry of Glycoproteins: Methods and Protocols*, Methods in Molecular Biology, vol. 2271, https://doi.org/10.1007/978-1-0716-1241-5_5, 

to reach the market, was a landmark in the field of Fc-engineered biopharmaceuticals [4]. Mogamulizumab is a humanized mAb with enhanced antibody-dependent cell-mediated cytotoxicity (ADCC) activity linked to optimized a-fucosylated glycoforms (Potelligent Technology) and indicated for patients with relapsed or refractory CCR4-positive adult T-cell leukemia-lymphoma. Since, two other glycoengineered antibodies have been approved in 2013 (obinutuzumab, Gazyvaro, Roche) [5] and in 2019 (benralizumab, Fasenra, Astra-Zeneca) [6], respectively. Benralizumab (MEDI-563) is a humanized IgG1 anti-IL-5Ra fucosylated antibody based on alpha 1,6-fucosyl transferase knock out CHO cells (also based on Potelligent Technology).

Obinutuzumab (GA101, Glycart Biotechnology) is a glycoengineered tumor-targeting anti-CD20 mAb with a modified crystallizable fragment (Fc) domain designed to increase the affinity for the FcγRIIIA/CD16 receptor, which was recently approved for clinical use in chronic lymphocytic leukemia (CLL) and follicular lymphoma [7]. Obinutuzumab is expressed from stable CHO K1 cell lines engineered to constitutively overexpress the heavy and light chains of as well as recombinant wild-type β-1,4-*N*-acetylglucosaminyltransferase III and wild-type Golgi α-mannosidase II using the glutamine synthetase expression system (Lonza) reduced fucose levels (<30%) [8].

In addition, a Fc glycoengineered antibody drug conjugate (ADC) based on Potelligent technology is also investigated in phase II pivotal and phase III clinical studies (belantamab mafodotin, GSK) [9]. A Biologics License Application (BLA) was filed in January 2020, with FDA priority reviews in Multiple Myeloma and in Prostate Cancer [10].

The development and optimization of antibody and related products rely on improving their analytical and functional characterization [11]. Structural insights can be obtained for intact mAbs by mass spectrometry (MS) [12], the higher mass accuracy provided by the more straightforward RP-HPLC-MS or HILIC-MS [13] analysis of their subunits remains valuable for example for Quality Control Labs and for comparability studies [14]. MAbs profiles can be simplified by reduction (yielding the light and heavy chains at ~25 and ~50 kDa, respectively) or by enzymatic treatments, such as *N*-deglycosylation (EndoS/IgGZERO or EndoS2/Glycinator) particularly useful for core afucosylation quantification [15, 16]carboxypeptidase B digestion or glutaminyl-peptide cyclotransferase treatment. Smaller mAb fragments can also be generated by papain digestion (producing ~50 kDa Fab/Fc fragments) or IdeS digestion (Fabricator, immunoglobulin degrading enzyme of *Streptococcus pyogenes*) followed by reduction with dithiothreitol (DTT, for Fc/2, LC, and Fd fragments of ~25 kDa) [17–19]. This approach has the advantages of being fast (requiring less than 2 h for the entire analysis, including digestion and RP-HPLC-MS analysis), informative, and inexpensive in terms of materials. Reduction experiments leading to individual light and heavy chains or IdeS treatment are also use for middle up and down mass spectrometry sequencing of IgGs as well as CE-MS analysis as orthogonal method to LC-MS [20].

### 1.1 Characterization of Obinutuzumab Subunits Under Reducing Conditions and EndosS2 Digestion (Middle-Level, 23–54 kDa Fragments) (Fig. 1)

Reducing treatments are a routine way to divide the analysis of mAbs, Fc-fusion proteins and ADCs into more manageable pieces. This middle-up strategy can be implemented on any current HPLC-MS instrumentation and is therefore available in most labs. Treatment mAbs with DTT or tris(2-carboxyethyl) phosphine (TCEP) fully reduces the remaining interchain disulfides and yields two main species: light chains, and heavy chains with different *N*-glycoforms. These species are stable in the denaturing organic solvent and can be successfully separated on a reversed phase column as illustrated here for obinutuzumab.

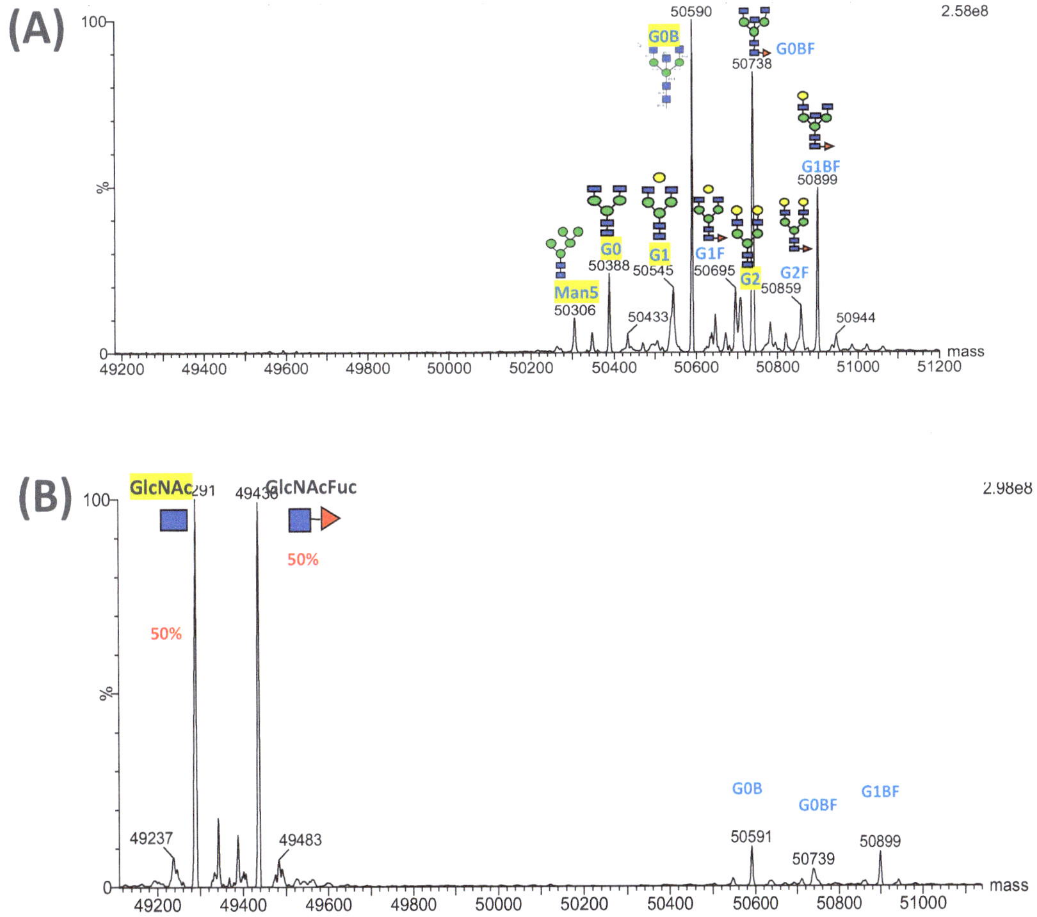


**Fig. 1** obinutuzumab (CHO, GnT III, and a-ManII GlycoMAb), reduction + EndoS2 (Roche/Glycart). (**a**) The glycoprofile of the heavy chains includes mainly biantennary complex structures with a third bisecting *N*-acetylglucosamine (G0B, G0BF, and G1BF). (**b**) After deglycosylation with Glycinator, the ratio of GlcNAc/GlcNAc-Fuc is 50/50 (Level of afucosylated glycans = 50%)

### *1.2 Characterization of Benralizumab Subunits After Enzymatic Cleavage (IdeS), Reduction and EndoS Deglycosylation (Middle-Level, 23–28 kDa Fragments) (Fig. 2)*

Downsized mAbs (or ADCs) can also be obtained by limited proteolytic cleavage under nondenaturing conditions in the hinge region of the heavy chain, yielding Fab or (Fab′)2, and Fc fragments, whose reduction (with DTT) produces even smaller fragments of approximately 25 kDa: the light chain and the two halves of the heavy chain (Fc/2 and Fd). Formerly conducted with proteases with a limited specificity, such as papain, pepsin, and endoprotease Lys-C, the enzymatic cleavage for middle-level analyses is currently mostly conducted using IdeS, a bacterial protease that specifically cleaves IgGs under the hinge region. The interest of IdeS has also been demonstrated for cysteine-linked ADCs [21]. The data can also be used for biosimilar comparability studies and Fc-fusion protein studies [22].

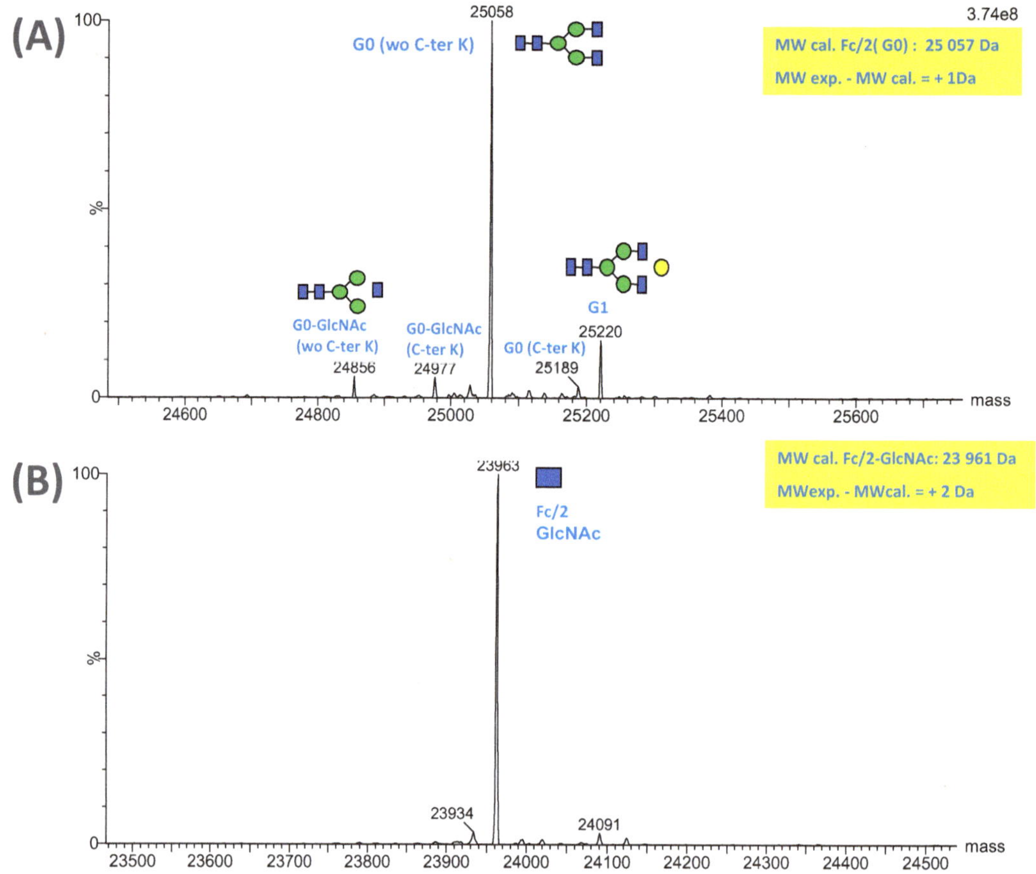


**Fig. 2** benralizumab (CHO, FUT8 −/−, Potteligent): reduction + IdeS + reduction + EndoS. (**a**) The glycoprofile of Fc/2 displays only afucosylated biantennary complex structures G0 and G1. (**b**) Deglycosylated Fc/2 with IgGzero (Level of afucosylated glycans = 100% confirmation)

**1.3 Characterization of mAb A (Produced in CHO Cells) and mAb B (Produced in CHO Cells) and Subunits Cultivated After Enzymatic Cleavage (IdeS), Reduction, and EndoS2 Deglycosylation (Middle-Level, 23–28 kDa Fragments) (Fig. 3a, b)**

*N*-glycosylation of recombinant IgGs produced in CHO cells can be metabolically modulated using kifunensine, an amannosidase I inhibitor, resulting in the production of antibodies with oligomannose-type *N*-glycans (*see* **Note 1**). Growing for 11 days in batch culture with a single treatment of kifunensine (60 ng/mL) is enough to elicit this effect without any significant impact on cell viability or antibody production. The resulting antibodies contained mainly oligomannose-type glycans and demonstrated increased ADCC activity and affinity for FcgRIIIA, but reduced C1q binding [23–26].

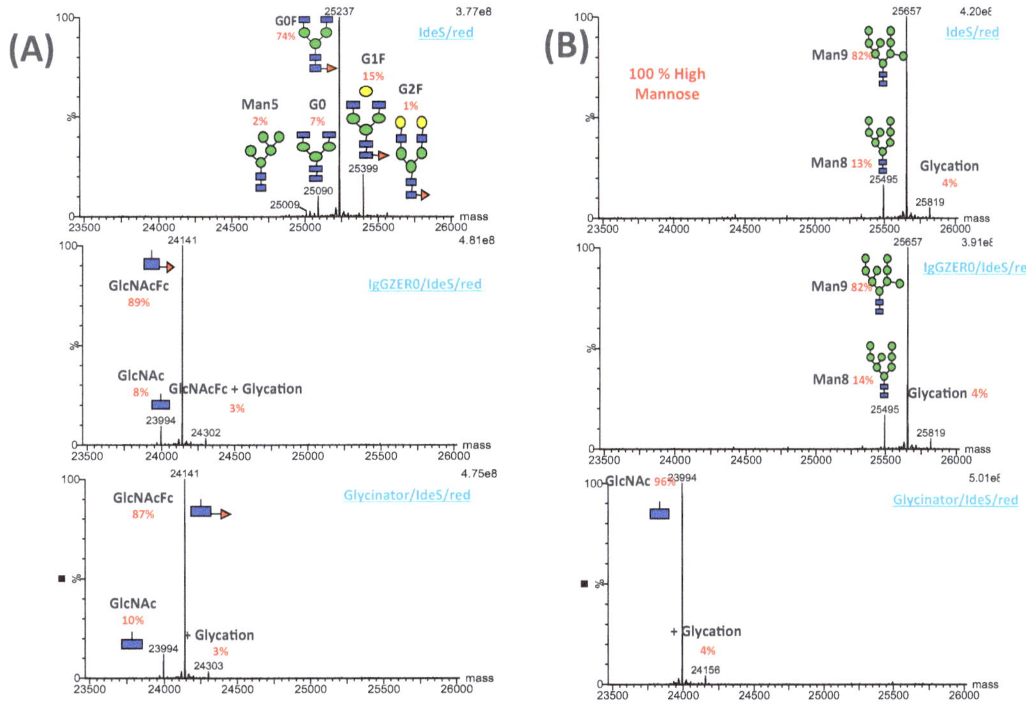


**Fig. 3** mAb A (CHO) vs. mAb B (mAb A, CHO + kifunensine). (**a**) The glycoprofile of Fc/2 of typical CHO cells produced mAb A displays mainly G0F and G1F and a smaller amount of G0, Man5, and G2F; Deglycosylated Fc/2 with IgGzero (Level of afucosylated glycans = 8–10%). (**b**) The glycoprofile of Fc/2 of CHO cells produced mAb B with culture medium completed with kifunensine displays mainly high mannose glycans; Deglycosylated Fc/2 with IgGzero (Level of afucosylated glycans = 100%). In both mAbs, around 3–4% glycated heavy chains are also detected

## 2 Materials

### 2.1 Chemicals

Ultrapure water produced from a Milli-Q Water System™(Millipore). All chemicals were of analytical grade. References are given as information and may be replaced by equivalent reagents.

1. Acetic acid 90%, (VWR, ref.: 20109.295).
2. Acetonitrile (Carlo Erba, ref.: 412342).
3. Cesium Iodide (CsI, Merck, ref.: 102861).
4. Dithiothreitol (DTT, Aldrich, ref.: 150460).
5. FabRICATOR 2000 units (IdeS, Genovis, ref.: A0-FR1-020) (*see* **Note 2**).
6. IgGZERO, 5000 units (EndoS, Genovis, ref.: A0-IZ1-050) (*see* **Note 3**).
7. GlycINATOR, 2000 units (EndoS2, Genovis, ref.: A0-GL1-020) (*see* **Note 4**).
8. Ethylenediamine tetraacetic acid (EDTA, ref.: ED2SS).
9. Guanidine hydrochloride 99% (Aldrich, ref.: 177253).
10. Trifluoroacetic Acid (TFA, Fluka, ref.: 91699 or).
11. Tris HCl (Trizma Base, Sigma, ref.: T6066).

### 2.2 mAbs

Obinutuzumab was purchased from Roche and benralizumab from MedImmune. MAbs A and B were produced by Pierre Fabre.

### 2.3 Ultra-performance Liquid Chromatography

1. LC equipment. Acquity™ UPLC system consisting in a Binary solvent manager, a sample manager, and a TUV detector (Waters).
2. Column Bioresolved RP mAb Polyphenyl 450 Å, 2.1 × 150 mm, 2.7 μm (Waters 186,008,946).
3. Mobile phases. Eluting solution A: MilliQ water + 0.05% TFA. Eluting solution B: Acetonitrile + 0.05% TFA.

### 2.4 Mass Analysis

Synapt G2Si™ Waters mass spectrometer equipped with an electrospray (ESI) source and a time-of-flight (TOF) analyzer.

Calibration and Lock Spray CsI solution: dissolve CsI in water/isopropanol, 50/50 (v/v) to have a 2 mg/mL solution for daily use.

After instrument conditioning with a mixture of eluting solutions A/B 50/50 (v/v) at 0.2 mL/min, calibrate the mass spectrometer by infusing the CsI calibration solution. CsI forms 12 charged clusters from 900 to 4000 *m/z*.

### 2.5 Reagents Preparation

1. 0.5 mL Eppendorf tubes.
2. Pipettes and corresponding tips.

3. Thermomixer + block tube 0.5 mL.
4. Reduction buffer. 6 M guanidine buffer pH 8 containing 2 mM EDTA and. 0.1 M Tris–HCl. For 10 mL: dissolve 0.121 g Tris HCl, 7.4 mg EDTA, 5.730 g guanidine hydrochloride, solubilize in 9 mL MilliQ water, adjust the pH at 8.0 with 6 N HCl and complete to 10 mL with MilliQ water.
5. Reducing reagent. Dissolve DTT in MilliQ water to have a 500 mM solution (77 mg/mL) for extemporaneous use.

## 3 Methods

### 3.1 General Principle

The mAbs are either reduced into two chains (light chain and heavy chain) or IdeS digested and reduced into three fragments (Fc/2, Fd, LC). Deglycosylation using EndoS/S2 can be done together with IdeS or sequentially. All enzymes work in combination so the IdeS digestion + EndoS/S2 deglycosylation can be performed in the same vial.

These mAbs subunits are adsorbed on the reversed phase column thanks to hydrophobic interactions. They are then eluted by increasing the amount of eluting solution B during the chromatographic gradient.

UV detection at 210 nm of the separated mAbs fragments. MS detection is achieved in parallel to check the identity of each fragment.

### 3.2 Sample Preparation

#### 3.2.1 Reduction

1. Add 25 μg of the mAb sample into an Eppendorf tube.
2. Dilute with reduction buffer to reach a volume of 23.5 μL.
3. Add 1.5 μL of 500 mM DTT. The final concentration of mAb is 1 mg/mL and DTT concentration is 30 mM.
4. Incubate for 45 min in the thermomixer at 56 °C under agitation (750 tr/min).
5. Quench the reaction by adding 1 μL acetic acid.

#### 3.2.2 IdeS Digestion and Reduction

1. Add 25 μg of the mAb sample into an Eppendorf tube.
2. Add 1.25 μL of FabRICATOR (IdeS, *see* **Note 2**) (1 unit of IdeS/μg of sample).
3. Complete to 10 μL with MilliQ Water.
4. Incubate for 30 min in the thermomixer at 37 °C under agitation (750 tr/min).
5. Dilute with reduction buffer to reach a volume of 23.5 μL.
6. Add 1.5 μL of 500 mM DTT. The final concentration of mAb is 1 mg/mL and DTT concentration is 30 mM.
7. Incubate for 45 min in the thermomixer at 56 °C under agitation (750 tr/min).

8. Quench the reaction by adding 1 μL acetic acid.
9. LC-ESI-TOF sample analysis: equilibrate the column by running through 95% solvent A at a flow rate of 0.3 mL/min during 10 min.
10. Set up the mass spectrometer and check the stable spray with elution buffer. The voltage applied to the capillary was set to 2500 V. Ions are scanned over a *m/z* range of 500–5000. Source and desolvation temperatures are set to 100 °C and 300 °C, respectively. Cone and source offset voltages are set to 45 V and 60 V, respectively. Nitrogen gas flow rates are set at 100 L/h for the cone and 1000 L/h for desolvation.
11. Inject the mAb sample preparation (5 μg) onto the column and simultaneously start both the chromatography gradient and the mass spectrometer data collection.
12. The analytical column is eluted typically at a flow rate of 0.3 mL/min by a four-steps linear gradient: (1) 5% B to 30% B in 8 min, (2) 30% B to 40% B in 32 min, (3) 40% B to 95% B in 10 min, (4) 95% B to 5% B in 2 min, followed by a 10 min equilibration step at 5% B.

### 3.3 Data Analysis

#### 3.3.1 MS Data Treatment

1. Using MassLynx™, open the Total Ion Chromatogram.
2. The lock mass correction factor is calculated from the MS signal of CsI solution infused within the lockspray (*m/z*: 1691.765) (*see* **Note 5**).
3. Combine spectra of each chromatographically separated peak.
4. Smooth and perform spectrum deconvolution using Maxent-1™.

#### 3.3.2 UV Data Treatment

1. Display the UV chromatogram at 210 nm in MassLynx™.
2. Integrate each chromatographic peak (reduced mAbs or IdeS digested and reduced mAbs).
3. From the result of integration (peak surfaces), calculate the percentage of each *N*-glycoforms.

## 4 Notes

1. EndoS2 (GlycINATOR®) is an IgG-specific endoglycosidase that hydrolyzes all glycoforms present at the Fc-glycosylation site. The enzyme acts on the chitobiose core and leaves the core GlcNAc intact.
2. After lock mass correction, measured masses should be within ±5 Da around theoretical masses calculated from the amino-acid sequence.

3. IdeS (FabRICATOR®) is a cysteine protease that digests antibodies at a specific site below the hinge, generating a homogenous pool of $F(ab')_2$ and Fc/2 fragments for Human IgG1-4, IgG from monkey, rat, rabbit and sheep.
4. EndoS (IgGZERO®) is an IgG-specific endoglycosidase acting on complex type *N*-glycans at the Fc-glycosylation site of IgG. The enzyme acts on the chitobiose core and leaves the core GlcNAc intact.
5. High mannose-type *N*-glycans contain from five to nine mannose residues and are found on antibodies produced in mammalian cells, yeast, insect cells, and plants, but only at a very low level in normal human antibodies. High mannose glycans on the Fc region of therapeutic IgG antibodies increase serum clearance in humans. As a consequence, only mammalian-based production systems are used for the manufacturing of approved biopharmaceuticals, which need near-human glycosylation.

## References


1. Beck A, Liu H (2019) Macro- and micro-heterogeneity of natural and recombinant IgG antibodies. Antibodies 8:18. https://doi.org/10.3390/antib8010018
2. Duivelshof BL, Jiskoot W, Beck A, Veuthey J-L, Guillarme D, D'Atri V (2019) Glycosylation of biosimilars: recent advances in analytical characterization and clinical implications. Anal Chim Acta 1089:1–18. https://doi.org/10.1016/j.aca.2019.08.044
3. De Leoz MLA, Duewer DL, Fung A, Liu L, Yau HK, Potter O, Staples GO, Furuki K, Frenkel R, Hu Y, Sosic Z, Zhang P, Altmann F, Nwald-Grube CG, Shao C, Zaia J, Evers W, Pengelley S, Suckau D, Wiechmann A, Resemann A, Jabs W, Beck A, Froehlich JW, Huang C, Li Y, Liu Y, Sun S, Wang Y, Seo Y, An HJ, Reichardt N-C, Ruiz JE, Archer-Hartmann S, Azadi P, Bell L, Lakos Z, An Y, Cipollo JF, Pucic-Bakovic M, Štambuk J, Lauc G, Li X, Wang PG, Bock A, Hennig R, Rapp E, Creskey M, Cyr TD, Nakano M, Sugiyama T, Leung P-KA, Link-Lenczowski P, Jaworek J, Yang S, Zhang H, Kelly T, Klapoetke S, Cao R, Kim JY, Lee HK, Lee JY, Yoo JS, Kim S-R, Suh S-K, de Haan N, Falck D, Lageveen-Kammeijer GSM, Wuhrer M, Emery RJ, Kozak RP, Liew LP, Royle L, Urbanowicz PA, Packer NH, Song X, Everest-Dass A, Lattová E, Cajic S, Alagesan K, Kolarich D, Kasali T, Lindo V, Chen Y, Goswami K, Gau B, Amunugama R, Jones R, CJM S, Kato K, Yagi H, Kondo S, Yuen CT, Harazono A, Shi X, Magnelli PE, Kasper BT, Mahal L, Harvey DJ, O'Flaherty R, Rudd PM, Saldova R, Hecht ES, Muddiman DC, Kang J, Bhoskar P, Menard D, Saati A, Merle C, Mast S, Tep S, Truong J, Nishikaze T, Sekiya S, Shafer A, Funaoka S, Toyoda M, de Vreugd P, Caron C, Pradhan P, Tan NC, Mechref Y, Patil S, Rohrer JS, Chakrabarti R, Dadke D, Lahori M, Zou C, Cairo C, Reiz B, Whittal RM, Lebrilla CB, Wu L, Guttman A, Szigeti M, Kremkow BG, Lee KH, Sihlbom C, Adamczyk B, Jin C, Karlsson NG, Örnros J, Larson G, Nilsson J, Meyer B, Wiegandt A, Komatsu E, Perreault H, Bodnar ED, Said N, Francois Y-N, Leize-Wagner E, Maier S, Zeck A, AJR H, Yang Y, Haselberg R, Yu YQ, Alley W, Leone JW, Yuan H, Stein SE (2020) NIST Interlaboratory Study on Glycosylation Analysis of Monoclonal Antibodies: Comparison of Results from Diverse Analytical Methods. Mol Cell Proteomics 19:11–30. https://doi.org/10.1074/mcp.RA119.001677
4. Beck A, Reichert JM (2012) Marketing approval of mogamulizumab: a triumph for glyco-engineering. MAbs 4:419–425. https://doi.org/10.4161/mabs.20996
5. Stoll DR, Harmes DC, Staples GO, Potter OG, Dammann CT, Guillarme D, Beck A (2018) Development of comprehensive online two-dimensional liquid chromatography/mass spectrometry using hydrophilic

interaction and reversed-phase separations for rapid and deep profiling of therapeutic antibodies. Anal Chem 90:5923–5929. https://doi.org/10.1021/acs.analchem.8b00776

6. Carter PJ, Lazar GA (2018) Next generation antibody drugs: pursuit of the "high-hanging fruit". Nat Rev Drug Discov 17:197–223. https://doi.org/10.1038/nrd.2017.227
7. Capuano C, Pighi C, Maggio R, Battella S, Morrone S, Palmieri G, Santoni A, Klein C, Galandrini R (2020) CD16 pre-ligation by defucosylated tumor-targeting mAb sensitizes human NK cells to γc cytokine stimulation via PI3K/mTOR axis. Cancer Immunol Immunother 69:501–512. https://doi.org/10.1007/s00262-020-02482-2
8. Mössner E, Brünker P, Moser S, Püntener U, Schmidt C, Herter S, Grau R, Gerdes C, Nopora A, van Puijenbroek E, Ferrara C, Sondermann P, Jäger C, Strein P, Fertig G, Friess T, Schüll C, Bauer S, Dal Porto J, Del Nagro C, Dabbagh K, Dyer MJS, Poppema S, Klein C, Umaña P (2010) Increasing the efficacy of CD20 antibody therapy through the engineering of a new type II anti-CD20 antibody with enhanced direct and immune effector cell–mediated B-cell cytotoxicity. Blood 115:4393–4402. https://doi.org/10.1182/blood-2009-06-225979
9. Tai Y-T, Mayes PA, Acharya C, Zhong MY, Cea M, Cagnetta A, Craigen J, Yates J, Gliddon L, Fieles W, Hoang B, Tunstead J, Christie AL, Kung AL, Richardson P, Munshi NC, Anderson KC (2014) Novel anti–B-cell maturation antigen antibody-drug conjugate (GSK2857916) selectively induces killing of multiple myeloma. Blood 123:3128–3138. https://doi.org/10.1182/blood-2013-10-535088
10. Beck A, D'Atri V, Ehkirch A, Fekete S, Hernandez-Alba O, Gahoual R, Leize-Wagner E, François Y, Guillarme D, Cianférani S (2019) Cutting-edge multi-level analytical and structural characterization of antibody-drug conjugates: present and future. Expert Rev Proteomics 16:337–362. https://doi.org/10.1080/14789450.2019.1578215
11. van der Burgt YEM, Kilgour DPA, Tsybin YO, Srzentić K, Fornelli L, Beck A, Wuhrer M, Nicolardi S (2019) Structural analysis of monoclonal antibodies by ultrahigh resolution MALDI in-source decay FT-ICR mass spectrometry. Anal Chem 91:2079–2085. https://doi.org/10.1021/acs.analchem.8b04515
12. Largy E, Cantais F, Van Vyncht G, Beck A, Delobel A (2017) Orthogonal liquid chromatography–mass spectrometry methods for the comprehensive characterization of therapeutic glycoproteins, from released glycans to intact protein level. J Chromatogr A 1498:128–146. https://doi.org/10.1016/j.chroma.2017.02.072
13. D'Atri V, Nováková L, Fekete S, Stoll D, Lauber M, Beck A, Guillarme D (2019) Orthogonal middle-up approaches for characterization of the glycan heterogeneity of Etanercept by hydrophilic interaction chromatography coupled to high-resolution mass spectrometry. Anal Chem 91:873–880. https://doi.org/10.1021/acs.analchem.8b03584
14. Ambrogelly A, Gozo S, Katiyar A, Dellatore S, Kune Y, Bhat R, Sun J, Li N, Wang D, Nowak C, Neill A, Ponniah G, King C, Mason B, Beck A, Liu H (2018) Analytical comparability study of recombinant monoclonal antibody therapeutics. MAbs 10:513–538. https://doi.org/10.1080/19420862.2018.1438797
15. Liu S, Zang L (2016) Rapid quantitation of monoclonal antibody N-glyco-occupancy and afucosylation using mass spectrometry. Anal Biochem 509:142–145. https://doi.org/10.1016/j.ab.2016.06.029
16. Upton R, Bell L, Guy C, Caldwell P, Estdale S, Barran PE, Firth D (2016) Orthogonal assessment of biotherapeutic glycosylation: a case study correlating N-glycan Core Afucosylation of Herceptin with mechanism of action. Anal Chem 88:10259–10265. https://doi.org/10.1021/acs.analchem.6b02994
17. Collin M (2001) EndoS, a novel secreted protein from *Streptococcus pyogenes* with endoglycosidase activity on human IgG. EMBO J 20:3046–3055. https://doi.org/10.1093/emboj/20.12.3046
18. Sjögren J, Struwe WB, Cosgrave EFJ, Rudd PM, Stervander M, Allhorn M, Hollands A, Nizet V, Collin M (2013) EndoS2 is a unique and conserved enzyme of serotype M49 group A streptococcus that hydrolyses N-linked glycans on IgG and α1-acid glycoprotein. Biochem J 455:107–118. https://doi.org/10.1042/BJ20130126
19. Sjögren J, Olsson F, Beck A (2016) Rapid and improved characterization of therapeutic antibodies and antibody related products using IdeS digestion and subunit analysis. Analyst 141:3114–3125. https://doi.org/10.1039/C6AN00071A
20. Giorgetti J, Beck A, Leize-Wagner E, François Y-N (2020) Combination of intact, middle-up and bottom-up levels to characterize 7 therapeutic monoclonal antibodies by capillary electrophoresis—mass spectrometry. J Pharm

Biomed Anal 182:113107. https://doi.org/10.1016/j.jpba.2020.113107

21. Beck A, Terral G, Debaene F, Wagner-Rousset-E, Marcoux J, Janin-Bussat M-C, Colas O, Dorsselaer AV, Cianférani S (2016) Cutting-edge mass spectrometry methods for the multi-level structural characterization of antibody-drug conjugates. Expert Rev Proteomics 13:157–183. https://doi.org/10.1586/14789450.2016.1132167
22. Beck A, Debaene F, Diemer H, Wagner-Rousset E, Colas O, Dorsselaer AV, Cianférani S (2015) Cutting-edge mass spectrometry characterization of originator, biosimilar and biobetter antibodies: biosimilar antibodies. J Mass Spectrom 50:285–297. https://doi.org/10.1002/jms.3554
23. Kanda Y, Yamada T, Mori K, Okazaki A, Inoue M, Kitajima-Miyama K, Kuni-Kamochi-R, Nakano R, Yano K, Kakita S, Shitara K, Satoh M (2007) Comparison of biological activity among nonfucosylated therapeutic IgG1 antibodies with three different N-linked Fc oligosaccharides: the high-mannose, hybrid, and complex types. Glycobiology 17:104–118. https://doi.org/10.1093/glycob/cwl057
24. Zhou Q, Shankara S, Roy A, Qiu H, Estes S, McVie-Wylie A, Culm-Merdek K, Park A, Pan C, Edmunds T (2008) Development of a simple and rapid method for producing non-fucosylated oligomannose containing antibodies with increased effector function. Biotechnol Bioeng 99:652–665. https://doi.org/10.1002/bit.21598
25. van Berkel PHC, Gerritsen J, Perdok G, Valbjørn J, Vink T, van de Winkel JGJ, Parren PWHI (2009) N-linked glycosylation is an important parameter for optimal selection of cell lines producing biopharmaceutical human IgG. Biotechnol Prog 25:244–251. https://doi.org/10.1002/btpr.92
26. Crispin M, Bowden TA, Coles CH, Harlos K, Aricescu AR, Harvey DJ, Stuart DI, Jones EY (2009) Carbohydrate and domain architecture of an immature antibody Glycoform exhibiting enhanced effector functions. J Mol Biol 387:1061–1066. https://doi.org/10.1016/j.jmb.2009.02.033

# Chapter 6

# Characterization of Glycosylated Proteins at Subunit Level by HILIC/MS

Valentina D'Atri and Davy Guillarme


## Abstract

Hydrophilic interaction chromatography (HILIC) coupled to mass spectrometry (MS) is considered as the reference analytical technique for glycans profiling, especially for the characterization of glycosylated protein therapeutics such as monoclonal antibodies (mAbs) and mAbs-related products. Although HILIC/MS is mainly known to profile enzymatically released and fluorescently labeled *N*-glycans, the recent commercialization of new widepore HILIC amide bonded stationary phases packed with sub-2 μm particles has allowed for remarkable separations also at the subunit level. Here, we describe a simple protocol to perform the mAb glycans profiling at subunit level by HILIC/MS.


**Key words** Hydrophilic interaction chromatography, Mass spectrometry, *N*-glycosylation, Biopharmaceutical proteins, Protocol, Monoclonal antibodies

## 1 Introduction

Glycosylation is reported as the most complex posttranslational modification (PTM) occurring in the expression of eukaryotic proteins [1]. It can greatly affects the structural heterogeneity of proteins and consequently their physical and functional properties such as solubility, conformation, folding, stability as well as their biological role [1]. For these reasons, glycosylation is of utmost importance in the field of biopharmaceutical drugs. In fact, the characterization of the glycans profile of protein therapeutics, such as monoclonal antibodies (mAbs) and mAbs-related products, is one of the most important critical quality attribute (CQA), due to the effects that glycans might have on the immunogenicity and clinical efficacy of these biopharmaceutical products [2].

*N*-linked glycosylation is the predominant type of glycosylation in protein therapeutics and it occurs on a conserved Asn residue through the Asn-X-Thr/Ser consensus sequence (where X is not a Pro). Major *N*-glycans components consists of a complex, biantennary and generally fucosylated main core including

Arnaud Delobel (ed.), *Mass Spectrometry of Glycoproteins: Methods and Protocols*, Methods in Molecular Biology, vol. 2271, https://doi.org/10.1007/978-1-0716-1241-5_6, 

N-acetylglucosamines and mannoses that could include up to 2 terminal galactoses (G0F, G1F, and G2F species). Minor forms might also consist of high-mannose (M5) or afucosylated species (G0, G1, and G2) and eventually include sialylation (G2S1, G2FS1, G2S2 and G2FS2 species). Figure 3 gives a synopsis of the typical *N*-glycoforms identified in protein therapeutics represented based on the Symbol Nomenclature for Graphical Representation of Glycans (SNFG) [3].

The analytical characterization of *N*-glycosylation in protein therapeutics is mainly performed by high-performance liquid chromatography coupled to spectroscopic detector (HPLC-UV and HPLC-FLD) or mass spectrometry (HPLC/MS) at different level of the analysis, including released glycans, glycopeptides (bottom-up level), protein subunits (middle-up level) and intact proteins [4]. Hydrophilic interaction chromatography (HILIC) is considered as the reference chromatographic technique at the released glycans level, although this approach requires the tedious chemical release, labelling and purification of the glycans prior their HILIC separation. On the other hand, reversed-phase liquid chromatography (RPLC) is mainly used at bottom-up level for peptides analysis. However, HILIC is also becoming a valid orthogonal technique to this approach thanks to its ability to separate the glycopeptides that would otherwise be poorly retained in RPLC [5, 6]. In this context, HILIC has also been reported as innovative approach for glycans profiling at middle-up level of analysis, after protein subunits generation obtained by combining enzymatic digestions and chemical reductions [7–9]. In fact, therapeutic antibodies treated with immunoglobulin-degrading enzyme of *Streptococcus pyogenes* (*IdeS* enzyme) are cleaved in two main protein subunits denoted as F$(ab')_2$ and Fc fragments. Indeed, six protein domains of around 25 kDa each (respectively two light chains (LC), Fd′ and Fc/2 fragments) can be obtained when *IdeS* digestion is followed by disulfide bond reduction, with glycosylation generally occurring on the Fc/2 domains [10]. The novelty of the HILIC analysis performed at middle-up level lies in the fact that protein subunits bearing different glycans are resolved on HILIC widepore (300 Å) sub-2-μm particles, amide bonded stationary phases based on the increase of hydrophilicity/polarity of each subunit that is mainly due to the size of the glycan attached on it. The same analysis performed by RPLC would result in the coelution of the subunits bearing the glycans, thus limiting the separation of the species at the chromatographic level [11].

Here, we present a protocol for the glycans profiling of therapeutic proteins performed at subunit level by HILIC/MS by using two representative commercial therapeutic mAbs, trastuzumab and rituximab. An overview of the workflow is represented in Fig. 1. The HILIC/MS chromatograms and deconvoluted mass spectra obtained from the analysis of trastuzumab and rituximab

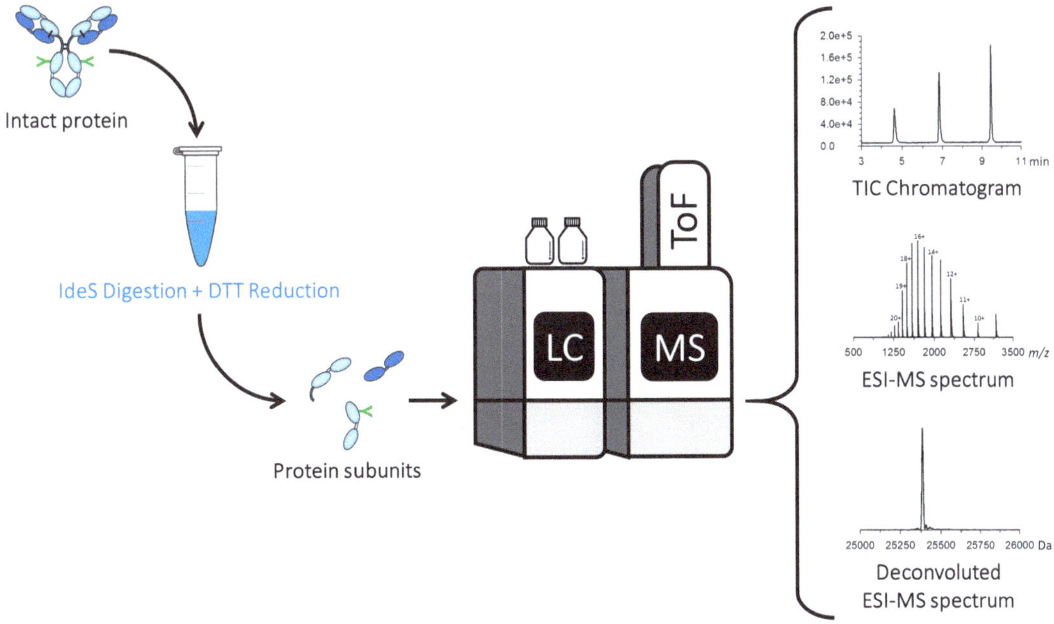


**Fig. 1** Strategy for HILIC/MS analysis of *N*-glycosylated protein subunits derived from *IdeS* digestion and DTT reduction

*N*-glycosylated protein subunits are depicted in Fig. 2, together with the graphic representation of each protein subunits eventually bearing a specific glycan. Detailed retention times and mass assignments are reported in Table 1.

Applying a middle-up approach streamlines the process of assessing glycosylation profiles, since a facilitated deconvolution of each peak and an easier assignment by MS are allowed by the increased chromatographic and mass spectrometric resolution obtained at subunit level, given that the analysis is conducted on subunits of around 25 kDa. After *IdeS* digestion and dithiothreitol (DTT) reduction, the structural heterogeneity of the mAbs is broken down and the three main protein subunits (LC, Fd′ and Fc/2) are resolved by hyphenating HILIC with MS, thanks to the wide pore HILIC stationary phase used here. For the characterization of trastuzumab and rituximab, Fig. 2 shows the total ion chromatograms (TIC) obtained from HILIC/MS analysis, with the light chains and the Fd′ fragments resolved as single peaks while the Fc/2 subunits are resolved as multiple peaks corresponding to individual Fc/2 protein subunits bearing different glycan moieties. Each peak is then deconvoluted (Fig. 2i–vii), identified against a theoretical molecular mass and assigned to a specific species, as reported in Table 1. Specifically, Fc/2 subunits bearing G0F, G1F, and G2F glycans species were identified for both trastuzumab and rituximab and Fc/2 bearing the G0 moiety was only identified for trastuzumab. Interestingly, other PTMs were highlighted, such

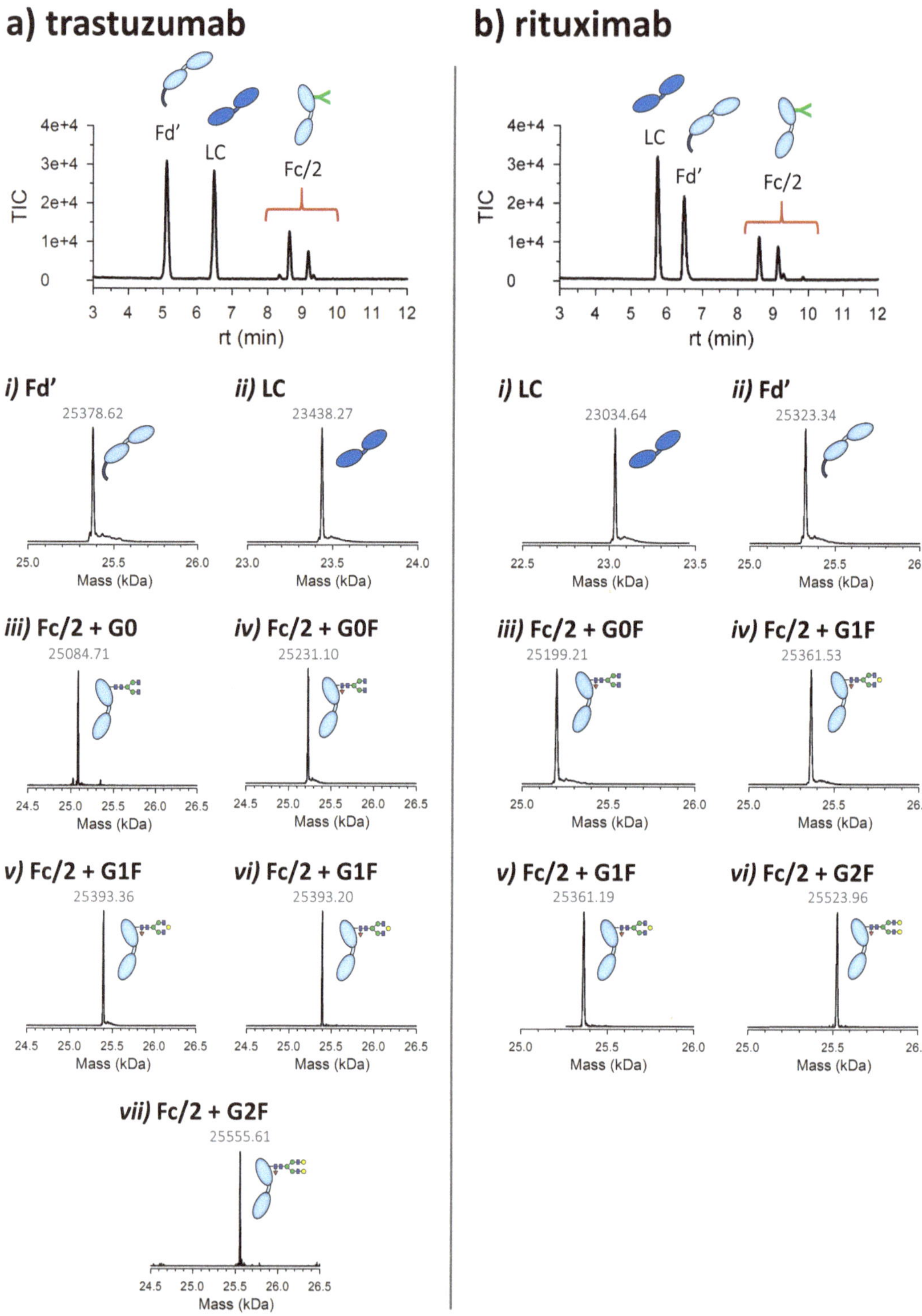


**Fig. 2** HILIC/MS analysis of trastuzumab (**a**) and rituximab (**b**). *N*-glycosylated protein subunits derived from *IdeS* digestion and DTT reduction. TIC chromatograms and deconvoluted mass spectra of each peak (i–vii). Detailed retention times and mass assignments are reported in Table 1

**Table 1**
**Trastuzumab and rituximab middle-up analysis performed by HILIC/MS. Protein subunits retention times and mass assignments**

| $T_r$ (min) | Assignment | Theoretical mass (Da) | Experimental mass (Da) | $\Delta m$ (Da) |
|---|---|---|---|---|
| *Trastuzumab* | | | | |
| 5.12 | Fd′ | 25379.59 | 25378.62 | 0.97 |
| 6.49 | LC | 23439.11 | 23438.27 | 0.84 |
| 8.35 | Fc/2—K + G0 | 25085.31 | 25084.71 | 0.60 |
| 8.63 | Fc/2—K + G0F | 25231.50 | 25231.10 | 0.40 |
| 9.18 | Fc/2—K + G1F | 25394.42 | 25393.36 | 1.06 |
| 9.34 | Fc/2—K + G1F | 25394.42 | 25393.20 | 1.22 |
| 9.89 | Fc/2—K + G2F | 25556.56 | 25555.61 | 0.95 |
| *Rituximab* | | | | |
| 5.74 | LC (Q/pE) | 23035.67 | 23034.64 | 1.03 |
| 6.49 | Fd′ (Q/pE) | 25324.48 | 25323.34 | 1.14 |
| 8.60 | Fc/2—K + G0F | 25200.22 | 25199.21 | 1.01 |
| 9.14 | Fc/2—K + G1F | 25362.36 | 25361.53 | 0.83 |
| 9.29 | Fc/2—K + G1F | 25362.36 | 25361.19 | 1.17 |
| 9.84 | Fc/2—K + G2F | 25524.50 | 25523.96 | 0.54 |

as the C-terminal Lys truncation for both trastuzumab and rituximab Fc/2 subunits and the pyroglutamic acid formation for the Fd′ and LC fragments of rituximab.

Thanks to the high selectivity of the HILIC separation, it is possible to directly assess qualitative differences in the mAbs glycosylation patterns and the application of the presented methods may be eventually used to support comparative analyses of glycans profiles, such as batch-to-batch controls or comparison between originator and biosimilars mAbs at the protein level [5, 8].

## 2 Materials

Follow appropriate laboratory practices and use LC-MS grade solvents and ultrapure water (18.2 MΩ·cm at 25 °C) to prevent contaminations in LC-MS. Prepare and store all reagents at room temperature, unless otherwise specified.

#### 2.1 Generation of Protein Subunits

1. Digestion buffer: 100 mM Tris–HCl, pH 7.5. Add about 5 mL ultrapure water to a 10 mL graduate cylinder. Weight 121 ± 1 mg TRIZMA base and transfer to the cylinder. Mix and adjust the pH with HCl. Complete up to 10 mL with ultrapure water. Store at 4 °C.
2. Reduction solution: 1 M dithiothreitol (DTT). Weight 31 ± 1 mg DTT and transfer to a 500 μL Eppendorf. Add 200 μL ultrapure water and homogenize the solution. Store at 4 °C.

#### 2.2 Hydrophilic Interaction Chromatography (HILIC) Coupled to Electrospray Ionization Mass Spectrometry (ESI-MS)

1. ACQUITY UPLC Glycoprotein BEH Amide, 300 Å, 1.7 μm column (Waters, Milford, MA, USA). Alternatively, the AdvanceBio Glycan Map, 1.8 μm column (Agilent, Santa Clara, CA, USA) can be used (*see* **Note 1**).
2. Mobile phase A (MPA): 0.08% TFA and 0.02% FA in water. Add 400 mL ultrapure water in a 500 mL graduate cylinder. Add 400 μL TFA by using a 1 mL micropipette. Add 100 μL FA by using a 200 μL micropipette. Complete to 500 mL with ultrapure water. Transfer to a 500 mL Duran bottle. Sonicate 5 min.
3. Mobile phase B (MPB): 0.08% TFA and 0.02% FA in ACN. Add 400 mL LC-MS grade ACN in a 500 mL graduate cylinder. Add 400 μL TFA by using a 1 mL micropipette. Add 100 μL FA by using a 200 μL micropipette. Complete to 500 mL with LC-MS grade ACN. Transfer to a 500 mL Duran bottle. Sonicate 5 min.
4. Weak wash solvent: 85:15 ACN–water. In a 200 mL graduate cylinder, add 30 mL of ultrapure water and complete up to 200 mL with LC-MS grade ACN. Transfer to a 250 mL Duran bottle.
5. Strong wash solvent: 40:60 ACN–water. In a 200 mL graduate cylinder, add 80 mL of LC-MS grade ACN and complete up to 200 mL with ultrapure water. Transfer to a 250 mL Duran bottle.
6. Seal wash: 10:90 methanol–water. In a 200 mL graduate cylinder, add 20 mL of LC-MS grade methanol and complete up to 200 mL with ultrapure water. Transfer to a 250 mL Duran bottle.
7. HPLC system: ACQUITY UPLC system (Waters, Milford, MA, USA) equipped with an autosampler, a binary pumping system, and a fixed loop injector (1 μL), and coupled to a fluorescence detector (FLD, $\lambda_{ex}$ at 280 nm and $\lambda_{em}$ at 360 nm). Typical HILIC elution parameters: flow rate: 0.4 mL/min; column temperature: 45 °C; first column equilibration time: 20 min at 85% MPB; gradient conditions: from 85% to 73% MPB in 0.2 min (fast high acetonitrile initial

gradient, *see* **Note 2**), from 73% to 65% MPB in 12 min (separation step), from 65% to 15% MPB in 0.3 min, 1 min at 15% MPB (washing step), from 15% to 85% MPB in 0.5 min, 9 min at 85% MPB (column reequilibration step).

8. Mass spectrometer: Quadrupole time-of-flight (Q-ToF) mass spectrometer (Xevo from Waters, Milford, MA, USA). Typical MS parameters: positive ion mode; capillary voltage: 3 kV; cone voltage: 30 V; source temperature: 150 °C; desolvation temperature: 500 °C; desolvation gas ($N_2$) flow rate: 1000 L/h; acquisition range: 400 to 4000 *m/z*; calibration of the instrument performed with sodium iodide (NaI, 2 μg/μL in 50% isopropanol).

## 3 Methods

### 3.1 Generation of Protein Subunits

1. In a 500 μL Eppendorf, add 100 μg of mAb, 100 U of immunoglobulin-degrading enzyme of *Streptococcus pyogenes* (*IdeS* enzyme, *see* **Note 3**), 10 μL of digestion buffer and complete to 90 μL with ultrapure water.
2. Vortex and incubate at 37 °C for 30 min.
3. After having performed the digestion, add 10 μL of DTT solution to the sample (*see* **Notes 4** and **5**). Vortex and incubate at 45 °C for 30 min.
4. Quench the reaction by acidification of the sample by adding 1% of TFA to the solution.
5. Transfer to HPLC vial equipped with a 100 μL insert (*see* **Note 6**).

### 3.2 HILIC-MS Data Treatment and Assessment of N-Glycan Profile

Mass spectra of mAb subunits separated by HILIC and obtained under electrospray conditions in positive ion mode typically exhibit multiply charged ions (the charge envelope) in charge states from +30 to +50 with mass range values within 4000 *m/z*.

1. Apply a mass deconvolution algorithm, to transform the multiply charged spectrum into a zero-charge, average molecular mass spectrum (deconvoluted mass spectrum) that represents the experimental mass of the mAb subunits (Fig. 1).
2. Identify the mAb subunits against an available theoretical molecular mass calculated, taking into account the mAb amino acids sequence and the possible presence of glycan moieties or others post-translational modifications (PTMs).
3. Reliable mAb amino acids sequences are available [12, 13] and can be used to compute the theoretical molecular mass of each mAb subunit through the use of specific software (*see* **Note 7**). Table 2 reports the amino acid sequences of trastuzumab and

**Table 2**
**Amino acid sequences of light chain and heavy chain of trastuzumab and rituximab. *N*-glycosylation site is in red and the *IdeS* proteases consensus site is in bold. Sequence source: DrugBank [13]**

| | Tratuzumab | Rituximab |
|---|---|---|
| Light Chain | DIQMTQSPSSLSASVGDRVTITCRA<br>SQDVN<br>TAVAWYQQKPGKAPKLLIYSASFL<br>YSGVPS<br>RFSGSRSGTDFTLTISSLQPEDFATY<br>YCQQ<br>HYTTPPTFGQGTKVEIKRTVAAPSV<br>FIFPP<br>SDEQLKSGTASVVCLLNNFYPREA<br>KVQWKV<br>DNALQSGNSQESVTEQDSKDSTYS<br>LSSTLT<br>LSKADYEKHKVYACEVTHQGLSSP<br>VTKSFN<br>RGEC | QIVLSQSPAILSASPGEKVTMTCRASSS<br>VS<br>YIHWFQQKPGSSPKPWIYATSNLASGV<br>PVR<br>FSGSGSGTSYSLTISRVEAEDAATYYCQ<br>QW<br>TSNPPTFGGGTKLEIKRTVAAPSVFIFPP<br>S<br>DEQLKSGTASVVCLLNNFYPREAKVQ<br>WKVD<br>NALQSGNSQESVTEQDSKDSTYSLSST<br>LTL<br>SKADYEKHKVYACEVTHQGLSSPVTK<br>SFNR<br>GEC |
| Heavy Chain | EVQLVESGGGLVQPGGSLRLSCAA<br>SGFNIK<br>DTYIHWVRQAPGKGLEWVARIYPT<br>NGYTRY<br>ADSVKGRFTISADTSKNTAYLQMN<br>SLRAED<br>TAVYYCSRWGGDGFYAMDYWGQ<br>GTLVTVSS<br>ASTKGPSVFPLAPSSKSTSGGTAAL<br>GCLVK<br>DYFPEPVTVSWNSGALTSGVHTFP<br>AVLQSS<br>GLYSLSSVVTVPSSSLGTQTYICNV<br>NHKPS<br>NTKVDKKVEPKSCDKTHTCPP**CPA**<br>**PELLG/G**<br>**PSVF**LFPPKPKDTLMISRTPEVTCV<br>VVDVS<br>HEDPEVKFNWYVDGVEVHNAKTK<br>PREEQYN<br>STYRVVSVLTVLHQDWLNGKEYK<br>CKVSNKA<br>LPAPIEKTISKAKGQPREPQVYTLPP<br>SREE<br>MTKNQVSLTCLVKGFYPSDIAVEW<br>ESNGQP<br>ENNYKTTPPVLDSDGSFFLYSKLTV<br>DKSRW<br>QQGNVFSCSVMHEALHNHYTQKS<br>LSLSPGK | QVQLQQPGAELVKPGASVKMSCKASG<br>YTFT<br>SYNMHWVKQTPGRGLEWIGAIYPGNG<br>DTSY<br>NQKFKGKATLTADKSSSTAYMQLSSLT<br>SED<br>SAVYYCARSTYYGGDWYFNVWGAGT<br>TVTVS<br>AASTKGPSVFPLAPSSKSTSGGTAALGC<br>LV<br>KDYFPEPVTVSWNSGALTSGVHTFPAV<br>LQS<br>SGLYSLSSVVTVPSSSLGTQTYICNVNH<br>KP<br>SNTKVDKKAEPKSCDKTHTCPP**CPAPE**<br>**LLG/**<br>**GPSVF**LFPPKPKDTLMISRTPEVTCVVV<br>DV<br>SHEDPEVKFNWYVDGVEVHNAKTKPR<br>EEQY<br>NSTYRVVSVLTVLHQDWLNGKEYKCK<br>VSNK<br>ALPAPIEKTISKAKGQPREPQVYTLPPS<br>RD<br>ELTKNQVSLTCLVKGFYPSDIAVEWES<br>NGQ<br>PENNYKTTPPVLDSDGSFFLYSKLTVD<br>KSR<br>WQQGNVFSCSVMHEALHNHYTQKSLS<br>LSPGK |

rituximab that have been used to compute the theoretical masses of each protein subunits. The glycosylation site is highlighted in red and the cleavage site of the *IdeS* enzyme is emphasized in bold to distinguish the two fragments deriving after the digestion, namely the Fd′ and the Fc/2 fragments, which respectively represent the N- and C-terminal portions of the heavy chains. Since glycosylation occurs on the Fc/2 domain, theoretical molecular masses of this subunit have to

| SNFG | ON Name | Shorthand name | Average mass shift (Da) |
|---|---|---|---|
| 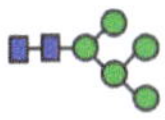 | M5 | M5 | 1217.09 |
| 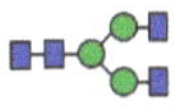 | A2 | G0 | 1299.19 |
| 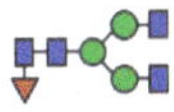 | FA2 | G0F | 1445.33 |
| 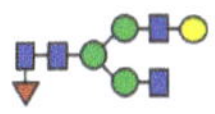 | FA2G1 | G1F | 1607.47 |
| 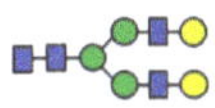 | A2G2 | G2 | 1623.47 |
| 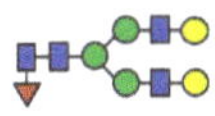 | FA2G2 | G2F | 1769.61 |
| 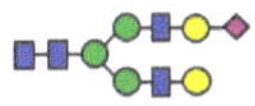 | A2G2S1 | G2S1 | 1914.73 |
| 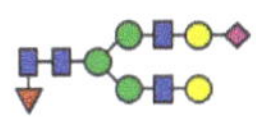 | FA2G2S1 | G2FS1 | 2060.87 |
| 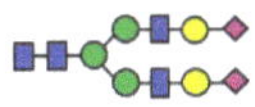 | A2S2S2 | G2S2 | 2205.98 |
| 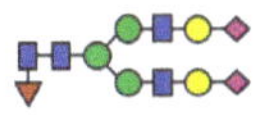 | FA2G2S2 | G2FS2 | 2352.12 |

**Fig. 3** Schematic representation of typical *N*-glycoforms identified in protein therapeutics along with their corresponding shorthand names based on the Oxford Notation (ON). Glycans represented based on the Symbol Nomenclature for Graphical Representation of Glycans (SNFG)

be evaluated by taking into account the typical mass shifts characteristic for each specific glycan moiety (Fig. 3). Additionally, other PTMs may be considered for the prediction of the theoretical molecular masses. Typical PTMs could include C-terminal Lys truncation (−K, −128.17 Da), pyroglutamic acid formation of the N-terminal Glu (E/pE, −18.02 Da) or Gln (Q/pE, −17.03 Da) residues and eventually Met or Trp oxidation (+15.99 Da) and Asn deamidation (+0.98 Da) (*see* **Note 8**).

## 4 Notes

1. Conditioning of previously unused HILIC columns should be performed through sequential injections of a representative protein sample until a stable chromatographic profile is achieved.
2. Under HILIC conditions, protein samples diluted in aqueous solutions could compromise the retention and generate breakthrough phenomena (peak distortion and band broadening) due to the strong eluotropic strength of the sample diluent. To prevent this issue, a fast initial gradient ramp that integrates a high percentage of ACN (85%) at the beginning of the method can be implemented to efficiently counterbalance the elutropic effect of the sample diluent [7–9, 14, 15].
3. *IdeS* proteases cleaves human IgG1-4 and chimeric IgG from monkey, rat, sheep and rabbit as well as Fc-fusion proteins. However, it has limited activity on murine IgG2a and IgG3, and IgG containing the LALA mutation (CPAPE**AA**G/GPSVF consensus site instead of CPAPE**LL**G/GPSVF). For digestion of these species, we recommend using *IdeZ* protease.
4. Be careful in performing the reactions in the proper order. Based on our experience, the *IdeS* digestion could be not effective or partially incomplete if the DTT reduction is performed before the *IdeS* digestion.
5. Amounts of protein and enzyme can be adapted based on specific needs by respecting the proportion of each component: 1 U of *IdeS* enzyme is able to digest 1 μg of IgG protein. Volumes could also be adapted based on the final required volume: final Tris–HCl and DTT concentrations are 10 mM and 100 mM, respectively.
6. Glass-silanized or polypropylene inserts should be preferred to avoid adsorption of the protein to vial wall [16].

7. MassLynx 4.1 software (Waters) was used to compute the theoretical molecular mass of each mAb subunit. Alternatively, the pI/Mw tool of ExPASy might be used free of charges [17].
8. The presence of Asn deamidation should be validated by performing peptide map analysis.

## References


1. Krištić J, Lauc G (2017) Ubiquitous importance of protein glycosylation. In: Lauc G (ed) High-throughput glycomics and glycoproteomics. Methods in molecular biology, vol 1503. Humana Press, New York, NY
2. Higel F, Seidl A, Sörgel F, Friess W (2016) N-glycosylation heterogeneity and the influence on structure, function and pharmacokinetics of monoclonal antibodies and Fc fusion proteins. Eur J Pharm Biopharm 100:94–100. https://doi.org/10.1016/j.ejpb.2016.01.005
3. Haltiwanger RS (2016) Symbol nomenclature for Glycans (SNFG). Glycobiology 26:217–217. https://doi.org/10.1093/glycob/cww005
4. Yang X, Bartlett MG (2019) Glycan analysis for protein therapeutics. J Chromatogr B Anal Technol Biomed Life Sci 1120:29–40. https://doi.org/10.1016/j.jchromb.2019.04.031
5. D'Atri V, Dumont E, Vandenheede I, Guillarme D, Sandra P, Sandra K (2017) Hydrophilic interaction chromatography for the characterization of therapeutic monoclonal antibodies at protein, peptide, and glycan levels. LC-GC Europe 30:424–434
6. Largy E, Cantais F, Van Vyncht G, Beck A, Delobel A (2017) Orthogonal liquid chromatography–mass spectrometry methods for the comprehensive characterization of therapeutic glycoproteins, from released glycans to intact protein level. J Chromatogr A 1498:128–146. https://doi.org/10.1016/j.chroma.2017.02.072
7. Periat A, Fekete S, Cusumano A, Veuthey J-L, Beck A, Lauber M, Guillarme D (2016) Potential of hydrophilic interaction chromatography for the analytical characterization of protein biopharmaceuticals. J Chromatogr A 1448:81–92. https://doi.org/10.1016/j.chroma.2016.04.056
8. D'Atri V, Fekete S, Beck A, Lauber M, Guillarme D (2017) Hydrophilic interaction chromatography hyphenated with mass spectrometry: a powerful analytical tool for the comparison of originator and biosimilar therapeutic monoclonal antibodies at the middle-up level of analysis. Anal Chem 89:2086–2092. https://doi.org/10.1021/acs.analchem.6b04726
9. D'Atri V, Fekete S, Stoll D, Lauber M, Beck A, Guillarme D (2018) Characterization of an antibody-drug conjugate by hydrophilic interaction chromatography coupled to mass spectrometry. J Chromatogr B 1080:37–41. https://doi.org/10.1016/j.jchromb.2018.02.026
10. Sjögren J, Olsson F, Beck A (2016) Rapid and improved characterization of therapeutic antibodies and antibody related products using IdeS digestion and subunit analysis. Analyst 141:3114–3125. https://doi.org/10.1039/c6an00071a
11. D'Atri V, Goyon A, Bobaly B, Beck A, Fekete S, Guillarme D (2018) Protocols for the analytical characterization of therapeutic monoclonal antibodies. III—Denaturing chromatographic techniques hyphenated to mass spectrometry. J Chromatogr B 1096:95–106. https://doi.org/10.1016/j.jchromb.2018.08.013
12. IMGT, the international ImMunoGeneTics information system. http://www.imgt.org. Accessed 1 Jun 2019
13. DrugBank. https://www.drugbank.ca. Accessed 1 Jun 2019
14. Bobály B, D'Atri V, Beck A, Guillarme D, Fekete S (2017) Analysis of recombinant monoclonal antibodies in hydrophilic interaction chromatography: a generic method development approach. J Pharm Biomed Anal 145:24–32. https://doi.org/10.1016/j.jpba.2017.06.016
15. Gargano AFG, Roca LS, Fellers RT, Bocxe M, Domínguez Vega E, Somsen GW (2018) Capillary HILIC-MS: a new tool for sensitive top-down proteomics. Anal Chem 90(11):6601–6609. https://doi.org/10.1021/acs.analchem.8b00382
16. Rodriguez-Aller M, Cusumano A, Alain B, Guillarme D, Fekete S (2016) Importance of vial shape and type on the reproducibility of size exclusion chromatography measurement of monoclonal antibodies. J Chromatogr B 1032:131–138. https://doi.org/10.1016/j.jchromb.2016.04.032
17. Compute pI/Mw tool. https://web.expasy.org/compute_pi/. Accessed 1 Jun 2019

# Chapter 7

# Analysis of Monoclonal Antibody Glycopeptides by Capillary Electrophoresis–Mass Spectrometry Coupling (CE-MS)

Josiane Saadé, Michael Biacchi, Jérémie Giorgetti, Antony Lechner, Alain Beck, Emmanuelle Leize-Wagner, and Yannis-Nicolas François

## Abstract

Glycosylation is a crucial posttranslational modification (PTM) that might affect the safety and efficacy of monoclonal antibodies (mAbs). Capillary electrophoresis–mass spectrometry (CE-MS) enables the characterization of the primary structure of mAbs. A bottom-up proteomic workflow is designed to provide detailed information about the glycosylation. In this chapter, we describe the validated experimental protocol applied for the characterization and relative quantification of mAbs *N*-glycosylation at the glycopeptide level.

**Key words** Monoclonal antibody, Glycopeptide, Glycosylation profiling, Structural characterization, Capillary electrophoresis–mass spectrometry

## 1 Introduction

Monoclonal antibodies (mAbs) are therapeutic proteins efficacious for various diseases including oncology, inflammatory diseases, organ transplantation, cardiology, viral infection, allergy, and tissue growth and repair [1–3]. MAbs are tetrameric glycoproteins having a molecular mass of approximately 150 kDa, composed of two heavy chains and two light chains, interlinked by several disulfide bonds, and having at least one conserved *N*-glycosylation site located in the fragment crystallizable Fc domain [4, 5]. Glycosylation forms are attached on an Asparagine residue enclosed in a specific amino acid sequence composed of Asn-X-Ser/Thr where X can be whichever amino acid except a Proline. Due to their heterogeneity caused by posttranslational modification (PTM), the characterization of mAbs structure has become a challenge for analytical sciences [2, 6–8]. Several PTM can be observed in biopharmaceuticals [9], including glycosylation and small chemical modifications such as methionine or tryptophan oxidation

Arnaud Delobel (ed.), *Mass Spectrometry of Glycoproteins: Methods and Protocols*, Methods in Molecular Biology, vol. 2271, https://doi.org/10.1007/978-1-0716-1241-5_7, 

| Graphical representation | Saccharide name | Abbr. | Formula | Average mass (Da) | Mass shift (Da) |
|---|---|---|---|---|---|
| | *N*-Acetyl-Glucosamine | N | C8H15NO6 | 221.2078 | 203.1925 |
| | Fucose | F | C6H12O5 | 164.1565 | 146.1412 |
| | Mannose | H | C6H12O6 | 180.1559 | 162.1406 |
| | Galactose | H | C6H12O6 | 180.1559 | 162.1406 |
| | *N*-Acetyl Neuraminic Acid | Sa | C11H19NO9 | 309.2699 | 291.2546 |
| | *N*-Glycolyl Neuraminic Acid | Sg | C11H19NO10 | 325.2693 | 307.2540 |

**Fig. 1** Graphical representation and corresponding names, abbreviations, formula, and average masses of the saccharides involved in the formation of *N*-glycans [18]

(+ 15.99 Da) and asparagine deamidation (+ 0.98 Da). Glycosylation depends on multiple factors like production system, selected clonal population, manufacturing process and may be genetically or chemically engineered [2]. The Fc glycosylation pattern is a critical PTM because it greatly affects the mAb characteristics such as solubility, stability, pharmacokinetic and pharmacodynamic properties, as well as in vivo efficacy [10–12]. As a consequence, the mAbs glycosylation profile is considered as a critical quality attribute (CQA) and must be thoroughly analyzed [13–17]. The main core of a glycan is usually composed of *N*-acetyl-glucosamines and mannoses residues, and then additional saccharides can be present, such as fucose, galactose, and sialic acid (Fig. 1) [18]. The complexity and heterogeneity of the glycosylation (Fig. 2) requires a number of orthogonal analytical techniques to be fully characterized. Separative techniques (liquid chromatography (LC), capillary electrophoresis (CE)) often coupled to spectrometric, amperometric and mass spectrometric detection have been described for the glyco-variants characterization at different levels (from released glycans to intact protein level) [19–21]. Reusch's group published two studies dealing with the analysis of Fc-glycosylation profiles, and comparing several separation methods hyphenated or not with mass spectrometry (MS) detection [21]. If all the methods showed excellent precision and accuracy, some differences were observed with regard to the detection and the quantitation of minor glycan species, such as sialylated glycans.

Concerning MS-based methods, a large panel was evaluated [20–23]. High-performance liquid chromatography coupled to mass spectrometry (HPLC-MS) plays a key role in the structural characterization. Especially, reversed-phase liquid chromatography-MS/MS (RPLC-MS/MS) is used at the bottom-up level to perform amino acid sequence verification and PTMs targeting. In

| | | theo M (glycan) | theo M (glycopeptide) | z=1 (+) | z=2 (+) | z=3 (+) | z=4 (+) |
|---|---|---|---|---|---|---|---|
| Peptide | EEQFNSTYR | 1172.510 | 1172.510 | 1173.510 | 587.255 | 391.837 | 294.127 |
| G0F [H3N4F1] | 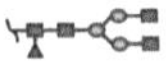 | 1444.534 | 2617.044 | 2618.044 | 1309.522 | 873.348 | 655.261 |
| G1F [H4N4F1] | 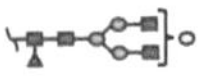 | 1606.587 | 2779.097 | 2780.097 | 1390.548 | 927.366 | 695.774 |
| G2F [H5N4F1] | 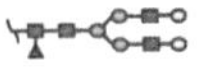 | 1768.640 | 2941.149 | 2942.149 | 1471.575 | 981.383 | 736.287 |
| G0F-N [H3N3F1] | 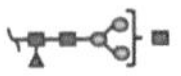 | 1241.455 | 2413.964 | 2414.964 | 1207.982 | 805.655 | 604.491 |
| G1F-N [H4N3F1] | 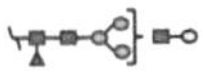 | 1403.507 | 2576.017 | 2577.017 | 1289.009 | 859.672 | 645.004 |
| G0 [H3N4] |  | 1298.476 | 2470.986 | 2471.986 | 1236.493 | 824.662 | 618.746 |
| G1 [H4N4] | 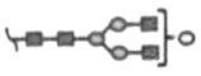 | 1460.529 | 2633.039 | 2634.039 | 1317.519 | 878.680 | 659.260 |
| G2 [H5N4] | 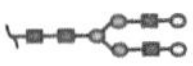 | 1622.582 | 2795.091 | 2796.091 | 1398.546 | 932.697 | 699.773 |
| G0-N [H3N3] | 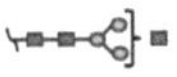 | 1095.397 | 2267.906 | 2268.906 | 1134.953 | 756.969 | 567.977 |
| G1-N [H4N3] |  | 1257.449 | 2429.959 | 2430.959 | 1215.980 | 810.986 | 608.490 |
| G1FS-N [H4N3FS1] | 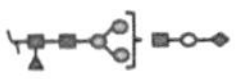 | 1710.598 | 2883.107 | 2884.107 | 1442.554 | 962.036 | 721.777 |
| G1FS [H4N4FS1] |  | 1913.677 | 3086.187 | 3087.187 | 1544.093 | 1029.729 | 772.547 |
| M5 [H5N2] | 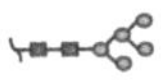 | 1216.423 | 2388.933 | 2389.933 | 1195.466 | 797.311 | 598.233 |
| M6 [H6N2] | 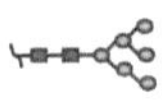 | 1378.476 | 2550.985 | 2551.985 | 1276.493 | 851.328 | 638.746 |

**Fig. 2** Overview of glycopeptide masses of natalizumab. For detailed information on saccharides involved in the glycan composition, refer to Fig. 1

order to enhance N-glycan characterization, peptide mapping is a method of choice to boost the different glycopeptides tracing and allow glycan structure analysis.

Regarding CE-MS coupling, Gennaro et al. described the development of capillary electrophoresis to electrospray ionization mass spectrometry (CE-ESI-MS) technology with online laser-induced fluorescence (LIF) detection that allows identification of major and minor glycan species observed in the routine CE-LIF assay [24]. More recently, CE-ESI-MS/MS methods with a sheathless interface have been developed to perform the characterization of mAbs in one injection including amino acid sequence, glycosylation characterization and other types of posttranslational modifications (methionine oxidation, asparagine deamidation, . . .) [14, 25, 26]. In 2018, Giorgetti et al. validated a CE-ESI-MS method in terms of robustness and reproducibility through the relative quantitation of glycosylation profiles for ten different mAbs produced in different cell lines [27]. Systematic comparison of the glycosylation patterns obtained for each mAb was compared with that obtained with the HILIC-FD reference method. Results obtained with the CE-ESI-MS approach and hydrophilic interaction liquid chromatography—Fluorescence Detector (HILIC-FD) showed very similar glycoprofiling, demonstrating the attractiveness of CE-ESI-MS method to characterize and quantify the glycosylation heterogeneity of a wide range of therapeutic mAbs, with high accuracy and precision. To enhance the repeatability for glycopeptides analysis with higher sensitivities by sheathless CE-ESI-MS/MS, Kammeijer et al. combined a Dopant Enriched Nitrogen Gaz (DEN-Gaz) to sheathless CE-ESI-MS/MS [28]. Moreover, the same group demonstrated CE-ESI-MS/MS technique as a promising tool for separating sialic acid linkage isomers [29]. Due to very identical molecular formulas and similar fragmentation patterns, the separation of isomeric sialylated glycopeptides remains challenging. Where conventional MS(/MS) approaches cannot resolve these isomers, CE enabled the baseline separation of sialylated glycopeptides due to a difference in their electrophoretic mobilities, correlated by a difference of acidity, as shown for the two representative compounds α 2,3-sialyllactose and α 2,6-sialyllactose.

In this chapter, we detail a CE-ESI-MS/MS method which enables the comprehensive characterization and relative quantification of mAbs *N*-glycosylation. As an example to describe this methodology, natalizumab (Tysabri®) results have been selected.

## 2 Materials

### *2.1 Reagents and Buffers*

1. 0.2% RapiGest solution: Add 500 μL of water to a 1 mg vial of RapiGest SF surfactant (Waters, Milford, MA, USA).

2. 0.5 μg/μL trypsin solution: dissolve sequencing grade trypsin (Promega, Madison, WI, USA) in reconstitution buffer to a final concentration of 0.5 μg/mL.
3. 200 mM ammonium acetate buffer, pH 4.0: Add about 10 mL of water to a 25 mL volumetric flask. Weigh 0.385 g of ammonium acetate and transfer to the flask. Add 0.996 mL of acetic acid (>99%). Make up to 25 mL with water.
4. 1 M ammonia solution: Add about 10 mL of water to a 25 mL volumetric flask. Add 1.56 mL of 30% ammonia (16.04 M). Make up to 25 mL with water.
5. 50 mM ammonium bicarbonate buffer, pH 8.0: Add about 10 mL of water to a 50 mL volumetric flask. Weigh 0.2085 g of ammonium bicarbonate and transfer to the flask. Add 0.05 mL of 1 M ammonia solution. Make up to 50 mL with water.
6. 500 mM DTT solution: Weigh 0.077 g of DTT and transfer to a 2.5 mL microtube. Make up to 1 mL with ammonium bicarbonate buffer (50 mM, pH 8.0).
7. 500 mM IAA solution: Weigh 0.092 g of IAA and transfer to a 2.5 mL microtube. Make up to 1 mL of ammonium bicarbonate buffer (50 mM, pH 8.0).
8. Calibration solution: Mix 5 μL of digest of β-galactosidase (8 μM) and 15 μL ammonium acetate 200 mM, pH 4.0 to obtain a final concentration of 2.0 μM.
9. Monoclonal antibody: Natalizumab EMA/FDA-approved formulations kindly provided by Pierre Fabre laboratories (Saint-Julien en Genevois, France). mAbs were stored at 4 °C.

### 2.2 Instrumentation

1. CESI8000 system (Sciex, Brea, CA, USA) was hyphenated to MS detection using a sheathless nanoelectrospray (nanoESI) interface.
2. Separations were performed using bare fused-silica capillaries (total length 100 cm; 30 mm i d.) with characteristic porous tip on their final end on 3 cm.
3. A second capillary (total length 80 cm; 50 mm i.d.) filled during experiments with 10% acetic acid background electrolyte (BGE) allows electric contact.
4. A Triple TOF 5600+ mass spectrometer (Sciex, Darmstadt, Germany) is equipped with a hybrid analyzer composed of a quadrupole followed by a time-of flight (TOF) analyzers.
5. Analyst software (Sciex, Darmstadt, Germany) to control the MS system.
6. PeakView software (Sciex, CA).
7. Dedicated software to process protein MS data (e.g., Biotools (Bruker, Germany)).

## 3 Methods

The following protocol details the different steps for the characterization of mAbs *N*-glycans using sheathless CE-ESI-MS instrument. Carry out all procedures at room temperature, unless otherwise specified.

### 3.1 Trypsin Digestion

1. Mix 10 μL of protein at 66.7 μM and 5 μL of ultrapure water in a 500 mL microtube to obtain a protein concentration of 45.6 μM (total volume 15 μL).
2. Add 15 μL of 0.2% RapiGest surfactant (*see* **Note 1**) to obtain mAb concentration of 22.2 μM. Heat sample at 40 °C for 10 min (total volume 30 μL).
3. Add 1.5 μL of 500 mM DTT solution to obtain a concentration of 25 mM (*see* **Note 2**). Heat sample at 95 °C for 5 min (total volume 31.5 μL).
4. Once cooled down to room temperature, alkylate the cysteines (Cys) to avoid the reformation of the disulfide bonds (*see* **Note 3**): add 0.65 μL of 500 mM IAA to obtain a final concentration of 10 mM. Leave sample at room temperature for 20 min in the dark (total volume 32.15 μL).
5. Add a first volume of 1 μL of trypsin (0.5 μg/μL). Leave at room temperature for 3 h (total volume 33.15 μL).
6. Add a second volume of 1 μL of trypsin (0.5 μg/μL). Heat sample overnight at 37 °C (total volume 34.15 μL).
7. Add 0.35 μL of formic acid (final concentration of 1% (v/v)) (*see* **Note 4**). Leave sample at room temperature for 2 h. The final concentration of mAb is 3 μg/μL (20 μM) (total volume 34.5 μL).
8. Before CE-MS analysis, mix 1 μL of mAb sample and 9 μL of 200 mM ammonium acetate, pH 4.0 to obtain a final protein concentration of 2 μM (*see* **Note 5**).

### 3.2 CE-ESI-MS Analysis

1. New capillaries are flushed at 75 psi (5.17 bars) for 10 min with methanol, then 10 min with 0.1 M sodium hydroxide, followed by 10 min with 0.1 M hydrochloric acid and water for 20 min.
2. Before each analysis, separation capillary is flushed at 75 psi (5.17 bars) for 10 min with the BGE composed of 10% acetic acid. The second capillary for maintaining the electric fields is flushed with the same BGE for 5 min at 75 psi (5.17 bars).
3. After each analysis, separation capillary is flushed at 50 psi (3.45 bars) for 5 min with methanol, then 5 min with 0.1 M sodium hydroxide, followed by 5 min with 0.1 M hydrochloric acid.

After each solvent, rinse the capillary with water at 75 psi (5.17 bars) for 3 min. The second capillary is flushed with the same BGE for 3 min at 75 psi (5.17 bars). A rinsing step of 5 min at 50 psi (3.45 bar) with ultrapure water is performed between each flush.

4. Add 10 μL of digested mAb sample (2 μM) in a microvial. Place it in the sample platform of the instrument.
5. Perform a hydrodynamic injection of 100 nL by applying 6 psi for 2 min (around 200 fmol injected).
6. Perform the separations using a voltage of +20 kV for 60 min (*see* **Note 6**).
7. Set ESI source parameters as follows: ESI voltage −1.75 kV while Gas Supplies (GS1 and GS2) were deactivated. Source heating temperature 150 °C and curtain gas value 5 (*see* **Note 7**). Mass/charge ($m/z$) range was 100–2000 in MS and 50–2000 in MS/MS.
8. Calibrate all spectra by external calibration using a digest of *β-galactosidase*.

### 3.3 MS and MS/MS Data Analysis

1. Use Analyst software to convert your MS-raw data from .wiff format to .mgf format (*see* **Note 8**).
2. Open the .mgf file with a dedicated software to perform the identification of the primary structure of the mAb. The mass tolerance for search algorithm identification was set to ±5 ppm for precursor ions and ±0.05 Da for fragment ions.
3. For the missing peptides not found by the search algorithm, calculate the theoretical monoisotopic masses using a fragment ion calculator (e.g., Proteomics Toolkit). Extract found masses from the MS-raw data using PeakView software and validate the presence of missing peptides using MS/MS data.
4. For glycopeptides, enter the theoretical modification in the search algorithm and perform the identification.
5. For the missing glycopeptides, follow the same protocol described in **step 3** with the masses of modified peptides (Fig. 2).
6. To perform the glycan profiling of each mAb, estimate relative occurrence levels from the sum of isotopic peak intensities, considering all charge states of the ion corresponding to one glycopeptide (Fig. 3a) [27]. Perform the comparison of all glycoforms abundance (Fig. 3b) (*see* **Note 9**).

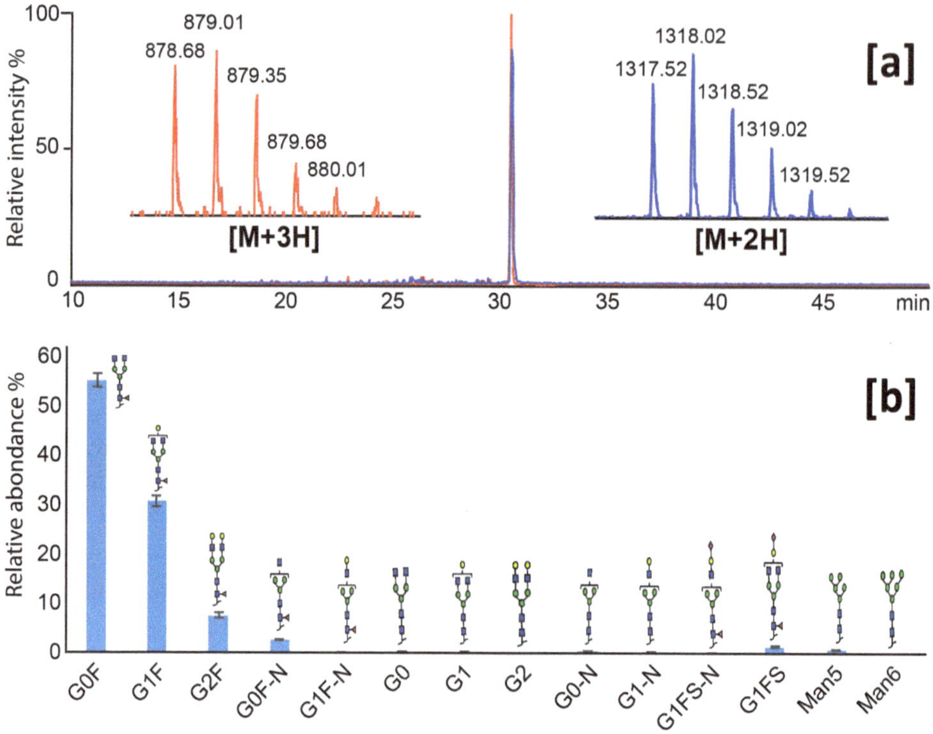


**Fig. 3** (**a**) Extracted ion electropherogram (EIE) of *m/z* ratios 878.68 and 1317.52 ([EEQFN*STYR] + G0F) and corresponding MS/MS fragmentation spectra (right-hand side). (**b**) Glycoforms relative abundance results obtained through the CE-ESI-MS data for the natalizumab Fc glycopeptide. (Copyright (2018) Elsevier)

## 4 Notes

1. RapiGest surfactant is a well-known denaturing agent used in this protocol. Other denaturing agents such as guanidine hydrochloride or urea could be used [30].
2. DTT is a well-known reducing agent used in this protocol. Other reducing agent such as TCEP (tris(2-carboxyethyl) phosphine) could be used [31].
3. Alkylation step may not necessary if a short digestion time is applied or acidic conditions sustained [31].
4. RapiGest, also known as sodium 3-[(2-methyl-2-undecyl-1,3-dioxolan-4-yl)methoxy]-1-propanesulfonate is an acid-cleavable anionic detergent used to enhance the enzymatic digestion of proteins. Addition of an acid is mandatory to cleave the surfactant promoting MS detection of peptides.
5. Dilution of mAb sample (3 μg/μL) in 200 mM ammonium acetate, pH 4.0 modifies the sample buffer and allows sample to be in favorable conditions to perform online preconcentration by transient isotachophoresis during the CE-ESI-MS analysis.

6. CE instrument allows for following the current profile during the separation. If the current drops, this indicates that you have a loss of separation and therefore a decrease in resolution. Among the most probable causes, the formation of air bubble at the capillary inlet, the clogging or rupture of the capillary have to be checked.
7. Due to the ultra-low flow rate obtained with the sheathless CE-ESI-MS interface, curtain gas must be less than 10 to get a stable spray.
8. Each mass spectrometer manufacturer requires obtaining raw data in a proprietary format (For example, .wiff for Sciex). However, analysis of this data implies the use of dedicated software. The most commonly used file format used in MS proteomics is the Mascot Generic Format (.mgf) file. The mgf file was developed by Matrix Science (London, UK), the maker of Mascot, the most widely used commercial search engine, but it is widely supported by many proteomics search engines. Then it allows a better flexibility for data treatment of obtained MS data.
9. Peak area or peak intensity of the different glycosylation can be displayed automatically with designed software as Skyline.

## Acknowledgments

The authors would like to thank Sciex and Bruker Daltonics for the fruitful collaboration and the support. The authors would like also to express their gratitude to Dr. E. Wagner-Rousset, MC. Janin-Bussat and O. Colas (Centre d'Immunologie Pierre Fabre, St Julien en Genevois, France) for helpful discussions. This work was supported by the CNRS (UMR 7140), the University of Strasbourg, and a doctoral fellowship from Bruker Daltonics and the General Council of Region Grand Est to JG.

### References

1. Beck A, Reichert JM (2012) Marketing approval of mogamulizumab a triumph for glyco-engineering. MAbs 4(4):419–425
2. Beck A, Wagner-Rousset E, Bussat MC et al (2008) Trends in glycosylation, glycoanalysis and glycoengineering of therapeutic antibodies and Fc-fusion proteins. Curr Pharm Biotechnol 9(6):482–501
3. Kaplon H, Reichert JM (2018) Antibodies to watch in 2018. MAbs 10(2):183–203
4. Berger M, Shankar V, Vafai A (2002) Therapeutic applications of monoclonal antibodies. Am J Med Sci 324(1):14–30
5. Beck A, Wagner-Rousset E, Ayoub D, Van Dorsselaer A, Sanglier-Cianférani S (2013) Characterization of therapeutic antibodies and related products. Anal Chem 85(2):715–736
6. Fekete S, Gassner AL, Rudaz S, Schappler J, Guillarme D (2013) Analytical strategies for the characterization of therapeutic monoclonal antibodies. Trac—Trend Anal Chem 42:74–83
7. Gahoual R, Beck A, Leize-Wagner E, François Y-N (2016) Cutting-edge capillary electrophoresis characterization of monoclonal antibodies and related products. J Chromatogr B 1032:61–78
8. Lechner A, Giorgetti J, Gahoual R, Beck A, Leize-Wagner E, François Y-N (2019) Insights from capillary electrophoresis approaches for characterization of monoclonal antibodies and

antibody drug conjugates in the period 2016–2018. J Chromatogr B 1122-1123:1–17
9. Walsh G, Jefferis R (2006) Post-translational modifications in the context of therapeutic proteins. Nat Biotechnol 24(10):1241–1252
10. Cymer F, Beck H, Rohde A, Reusch D (2018) Therapeutic monoclonal antibody N-glycosylation – structure, function and therapeutic potential. Biologicals 52:1–11
11. Higel F, Sandl T, Kao C-Y et al (2019) N-glycans of complex glycosylated biopharmaceuticals and their impact on protein clearance. Eur J Pharm Biopharm 139:123–131
12. Beck A, Liu H (2019) Macro- and micro-heterogeneity of natural and recombinant IgG antibodies. Antibodies 8(1):18
13. Ayoub D, Jabs W, Resemann A et al (2013) Correct primary structure assessment and extensive glyco-profiling of cetuximab by a combination of intact, middle-up, middle-down and bottom-up ESI and MALDI mass spectrometry techniques. MAbs 5(5):699–710
14. Gahoual R, Burr A, Busnel JM et al (2013) Rapid and multi-level characterization of trastuzumab using sheathless capillary electrophoresis-tandem mass spectrometry. MAbs 5(3):479–490
15. Reusch D, Tejada ML (2015) Fc glycans of therapeutic antibodies as critical quality attributes. Glycobiology 25(12):1325–1334
16. Schiestl M, Stangler T, Torella C, Čepeljnik T, Toll H, Grau R (2011) Acceptable changes in quality attributes of glycosylated biopharmaceuticals. Nat Biotechnol 29:310
17. Largy E, Cantais F, Van Vyncht G, Beck A, Delobel A (2017) Orthogonal liquid chromatography–mass spectrometry methods for the comprehensive characterization of therapeutic glycoproteins, from released glycans to intact protein level. J Chromatogr A 1498:128–146
18. D'Atri V, Goyon A, Bobaly B, Beck A, Fekete S, Guillarme D (2018) Protocols for the analytical characterization of therapeutic monoclonal antibodies. III—denaturing chromatographic techniques hyphenated to mass spectrometry. J Chromatogr B 1096:95–106
19. Zhang L, Luo S, Zhang B (2015) Glycan analysis of therapeutic glycoproteins. MAbs:1–11
20. Dotz V, Haselberg R, Shubhakar A et al (2015) Mass spectrometry for glycosylation analysis of biopharmaceuticals. TrAC Trends Anal Chem 73:1–9
21. Reusch D, Haberger M, Falck D et al (2015) Comparison of methods for the analysis of therapeutic immunoglobulin G fc-glycosylation profiles-part 2: mass spectrometric methods. MAbs 7(4):732–742
22. Banazadeh A, Veillon L, Wooding KM, Zabet-moghaddam M, Mechref Y (2017) Recent advances in mass spectrometric analysis of glycoproteins. Electrophoresis 38(1):162–189
23. Xiao H, Suttapitugsakul S, Sun F, Wu R (2018) Mass spectrometry-based chemical and enzymatic methods for global analysis of protein glycosylation. Acc Chem Res 51 (8):1796–1806
24. Gennaro LA, Salas-Solano O (2008) On-line CE-LIF-MS technology for the direct characterization of N-linked glycans from therapeutic antibodies. Anal Chem 80(10):3838–3845
25. Gahoual R, Busnel J-M, Beck A, François Y-N, Leize-Wagner E (2014) Full antibody primary structure and microvariant characterization in a single injection using transient isotachophoresis and sheathless capillary electrophoresis--tandem mass spectrometry. Anal Chem 86 (18):9074–9081
26. Gahoual R, Beck A, François Y-N, Leize-Wagner E (2016) Independent highly sensitive characterization of asparagine deamidation and aspartic acid isomerization by sheathless CZE-ESI-MS/MS. J Mass Spectrom 51 (2):150–158
27. Giorgetti J, D'Atri V, Canonge J et al (2018) Monoclonal antibody *N*-glycosylation profiling using capillary electrophoresis—mass spectrometry: assessment and method validation. Talanta 178(Supplement C):530–537
28. Kammeijer GSM, Kohler I, Jansen BC et al (2016) Dopant enriched nitrogen gas combined with Sheathless capillary electrophoresis–electrospray ionization-mass spectrometry for improved sensitivity and repeatability in Glycopeptide analysis. Anal Chem 88 (11):5849–5856
29. Kammeijer GSM, Jansen BC, Kohler I et al (2017) Sialic acid linkage differentiation of glycopeptides using capillary electrophoresis—electrospray ionization—mass spectrometry. Sci Rep 7(1):3733
30. Dada OO, Zhao YM, Jaya N, Salas-Solano O (2017) High-resolution capillary zone electrophoresis with mass spectrometry peptide mapping of therapeutic proteins: peptide recovery and post-translational modification analysis in monoclonal antibodies and antibody-drug conjugates. Anal Chem 89 (21):11236–11242
31. Bobaly B, D'Atri V, Goyon A et al (2017) Protocols for the analytical characterization of therapeutic monoclonal antibodies. II—enzymatic and chemical sample preparation. J Chromatogr B 1060:325–335

# Chapter 8

# Enrichment of Intact Glycopeptides Using Strong Anion Exchange and Electrostatic Repulsion Hydrophilic Interaction Chromatography

**Abel Bermudez and Sharon J. Pitteri**

## Abstract

Glycosylation is a biologically important and complex protein posttranslational modification. The emergence of glycoproteomic technologies to identify and characterize glycans on proteins has the potential to enable a better understanding the role of glycosylation in biology, disease states, and other areas of interest. In particular, the analysis of intact glycopeptides by mass spectrometry allows information about glycan location and composition to be ascertained. However, such analysis is often complicated by extensive glycan diversity and the low abundance of glycopeptides in a complex mixture relative to nonglycosylated peptides. Enrichment of glycopeptides from a protein enzymatic digest is an effective approach to overcome such challenges. In this chapter, we described a glycopeptide enrichment method combining strong anion exchange, electrostatic repulsion, and hydrophilic interaction chromatography (SAX-ERLIC). Following enzymatic digestion of proteins into peptides, SAX-ERLIC is performed by solid phase extraction to enrich glycopeptides from biological samples with subsequent LC-MS/MS analysis. Glycopeptide data generated using the SAX-ERLIC enrichment yields a high number of total and unique glycopeptide identifications which can be mapped back to proteins. The enrichment strategy is robust, easy to perform, and does not require cleavage of glycans prior to LC-MS/MS analysis.

**Key words** Glycopeptide enrichment, Glycoproteomics, LC-MS, Protein glycosylation

## 1 Introduction

Covalent attachment of glycans to proteins is a common type of post-translational modification that has important downstream effects on a variety of cellular functions as well as health and disease [1, 2]. Despite its biological importance, protein glycosylation remains understudied and poorly understood due to substantial analytical challenges in the complexity of the modification. First, glycosylation is a nonstoichiometric modification with variable occupancy at a given glycosite which can be dynamic and change with biological condition. Second, site heterogeneity is common where a given glycosite can be occupied by any number of different

Arnaud Delobel (ed.), *Mass Spectrometry of Glycoproteins: Methods and Protocols*, Methods in Molecular Biology, vol. 2271, https://doi.org/10.1007/978-1-0716-1241-5_8, 

glycoforms (e.g., high mannose, complex branched, hybrid structures). Third, in a typical protein tryptic digest, glycopeptides comprise a minor component of the overall mixture compared to nonglycopeptides, thereby creating a dynamic range challenge which is further amplified in a complex mixture. Fourth, glycopeptides can suffer from poor ionization efficiency in the presence of coeluting nonglycosylated peptides hindering LC-MS/MS analysis. To help overcome some of these challenges, enrichment of glycopeptides from a complex mixture of total peptides is a reasonable strategy to simplify the mixture by reducing the background of nonglycopeptides. Enrichment aids the analysis of intact glycopeptides by LC-MS/MS thereby retaining information about glycosylation site and composition. Such analysis allows systematic studies of protein glycosylation in biological and clinical samples.

Examples of common approaches for enrichment of glycoproteins and glycopeptides include the use of hydrazide chemistry or titanium dioxide ($TiO_2$). The hydrazide chemistry-based approach typically utilizes a solid support that is functionalized with hydrazide to capture glycans through the formation of hydrazone bonds [3–6]. This method provides outstanding specificity for glycoproteins, and results in the identification of thousands of novel glycosites. However, this approach requires the release of the glycan from the peptides, usually by PNGase F, and subsequent analysis of deglycosylated peptides for identification of glycosylation sites—thereby, losing important information about the glycan structures that occupied the glycosylation site. $TiO_2$ has also been shown to be highly selective for capture and enrichment of sialylated glycopeptides, but also typically involves the cleavage of the glycan from the peptide [7–12].

Other types of glycopeptide enrichment methods have been developed that allow subsequent analysis of intact glycoproteins or glycopeptides, allowing information about glycosite localization, occupancy, and glycan composition to be obtained [13–17]. Some of the most common enrichment strategies of this type include lectin affinity chromatography which uses proteins (lectins) with binding affinities toward specific carbohydrate moieties to capture glycosylated species at the protein or peptide level [18–23]. However, the typical lectin density is often insufficient for efficient binding of glycopeptides with only a single glycan structure (binding site) on each peptide. Hydrophilic interaction chromatography (HILIC) chromatography has also been use for glycopeptide enrichment. HILIC-based approaches separate glycopeptides based on hydrophobicity and glycan size, and in general show less bias for specific glycan types compared to lectin affinity and $TiO_2$ [24–27].

Electrostatic repulsion hydrophilic interaction liquid chromatography (ERLIC) is a method that combines hydrophilic interaction and electrostatic repulsion [28, 29]—essentially combining the characteristics of HILIC and ion exchange chromatography in a mixed mode. The hydrophilic interactions are controlled by the percentage of organic solvent versus water in the mobile phase, whereas the electrostatic interactions are controlled by the charge of the media. At high organic solvent in the mobile phase, hydrophilic interactions are stronger than electrostatic interactions so both positive and negative species can be retained. The negative charges on sialic acid sugars and the hydrophilic nature of the glycans promote their retention on an anion exchange (positively charged) column. In high aqueous solvent, hydrophilic species, including glycopeptides are eluted. ERLIC has been demonstrated as an effective method for the enrichment of phosphopeptides and glycopeptides and an attractive alternative to traditional HILIC strategies which show lower selectivity [7, 28, 30–35].

In this chapter, we describe a method using ERLIC with a strong anion exchange resin (SAX-ERLIC) using solid phase extraction (SPE) to capture and enrich glycopeptides from complex mixtures. The general principle of this method is shown in Fig. 1. At high organic mobile phase, the positively charged media interacts with the negative charges on sialic acids. Glycans on glycopeptides are highly hydrophilic and acidic which further allows for their retention on the SPE cartridge at high organic. At high organic, most nonglycosylated (and nonphosphorylated) peptides are eluted. When the mobile phase is changed to a higher concentration of aqueous solution, the hydrophilic interaction of the glycopeptides predominates, allowing their elution from the media. A significant advantage of ERLIC is that the use of charged chromatography media may select a wider range of glycopeptides than other methods, thereby reducing the enrichment bias due to glycan or peptide size. Furthermore, the glycopeptides are eluted intact, thereby allowing analysis of the peptide sequence and glycan structure simultaneously by LC-MS/MS.

The SAX-ERLIC method described in this chapter is versatile and can be used for enriching both N- and O-linked glycopeptides from a variety of biological and clinical sample types. The ability to study intact glycopeptides allows one to obtain both protein and glycosylation information simultaneously. The method is easy to use with a variety of practical forms of implementation. SAX-ERLIC is a practical, simple, and robust reasonable means of enriching glycopeptides from biological and clinical samples such as serum/plasma, tissue, and cells. SAX-ERLIC does not require *N*-glycan cleavage prior to LC-MS/MS analysis and thus retains site-specific information and glycosite mapping, thereby enabling biological research and systematic studies of protein glycosylation with unprecedented coverage and depth. Also, SAX-ERLIC method allows a high number of total and unique glycopeptide identifications, some of the highest ever reported among comparable studies [29].

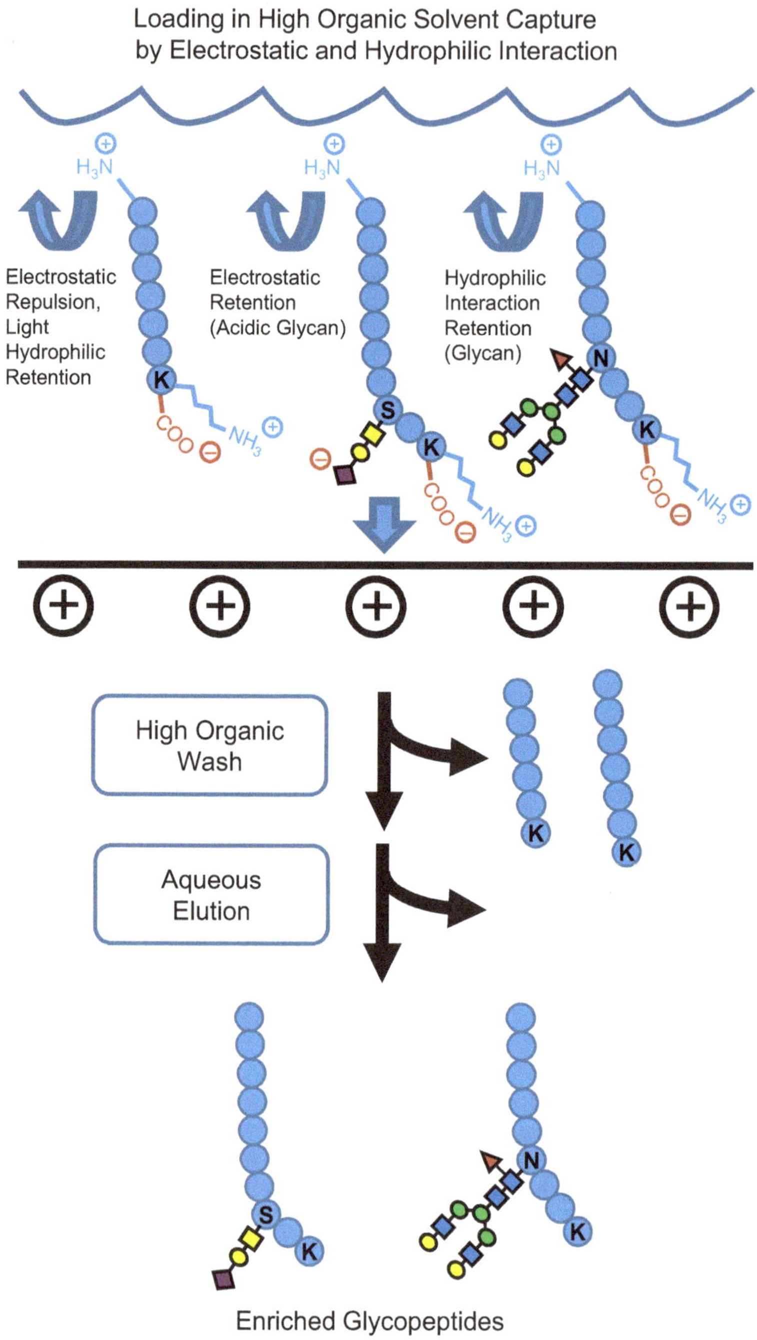


**Fig. 1** General principle of SAX-ERLIC. Peptides are loaded onto an anion exchange column (positive charge) in high organic solvent. Glycopeptides are enriched in the aqueous elution

From a practical perspective, SAX-ERLIC is relatively simple to perform in the laboratory, and does not require additional desalting steps which minimizes sample loss and experimental variability. Furthermore, N- and O-linked glycopeptides could be isolated simultaneously, thereby allowing for concurrent studies of both glycosylation types. One potential pitfall to this approach is that O-linked glycans are typically smaller and possibly less hydrophilic than N-linked glycans. In cases where O-linked glycopeptides have small glycan moieties and have no sialic acid, the SAX-ERLIC may be less effective at capturing these species. Sample complexity is a major technical challenge for analysis of glycopeptides which are a minor species in a complex mixture. SAX-ERLIC can be used to separate glycopeptides from other peptides.

## 2 Materials

### *2.1 Sample Homogenization (See Note 1)*

1. 5× Homogenization Buffer: 1.0 M Tris–HCl, pH 8.0, 0.5 Ethylenediaminetetraacetic acid (EDTA) (*see* **Note 2**).
2. Urea Buffer: 8 M urea. Weigh 12 g of urea and add 16 mL LC-MS grade water (*see* **Note 3**).
3. Lysis Buffer: Combine 4 parts Urea Buffer with 1 part 5× Homogenization Buffer and 1 cOmplete, Mini, EDTA-free Protease Inhibitor tablet (Sigma Aldrich).
4. Pierce BCA Protein Assay Kit (Thermo Fisher Scientific) or similar.
5. Falcon Tubes 5 mL (Corning) or similar.

### *2.2 Sample Digestion*

1. Ammonium Bicarbonate Buffer: Weight 39.5 mg ammonium bicarbonate and add 10 mL of LC-MS grade water for a concentration of 50 mM ammonium bicarbonate.
2. Trypsin Solution: Reconstitute 20 μg lyophilized trypsin enzyme in 50 mM ammonium bicarbonate buffer for a final concentration of 0.2 μg/μL trypsin.
3. Reduction Solution: Weight 25.0 mg Tris(2-carboxyethyl) phosphine (TCEP) and add 500 μL of LC-MS grade water to make a 200 mM TCEP.
4. Alkylation Solution: Weight 7.4 mg iodoacetamide and add 200 μL of LC-MS grade water to make 200 mM iodoacetamide.

### *2.3 SAX-ERLIC*

1. SOLA SAX Solid-Phase-Extraction (SPE) Columns (PN#60109-003, Thermo Fisher Scientific).
2. Disposable Flow Control Valve Liners for the Visiprep DL (Supelco).

3. Safe-Lock centrifuge tubes (1.5 and 2.0 mL).
4. Conical Centrifuge 15 mL tubes.
5. LC-MS grade acetonitrile.
6. 100 mM triethylammonium acetate: dilute 1 M stock solution by 10× using LC-MS grade water (*see* **Note 4**).
7. 1% trifluoroacetic acid (TFA) in water.
8. 95% acetonitrile with 1% TFA in water.
9. 50% acetonitrile with 0.1% TFA in water.
10. 5% acetonitrile with 0.1% TFA in water.

#### 2.4 Reversed-Phase Chromatography and Mass Spectrometry

1. Reversed-Phase Buffer A: 0.1% formic acid in LC-MS grade water.
2. Reversed-Phase Buffer B: 0.1% formic acid in LC-MS grade acetonitrile.
3. C18 trap column (Thermo Fisher Scientific Acclaim PepMap 100 or similar).
4. C18 analytical column (Picofrit 75 μM ID, New Objective, packed to 25 cm long with Magic C18 AQ resin, or similar).

#### 2.5 Instrumentation

1. PRO250 Homogenizer (ProScientific) or similar.
2. Brason SLPe Digital Sonifier (Fisher Scientific) or similar.
3. Visiprep SPE Vacuum manifold or similar.
4. Acid-Resistant CentriVap Centrifugal Concentrator (Labconco) or similar.
5. Ultimate 3000 RSLCnano System (Thermo Fisher Scientific) or similar.
6. LTQ Orbitrap Elite Mass Spectrometer (Thermo Fisher Scientific) or similar.
7. Byonic Software (Protein Metrics Inc.) or similar.

## 3 Methods

The protocol described here is optimized for 500 μg of starting protein material from biological or clinical samples.

#### 3.1 Sample Homogenization

1. Fill a 100 mL beaker with ice and deionized water for cooling samples during homogenization (*see* **Note 5**).
2. Rinse homogenizer probe with water and ethanol.
3. Set homogenizer to level 4.
4. Transfer tissue sample into a 5.0 mL Falcon tube and immediately add 800 μL (*see* **Note 6**) of Lysis Buffer.

5. Place the sample on ice and insert the homogenizer probe into the sample solution not touching the bottom of the Falcon tube. Move the probe up and down during homogenization.
6. Homogenize the sample for 20 s, followed by 30 s resting on ice. Repeat this cycle two additional times. If tissue is not completely homogenized, repeat this process again until the tissues has been completely homogenized.
7. Clean the probe by rinsing with water and ethanol between samples and after use. Ensure that no residue is stuck between the probe's blades.
8. Following homogenization, sonicate samples using the probe sonicator in a continuous mode with the amplitude set at 40%. Sonicate for 15 s (*see* **Note 7**) on ice, followed by resting on ice for 30 s. Note: Keep the sample on ice at all times and do not let the sample heat up.
9. Transfer lysed samples to a 2.0 mL Eppendorf tube. Centrifuge samples for 10 min at 14,000 × *g* at 4 °C. Collect the supernatant and transfer to a 1.5 mL Eppendorf tube.
10. Use a BCA Protein Assay kit to measure the protein concentration in the sample.
11. For disulfide bond reduction, aliquot 500 μg of protein into a new 1.5 mL Eppendorf tube and bring the sample volume to 120 μL using 50 mM ammonium bicarbonate. Add 6 μL of Reduction Solution. Incubate samples for 2 h at room temperature (*see* **Note 8**).
12. Alkylate proteins by adding 9 μL Alkylation Solution and incubate sample for 45 min in the dark at room temperature.

### 3.2 Trypsin Digestion

1. Dilute the urea concentration in the sample to less than 500 mM using 50 mM ammonium bicarbonate (*see* **Note 9**).
2. Add trypsin at a ratio of 1:30 trypsin to protein. Incubate sample at 37 °C overnight without shaking.
3. Concentrate sample volume down to 200 μL (if needed) using a speed vacuum.

### 3.3 Strong Anion Exchange and Electrostatic Repulsion Hydrophilic Interaction Liquid Chromatography (SAX-ERLIC)

Precondition, load, wash, and elution steps for a single SPE cartridge are shown in Fig. 2.

1. Open the SPE vacuum valves and insert disposable liners inside individual wells on the SPE vacuum manifold. Close the SPE valves.
2. Place 15 mL conical tubes inside the SPE vacuum manifold to collect waste solvents.
3. Place a SOLA SAX SPE cartridge for each sample on top of each disposable liner.

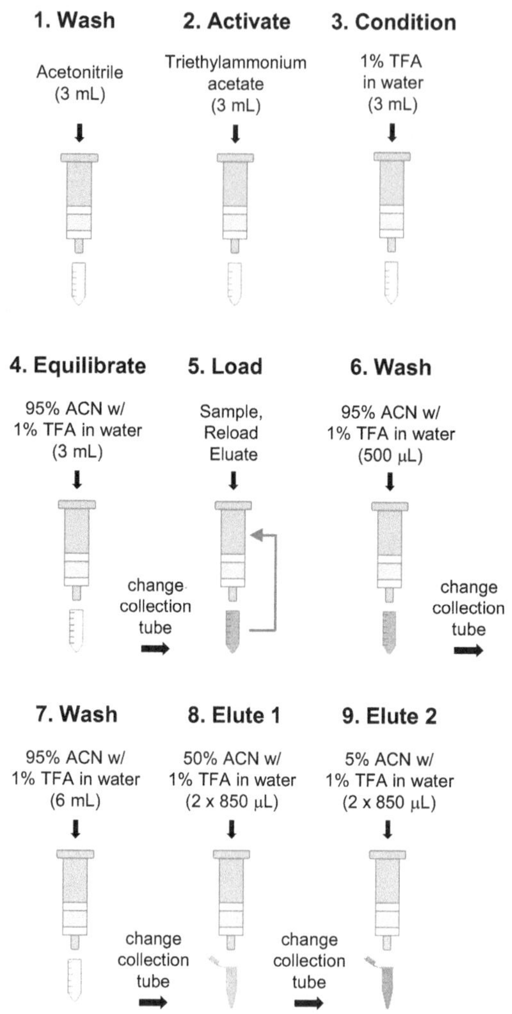


**Fig. 2** Overview of SAX-ERLIC workflow

4. Precondition SAX-ERLIC SPE cartridges as follows (*see* **Note 10**):
   (a) Using a pipette, add 3 mL of acetonitrile to the SPE cartridge, 1 mL at the time (*see* **Note 11**). Start adjusting the flow rate by opening the SPE valves slowly to an approximate flow rate of 1 mL/min. Do not let the SPE cartridges dry out during the glycopeptide enrichment.
   (b) Activate the SPE cartridge with 3 mL of 100 mM triethylammonium acetate.
   (c) Condition the SPE cartridge with 3 mL of 1% TFA in water.
   (d) Equilibrate the SPE cartridge with 3 mL of 95% acetonitrile with 1% TFA in water. Leave approximately 50 μL of 95% acetonitrile with 1% TFA on the SPE cartridge by closing the SPE valve.

5. Transfer the tryptic peptide sample solution into a 15 mL conical tube and adjust the organic solvent content of the solution by adding 3 mL of 95% acetonitrile with 1% TFA in water (*see* **Note 12**).
6. Remove the waste 15 mL conical tube from the SPE vacuum manifold and replace with a new 15 mL conical tube to collect the sample flow through.
7. Pipet the digestion solution (sample) on the SOLA-SAX SPE cartridge, 1 mL at the time at a flow rate of 0.5 mL/min.
8. Collect the sample flow through and reload onto the SOLA-SAX SPE cartridge as in the previous step. If necessary, apply gentle pressure on the SOLA-SAX SPE cartridge using a pipette bulb or syringe to elute the digestion solution.
9. Wash the cartridge with 500 μL of 95% acetonitrile with 1% TFA in water. Nonglycosylated peptides will flow through the SOLA-SAX SPE cartridge into the conical tube.
10. Remove the 15 mL conical tube containing the flow through and replace with a new 15 mL conical tube.
11. Wash the cartridge with an additional 6 mL of 95% acetonitrile with 1% TFA in water.
12. Remove the 15 mL conical collection tube and replace them with a 2 mL Eppendorf tube to collect the 50% acetonitrile elution fraction.
13. Slowly start eluting glycopeptides with two 850 μL aliquots of 50% acetonitrile with 0.1% TFA in water into the 2 mL Eppendorf tube.
14. Remove the 2.0 mL Eppendorf tube containing the 50% acetonitrile enriched glycopeptide fraction and replace with a new 2.0 mL Eppendorf tube labeled 5% acetonitrile fraction.
15. Elute larger glycopeptides with two 850 μL aliquots of 5% acetonitrile with 0.1% TFA in water.
16. Concentrate eluate from 50% and 5% acetonitrile fractions using a speed vacuum, combine both fractions, dry down, and store at −80 °C if needed.

### 3.4 Reverse-Phase Liquid Chromatography–Tandem Mass Spectrometry Analysis

Dried glycopeptides are reconstituted in 20 μL of 0.1% formic acid in water prior to LC-MS/MS analysis. The analysis described below is performed on a LTQ-Orbitrap Elite mass spectrometer coupled with a nanoLC. The reversed-phase liquid chromatography gradient for glycopeptide separation is shown in Table 1.

1. Load 4 μL of each glycopeptide enriched sample onto a 20 μL sample loop.
2. Load the sample from the loop onto a C18 trap column using 2% Reversed-Phase Buffer B for 10 min at a flow rate of 5 μL/min.

**Table 1**
**Nano-flow reverse-phased liquid chromatography method**

| Time (min) | Flow rate μL/min | Buffer B (%) | RP chromatography |
|---|---|---|---|
| 0 | 0.5 | 2.0 | Starts loading on C18 trap |
| 10 | 0.5 | 2.0 | Gradient starts |
| 125 | 0.5 | 22.0 | |
| 130 | 0.5 | 85.0 | Gradient ramps to high organic solvent % |
| 140 | 0.5 | 85.0 | |
| 140.1 | 0.5 | 2.0 | Equilibration begins |
| 150 | 0.5 | 2.0 | |

3. Glycopeptides are separated based on their hydrophobicity on a 25 cm long C18 column.
4. A LTQ-Orbirap Elite mass spectrometer is used to acquire spectra utilizing two complementary methods of fragmentations.
   (a) Higher Energy Collision Dissociation (HCD) for obtaining glycan fragmentation; and
   (b) Electron transfer dissociation (ETD) to acquire peptide backbone fragmentation.

   Peptides are first fragmented by HCD. If glycan oxonium ions (singly charged glycan fragment ions such as *m/z*: 138.0545, 204.0867, 274.0900, 292.0800, and 366.1396) are observed (indicating a glycopeptide), a subsequent scan with electron transfer dissociation is triggered (method known as HCD-product dependent-ETD [36]) to obtain peptide fragmentation.
5. 1.8 kV is supplied to the mobile phase to protonate glycopeptides and the heated capillary temperature is set to 280 °C for solvent desolvation.
6. The FT resolution on the LTQ-Orbitrap Elite mass spectrometer is set to 30,000, scanning within mass scan range 400–1800 *m/z* for MS, followed by an MS/MS on the top 10 most abundant ions with +2 and higher charge state. +4 charge state is used as the default for HCD selection with a normalized collision energy of 35.0 and 15,000 resolution for the MS/MS scans. Dynamic exclusion is enabled with repeat count of 3 and exclusion duration of 30 s. Precursor ion isolation width is set to 4.0 *m/z* and the FT HCD scan window is fixed at 100 *m/z*.

### 3.5 Glycoprotein Identification

LC-MS raw data is searched using Byonic software from Protein Metrics Inc. for glycoprotein identification. Suggested search parameters are listed here.

Protein Database: Swiss-Prot or similar containing species of interest. The database can be customized and/or also created from shotgun proteomics analysis of the sample of interest.

*N*-Glycan Library: Curated glycan library from Byonic or custom glycan database.

Sample Digestion

1. Cleavage sites: RK.
2. Cleavage side: C-terminal.
3. Digestion Specificity: Fully specific (fastest).
4. Missed Cleavages: 2.

Instrument Parameters

1. Precursor mass tolerance: 10 ppm.
2. Fragmentation type: QTOF/HCD.
3. Fragment mass tolerance: 20 ppm.
4. Recalibration lock mass: None.

Modifications

1. Carbamidomethyl/+57.021464@C fixed.
2. Oxidation/+15.9949@M common1.
3. Deamidated/+0.984016@N rare2.
   Total common max: 2.
   Minimum mass: −40.
   Maximum mass: 100.

Spectrum Input Options

1. Charge states apply to unassigned spectra: 2, 3, 4, 5.
2. Precursor isotope off by x: Too high or low (narrow).
3. Maximum precursor mass: 10,000.
4. Precursor and charge assignments: Compute from MS1.
5. Maximum # of precursors per scan: 2.
6. Smoothing width (*m/z*): 0.01.

Peptide Output Options

1. Automatic score cut: enabled.
2. Show all N-glycopeptides: enabled.

Protein Output Options

1. Protein FDR: 1% FDR (or 20 reverse count).

   Data can be filtered by a score cut off greater than 150. Manual checking of the identifications to determine quality and most appropriate score cutoffs for specific experiments are recommended.

## 4 Notes

1. The protocol described here is for tissue samples. Other sample types can be used.
2. Always prepare solutions using ultrapure LC-MS grade solvents (e.g., water, acetonitrile) and store reagents as stated in this protocol.
3. Urea solution should be made fresh, ideally on the day of homogenization.
4. Triethylammonium acetate buffer should be made fresh on the day of use.
5. Keep samples on ice all the time during the lysing process to prevent proteolysis.
6. The amount of lysis buffer should be optimized to result in a protein concentration greater than 4 mg/mL following homogenization.
7. If 15 s of sonication is too long and the sample heats up, reduce the sonication time to 10 s.
8. Do not heat up samples during reduction step if lysis buffer contains urea. Increasing temperature may result in carbamylation on the side chains of lysine and arginine residues possibly preventing proteins from enzymatic digestion.
9. Samples can also be desalted using a desalting spin column.
10. Subsequent SAX-ERLIC steps are given for a single SPE cartridge but can be applied to each cartridge in the manifold before moving to the next step.
11. Gravity will pull the acetonitrile through the cartridge, no need to use a vacuum pump.
12. If precipitation occurs, centrifuge the sample and proceed with supernatant only.

### References


1. Varki A, Cummings RD et al (eds) (2015) Essentials of glycobiology. Cold Spring Harbor (NY), New York
2. Reily C, Stewart TJ, Renfrow MB, Novak J (2019) Glycosylation in health and disease. Nat Rev Nephrol 15(6):346–366. https://doi.org/10.1038/s41581-019-0129-4
3. Zhu J, Sun Z, Cheng K, Chen R, Ye M, Xu B, Sun D, Wang L, Liu J, Wang F, Zou H (2014) Comprehensive mapping of protein

N-glycosylation in human liver by combining hydrophilic interaction chromatography and hydrazide chemistry. J Proteome Res 13 (3):1713–1721. https://doi.org/10.1021/pr401200h

4. Zhang H, Li XJ, Martin DB, Aebersold R (2003) Identification and quantification of N-linked glycoproteins using hydrazide chemistry, stable isotope labeling and mass spectrometry. Nat Biotechnol 21(6):660–666. https://doi.org/10.1038/nbt827
5. Liu T, Qian WJ, Gritsenko MA, Camp DG 2nd, Monroe ME, Moore RJ, Smith RD (2005) Human plasma N-glycoproteome analysis by immunoaffinity subtraction, hydrazide chemistry, and mass spectrometry. J Proteome Res 4(6):2070–2080. https://doi.org/10.1021/pr0502065
6. Li X, Jiang J, Zhao X, Wang J, Han H, Zhao Y, Peng B, Zhong R, Ying W, Qian X (2013) N-glycoproteome analysis of the secretome of human metastatic hepatocellular carcinoma cell lines combining hydrazide chemistry, HILIC enrichment and mass spectrometry. PLoS One 8(12):e81921. https://doi.org/10.1371/journal.pone.0081921
7. Magdeldin S, Yamamoto K, Yoshida Y, Xu B, Zhang Y, Fujinaka H, Yaoita E, Yates JR 3rd, Yamamoto T (2014) Deep proteome mapping of mouse kidney based on OFFGel prefractionation reveals remarkable protein post- translational modifications. J Proteome Res 13 (3):1636–1646. https://doi.org/10.1021/pr401122m
8. Zhang B, Sheng Q, Li X, Liang Q, Yan J, Liang X (2011) Selective enrichment of glycopeptides for mass spectrometry analysis using C18 fractionation and titanium dioxide chromatography. J Sep Sci 34(19):2745–2750. https://doi.org/10.1002/jssc.201100427
9. Zhu J, Wang F, Cheng K, Dong J, Sun D, Chen R, Wang L, Ye M, Zou H (2013) A simple integrated system for rapid analysis of sialic-acid-containing N-glycopeptides from human serum. Proteomics 13(8):1306–1313. https://doi.org/10.1002/pmic.201200367
10. Palmisano G, Parker BL, Engholm-Keller K, Lendal SE, Kulej K, Schulz M, Schwammle V, Graham ME, Saxtorph H, Cordwell SJ, Larsen MR (2012) A novel method for the simultaneous enrichment, identification, and quantification of phosphopeptides and sialylated glycopeptides applied to a temporal profile of mouse brain development. Mol Cell Proteomics 11(11):1191–1202. https://doi.org/10.1074/mcp.M112.017509
11. Palmisano G, Lendal SE, Larsen MR (2011) Titanium dioxide enrichment of sialic acid-containing glycopeptides. Methods Mol Biol 753:309–322. https://doi.org/10.1007/978-1-61779-148-2_21
12. Larsen MR, Jensen SS, Jakobsen LA, Heegaard NH (2007) Exploring the sialiome using titanium dioxide chromatography and mass spectrometry. Mol Cell Proteomics 6 (10):1778–1787. https://doi.org/10.1074/mcp.M700086-MCP200
13. Kirwan A, Utratna M, O'Dwyer ME, Joshi L, Kilcoyne M (2015) Glycosylation-based serum biomarkers for cancer diagnostics and prognostics. Biomed Res Int 2015:490531. https://doi.org/10.1155/2015/490531
14. Wuhrer M, Catalina MI, Deelder AM, Hokke CH (2007) Glycoproteomics based on tandem mass spectrometry of glycopeptides. J Chromatogr B Analyt Technol Biomed Life Sci 849 (1-2):115–128. https://doi.org/10.1016/j.jchromb.2006.09.041
15. Huang BY, Yang CK, Liu CP, Liu CY (2014) Stationary phases for the enrichment of glycoproteins and glycopeptides. Electrophoresis 35 (15):2091–2107. https://doi.org/10.1002/elps.201400034
16. Chen CC, Su WC, Huang BY, Chen YJ, Tai HC, Obena RP (2014) Interaction modes and approaches to glycopeptide and glycoprotein enrichment. Analyst 139(4):688–704. https://doi.org/10.1039/c3an01813j
17. Ongay S, Boichenko A, Govorukhina N, Bischoff R (2012) Glycopeptide enrichment and separation for protein glycosylation analysis. J Sep Sci 35(18):2341–2372. https://doi.org/10.1002/jssc.201200434
18. Lis H, Sharon N (1993) Protein glycosylation. Structural and functional aspects. Eur J Biochem 218(1):1–27
19. Varki A, Cummings RD, Esko JD, Freeze HH, Stanley P, Bertozzi CR, Hart GW, Etzler ME (2009) Essentials of glycobiology. Essentials of glycobiology, 2nd edn. Cold Spring Harbor Laboratory Press, Cold Spring Harbor (NY), New York
20. Young NM, Brisson JR, Kelly J, Watson DC, Tessier L, Lanthier PH, Jarrell HC, Cadotte N, St Michael F, Aberg E, Szymanski CM (2002) Structure of the N-linked glycan present on multiple glycoproteins in the gram-negative bacterium, campylobacter jejuni. J Biol Chem 277(45):42530–42539. https://doi.org/10.1074/jbc.M206114200
21. Qiu R, Regnier FE (2005) Use of multidimensional lectin affinity chromatography in differential glycoproteomics. Anal Chem 77 (9):2802–2809. https://doi.org/10.1021/ac048751x

22. Geng M, Zhang X, Bina M, Regnier F (2001) Proteomics of glycoproteins based on affinity selection of glycopeptides from tryptic digests. J Chromatogr B Biomed Sci Appl 752 (2):293–306
23. Alvarez-Manilla G, Warren NL, Atwood J 3rd, Orlando R, Dalton S, Pierce M (2010) Glycoproteomic analysis of embryonic stem cells: identification of potential glycobiomarkers using lectin affinity chromatography of glycopeptides. J Proteome Res 9(5):2062–2075. https://doi.org/10.1021/pr8007489
24. Picariello G, Ferranti P, Mamone G, Roepstorff P, Addeo F (2008) Identification of N-linked glycoproteins in human milk by hydrophilic interaction liquid chromatography and mass spectrometry. Proteomics 8 (18):3833–3847. https://doi.org/10.1002/pmic.200701057
25. Neue K, Mormann M, Peter-Katalinic J, Pohlentz G (2011) Elucidation of glycoprotein structures by unspecific proteolysis and direct nanoESI mass spectrometric analysis of ZIC-HILIC-enriched glycopeptides. J Proteome Res 10(5):2248–2260. https://doi.org/10.1021/pr101082c
26. Mysling S, Palmisano G, Hojrup P, Thaysen-Andersen M (2010) Utilizing ion-pairing hydrophilic interaction chromatography solid phase extraction for efficient glycopeptide enrichment in glycoproteomics. Anal Chem 82(13):5598–5609. https://doi.org/10.1021/ac100530w
27. Kondo A, Thaysen-Andersen M, Hjerno K, Jensen ON (2010) Characterization of sialylated and fucosylated glycopeptides of beta2-glycoprotein I by a combination of HILIC LC and MALDI MS/MS. J Sep Sci 33 (6-7):891–902. https://doi.org/10.1002/jssc.200900802
28. Alpert AJ (2008) Electrostatic repulsion hydrophilic interaction chromatography for isocratic separation of charged solutes and selective isolation of phosphopeptides. Anal Chem 80(1):62–76. https://doi.org/10.1021/ac070997p
29. Totten SM, Feasley CL, Bermudez A, Pitteri SJ (2017) Parallel comparison of N-linked Glycopeptide enrichment techniques reveals extensive Glycoproteomic analysis of plasma enabled by SAX-ERLIC. J Proteome Res 16 (3):1249–1260. https://doi.org/10.1021/acs.jproteome.6b00849
30. Sok Hwee Cheow E, Hwan Sim K, de Kleijn D, Neng Lee C, Sorokin V, Sze SK (2015) Simultaneous enrichment of plasma soluble and extracellular vesicular glycoproteins using prolonged ultracentrifugation-electrostatic repulsion-hydrophilic interaction chromatography (PUC-ERLIC) approach. Mol Cell Proteomics 14(6):1657–1671. https://doi.org/10.1074/mcp.O114.046391
31. Pagel O, Loroch S, Sickmann A, Zahedi RP (2015) Current strategies and findings in clinically relevant post-translational modification-specific proteomics. Expert Rev Proteomics 12(3):235–253. https://doi.org/10.1586/14789450.2015.1042867
32. Kuo CW, Wu IL, Hsiao HH, Khoo KH (2012) Rapid glycopeptide enrichment and N-glycosylation site mapping strategies based on amine-functionalized magnetic nanoparticles. Anal Bioanal Chem 402(9):2765–2776. https://doi.org/10.1007/s00216-012-5724-1
33. Hao P, Zhang H, Sze SK (2011) Application of electrostatic repulsion hydrophilic interaction chromatography to the characterization of proteome, glycoproteome, and phosphoproteome using nano LC-MS/MS. Methods Mol Biol 790:305–318. https://doi.org/10.1007/978-1-61779-319-6_23
34. Darula Z, Sarnyai F, Medzihradszky KF (2016) O-glycosylation sites identified from mucin core-1 type glycopeptides from human serum. Glycoconj J 33(3):435–445. https://doi.org/10.1007/s10719-015-9630-6
35. Cao L, Yu L, Guo Z, Li X, Xue X, Liang X (2013) Application of a strong anion exchange material in electrostatic repulsion-hydrophilic interaction chromatography for selective enrichment of glycopeptides. J Chromatogr A 1299:18–24. https://doi.org/10.1016/j.chroma.2013.05.037
36. Saba J, Dutta S, Hemenway E, Viner R (2012) Increasing the productivity of glycopeptides analysis by using higher-energy collision dissociation-accurate mass-product-dependent electron transfer dissociation. Int J Proteomics 2012:560391. https://doi.org/10.1155/2012/560391

# Chapter 9

# In-Depth Glycan Characterization of Therapeutic Glycoproteins by Stepwise PGC SPE and LC-MS/MS

**Myung Jin Oh, Youngsuk Seo, Unyong Kim, and Hyun Joo An**

## Abstract

Glycosylation of biologics, an important factor in pharmacological functions such as efficacy, safety, and biological activity, is easily affected by subtle changes in the cellular environment. Therefore, comprehensive and in-depth glycan characterization of therapeutic glycoproteins should be performed to ensure product quality and process consistency, but it is analytically challenging due to glycan microheterogeneity occurring in the glycan biosynthesis pathway. LC-based chromatographic separation combined with mass spectrometry (MS) has been widely used as a prominent tool for the qualitative and quantitative analysis of glycosylation of therapeutic glycoproteins. However, prior to LC/MS analysis, glycans are selectively captured and fractionated by solid-phase extraction (SPE) utilizing physicochemical characteristics for comprehensive characterization of a wide range of glycan heterogeneity on glycoengineered therapeutic proteins. In particular, porous graphitized carbon (PGC) SPE has been employed as a useful technique for the fractionation of native glycans having different sizes and polarities. Here, we describe a systematic method for comprehensive glycan characterization of therapeutic proteins using stepwise PGC SPE and LC/MS.

**Key words** Glycan, PGC, SPE, Fractionation, Biotherapeutics, Mass spectrometry

## 1 Introduction

Glycosylation, one of the most common and complex posttranslational modifications (PTMs), significantly impacts on physicochemical properties, bioactivity, stability, and immunogenicity of biopharmaceutical products [1]. Since glycosylation is influenced by host cells and fermentation conditions, it should be thoroughly monitored throughout the life cycle of the drug from development to routine manufacturing to ensure safety, product consistency, and quality [2]. Glycans in therapeutic glycoproteins exist in various forms and they may be neutral, acidic (sialylated, phosphorylated, sulfated, or *O*-acetylated), highly branched, and polylactosaminylated [3, 4]. To date, there is no single assay platform for glycan simultaneous detection of highly glycosylated biotherapeutics

Arnaud Delobel (ed.), *Mass Spectrometry of Glycoproteins: Methods and Protocols*, Methods in Molecular Biology, vol. 2271, https://doi.org/10.1007/978-1-0716-1241-5_9, 

containing various glycan forms. Therefore, analytical platform capable of analyzing a variety of glycans with various physicochemical and pharmacological properties is definitely needed for comprehensive characterization of therapeutic glycoproteins. Chromatographic separation techniques such as high-performance liquid chromatography (HPLC) and capillary electrophoresis (CE) coupled with mass spectrometry have emerged as a premier tool for glycan analysis [5]. However, due to glycan heterogeneity and complexity, comprehensive glycan analysis often requires additional fractionation beyond of what chromatography can provide. Consequently, SPE-based prefractionation is often adapted to reduce glycan complexity by capturing of target glycans based on physicochemical properties [3, 6].

SPE is one of the representative separation techniques to isolate analytes of interest from sample mixtures [7]. The PGC cartridge consisting of flat layers of hexagonally arranged carbon atoms has been applied to capture native glycans prior to LC/MS or CE-MS analysis [8, 9]. All types of glycans including neutral and acidic moieties are tightly bound to PGC [10]. In particular, low polar glycans such as high-mannose and nonacidic glycans can be separated early from the PGC by organic and aqueous solvent, while high polar glycans containing sialylation or acetylation group are eluted later [11–13]. Therefore, glycans with different sizes and polarities can be fractionated via PGC by adjusting elution solution of organic solvents and acid [6].

This chapter will introduce comprehensive glycan profiling of therapeutic glycoproteins using PGC SPE combined with LC/MS analysis. Stepwise SPE fractionation based on glycan size and polarity (acidity, p$K_a$) on PGCs was used to capture neutral and acidic glycans containing phosphorylated or sialylated groups, respectively prior to LC/MS. This analytical strategy can be applied to other therapeutic glycoproteins with complex glycoform [12, 14, 15] as well as biological samples such as serum [16], saliva [17], tissue [18], and cell [10].

## 2 Materials

### 2.1 Samples

1. Infliximab (Recombinant monoclonal antibody) from Janssen.
2. Agalsidase-beta (Recombinant human alpha-galactosidase) from Genzyme.

### 2.2 Detergent Removal

1. Protein concentrator for 10,000 MWCO (Molecular Weight Cut-Off).
2. Washing solution: deionized water.
3. Centrifuge.

### 2.3 Release of N-Glycans

1. Enzyme digestion solution: 200 mM ammonium bicarbonate ($NH_4HCO_3$) and 10 mM dithiothreitol (DTT).
2. Peptide-*N*-glycosidase F (PNGasF).

### 2.4 Release of N-Glycans

1. SPE graphitized carbon cartridges (250 mg, 6 mL).
2. SPE activation solution: deionized water and 80% acetonitrile (ACN) in 0.1% trifluoro acetic acid (TFA) (v/v).
3. SPE Washing solution: deionized water.
4. Elution solutions:

| Monoclonal antibody (infliximab) | Lysosomal enzyme |
|---|---|
| (a) 10% ACN in $H_2O$ | (a) 20% ACN in $H_2O$ |
| (b) 20% ACN in $H_2O$ | (b) 10% ACN in $H_2O$ with 0.05% TFA |
| (c) 40% ACN in $H_2O$ with 0.05% TFA | (c) 40% ACN in $H_2O$ with 0.05% TFA |

### 2.5 LC Separation and MS Analysis of Glycans

1. The nano-LC Hypercarb porous graphitized carbon (PGC) chip (a 9 × 0.075 mm i.d. enrichment column and a 43 × 0.075 mm i.d. analytical column, both packed with 5 μm PGC as the stationary phase, with an integrated nano-ESI spray tip).
2. LC Solutions.

| Monoclonal antibody (infliximab) | Lysosomal enzyme |
|---|---|
| (a) 3% ACN in $H_2O$ with 0.1% formic acid (FA) (v/v) | (a) 3% ACN in $H_2O$ with 0.5% FA (v/v) |
| (b) 90% ACN in $H_2O$ with 0.1% FA (v/v) | (b) 90% ACN in $H_2O$ with 0.5% FA (v/v) |

## 3 Methods

The whole procedure from purification of a therapeutic glycoprotein to SPE fractionation of glycans is outlined in Fig. 1. Initially, drug excipients such as surfactants, stabilizer, and buffers were removed by spin-filtering the glycoprotein solution using MWCO. *N*-glycans on a therapeutic protein were enzymatically released by PNGase F digestion. Prior to LC/MS analysis, the liberated *N*-glycans were fractionated through PGC SPE according to their molecule size and polarity to reduce the sample complexity. In particular, this chapter focuses on the PGC SPE procedure for glycan analysis of therapeutic mAb and lysosomal enzyme, respectively containing different types of glycans.

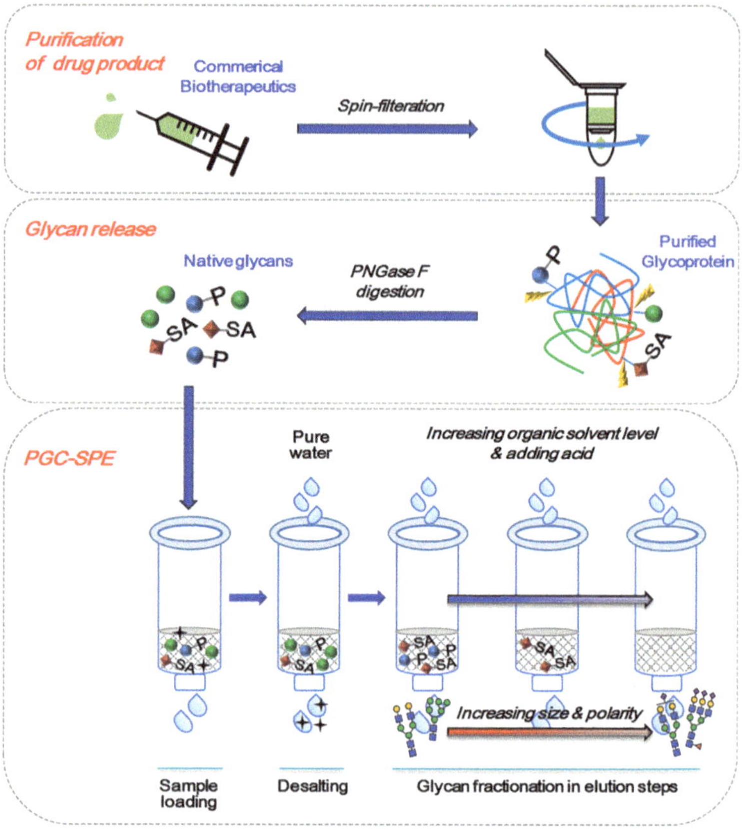


**Fig. 1** Experimental workflow for stepwise PGC SPE fractionation of various glycans. Prior to the PGC SPE, *N*-glycans were enzymatically released from a therapeutic glycoprotein purified by spin filtration

### 3.1 Detergent Removal

1. Rinse the spin concentrator with deionized water to remove trace additives such as glycerin and sodium azide within the concentrators (*see* **Note 1**).
2. After prerinsing the spin concentrator with water, solution of biological product (100 μg/500 mL) is loaded into the sample chamber.
3. Place the concentrator assembly into the rotor with a proper counterbalance, then centrifuge at 12,000 × *g* for 10 min to until the desired concentration factor is achieved.
4. Discard filtered solution in a collection tube.
5. Dilute the sample to the original volume with deionized water.
6. This filtration step (**steps 3–6**) was repeated three times.
7. Recover the concentrated and desalted solution from the sample chamber and then dry the solution under vacuum.

### *3.2 Enzymatic Release of N-Glycans*

1. Add sample (100 μg/50 μL) to 50 μL of digestion solution (*see* **Note 2**).
2. Heat the reaction mixture by alternating between a 100 and 25 °C water bath for six cycles of 20 s each (*see* **Note 3**).
3. After cooling at room temperature, add 2 μL of PNGaseF and incubate the mixture (pH 7.5) at 37 °C for 16 h (*see* **Note 4**).

### *3.3 Purification and Fractionation of Neutral and Sialylated Glycans on mAb Using PGC SPE*

1. Wash the PGC cartridge sequentially with deionized water and 80% ACN in 0.1% aqueous TFA (v/v) and then deionized water.
2. Load the glycan solutions to the PGC cartridge and wash subsequently with 6 mL of deionized water. The flow rate during the SPE extraction should be kept at 200 μL/min after the sample loading to maximize the interaction of glycans with the stationary phase of the cartridge.
3. After the wash, elute the glycans with 6 mL of 20% ACN in $H_2O$ (v/v) and 6 mL of 40% ACN with 0.05% TFA in $H_2O$ (v/v), respectively (*see* **Note 5**).
4. Collect each fraction and dry under vacuum.
5. Reconstitute the fractions in 10–30 μL of nano pure water prior to MS analysis.

### *3.4 Purification and Fractionation of Neutral, Sialylated, and Phosphorylated Glycans on Lysosomal Enzyme Using PGC SPE*

1. Stabilization, sample loading, and purification of PGC SPE cartridge were conducted in the same manner as the mAb.
2. First, elute preferentially neutral glycans from the PGC using 6 mL of 20% ACN in water (v/v).
3. Selectively fractionate phosphorylated glycans by 6 mL of 10% ACN in water with 0.05% TFA (v/v).
4. Lastly, collect the sialylated glycans remained on the PGC cartridge by 6 mL of 40% ACN in 0.05% TFA (v/v) (*see* **Note 5**).
5. Dry all SPE fractions under vacuum.

### *3.5 PGC-LC/MS Analysis*

1. Purified glycans are chromatographically separated and analyzed using a PGC chip nano-LC/Q-TOF MS system.
2. The mAb glycan elution gradient is delivered at 0.3 μL/min using solutions of: (A) 3.0% ACN with 0.1% FA (v/v) in $H_2O$, and (B) 90.0% ACN with 0.1% FA (v/v) in $H_2O$, at the following proportions and time points: 6% B, 0–2.5 min; 6–16% B, 2.5–20 min; 16–30% B, 20–38 min; and 30–99.9% B, 38–42 min. Remaining nonglycan compounds are flushed out with 99.9% B for 10 min. Finally, the column is reequilibrated with 5% B for 10 min. In the case of a lysosomal enzyme, neutral and phosphorylated glycan fractions are chromatographically separated in the applied gradient, individually:

3% B, 0–4 min; 3–5% B, 4–6 min; 5–20% B, 6–25 min; 20–100% B, 25–30 min. Sialylated glycan fraction is analyzed in the LC gradient: 3% B, 0–4 min; 3–10% B, 4–6 min; 10–40% B, 6–25 min; 40–100% B, 25–30 min. More acidic solutions of (A) 3% ACN with 0.5% FA (v/v) in $H_2O$ and (B) 90% ACN with 0.5% FA (v/v) in $H_2O$ are used for LC/MS analysis of glycans with acidic functional groups.

3. Acquire MS spectra in the positive ion detection mode over a mass range of $m/z$ 500–2000 with an acquisition time of 1.5 s per spectrum.

### 3.6 MS Data Analysis

1. After data acquisition, identify glycans within 20 ppm accurate mass criteria using the Molecular Feature Extractor algorithm included in the Mass-Hunter Qualitative Analysis software.
2. Filter MS peaks with a signal-to-noise ratio of 5.0 and parse them into individual ion species. Using expected isotopic distribution, charge-state information and retention time, sum together all ion species associated with a single compound (e.g., the doubly protonated ion, the triply protonated ion, and all associated isotopologs), and calculate the neutral mass of the compound.
3. Use computerized algorithms to identify *N*-glycan compositions by accurate mass. Compare deconvoluted experimental masses against theoretical glycan masses using a mass error tolerance of 5 ppm. On the basis of known biosynthetic pathways and glycosylation patterns, glycan compositions containing hexose (Hex), *N*-acetylhexosamine (HexNAc), fucose (Fuc), *N*-acetylneuraminic acid (NeuAc), and *N*-glycolyneuraminic acid (NeuGc), and phosphate (P) can be determined [19, 20].

### 3.7 Glycan Profiling

1. **Neutral and sialylated glycans on mAb**. In order to demonstrate the glycan profiling of therapeutic mAb using PGC SPE and LC/MS analysis, infliximab produced in SP2/O cell line, representative commercial therapeutic mAb, was used. Glycans of infliximab were fractionated according to their size and polarity by PGC SPE. Representative extracted compound chromatograms (ECCs) of *N*-glycans identified in 20% ACN and 40% ACN with 0.05% TFA fraction, respectively were shown in Fig. 2. Indeed, glycan size as well as glycan polarity increases together as ACN progressed from 20% to 40%. To be specific, high mannose and small neutral complex/hybrid type glycans (blue color) were found in the 20% ACN elution (Fig. 2a), while sialylated complex and hybrid type glycans (red color) were mainly detected in 40% ACN solution with 0.05% TFA (Fig. 2b). Each of the identified glycan compositions included two or more peaks corresponding to structural

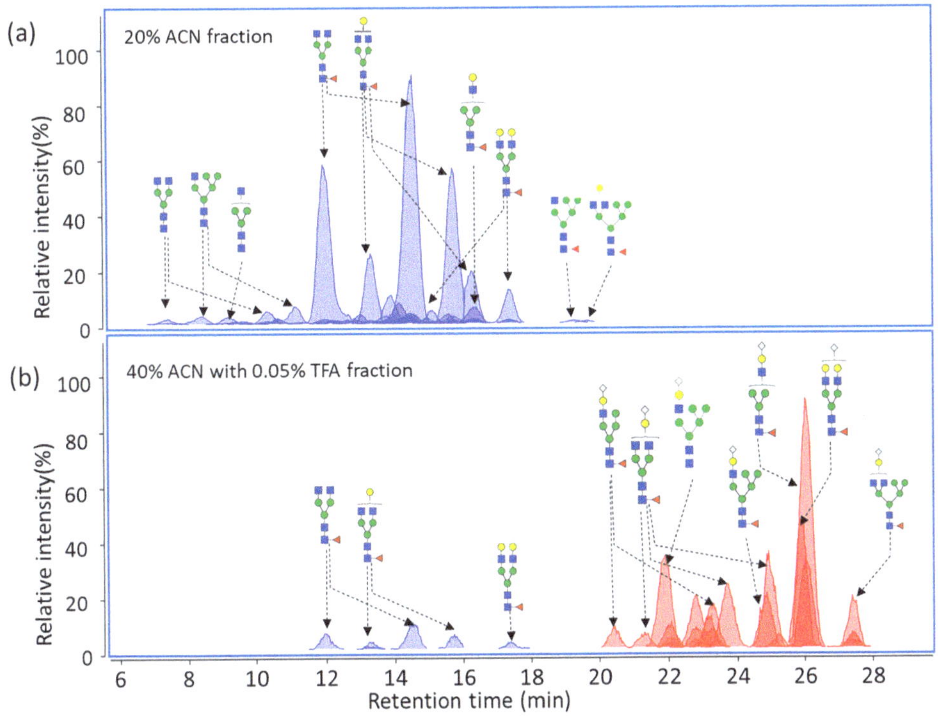


**Fig. 2** Representative extracted compound chromatograms (ECCs) of mAb *N*-glycans obtained from two PGC SPE fractions. (**a**) The first fraction by 20% ACN in $H_2O$, (**b**) the second fraction by 40% ACN in $H_2O$ with 0.05% TFA. Chromatograms are color-coded: blue for neutral glycan and red for sialylated glycan

and/or linkage isomers. All *N*-glycans found in PGC fractions were listed in Table 1 with glycan composition, accurate mass, and normalized intensity. Normalized intensity of individual glycan composition was calculated by dividing the peak intensity of a single glycan into the sum of peak intensities of all glycans found in each SPE fraction. Finally, 19 neutral *N*-glycans and 7 sialylated *N*-glycan compositions were obtained in LC/MS profile from combined the 20% and 40% ACN SPE fractionation.

2. **Neutral, sialylated, and phosphorylated glycans on a lysosomal enzyme.** Fig. 3 is the representative ECCs of three glycan fractions obtained via the stepwise PGC SPE in glycan analysis of a therapeutic lysosomal enzyme. Neutral, phosphorylated, and sialylated glycans were successfully captured by SPE solutions with different proportions of ACN, pure water, and TFA [6]. Numerous glycans of each SPE fraction were reseparated and determined by PGC-LC/MS analysis. To be specific, high mannose and neutral complex-type glycans were observed in the SPE eluent of 20% ACN in water (v/v) (Fig. 3a). In addition, we found that phosphorylated high mannose and hybrid-type glycans were collected altogether in

**Table 1**
**The list of total glycans on infliximab found in each PGC SPE fraction**

| Molecular mass [M] | Composition[a] | Glycan type[b] | AVE[c] | SE[d] |
|---|---|---|---|---|
| *The first eluent by 20% ACN in $H_2O$* | | | | |
| 910.3272 | 3_2_0_0_0 | Core | 0.23 | 0.01 |
| 1072.379 | 4_2_0_0_0 | HM | 0.29 | 0.08 |
| 1113.405 | 3_3_0_0_0 | C | 1.57 | 0.07 |
| 1234.431 | 5_2_0_0_0 | HM | 1.12 | 0.11 |
| 1259.466 | 3_3_1_0_0 | C–F | 4.74 | 0.14 |
| 1275.462 | 4_3_0_0_0 | CH | 1.13 | 0.21 |
| 1316.488 | 3_4_0_0_0 | C | 3.07 | 0.37 |
| 1421.520 | 4_3_1_0_0 | CH–F | 3.13 | 0.22 |
| 1437.515 | 5_3_0_0_0 | H | 2.61 | 0.21 |
| 1462.548 | 3_4_1_0_0 | C–F | 44.26 | 0.20 |
| 1478.540 | 4_4_0_0_0 | CH | 1.55 | 0.04 |
| 1583.572 | 5_3_1_0_0 | H–F | 1.07 | 0.03 |
| 1599.568 | 6_3_0_0_0 | H | 1.31 | 0.08 |
| 1624.601 | 4_4_1_0_0 | CH–F | 29.16 | 0.40 |
| 1640.593 | 5_4_0_0_0 | CH | 0.35 | 0.05 |
| 1745.625 | 6_3_1_0_0 | H–F | 0.45 | 0.03 |
| 1786.653 | 5_4_1_0_0 | CH–F | 3.67 | 0.41 |
| 1948.706 | 6_4_1_0_0 | H–F | 0.29 | 0.02 |
| *The second eluent by 40% ACN in $H_2O$ with 0.05% TFA* | | | | |
| 1462.548 | 3_4_1_0_0 | C–F | 3.55 | 0.25 |
| 1624.596 | 4_4_1_0_0 | CH–F | 1.84 | 0.03 |
| 1786.651 | 5_4_1_0_0 | CH–F | 0.36 | 0.09 |
| 1728.610 | 4_3_1_0_1 | H–F–S | 33.76 | 2.56 |
| 1890.664 | 5_3_1_0_1 | H–F–S | 5.26 | 0.97 |
| 1906.659 | 6_3_0_0_1 | H–S | 10.62 | 0.87 |
| 1931.692 | 4_4_1_0_1 | CH–F–S | 18.53 | 0.65 |
| 2052.718 | 6_3_1_0_1 | CH–F–S | 5.46 | 0.61 |
| 2093.743 | 5_4_1_0_1 | H–F–S | 15.67 | 0.61 |
| 2255.801 | 6_4_1_0_1 | H–S | 4.96 | 0.23 |

[a]Composition: Hex_HexNAc_Fuc_NeuAc_NeuGc
[b]Glycan type: *HM* high mannose, *H* hybrid type, *C* complex type, *CH* complex or hybrid type, *F* fucosylated, *S* sialylated
[c]*AVE* average
[d]*SE* standard error

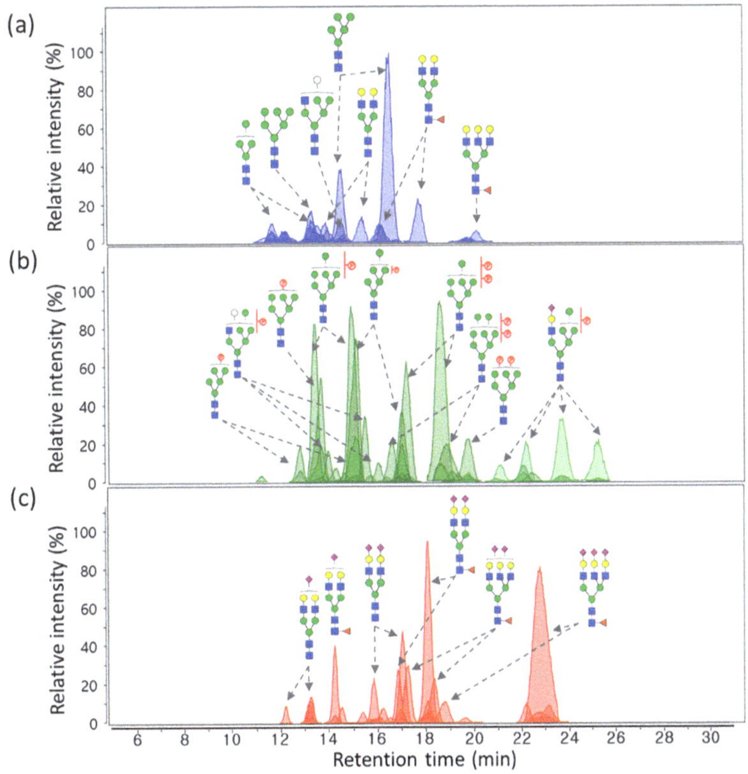


**Fig. 3** Representative extracted compound chromatograms (ECCs) of three glycan fractions of lysosomal enzyme obtained from PGC SPE. (**a**) The first SPE fraction by 20% ACN in $H_2O$, (**b**) the second SPE fraction by 10% ACN in $H_2O$ with 0.05% TFA, and (**c**) the third SPE fraction by 40% ACN in $H_2O$ with 0.05% TFA

10% ACN in water with 0.05% TFA (v/v) fraction (Fig. 3b). Lastly, sialylated glycans containing bi- and trisialylated complex type as major components were eluted only by 40% ACN solution with 0.05% TFA (v/v) (Fig. 3c).

## 4 Notes

1. Excipients present in pharmaceutical formula should be removed prior to LC/MS because they interrupt glycan analysis. Excipient removal can be performed by various tools, depending on the specific physicochemical characteristics, but in this experiment, ultrafiltration membrane was applied to remove polysorbates or other surfactants and enrich the target therapeutic glycoprotein. Note that MWCO filter should not exceed half of the molecular mass of target biotherapeutic protein to minimize possible sample loss.

2. Ammonium bicarbonate buffer containing dithiothreitol (DTT) was used for protein denaturation. The final concentration used in the denaturation process is 100 mM ammonium bicarbonate buffer and 5 mM DTT. The denaturation solution was prepared in 2× concentration of ammonium bicarbonate and DTT and the denaturation solution and sample were mixed in 1:1 (vol: vol) to the final concentration.
3. It is best to perform the protein denaturation using the cycling of hot water and room temperature water to prevent protein aggregation.
4. For complete deglycosylation, a sufficient amount of PNGase F and incubation time is required depending on protein structure with glycosylation sites. In the presence of α1,3-core fucosylated *N*-glycan producing in plant or insect glycoproteins, an alternate enzyme such as PNGase A should be used instead of PNGase F, which can no longer function.
5. For sophisticated fractionation of diverse glycans, the volume of SPE elution solution as well as the ratio of solution components such as organic (ACN) and acid (TFA) can be changed.

## Acknowledgments

The authors are grateful for the support provided by the Ministry of Food and Drug Safety (19172MFDS169) and the National Research Foundation (NRF-2015M3A9B5053960 for H. J. An).

### References

1. Jefferis R (2009) Glycosylation as a strategy to improve antibody-based therapeutics. Nat Rev Drug Discov 8(3):226–234. https://doi.org/10.1038/nrd2804
2. del Val IJ, Kontoravdi C, Nagy JM (2010) Towards the implementation of quality by design to the production of therapeutic monoclonal antibodies with desired glycosylation patterns. Biotechnol Prog 26(6):1505–1527. https://doi.org/10.1002/btpr.470
3. Oh MJ, Hua S, Kim U, Kim HJ, Lee J, Kim J-H, An HJ (2016) Analytical detection and characterization of biopharmaceutical glycosylation by MS. Bioanalysis 8(7):711–727. https://doi.org/10.4155/bio.16.20
4. Hua S, Oh MJ, Ozcan S, Seo YS, Grimm R, An HJ (2015) Technologies for glycomic characterization of biopharmaceutical erythropoietins. Trac—Trend Anal Chem 68:18–27. https://doi.org/10.1016/j.trac.2015.02.004
5. Hajba L, Szekrenyes A, Borza B, Guttman A (2018) On the glycosylation aspects of biosimilarity. Drug Discov Today 23(3):616–625. https://doi.org/10.1016/j.drudis.2018.01.009
6. Seo Y, Oh MJ, Park JY, Ko JK, Kim JY, An HJ (2019) Comprehensive characterization of biotherapeutics by selective capturing of highly acidic Glycans using stepwise PGC-SPE and LC/MS/MS. Anal Chem 91(9):6064–6071. https://doi.org/10.1021/acs.analchem.9b00603
7. Tian YA, Zhou Y, Elliott S, Aebersold R, Zhang H (2007) Solid-phase extraction of N-linked glycopeptides. Nat Protoc 2(2):334–339. https://doi.org/10.1038/nprot.2007.42
8. Bapiro TE, Richards FM, Jodrell DI (2016) Understanding the complexity of porous graphitic carbon (PGC) chromatography:

modulation of Mobile-stationary phase interactions overcomes loss of retention and reduces variability. Anal Chem 88(12):6190–6194. https://doi.org/10.1021/acs.analchem.6b01167

9. Abrahams JL, Campbell MP, Packer NH (2018) Building a PGC-LC-MS N-glycan retention library and elution mapping resource. Glycoconj J 35(1):15–29. https://doi.org/10.1007/s10719-017-9793-4
10. Hua S, Saunders M, Dimapasoc LM, Jeong SH, Kim BJ, Kim S, So M, Lee KS, Kim JH, Lam KS, Lebrilla CB, An HJ (2014) Differentiation of cancer cell origin and molecular subtype by plasma membrane N-glycan profiling. J Proteome Res 13(2):961–968. https://doi.org/10.1021/pr400987f
11. Seo Y, Kim U, Oh MJ, Yun NY, An HJ (2015) MS platform for erythropoietin glycome characterization. Mass Spectrom Lett 6(3):53–58. https://doi.org/10.5478/Msl.2015.6.3.53
12. Oh MJ, Hua S, Kim BJ, Jeong HN, Jeong SH, Grimm R, Yoo JS, An HJ (2013) Analytical platform for glycomic characterization of recombinant erythropoietin biotherapeutics and biosimilars by MS. Bioanalysis 5 (5):545–559. https://doi.org/10.4155/Bio.12.327
13. Ruhaak LR, Deelder AM, Wuhrer M (2009) Oligosaccharide analysis by graphitized carbon liquid chromatography-mass spectrometry. Anal Bioanal Chem 394(1):163–174. https://doi.org/10.1007/s00216-009-2664-5
14. Kim U, Oh MJ, Seo Y, Jeon Y, Eom JH, An HJ (2017) Sensitive and comprehensive analysis of O-glycosylation in biotherapeutics: a case study of novel erythropoiesis stimulating protein. Bioanalysis 9(18):1373–1383. https://doi.org/10.4155/bio-2017-0085
15. Seo Y, Park GM, Oh MJ, An HJ (2017) Investigation of O-glycosylation heterogeneity of recombinant coagulation factor IX using LC-MS/MS. Bioanalysis 9(18):1361–1372. https://doi.org/10.4155/bio-2017-0086
16. Ozcan S, Barkauskas DA, Renee Ruhaak L, Torres J, Cooke CL, An HJ, Hua S, Williams CC, Dimapasoc LM, Han Kim J, Camorlinga-Ponce M, Rocke D, Lebrilla CB, Solnick JV (2014) Serum glycan signatures of gastric cancer. Cancer Prev Res (Phila) 7(2):226–235. https://doi.org/10.1158/1940-6207.CAPR-13-0235
17. Kim BJ, Han C, Moon H, Kwon J, Jang IS, Lim SK, Park KW, Choi JS, An HJ (2018) Monitoring of post-mortem changes of saliva N-glycosylation by nano LC/MS. Anal Bioanal Chem 410(1):45–56. https://doi.org/10.1007/s00216-017-0702-2
18. Ji IJ, Hua S, Shin DH, Seo N, Hwang JY, Jang IS, Kang MG, Choi JS, An HJ (2015) Spatially-resolved exploration of the mouse brain glycome by tissue glyco-capture (TGC) and nano-LC/MS. Anal Chem 87(5):2869–2877. https://doi.org/10.1021/ac504339t
19. Aldredge D, An HJ, Tang N, Waddell K, Lebrilla CB (2012) Annotation of a serum N-glycan library for rapid identification of structures. J Proteome Res 11(3):1958–1968. https://doi.org/10.1021/pr2011439
20. Kronewitter SR, An HJ, de Leoz ML, Lebrilla CB, Miyamoto S, Leiserowitz GS (2009) The development of retrosynthetic glycan libraries to profile and classify the human serum N-linked glycome. Proteomics 9 (11):2986–2994. https://doi.org/10.1002/pmic.200800760

# Chapter 10

# Site-Specific *N*-glycosylation Analysis of Recombinant Proteins by LC/MS$^E$

**Kevin Canis, Estelle Garénaux, and Jean-François Boe**

## Abstract

The glycosylation process is extremely heterogeneous, dynamic, and complex compared with any other post-translational modification of protein. In the context of recombinant glycoproteins, glycosylation is a critical attribute as glycans could dramatically alter protein functions and properties including activity, half-life, in vivo localization, stability, and, last but not least, immunogenicity. Liquid chromatography combined to mass spectrometry constitutes the most powerful analytical approach to achieve the comprehensive glycan profile description or comparison of glycoproteins. This chapter details a versatile yet straightforward LC-MS approach for sample preparation, analysis, and data interpretation, enabling the evaluation of site-specific *N*-glycosylation of recombinant glycoproteins.

**Key words** Recombinant glycoprotein, Microheterogeneity, Mass spectrometry, Liquid chromatography, Site-specific glycosylation

## 1 Introduction

Glycosylation is the most complex and diverse post-translational modification of proteins. The intracellular machinery underlying this process is very dynamic thus the resulting glycosylation depends not only on the expression cell type but also on its biological status and local environmental conditions. In nature, glycan's structural heterogeneity reflects a wide diversity of functions at each molecular, cellular, and tissular levels including protein functions and activity, tissue adhesion, metastasis, and cell-to-cell or cell-to-matrix interactions [1, 2].

As most therapeutics bioproducts are glycoproteins, glycosylation is conventionally a central feature to control. Glycosylation of biotherapeutics plays a predominant role on their pharmacokinetics, pharmacodynamics, stability, and immunogenicity. Beyond the expression cell type, the growth conditions must be finely regulated within the bioreactor to ensure consistency of the production.

Arnaud Delobel (ed.), *Mass Spectrometry of Glycoproteins: Methods and Protocols*, Methods in Molecular Biology, vol. 2271, https://doi.org/10.1007/978-1-0716-1241-5_10, © Springer Science+Business Media, LLC, part of Springer Nature 2021

Selected key glycosylation attributes of the final recombinant product must be monitored to guarantee its safety and efficacy [3, 4].

Unlike nucleic acids and proteins, biosynthesis of carbohydrates is a non-template driven process that leads to significant structural heterogeneity. Comprehensive glycomics study of glycoproteins traditionally includes three main levels of investigation:

1. Monosaccharide level: determination of carbohydrate content in terms of neutral, amino, and acidic monosaccharides.
2. Oligosaccharide level: qualitative and quantitative description of each individual species of the glycan population.
3. Glycopeptide level: determination of the glycan distribution at each glycosylation site along the amino acid sequence (microheterogeneity).

Considering the complexity and versatility of glycosylation, characterisation and monitoring are a challenging task requiring complementary analytical approaches. Mass spectrometry (MS) technologies offer strong advantages with high sensitivity, acquisition of fragmentation data, and high performances on complex samples when coupled to separative systems such as liquid chromatography (LC) [5]). Examples of combination of LC and MS-based glycoproteomics approaches are available from a growing number of publications [6–8].

In this chapter, we cover the study of site microheterogeneity by investigating *N*-glycosylation at the glycopeptide level. Based on a peptide mapping approach, we propose a robust procedure involving little protein chemistry and relatively common LC-$MS^E$ instrumentation. The procedure consists in three successive phases, respectively in silico sequence investigation, sample preparation/ analysis, and data interpretation. The whole process is illustrated using data produced from LC-$MS^E$ analysis of human coagulation factor VIII.

## 2 Materials

### 2.1 Equipment and Consumables

1. Cooled centrifuge.
2. Vortex.
3. Thermostated water or dry bath.
4. ThermoFisher Zeba spin columns, 0.5 mL 7K MWCO.
5. Lo-bind microcentriguge eppendorf tubes.

### 2.2 Reagents

We recommend to prepare all the following solutions extemporaneously.

1. *Processing buffer*: 50 mM Ammonium bicarbonate pH 7.8 solution. Add 198 mg Ammonium bicarbonate in 50 mL of ULC/MS grade $H_2O$ into a centrifuge tube. Vortex until dissolved.
2. *Reduction buffer*: 20 mM dithiothreitol (DTT) in processing buffer. Add 3.1 mg DTT in 1 mL of processing buffer into a microcentrifuge tube. Vortex until dissolved.
3. *Alkylation buffer*: 200 mM iodoacetamide (IAM) in processing buffer. Add 37.0 mg IAM in 1 mL of processing buffer into a microcentrifuge tube. Vortex until dissolved and keep protected from light.
4. *Quenching buffer*: 200 mM DTT in denaturation solution. Add 3.1 mg DTT in 100 μL of processing buffer into a microcentrifuge tube. Vortex until dissolved.
5. *LC-MS Solvent A*: 0.1% trifluoroacetic acid (TFA) in $H_2O$. Add 1 mL of TFA into 1 L of $H_2O$.
6. *LC-MS Solvent B*: 0.1% TFA in ACN/$H_2O$ (90/10). Add 1 mL of TFA into 900 mL ACN with 100 mL $H_2O$.
7. *MS Calibrant solution*: 2 mg/mL sodium iodide (NaI). To 40 mg of NaI, add 20 mL $H_2O$ and 20 mL isopropanol. Vortex until dissolved.
8. *MS Lockspray™ solution*: 1 μg/mL Leu-Enkephaline. To 10 mg of Leu-Enkephaline (Leucine Enkephalin acetate salt hydrate), add 10 mL $H_2O$ to prepare a Stock solution at 1 mg/mL. Vortex until dissolved. Dilute 200 μL of 1 mg/mL Leu-Enkephaline solution with 100 mL $H_2O$, 100 mL ACN, and 200 μL glacial acetic acid.
9. Formic acid (FA).
10. Trypsin (Sequencing Grade, Promega Product No. V5111).

## 2.3 Equipment

### 2.3.1 Hardware

The LC-MS system is a Waters Acquity™ UPLC I-Class integrated chromatography system coupled to a Waters Xevo G2-S™ Q-TOF mass spectrometer. The UPLC is composed of a Binary Solvent Manager, a Sample Manager, and a Column Manager. The Xevo G2-S Q-Tof is a hybrid, quadrupole, orthogonal acceleration, time-of-flight (TOF) mass spectrometer fitted with an ElectroSpray source and a LockSpray™ source.

### 2.3.2 Software

The LC-MS system is operated by Waters MassLynx™ software version 4.1. Interpretation of the raw data is aided by the use of the Waters BioLynx™ software. The use of Microsoft Excel™ is recommanded to produce lists of target ions.

#### 2.3.3 Column

Reverse-phase Waters Acquity UPLC™ BEH $C_{18}$ column, 1.7 μm particle diameter, 2.1 mm × 150 mm (Product No. 186002353).

## 3 Methods

On overview of the method workflow is presented in Figure 1.

### 3.1 In Silico Analysis

The first phase of the study consists in carefully examining the protein amino acid sequence and its expression system. This preparative work is crucial and should never be overlooked as it directly determines the quality of the study outcome. Before starting, we advise to refer to the expert MIRAGE guideline series, discussing critical aspects of sample preparation and analysis in the context of glycomics studies [9–11].

The assistance of appropriate open-source tools and resources remains very valuable to anticipate the complexity of recombinant glycoproteins. In this regard, the portals held by ExPASy (www.expasy.org) and Consortium of Functional Glycomics

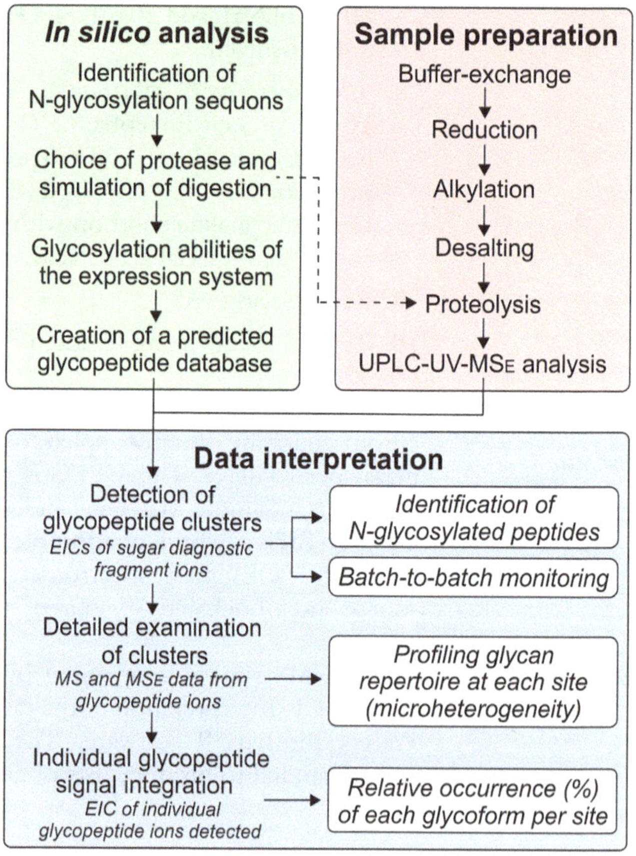


**Fig. 1** Overview of the experimental procedure

(www.functionalglycomics.org) are suggested here as they offer a large set of complementary prediction tools and comprehensive databases. However, despite the availability of a large number of algorithms for glycopeptide analysis developed in recent years [12], the following procedure assumes a minimal use of automated data interpretation and instead encourages manual targeted investigation.

#### 3.1.1 Amino Acid Sequence Examination

1. Examine the protein amino acid sequence to identify the position of potential *N*-glycosylation consensus sequons, that is, Asn-Xxx-Ser/Thr tripeptide sequence were Xxx is any amino acid except Pro. Using most MS instrument manufacturer software (in our case Waters MassLynx™) or open-access resources such as NetNGlyc [13], the protein can be readily screened to highlight putative *N*-glycosylation sites along the sequence.
2. For each site identified, examine the amino acid sequence surrounding the *N*-glycosylation sequon for proteases cleavage sites. An overview of common enzymes and chemicals used for proteomics studies is given in **Note 1** and further discussed in Ref. 14.
3. Perform in silico digestions with the most relevant protease (s) selected using Waters MassLynx™ or open-access resources such as PeptideMass [15]. Peptide sequences covering each potential *N*-glycosylation sequons should be examined.
4. Select the protease to be used taking the following rules into account:
   (a) *The size of the resulting peptides*: best MS sensitivity and resolution are generally achieved for glycopeptides ranging from 1500 to 5000 Da.
   (b) *The protease specificity and efficiency*: limiting non-specific or missed cleavages improves the detection of low-abundancy glycopeptides as well as reproducibility.
   (c) *The position of the glycosylation site within the peptide*: enzymatic proteolysis may not occur efficiently at the vicinity of glycosylated Asn residues. At least two amino acids between the cleavage site and the glycosylated Asn is preferable.
   (d) *The hydrophobicity of the resulting peptides*: our strategy takes advantage of reverse-phase chromatography for its capacity of eluting closely the glycoforms of a single peptide. Better retention improves this relative selectivity.
5. Consider the use of several proteases if no single enzyme allows to generate a set of suitable peptides for each *N*-glycosylation sequon. Proteases can be used either combined sequentially or independently in separate assays.

6. Once the protease is selected, use the appropriate in silico digestion prepared in **step 3** to generate a list containing each potentially *N*-glycosylated peptides together with their respective monoisotopic molecular weight.

#### 3.1.2 Glycosylation Properties of the Expression System

The objective is to identify a set of *N*-glycan candidates which are most likely expressed on the recombinant protein.

1. A literature survey is the first task to undertake. In addition the general knowledge of glycan biosynthetic pathways, the glycosylation capacities of many expression cell types are well documented [16].
2. The cell and tissue glycan profiling database hosted by the Consortium of Functional Glycomics (CFG) together with the glycomics links contained in the portal can also be a good starting point to gather useful information on most common expression systems.
3. Any set of experimental data generated from the recombinant product should be included in the investigation, such as *N*-glycan profiling by MALDI-MS [17] or LC-FLR-MS [18, 19]. These are further discussed in Subheading 4.
4. Once an expected glycan repertoire is defined, select a set of three to five *N*-glycan structures on the basis of their relative abundancy. These glycans will be used to probe the MS data and facilitate the data interpretation. It is advisable to choose *N*-glycan species structurally different from each other to maximize hits during data processing.
5. Determine the monoisotopic molecular weight of the selected *N*-glycans using the monosaccharide residual masses reported in Table 1. The use of at least three decimals for calculations is recommended.

**Table 1**
**Monosaccharides commonly found on therapeutic proteins and corresponding residual monoisotopic masses [21]**

| Class | Common species and corresponding key | Residual monoisotopic mass (Da) |
|---|---|---|
| Pentose | Xylose ☆ | 132.042 |
| Deoxyhexose | Fucose △ | 146.058 |
| Hexose | Mannose ○<br>Glucose ○<br>Galactose ○ | 162.053 |
| Hexuronic acid | Glucuronic acid ◇ | 176.032 |
| N-acetylhexosamine | N-acetylglucosamine □<br>N-acetylgalactosamine □ | 203.079 |
| Sialic acid | N-acetylneuraminic acid ◇<br>N-glycolylneuraminic acid ◇ | 291.095<br>307.090 |

6. Alternatively, open-access software such as GlycoWorkbench [20] calculates the monoisotopic molecular mass of glycans from their carbohydrate content. Additionally, the software offers useful features to build up structures and provide support to MS data interpretation.

#### 3.1.3 Creation of the Glycopeptide Database

1. Combining the monoisotopic masses predicted for both the peptide and the glycan moieties, produce a list containing masses of likely occurring glycopeptides.
2. Calculate the *m/z* values of relevant charge states for each glycopeptide using $M_H^+ = 1.0078$. Usually, $[M + H]^+$ to $[M + 4H]^{4+}$ ions are appropriate but depends on glycopeptide properties (size, amino acid composition, sialic acids, etc.).
3. An example of calculation is given in Table 2.
   (a) Peptide: human factor VIII amino acids 1804-1813 (KN FVKPNETK).
   (b) Expression system: human liver and endothelial cells.
   (c) *N*-glycan candidate: monosialylated biantennary core fucosylated structure (composition $Hex_5HexNAc_4dHex_1$-$NeuAc_1$, FA2G2S1).

### 3.2 Sample Preparation and Analysis

The following procedure applies to glycoprotein samples at a concentration of 1 mg/mL in formulation buffer containing moderate amounts of salts and low molecular weight compounds (amino acids, mono or disaccharides, etc.). Variations of the protein concentration, high content of formulation compounds, potential interferents or presence of polymers/surfactants may require adjustments as discussed in **Note 2**.

#### 3.2.1 Sample Deformulation

1. Set the centrifuge at 8 °C and prepare a Zeba™ spin desalting column (0.5 mL, 7K MWCO) in a collection tube (*see* **Note 2**).
2. Remove the storage solution from the column by centrifugation for 1 min at 1500 × *g*.
3. Condition the resin by addition of 300 μL of processing buffer followed by centrifugation for 1 min at 1500 × *g* (*see* **Note 3**). This step is repeated two additional times.
4. Transfer the desalting column into a clean collection tube and load a 100 μL aliquot of the sample solution (i.e., 100 μg of glycoprotein sample) onto the resin.
5. Collect the buffer-exchanged sample by centrifugation for 2 min at 1500 × *g*.

**Table 2**
**$[M + H]^{+}$ to $[M + 4H]^{4+}$ glycopeptide *m/z* determination**

| Monoisotopic mass (Da) | | | Glycopeptide charge state and *m/z* | | | |
|---|---|---|---|---|---|---|
| Human FVIII aa 1804-1813 | FA2G2S1 N-glycan | Resulting glycopeptide | 1+ | 2+ | 3+ | 4+ |
| 1203.661 | 2077.749 | 3263.396 | 3264.404 | 1632.706 | 1088.807 | 816.857 |

*3.2.2 Sample Reduction*

1. Transfer a 50 μL aliquot of the buffer-exchanged sample into a clean collection tube. The remaining buffer-exchanged sample can be stored as back-up in a deep freezer.
2. Add 30 μL of processing buffer followed by 10 μL of reduction buffer (*see* **Note 4**).
3. Vortex and incubate the sample for 60 min at 37 °C. Following the incubation, leave the sample to cool down at room temperature for 5 min.

*3.2.3 Sample Alkylation*

1. Add 10 μL of alkylation buffer into the reduced sample (*see* **Note 5**).
2. Vortex and incubate the sample for 30 min at room temperature protected from light.
3. Following the incubation, quench the reaction by adding 5 μL of quenching buffer. Vortex and leave to incubate at room temperature for 10 min.

*3.2.4 Sample Desalting*

1. Set the centrifuge at 8 °C and prepare a Zeba™ spin desalting column (0.5 mL, 7K MWCO) in a collection tube.
2. Remove the storage solution from the column by centrifugation for 1 min at 1500 × *g*.
3. Condition the resin by addition of 300 μL of processing buffer followed by centrifugation for 1 min at 1500 × *g*. This step should be repeated two additional times.
4. Transfer the desalting column into a clean collection tube and load a 100 μL aliquot of the sample solution onto the resin.
5. Collect the desalted sample by centrifugation for 2 min at 1500 × *g*.

*3.2.5 Proteolytic Digestion*

1. Prepare the trypsin solution (*see* **Note 1**) by resuspending 20 μg of Sequencing Grade Trypsin in 200 μL of processing buffer. Vortex until dissolved.
2. Add 25 μL of trypsin solution into the desalted sample (1/20e *weight-to-weight* enzyme–substrate ratio).
3. Vortex and incubate the sample for 4 h at 37 °C. Following incubation, quench the reaction by addition of 1 μL acetic acid, vortex the sample and leave to cool down for 5 min at room temperature before transfer into an injection vial for analysis.

*3.2.6 Glycopeptide Enrichment (optional)*

It is well established that glycopeptides exhibit poor ionization efficiency compared to their non-glycosylated counterparts [22]. In recent years various strategies focusing on glycopeptides purification or enrichment have been developed. These separative techniques generally rely on glycans hydrophilicity, charge, or affinity to specific targeted terminal glycoepitopes.

**Table 3**
**Liquid chromatography solvent gradient**

| Time (min) | % solvent B |
|---|---|
| 0 | 0 |
| 1.0 | 0 |
| 61.0 | 40 |
| 81.0 | 60 |
| 81.1 | 100 |
| 86.0 | 100 |
| 86.1 | 0 |
| 90.0 | 0 |

Enriching glycopeptides can greatly improve detection of glycopeptides in the context of site-specific N-glycosylation analyses. A comprehensive review of these methods cannot be proposed in these pages, but consult **Note 6** for some examples. Glycopeptide enrichment can also be further explored in Ref. 12, 23.

#### *3.2.7 LC-MS Analysis*

##### Liquid Chromatography

Solvent A: 0.1% TFA in $H_2O$.

Solvent B: 0.1% TFA in ACN/$H_2O$ (90/10).

Column: Acquity BEH $C_{18}$, 1.7 μm, 2.1 mm × 150 mm (*see* **Note 7**).

Flow rate: set at 300 μL/min.

Chromatographic gradient: *see* Table 3.

Column temperature: set at 60 °C.

Autosampler temperature: set at 8 °C.

Injection volume: 1–5 μL.

##### Mass Spectrometry

Calibration is performed using 2 mg/mL Sodium iodide and LockSpray™ is recorded using 1 μg/mL Leu-Enkephaline.

MS acquisition is acquired using two separate channels: Function 1 records data at low-collision energy (MS) and Function 2 records data at high-collision energy ($MS^E$). Detailed experimental conditions are given in Table 4.

### *3.3 Data Interpretation*

Proteolytic digests of large glycoproteins produce a very complex mixture of components reflected in the LC-MS data by a great variety of signals of diverse intensities. As an example, the total ion chromatograms (TIC) acquired during the LC-MS analysis of Human coagulation Factor VIII is shown in Fig. 2, top panel. Generally, glycopeptide signals are very low and not identifiable on TIC while intense signals correspond to non-modified peptides.

**Table 4**
**Mass spectrometry experimental conditions**

| Parameter | Value |
|---|---|
| Scanning mode | Positive electrospray $MS^E$ |
| *m/z* range | 100–3000 |
| Capillary | 3.0 kV |
| Sampling cone | 30 V |
| Function 1 collision energy | 4.0 eV |
| Function 2 collision energy | 15–50 eV |
| Calibration | Better than 10 ppm |
| LockSpray™ | Real-time correction |

#### 3.3.1 Detection of Glycopeptide-Related Clusters

Glycans produce very distinctive ions when subjected to fragmentation [9]. At high-collision energy, sugar loss from the glycan moieties of glycopeptides produces diagnostic ions that act as a diagnostic signature. Taking advantage of these ions, $MS^E$ data are filtered to highlight glycopeptide signals.

Reverse-Phase chromatography of a proteolytic digest separates the mixture mostly on the basis of the peptide moieties. In this procedure, we take advantage of this property to achieve a moderate separation of the glycoforms of each individual peptide sequence. Consequently, MS signals relevant to glycoforms of the same peptide are grouped within clusters.

1. Select a series of appropriate sugar diagnostic fragment ions based on the *N*-glycan species selected during the in silico investigation (*see* examples in Table 5). If necessary, GlycoWorkbench software provides useful features to anticipate fragmentation of definite glycan structures [20].
2. Using the selected diagnostic fragment ions, produce an extracted ion chromatograms (EIC) of the high-collision energy channel. The combination of appropriate fragment ions is recommended to improve signal sensitivity but also to target specific glycan species.
3. The extraction window used to produce the EIC should be well chosen to maximize both signal detection and specificity (*see* **Note 8**). Also, the selection of appropriate isotopes is also important to maximise EIC signal sensitivity (*see* **Note 9**).
4. As a result, EIC signals would generally appear as clusters (Fig. 2, *bottom panel*), each consisting in the different glycoforms of individual peptide sequence.

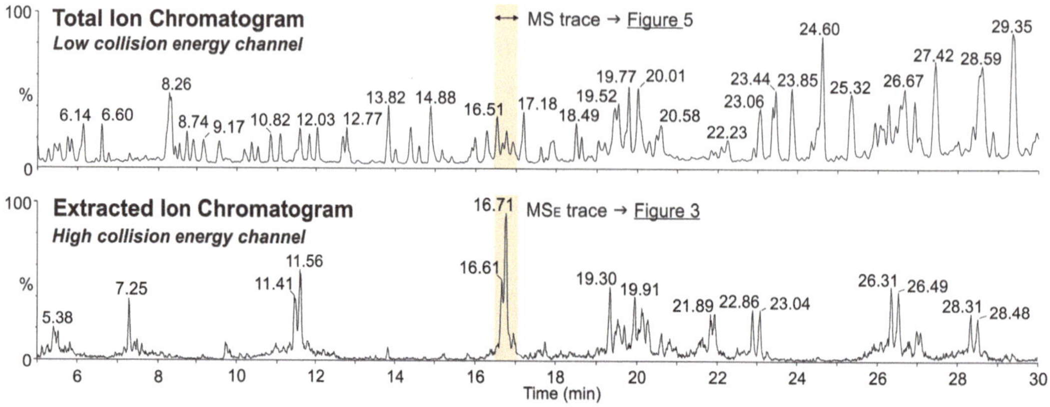


**Fig. 2** UPLC-MS$^E$ results produced from reduced, alkylated and trypsin-digested human coagulation Factor VIII. *Top panel*: Total Ion Chromatogram (low collision energy channel) acquired from 0 to 30 min. *Bottom panel*: Extracted Ion Chromatogram (high collision energy channel) generated using a combination of sugar diagnostic fragment ions (*m/z* 204.087 + 274.088 + 366.140 + 657.236)

**Table 5**
**Examples of common diagnostic fragments observed from mammalian glycans**

| Nature of fragment ion | *Associated N-glycan motif* | Monoisotopic *m/z* value |
|---|---|---|
| HexNAc | *Complex-type glycans* | 204.087 |
| NeuAc | *Sialylated glycans* | 274.088–292.103 |
| NeuGc | *Sialylated glycans* | 308.262 |
| HexNAc–Hex | *LacNAc antennae* | 366.140 |
| HexNAc–HexNAc | *LacdiNAc antennae* | 401.166 |
| Hex–HexNAc–Fuc | *Fucosylated antennae* | 512.198 |
| Hex–HexNAc–NeuAc | *Sialylated antennae* | 657.236 |

*3.3.2 Identification of the Peptide Sequences*

Analysis of individual clusters is initiated with the objective of identifying their associated peptide sequence. The following section focuses on the EIC cluster detected at 16.7 min in Figure 2, bottom panel.

1. Using the high-collision energy channel, combine scans corresponding to the whole cluster. The resulting combined MS$^E$ spectrum displays multiple signals, including fragment ions related to each glycan, glycopeptide, and peptide moiety (Fig. 3).
2. Examine the MS$^E$ spectrum: first the detection of low mass range $[M + H]^+$ monosaccharide diagnostic ions should confirm the presence of glycan fragments (Fig. 3).

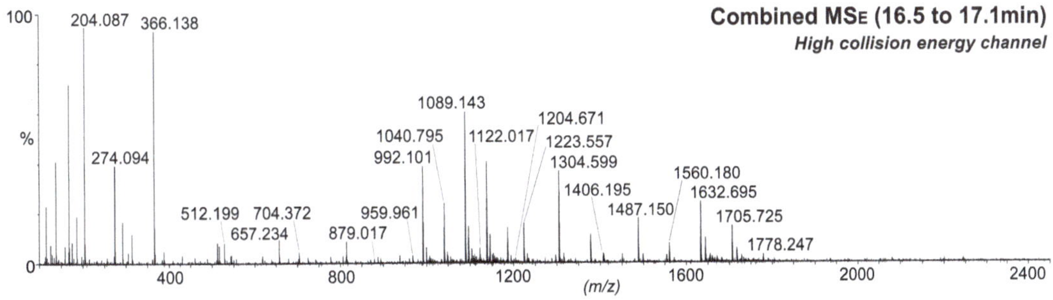


| m/z value | Charge | Assignement |
|---|---|---|
| 204.087 | 1 | HexNAc (1) |
| 274.094 | 1 | NeuAc (1) |
| 366.138 | 1 | Hex(1) HexNAc(1) |
| 512.199 | 1 | dHex (1) Hex(1) HexNAc(1) |
| 657.234 | 1 | Hex(1) HexNAc(1) NeuAc(1) |
| 704.372 | 2 | Peptide + HexNAc(1) |
| 777.413 | 2 | Peptide + dHex(1) HexNAc(1) |
| 879.017 | 2 | Peptide + dHex(1) HexNAc(2) |
| 959.961 | 2 | Peptide + dHex(1) Hex(1) HexNAc(2) |
| 1040.993 | 2 | Peptide + dHex(1) Hex(2) HexNAc(2) |
| 1122.017 | 2 | Peptide + dHex(1) Hex(3) HexNAc(2) |
| 1204.671 | 1 | Peptide KNFVKPNETK |
| 1223.557 | 2 | Peptide + dHex(1) Hex(3) HexNAc(3) |
| 1304.599 | 2 | Peptide + dHex(1) Hex(4) HexNAc(3) |
| 1406.195 | 2 | Peptide + dHex(1) Hex(4) HexNAc(4) |
| 1487.150 | 2 | Peptide + dHex(1) Hex(5) HexNAc(4) |
| 1560.180 | 2 | Peptide + dHex(0) Hex(5) HexNAc(4) NeuAc(1) |
| 1632.695 | 2 | Peptide + dHex(1) Hex(5) HexNAc(4) NeuAc(1) |
| 1705.725 | 2 | Peptide + dHex(2) Hex(5) HexNAc(4) NeuAc(1) |
| 1778.247 | 2 | Peptide + dHex(1) Hex(5) HexNAc(4) NeuAc(2) |

**Fig. 3** *Top panel*: Combined mass spectrum (high collision energy channel) recorded between 16.5 and 17.1 min. *Bottom panel*: Summary of data interpretation highlighting the identification of fragment ions consistent with the peptide (in red), glycan (in blue), and glycopeptide moieties (in green)

3. The mid- to high-mass range region should display a series of ions related altogether by mass shifts consistent with monosaccharide residues and thus consistent with glycopeptide fragments.
4. If necessary, a deconvolution algorithm such as Waters MaxEnt 3™ can be applied on the combined spectrum to ease the peak assignment (Fig. 4) (*see* **Note 10**).
5. Starting from the highest *m/z* glycopeptide ion, successively assign the related ions by trimming off monosaccharide units (Table 1) and using your knowledge of mammalian glycan biosynthetic pathways (*see* **Note 11**).
6. Once the last residues of the *N*-glycan chitobiose trimannosyl core have been removed, the *m/z* ions consistent with the

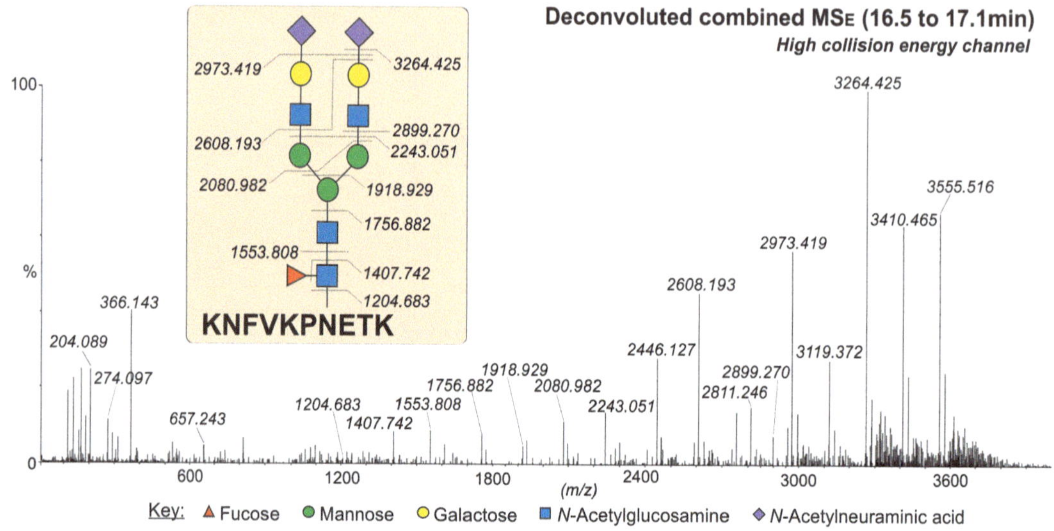


**Fig. 4** Deconvoluted mass spectrum generated from the data shown in Fig. 3

peptide itself should be identified. Based on the *m/z* value monitored, determine the molecular weight of the peptide moieties then identify it using the list prepared in Subheading 3.1.

7. The confirmation of the peptide identity allows to firmly assign the composition of at least one full glycopeptide structure. For example in Fig. 3, the $[M + 2H]^{2+}$ fragment ions at *m/z* 1632.695 correspond to KNFVKPNETK substituted to a glycan composition $dHex_1Hex_5HexNAc_4NeuAc_1$.

*3.3.3 Determination of the Glycan Repertoire*

Once the peptide sequence is identified, the exhaustive description of its associated glycan repertoire is undertaken.

1. Combine MS spectrum corresponding to cluster of interest in the low-collision energy channel. The resulting MS spectrum generated should include signals consistent with all the glycoforms of the identified peptide (Fig. 5).
2. In our example, two series of signals glycopeptides are observed: the first one is composed of $[M + 3H]^{3+}$ ions (*m/z* range 900–1300) and the second one of $[M + 2H]^{2+}$ ions (*m/z* range 1500–1900). Give priority to the most intense set of signals for assignment.
3. Using information gathered from the previous section, assign all signals consistent with glycopeptides, bearing in mind that glycopeptide species are related altogether by mass shifts consistent with monosaccharide residues as found in Table 1.
4. As in the previous section, a deconvolution algorithm could be applied to ease peak assignment although loss of signals could occur and alter final results (*see* **Note 12**).

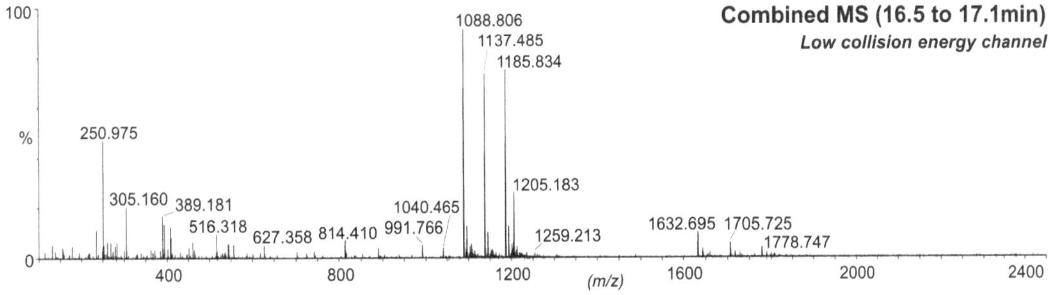


| m/z value | Charge | Assignement |
|---|---|---|
| 991.766 | 3 | KNFVKPNETK + dHex(1) Hex(5) HexNAc(4) NeuAc(0) |
| 1040.465 | 3 | KNFVKPNETK + dHex(0) Hex(5) HexNAc(4) NeuAc(1) |
| 1088.806 | 3 | KNFVKPNETK + dHex(1) Hex(5) HexNAc(4) NeuAc(1) |
| 1137.485 | 3 | KNFVKPNETK + dHex(2) Hex(5) HexNAc(4) NeuAc(1) |
| 1185.834 | 3 | KNFVKPNETK + dHex(1) Hex(5) HexNAc(4) NeuAc(2) |
| 1205.183 | 3 | KNFVKPNETK + dHex(2) Hex(5) HexNAc(5) NeuAc(1) |
| 1259.213 | 3 | KNFVKPNETK + dHex(2) Hex(6) HexNAc(5) NeuAc(1) |
| 1632.695 | 2 | KNFVKPNETK + dHex(1) Hex(5) HexNAc(4) NeuAc(1) |
| 1705.725 | 2 | KNFVKPNETK + dHex(2) Hex(5) HexNAc(4) NeuAc(1) |
| 1778.747 | 2 | KNFVKPNETK + dHex(1) Hex(5) HexNAc(4) NeuAc(2) |

**Fig. 5** *Top panel*: Combined mass spectra (low collision energy channel) recorded between 16.5 and 17.1 min. *Bottom panel*: Summary of data interpretation highlighting the identification of molecular ions related to the glycoforms of the peptide identified

#### 3.3.4 Evaluation of Peptide's Microheterogeneity

The appropriate integration of EIC signals allows to evaluate the relative occurrence of each glycopeptide species, making the crude assumption that each molecule has a similar MS response (*see* **Note 12**).

1. Produce the EIC specific to each individual glycopeptide assigned using the low-collision energy channel. The combination of at least two relevant charge states is recommended to improve the accuracy of the evaluation (Fig. 6).
2. For each EIC produced, integrate the signal, record the peak area and proceed to the semi-quantitative evaluation of each glycoform using a spreadsheet software.

$$\%\text{Glycoform X} = \frac{\text{Area Glycoform X}}{\sum \text{Area of detected Glycoforms}}$$

#### 3.3.5 Sample Monitoring/Batch-to-Batch Comparison

Providing that a reference batch has been preliminarily characterized, comparing glycosylation of recombinant protein batches can be readily achieved by using appropriate sugar diagnostic fragment ions to generate EICs. Figure 7 illustrates a straightforward approach for batch-to-batch monitoring of a same recombinant protein produced in different experimental conditions, as exemplified by the different glycopeptide cluster highlighted in orange.

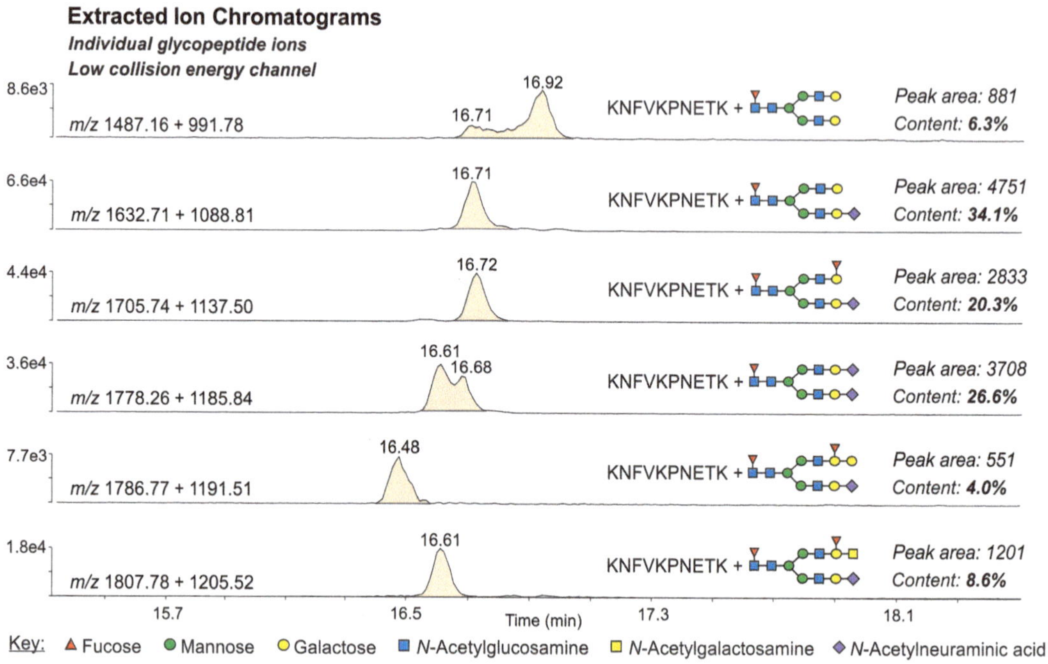


**Fig. 6** Extracted ion chromatograms (low collision energy channel) produced using individual molecular ions diagnostic to glycopeptides detected

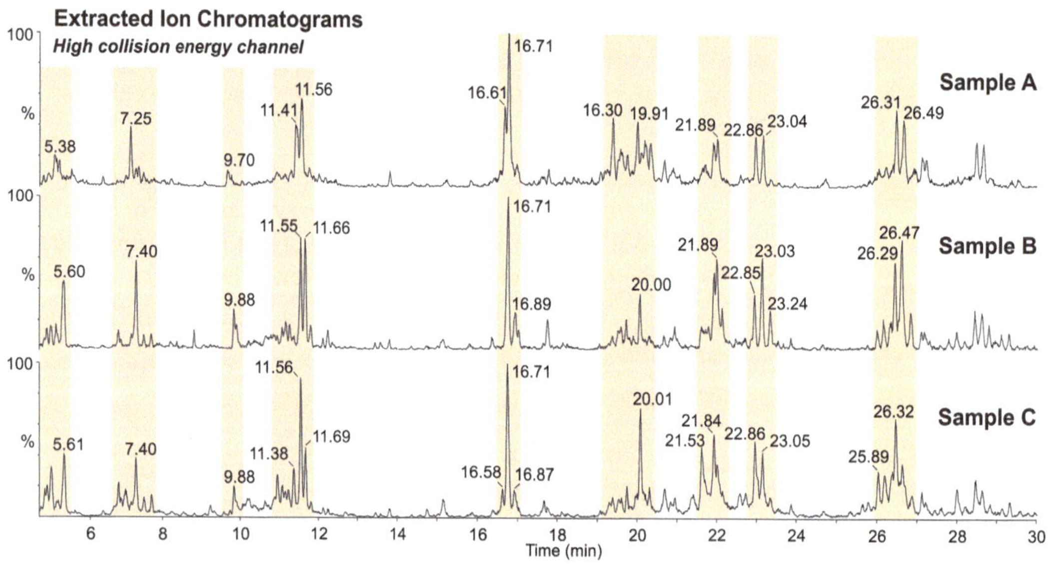


**Fig. 7** Extracted ion chromatograms (high collision energy channel) obtained from the UPLC-UV-MS$^E$ analysis of reduced, alkylated, and human coagulation factor VIII expressed from three different expression systems

### 3.4 Complementary MS-Based Approaches

This last section briefly discusses the advantages of combining complementary glycomic approaches when aiming at the comprehensive description of a glycoprotein glycan repertoire. Complementary approaches particularly apply to the context of recombinant therapeutic protein characterization at early stage of development and production.

A straightforward approach could consist in performing protein intact or reduced mass measurement before and after PNGase F treatment. Doing so provides preliminary information on the type of *N*-glycans substituted to the protein and may point out the subunit to focus on in the case of multimeric proteins.

In terms of complementarity to site-specific *N*-glycosylation data, analytical approaches at the oligosaccharide level are of primary interest. The most powerful technique for describing an oligosaccharide population consists in analyzing permethylated glycans by MALDI-TOF/TOF analysis [17]. Permethylation significantly enhances detection sensitivity but, more interestingly, will confer predictable fragmentation pathways to oligosaccharides. Taking advantage of these fragmentation rules while interpreting MS/MS spectra helps evidencing the presence of biologically relevant epitopes such as ABO blood groups, GalαGal, Lewis motif, or LacDiNac elongation.

Working with permethylated *N*-glycans also allows to get one step deeper into structural characterization through preparation of partially methylated alditol acetates (PMAA) from permethylated glycans. By analyzing PMAA by GC-MS, inter-monosaccharide linkages can be characterized, which provides precious supportive data for deciphering of oligosaccharide structures by MS/MS [24].

A final effective approach for *N*-glycan profiling is HILIC LC-FLR-MS analysis of fluorophore-labeled glycans. Such application allows to separate, identify and achieve relative quantification of individual *N*-glycan species using a minute amount of glycoprotein sample [18, 19].

## 4 Notes

1. The most useful and commonly used proteases and chemicals are summarized in Table 6. Sequencing Grade products are always recommended when undertaking proteomic analysis. Also, pay a particular attention to their degree of specificity. Manufacturer's instructions should include best digestion buffer composition and pH, as well as tolerance toward salts or other potential interferent. Expert comments are available in [14] .
2. Alternatives to gel filtration cartridges are available to achieve the deformulation of a sample solution. When choosing,

**Table 6**
**Proteases and chemicals most commonly used for peptide mapping approaches**

| Enzyme/Chemical | Specificity[a] | Suggestion |
|---|---|---|
| Trypsin | C-terminus of Lys and Arg | Promega prod. No. V5111 |
| Chymotrypsin | C-terminus of Phe, Trp, and Tyr | Promega prod. No. V1061 |
| Lys-C | C-terminus of Lys | Promega prod. No. V1071 |
| V8 protease (Glu-C) | C-terminus of Asp and Glu | Promega prod. No. V1651 |
| Cyanogen bromide | C-terminus of Met | Sigma-Aldrich prod. No. C91492 |

[a]The enzyme specificities is broadly described here and can be subjected to sequence specificities. The reader must refer to manufacturer's instructions for complementary information

carefully consider the desired molecular weight cutoff, the sample volume capacity and the possible resulting dilution factor.

If the sample formulation contains high concentration of low molecular weight compounds, dialysis cassettes offer the most powerful option in terms of deformulation efficiency (e.g., Thermo Fisher Slide-A-Lyzer™ cassettes). Look carefully for protein precipitation while performing buffer exchange and remember that sample dilution might not be neglectable. If the initial protein concentration is low and the sample needs to be concentrated while buffer-exchanged, then filter centrifugal devices should be preferred (e.g., Merck™ Amicon™ Ultra Centrifugal Filters).

Finally, if the formulation contains surfactants, the use of specific detergent removal system is recommended as such compounds prove difficult to remove using the aforementioned techniques (e.g., DetergentOUT™ Tween® Removal Spin Column).

3. The efficiency of protein unfolding can be enhanced by performing reduction and alkylation in presence of a chaotropic reagent, such as guanidine hydrochloride or urea. Such reagent might interfere with the proteolytic enzyme, therefore refer to manufacturer's instructions to assess the enzyme compatibility. If necessary, perform a buffer exchange prior to proteolysis to switch into a compatible digestion buffer.

4. The prevention of any alteration of the glycoprotein such as deamidation or oxidation is a prerequisite to generate high quality results. In general, high temperature, alkaline pH, and long incubation time favor such alterations. If significant levels of alteration is observed using the standard procedure, the use of tris(2-carboxyethyl)phosphine (TCEP) as reducing agent is an interesting alternative. Compared with DTT, TCEP has stronger reducing properties and is effective within a larger pH range.

**Table 7**
**Common lectins available in view of glycopeptide affinity chromatography**

| Lectin | Specificity[a] | Note |
|---|---|---|
| *Ulex europaeus UEA-I* | *Anti-O(H)* | *Preliminary desialylation of glycans required* |
| *Griffonia simplicifolia GS-I* | *Anti-B* | |
| *Helix pomatia HPA* | *Anti-A* | |
| *Sambucus nigra SNA-I* | *NeuAc$_{\alpha 2,6}$Gal* | *n/a* |

[a]Lectin specificity is broadly described here and can be subjected to sequence specificities. Refer to manufacturer for further information

5. As for alterations mentioned in **Note 4**, overalkylation of the glycoprotein must be avoided. Excessive incubation time and concentration of reactive are common sources of overalkylation. If necessary, many alternatives to Iodoacetamide are available, some of which could be based on a more favorable chemistry for your sample (iodoacetic acid, 4-vinylpyridine, *N*-ethylmaleimide, etc.).
6. Enrichment of glycopeptides can greatly improve their detection by MS and ease data interpretation. Enrichment is better implemented once the peptide/glycopeptide mixture is produced . Such techniques are based on either general properties of glycans (hydrophilicity, charge) or on affinity towards specific glycoepitopes through the use of lectins.

   Based on hydrophilicity of glycans, hydrophilic interaction liquid chromatography—solid phase extraction (HILIC SPE) is probably the most versatile method available to achieve glycopeptide enrichment disregarding their glycan component [25].

   In contrast, if a focus is given to specific glycan attributes, other chromatographic methods are investigated: for example, titanium oxide (TiO2) resin allows enrichment of sialylated glycopeptides by retention based on the charges carried by sialic acids [26]. Other glycoepitopes can be targeted by affinity chromatography using lectins. Examples of a few effective lectins are shown in Table 7.
7. In our procedure, we take advantage of hydrophobic properties of the reverse-phase $C_{18}$ chromatography to obtain a moderate separation of the glycoforms of each individual peptide sequence.
8. The extraction window used to produce EIC should ideally be as narrow as possible to improve signal specificity. However, mass accuracy must be considered to avoid a significant

sensitivity decrease while generating EIC. Considering the mass accuracy of our instrument, best results are achieved with an absolute 0.01 amu window for sugar diagnostic ions.

9. When targeting glycopeptides, EIC sensitivity is improved by extracting the *m/z* related to the major isotopic signal of the isotopic massif instead of the *m/z* from the monoisotopic signal. These isotopes corresponding to +1, +2, or +3 amu are more abundant for biomolecules above 2000 Da such as glycopeptides.

10. The intensity of minor signals observed on the MS spectrum can be significantly reduced or even lost following deconvolution. While it may not affect the identification of fragments at high-collision energy, a loss of signal will result in an incomplete description of glycopeptide repertoire when interpreting data at low-collision energy.

11. When tentatively identifying the peptide related to each cluster, keep in mind that any *N*-glycan is characterized by its constitutive pentasaccharide core ($GlcNAc_2Man_3$). Identifying the corresponding ions will lead you to the peptide backbone. The *N*-glycan core indeed produces a series of successive mass decrements of 162, 162, 162, 203, and 203, down to the peptide backbone itself. Remember that the proximal GlcNAc residue can be fucosylated. Also, it should be remembered that the "naked" peptide moieties commonly display a lower charge state than their glycosylated counterparts.

12. It is understood that such approach is not truly quantitative due to differential ionization and transmission of glycopeptide species, possible variations of ionization source regime, or MS signal suppression if eluted together with highly ionizable molecules. However, it remains a fast and reasonably reliable way to assess microheterogeneity and to compare the glycoforms distribution during batch-to-batch monitoring.

## Acknowledgments

The authors thank Baxalta Innovations GmbH for supporting the experimental work reported in these pages and for acknowledging the use of data to illustrate the chapter. The authors also thank Dr Cedric Atmanene and Mr Jeremy Pichot for their valuable contribution to the experimental work and the edition of the manuscript.

## References


1. Varki A (2017) Biological roles of glycans. Glycobiology 27(1):3–49
2. Varki A, Cummings RD, Esko JD, Stanley P, Hart GW, Aebi M, Darvill AG, Kinoshita T, Packer NH, Prestegard JH, Schnaar RL, Seeberger PH (2015–2017) Essentials of glycobiology, 3rd edn. Cold Spring Harbor Laboratory Press, New York
3. Zupke C, Brady LJ, Slade PG, Clark P, Caspary RG, Livingston B, Taylor L, Bigham K, Morris AE, Bailey RW (2015) Real-time product attribute control to manufacture antibodies with defined N-linked glycan levels. Biotechnol Prog 31:1433–1441
4. Sha S, Agarabi C, Brorson K, Lee DY, Yoon S (2016) N-glycosylation design and control of therapeutic monoclonal antibodies. Trends Biotechnol 34(10):835–846
5. Ruhaak LR, Xu G, Li Q, Goonatilleke E, Lebrilla CB (2018) Mass spectrometry approaches to glycomic and glycoproteomic analyses. Chem Rev 118(17):7886–7930
6. Oh MJ, Hua S, Kim U, Kim HJ, Lee J, Kim JH, An HJ (2016) Analytical detection and characterization of biopharmaceutical glycosylation by MS. Bioanalysis 8(7):711–727
7. Canis K, Anzengruber J, Garenaux E, Feichtinger M, Benamara K, Scheiflinger F, Savoy LA, Reipert BM, Malisauskas M (2018) In-depth comparison of N-glycosylation of human plasma-derived factor VIII and different recombinant products: from structure to clinical implications. J Thromb Haemost 16:1592–1603
8. Stavenhagen K, Gahoual R, Dominguez Vega E, Palmese A, Hipgrave Ederveen AL, Cutillo F, Palinsky W, Bierau H, Wuhrer M (2019) Site-specific N- and O-glycosylation analysis of atacicept. MAbs 11(6):1053–1063
9. Struwe WB, Agravat S, Aoki-Kinoshita KF, Campbell MP, Costello CE, Dell A, Feizi T, Haslam SM, Karlsson NG, Khoo KH, Kolarich D, Liu Y, McBride R, Novotny MV, Packer NH, Paulson JC, Rapp E, Ranzinger R, Rudd PM, Smith DF, Tiemeyer M, Wells L, York WS, Zaia J, Kettner C (2016) The minimum information required for a glycomics experiment (MIRAGE) project: sample preparation guidelines for reliable reporting of glycomics datasets. Glycobiology 26(9):907–910
10. Campbell MP, Abrahams JL, Rapp E, Struwe WB, Costello CE, Novotny M, Ranzinger R, York WS, Kolarich D, Rudd PM, Kettner C (2019) The minimum information required for a glycomics experiment (MIRAGE) project: LC guidelines. Glycobiology 29(5):349–354
11. Kolarich D, Rapp E, Struwe WB, Haslam SM, Zaia J, McBride R, Agravat S, Campbell MP, Kato M, Ranzinger R, Kettner C, York WS (2013) The minimum information required for a glycomics experiment (MIRAGE) project: improving the standards for reporting mass-spectrometry-based glycoanalytic data. Mol Cell Proteomics 12:991–995
12. Cao L, Qu Y, Zhang Z, Wang Z, Prykova I, Wu S (2016) Intact glycopeptide characterization using mass spectrometer. Expert Rev Proteomics 13(5):513–522
13. Gupta R, Jung E, Brunak S (2004) Prediction of N-glycosylation sites in human proteins. 46:203–206
14. Giansanti P, Tsiatsiani L, Low TY, Heck A (2016) Six alternative proteases for mass spectrometry-based proteomics beyond trypsin. Nat Protoc 11:993–1006
15. Wilkins MR, Lindskog I, Gasteiger E, Bairoch A, Sanchez JC, Hochstrasser DF, Appel RD (1997) Detailed peptide characterisation using PeptideMass—a world-wide web accessible tool. Electrophoresis 18 (3-4):403–408
16. Goh JB, Ng SK (2018) Impact of host cell line choice on glycan profile. Crit Rev Biotechnol 38(6):851–867
17. Harvey DJ (2018) Analysis of carbohydrates and glycoconjugates by matrix-assisted laser desorption/ionization mass spectrometry: An update for 2013-2014. Mass Spectrom Rev 37 (4):353–491
18. Lauber MA, Yu Y-Q, Brousmiche DW, Hua Z, Koza S, Guthrie E, Magnelli P, Taron CH, Fountain KH (2015) Rapid preparation of released N-glycans for HILIC analysis using a labelling reagent that facilitates sensitive fluorescence and ESI-MS detection. Anal Chem 87:5401–5409
19. Melmer M, Stangler T, Schiefermeier M, Brunner W, Toll H, Rupprechter A, Lindner W, Premstaller A (2010) HILIC analysis of fluorescence-labeled N-glycans from recombinant biopharmaceuticals. Anal Bioanal Chem 398:905–914
20. Ceroni A, Maass K, Geyer H, Geyer R, Dell A, Haslam S (2008) GlycoWorkbench: a tool for the computer-assisted annotation of mass spectra of glycans. J Proteome Res 7 (4):1650–1659
21. Varki A, Cummings RD, Aebi M, Packer NH, Seeberger PH, Esko JD, Stanley P, Hart G, Darvill A, Kinoshita T, Prestegard JJ, Schnaar RL, Freeze HH, Marth JD, Bertozzi CR, Etzler ME, Frank M, Vliegenthart JF,

Lütteke T, Perez S, Bolton E, Rudd P, Paulson J, Kanehisa M, Toukach P, Aoki-Kinoshita KF, Dell A, Narimatsu H, York W, Taniguchi N, Kornfeld S (2015) Symbol nomenclature for graphical representation of Glycans. Glycobiology 25:1323–1324

22. Stavenhagen K, Hinneburg H, Thaysen-Andersen M, Hartmann L, Varón Silva D, Fuchser J, Kaspar S, Rapp E, Seeberger PH, Kolarich D (2013) Quantitative mapping of glycoprotein micro-heterogeneity and macro-heterogeneity: an evaluation of mass spectrometry signal strengths using synthetic peptides and glycopeptides. J Mass Spectrom 48 (6):627–639
23. Zhu R, Zacharias L, Wooding KM, Peng W, Mechref Y (2017) Glycoprotein enrichment analytical techniques: advantages and disadvantages. Methods Enzymol 585:397–429
24. Geyer R, Geyer H (1994) Saccharide linkage analysis using methylation and other techniques. Methods Enzymol 230:86–108
25. Jensen PH, Mysling S, Højrup P, Jensen ON (2013) Glycopeptide enrichment for MALDI-TOF mass spectrometry analysis by hydrophilic interaction liquid chromatography solid phase extraction (HILIC SPE). Methods Mol Biol 951:131–144
26. Larsen MR, Jensen SS, Jakobsen LA, Heegaard NH (2007) Exploring the sialiome using titanium dioxide chromatography and mass spectrometry. Mol Cell Proteomics 6:1778–1787

# Chapter 11

# Mapping *O*-glycosylation Sites Using OpeRATOR and LC-MS

**Maria Nordgren, Andreas Nägeli, Helén Nyhlén, and Jonathan Sjögren**

## Abstract

*O*-glycosylation is a difficult posttranslational modification to analyze. *O*-glycans are labile and often cluster making their analysis by LC-MS very challenging. OpeRATOR is an *O*-glycan specific protease that cleaves the protein backbone N-terminally of glycosylated serine and threonine residues. This enables the generation of glycopeptides of suitable size for mapping *O*-glycosylation sites in detail by bottom-up LC-MS analysis. In this chapter we demonstrate a simple workflow for in-depth analysis of *O*-glycosylation sites on heavily glycosylated proteins using OpeRATOR digestion and HILIC-MS/MS analysis.

**Key words** *O*-glycosylation, OpeRATOR, *O*-glycan specific protease, Glycoproteomics, Glycopeptides, Mass spectrometry, LC-MS

## 1 Introduction

Protein glycosylation is an important posttranslational modification that may impact the structure and function of the glycoprotein and in this way affect cell signaling, transportation, receptor binding, adherence, inflammation, and much more [1]. The successful development of proteins as therapeutic drugs has spurred great interest in the detailed characterization of glycosylation profiles of biopharmaceuticals. The most common and well-studied type of glycosylation is carbohydrates attached to asparagine, the so-called *N*-linked glycosylation, whereas sugar residues attached to serine or threonine, called *O*-linked glycosylation, is far less studied. Nevertheless, *O*-glycans are present on biotherapeutics and characterization is of importance during development and quality control. An example is etanercept, one of the best-selling biologics currently on the market. It is an Fc-fusion molecule consisting of the extracellular domain of the TNF receptor II fused to an IgG1 Fc domain with the C-terminal domain of the receptor carrying 13 *O*-glycosylation sites [2].

The study of *O*-glycans has long been challenging due to the lack of an *O*-glycan consensus motif, density and clustering of *O*-glycans, and the lack of specific enzymatic tools for *O*-glycans

Arnaud Delobel (ed.), *Mass Spectrometry of Glycoproteins: Methods and Protocols*, Methods in Molecular Biology, vol. 2271, https://doi.org/10.1007/978-1-0716-1241-5_11, 

[3, 4]. The approaches for studying *O*-glycosylation include identification of glycosylated proteins using lectins or antibodies, analysis of released glycans (glycomics) or glycopeptides (glycoprotemics). A multicenter study of methods for *O*-glycan analysis concluded that the most reliable methods included mass spectrometric analysis of permethylated reduced glycans in positive mode, native reduced glycans in negative mode and LC-MS analysis of glycopeptides [5]. However, released glycan approaches often include tedious chemical methods for chemical release of the glycans and all information on the origin of specific glycans is lost. Glycoproteomics on the other hand is able to deliver site-specific information in some cases. However, traditional sample preparation methods using trypsin or Lys-C are often complicated by the clustering of the mucin-type *O*-glycans resulting in large chunks of glycopeptides with many *O*-glycosylation sites present. The poor ionization efficiency of heavily glycosylated peptides limits the sensitivity of such approaches and site-specific identification of *O*-glycans becomes challenging. Due to the lability of *O*-glycans, standard collision-based fragmentation techniques such as CID or HCD are ill-suited to this type of analysis and high-end instrumentation capable of alternative fragmentations (ETD, EThcD, and/or UVPD) is necessary [6, 7].

Recently, a novel *O*-glycan specific protease (OpeRATOR®, Genovis AB) originating from the mucin-degrading gut bacterium *Akkermansia muciniphila* has been brought to the market. OpeRATOR binds to *O*-glycosylated structures and hydrolyzes the peptide bond N-terminally to the *O*-glycosylated serine or threonine residue. The enzyme is specific for core 1 mucin-type *O*-glycans and shows increased activity on desialylated glycans. This allows digestion of heavily *O*-glycosylated proteins into peptides of suitable size for bottom-up proteomics applications and enables easier mapping of *O*-glycosylation sites based on identification of digestion sites rather than ETD fragmentation. A solid-phase method based on OpeRATOR digestion has been employed for in-depth analysis of *O*-glycosylation of Zika virus proteins [8] and for mapping over 3000 new *O*-glycosylation sites in human kidney tissues, T-cells, and serum [9]. In the same study a comparison of healthy and tumour kidney tissue was carried out, indicating OpeRATOR as a promising tool for clinical diagnostics.

In this chapter we outline the sample preparation and analytical workflows for obtaining site-specific information on the *O*-glycosylation of a biopharmaceutical using plasma derived human C1 inhibitor as an example. Briefly, samples were treated with OpeRATOR for digestion at *O*-glycosylation sites, a broad acting sialidase mix (SialEXO) to remove sialic acids and increase the performance of OpeRATOR, and PNGaseF to remove *N*-glycans. After graphite enrichment the resulting *O*-glycosylated peptides were analyzed by HILIC-MS/MS using an ESI-Q-TOF instrument. The OpeRATOR digestion sites were identified and allowed for mapping of the

*O*-glycosylation sites. Taken together this method presents a new workflow for site-specific analysis of *O*-glycosylated proteins with applications in characterization of biopharmaceuticals as well as basic research.

## 2 Materials

### 2.1 Protein Buffer Exchange

1. Microspin desalting column 40K, sample size 50–200 μl (cat # 87766, Thermo Scientific).
2. 20 mM Tris buffer, pH 7.5.
3. Centrifuge with fixed angle rotor that can accommodate 1.5–2 ml centrifuge tubes.

### 2.2 O-Glycoprotein Digestion

1. OpeRATOR™ 2000 U (cat # G2-OP1-020, Genovis AB). Endo-O-protease (*Akkermansia muciniphila*) recombinant, expressed in *E. coli.* Digest the protein backbone N-terminally to a desialylated core 1 *O*-glycan.
2. SialEXO™ 2000 U (cat # G1-SM1-020, Genovis AB). Mix of two neuraminidases (*Akkermansia muciniphila*) recombinant, expressed in *E. coli.* Cleaves all nonreducing terminal sialic acid residues α (2-3,6,8).
3. PNGaseF (cat # F8435, Sigma Aldrich). *N*-glycosidase (*Elizabethkingia meningoseptica*) recombinant, expressed in *E. coli.*
4. Human C1-esterase inhibitor (commercially available drug), glycoprotein derived from pooled human plasma [10].
5. 20 mM Tris buffer pH 7.0.
6. Incubator or water bath for 37 °C.

### 2.3 Sample Cleanup Using Graphite Spin Columns

1. Graphite spin columns™ (cat # 88302, Pierce).
2. 1 M $NH_4OH$ Approximate strength 56.6% (14.5 M). Prepare 1 M by the addition of 0.6 ml 56.6% strength to 8.1 ml MQ.
3. MQ grade water prepared from purification of deionized water with a PureLAB Classic ELGA system.
4. Acetonitrile, LC-MS grade.
5. 1.0% trifluoroacetic acid (TFA), LC-MS grade. Dilute in MQ to 1% (v/v).
6. 2.5% TFA, LC-MS grade. Dissolve in MQ to 2.5% (v/v).
7. 50% acetonitrile containing 0.1% formic acid (FA), LC-MS grade. Add 5 ml acetonitrile and 10 μl FA in 5 ml MQ (v/v).
8. Centrifuge with fixed angle rotor that can accommodate 1.5–2 ml centrifuge tubes.
9. Optional: vacuum centrifuge that can accommodate 1.5–2 ml centrifuge tubes.

#### *2.4 O-Glycopeptide Separation by Hydrophilic Interaction Chromatography and Tandem MS Analysis of O-Glycopeptides Using Collision Energy Stepping CID Fragmentation Technique*

1. Agilent UHPLC system equipped with a 1290 Infinity binary pump, 1290 high performance sampler with thermostat, 1290 temperature-controlled column compartment and 1260 multi wavelength detector.
2. Bruker Daltonics HyStar software version 3.2 for instrument control.
3. Acquity UPLC Glycoprotein Amide Column, 300 Å, 1.7 μm, 2.1 mm × 150 mm (cat # 186007963 Waters) (*see* **Note 1**).
4. KrudKatcher Ultra HPLC In-Line filter, 2.0 μm Depth Filter × 0.004 in. (cat # AF0-8497, Phenomenex) (*see* **Note 2**).
5. 0.5% FA in MQ.
6. 0.5% FA in 95% acetonitrile.
7. Hybride Quadrupole Time-Of-Flight mass spectrometer equipped with Apollo II Electrospray Ion Source, Impact II (Bruker).
8. Infusion syringe pump.
9. Bruker otof Control version 4.0 for instrument control.
10. Bruker Compass DataAnalysis version 4.4.
11. Biopharma Compass version 2.0.1519 with GlycoQuest integrated search engine for glycans.
12. ESI-L Low Concentration Tuning Mix (cat # G1969-85000 Agilent Technologies).

## 3 Methods

#### *3.1 Protein Buffer Exchange*

1. Reconstitute 500 U of the commercially available C1-inhibitor drug in 1000 μl of MQ to a concentration of 40 mg/ml [10].
2. A buffer exchange step is required to replace the C1-inhibitor drug formulation with a buffer that is compatible with the enzymatic activities.
3. Break off the bottom closure of the micro spin desalting column. Loosen the lid (do not remove the lid).
4. Place the column in a collection tube (1.5–2 ml) and centrifuge at 1500 × *g* for 1 min to remove the storage solution.
5. Discard the flow-through and place the column in the collection tube.
6. Add 300 μl 20 mM Tris pH 7.5 buffer on top of the resin. Centrifuge the column at 1500 × *g* for 1 min and discard the flow-through.
7. Repeat **steps 5** and **6** two more times. Last spin for 2 min.
8. Blot the bottom of the column to remove excess liquid. Place the column in a new collection tube (1.5–2 ml).

9. Apply the protein solution on top of the resin (100–200 μl).
10. Centrifuge at 1500 × *g* for 2 min to recover the flow-through with antibody in TBS buffer.

### 3.2 O-Glycoprotein Digestion

OpeRATOR is an *O*-glycan specific protease that catalyzes the hydrolysis of peptide bonds on *O*-glycosylated proteins, N-terminally to the glycosylated serine and threonine residues. OpeRATOR activity is significantly enhanced when the sialic acids are removed. SialEXO™, is a mix of two sialidases for complete removal of sialic acids (α2-3, α2-6, and α2-8 linkages). PNGaseF (*see* **Note 3**) is added to hydrolyze all *N*-glycans to reduce complexity and enable improved detection of the *O*-glycopeptides. The N-terminal domain of the C1-inhibitor is heavily *O*-glycosylated with glycans of mainly core 1 structure. From this domain, many *O*-glycopeptides can be generated using the OpeRATOR enzyme only (Fig. 1) (*see* **Notes 4** and **5**).

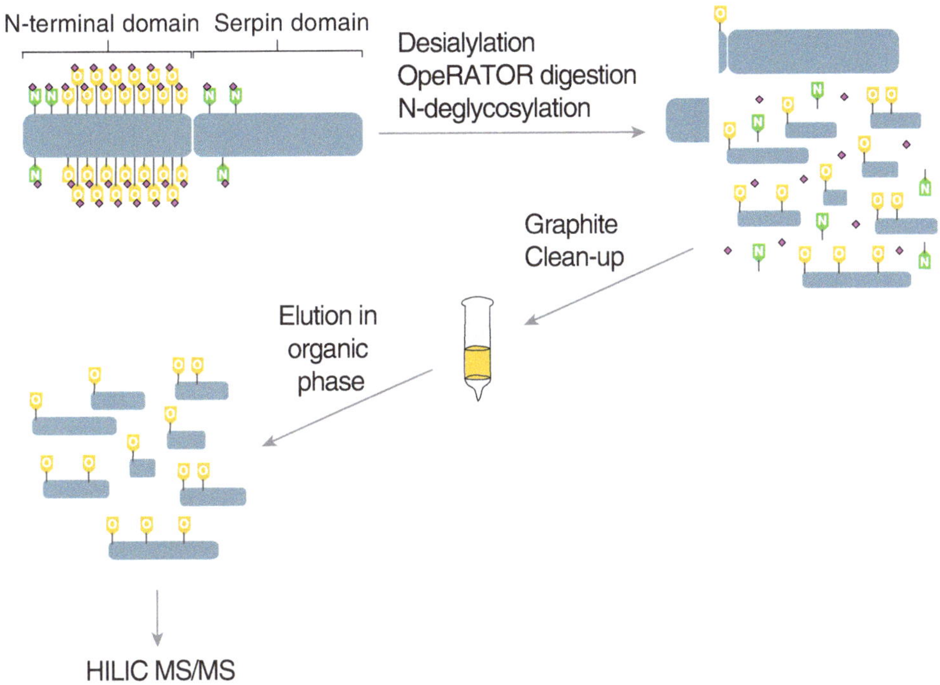


**Fig. 1** Workflow for site specific determination of *O*-glycans on the C1 inhibitor protein. The sialic acids and *N*-glycans of the glycoprotein are enzymatically removed whilst OpeRATOR digests the protein N-terminally of the *O*-glycosylated sites. After sample cleanup on graphite spin columns, the *O*-glycopeptides are separated on a HILIC column and analyzed with tandem MS using a collision energy stepping CID fragmentation technique. The overlapping peptides generated from OpeRATOR digestion will provide information on the modified sites

1. Reconstitute OpeRATOR enzyme 2000 U in 50 μl MQ to 40 U/μl.
2. Reconstitute SialEXO enzyme 2000 U in 50 μl MQ to 40 U/μl.
3. Reconstitute PNGaseF enzyme 300 U in 150 μl MQ to 2 U/μl.
4. Prepare digestion reactions of 100 μg by adding: 37.5 μl 20 mM Tris buffer pH 7.5, 2.5 μl C1-inhibitor (40 mg/ml), 2.5 μl SialEXO (40 U/μl), 2.5 μl OpeRATOR (40 U/μl), and 5 μl PNGaseF (2 U/μl) (*see* **Note 3**), resulting in a final concentration of 2 mg/ml C1 inhibitor, 1 U/μg of SialEXO, 1 U/μg OpeRATOR, and 0.1 U/μg of PNGaseF.
5. Incubate the samples at 37 °C overnight.

### *3.3 Sample Cleanup Using Graphite Resin*

Prior to analysis by LC-MS, sample clean-up is done on Graphite Spin columns. The graphite resin efficiently captures hydrophilic peptides and remove hydrophobic peptides, salts and other contaminants that interfere with LC-MS analysis (*see* **Note 6**).

1. Remove top and bottom cap of the spin column and place into a 1.5 ml collection tube.
2. Centrifuge the column at 2000 × *g* for 1 min to remove the storage solution.
3. Discard the flow-through.
4. Place the column in the collection tube.
5. Prepare the graphite by addition of 100 μl 1 M $NH_4OH$ on top of the resin and reseal the column with the lid.
6. Centrifuge the column at 2000 × *g* for 1 min and discard the flow-through.
7. Repeat the **steps 5** and **6** one more time.
8. Activate the graphite with 100 μl 100% acetonitrile.
9. Centrifuge the column at 2000 × *g* for 1 min and discard the flow-through.
10. Equilibrate the column by addition of 100 μl 1% TFA on top of the resin and reseal the column with the lid.
11. Centrifuge the column at 2000 × *g* for 1 min to remove the solution.
12. Repeat the **steps 8** and **9** one more time.
13. Dilute the entire 50 μl volume of the digested protein sample with 50 μl 2.5% TFA. Verify that pH is between 2.0 and 2.5 using pH indicator strips.
14. Apply sample (50 μl digest + 50 μl 2.5% TFA) and incubate for 10 min with periodic vortex mixing.
15. Centrifuge the column at 2000 × *g* for 3 min to remove the flow-through fraction.

16. Wash the resin by addition of 100 μl 1% TFA.
17. Centrifuge the column at 2000 × *g* for 1 min to remove the solution.
18. Repeat the **steps 16** and **17** one more time.
19. Elute *O*-glycopeptides with 50 μl 0.1% FA in 50% acetonitrile (*see* **Note 7**).
20. Centrifuge the column at 2000 × *g* for 1 min to remove the solution.
21. Reapply the eluted volume to the resin and repeat centrifugation three more times (*see* **Note 8**).
22. The organic content of the samples needs to be increased for retention on the HILIC column. Two options are described below (*see* **Note 9**).
23. Option 1: dry the eluted *O*-glycopeptides in a vacuum centrifuge/concentrator (*see* **Note 10**). Dissolve the peptides in 10 μl MQ (*see* **Note 11**), then add 40 μl 100% ACN + 0.25 μl of 100% FA to reach an acetonitrile concentration high enough for binding to the HILIC column [80% ACN 0.5% FA].
24. Option 2: adjust the eluted sample to approximate starting conditions of HILIC LC. To 20 μl of eluted sample add 30 μl of 100% ACN and 0.25 μl 100% FA [80% ACN 0.5% FA]. At a 20 μl injection this is enough organic phase for this sample to retain the material on this HILIC column (*see* **Note 12**).

### 3.4 Hydrophilic Interaction Chromatography and Tandem MS Analysis of O-Glycopeptides Using Stepping Collision Energy CID Fragmentation Technique

Separation of the purified and concentrated *O*-glycopeptides is performed on a hydrophilic interaction column where the polar peptides are retained and eluted in a gradient of increasing water. In this experiment a Waters Acquity UPLC Glycoprotein Amide Column 300 Å is used (*see* **Note 1**). The *O*-glycopeptides are analysed by MS and MS/MS using an Impact II Q-TOF mass spectrometer (Bruker) interfaced with a standard electrospray ionization source. In order to gain information on both the glycan structures and the peptide sequences, fragmentation spectra were acquired at different collision energies [11]. From the resulting mass spectra, glycopeptide spectra were identified based on the presence of diagnostic oxonium ions and searched for composition matches against a glycan database. The acquired glycan information is combined with peptide matches from a theoretical digest of the C1-inhibitor protein using N-terminal digestion to serine and threonine (Fig. 2). As a result of the OpeRATOR enzyme specificity, the *O*-glycosylation sites can be mapped based on the digestion sites (Fig. 3). Each identified *O*-glycopeptide carries a core 1 *O*-glycan at the N-terminal serine or threonine residue.

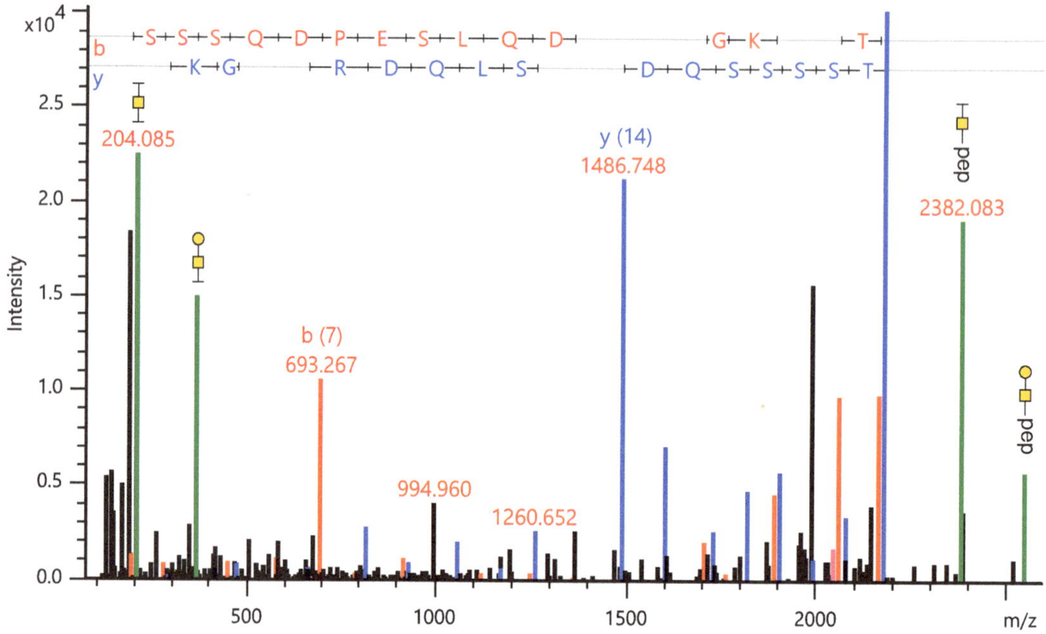


**Fig. 2** Example of combined results from theoretical digest peptide annotation and GlycoQuest glycan annotation. From the acquired data, the *O*-glycosylated peptide TSSSSQDPESLQDRGEGKVAT (27-47) HexNAcHex could be identified, and as a result of OpeRATOR specificity, the N-terminal threonine was defined as the site of modification. The identified oxonium ions are those of HexNAcHex (*m*/*z* 366.138) and (HexNAc *m*/*z* 204.087)

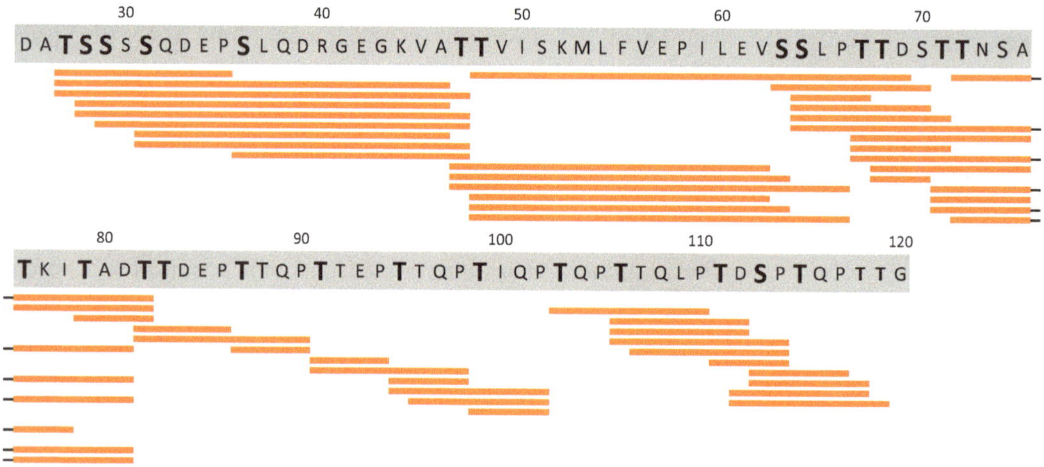


**Fig. 3** Peptide map of the *O*-glycosylation sites of C1 inhibitor. The identified *O*-glycosylation sites are shown in bold text. The two most C-terminal sites (T118, T119) could only be inferred from the digestion pattern without direct identification of a peptide containing the glycosylated amino acid. The sequence variation of *O*-glycopeptides is a result of the heterogeneous *O*-glycosylation

1. Equilibrate the column with the starting conditions of the gradient, 10% mobile phase A (0.5% FA in MQ), 90% mobile phase B (0.5% FA in 95% ACN 5% MQ), using a flow of 0.2 ml/min and a column temperature of 40 °C.
2. The gradient applied is 85 min. Hold 10% A for 0.5 min and increase to 50% A for 70 min. Wash the column at 95% A for 1 min and reequilibrate the column at 10% A for a minimum of 10 min.
3. Inject 20 μl of the prepared sample (*see* **Note 12**).
4. Queued samples are kept at 4 °C in the autosampler.
5. The LC system is connected to the mass spectrometer using the Apollo II ESI sprayer.
6. Calibrate the MS using the ESI-L Low Concentration Tuning Mix G1969-85000 Agilent Technologies at a rate of 180 μl/h using an infusion pump.
7. Applied settings for MS and MS/MS fragmentation analysis using the defined instrumentation (Table 1) (*see* **Note 14**).

**Table 1**
**Applied settings for MS and MS/MS fragmentation**

| | |
|---|---|
| Mode | Positive ion mode |
| Drying gas | Nitrogen |
| ESI source temperature | 220 °C |
| ESI source voltage | 4.5 kV |
| Nebulization gas pressure | 1.8 Bar |
| Nebulization gas flow rate | 8.0 l/min |
| MS spectra rate | 2 Hz |
| Collision gas | Nitrogen |
| Precursor selection range | 300–3000 m/z |
| Quadrupole ion energy | 5.0 eV |
| Collision cell energy | 7.0 eV |
| Isolation and fragmentation | Mass 300–1300, 23–65 eV<br>Increasing with *m/z* and lower charge state |
| Stepping | RF 800 Vpp, 100 μs<br>RF 2000 Vpp, 140 μs<br>Peptide fragmentation:<br>100% of collision energy, 80% of the time<br>glycan fragmentation:<br>50% of collision energy, 20% of the time<br>(*see* **Note 12**) |

8. Convert the acquired MS/MS data to .xml files in DataAnalysis v.3.0 and import into Biopharma Compass as singly charges masses. The Spectra Classifier based on diagnostic MS/MS spectra features, detects glycopeptide spectra and calculates the mass of the peptide moiety.
9. Applied classification parameters as follows. The *m/z* Da signals of the expected oxonium ions is added in the search: H-163.0601, N-204.086649, HN-366.139472. Additional ions that could be present S-$H_2O$-274.092128, S-292.102693, HNS-657.234889, HS-454.155516 (if missed by the sialidase), HH-325.113471, HHN-528.192296 and F-147.065185. Accept a distance tolerance of 0.02 Da.
10. The classified glycopeptide spectra are searched against a glycan databank for structure assignment. Here against Glycome DB (www.glycome-db.org) using the search engine GlycoQuest, which is integrated in Biopharma Compass. Apply free end and a glycan composition restriction of H 0-3, N 1-3, S 0-2, F 0-2, MS tolerance (*m/z*) 10 ppm and MS/MS tolerance (*m/z*) 0.05 Da.
11. All glycopeptide spectra with corresponding calculated peptide mass from the glycopeptide classification are searched for matches against the known protein sequence (*see* **Note 15**). If the endo-O-protease is not already available in the enzyme list, add a new enzyme with N-terminal digestion to serine and threonine residues. The number of missed cleavages needs to be considered based on the protein sequence as *O*-glycans are often present in serine and threonine rich stretches but there is not an *O*-glycan at every site.
12. The theoretical digest and glycan search results are combined to a glycopeptide result.
13. The MS data is processed using DataAnalysis v.3.0. The acquired intact glycopeptide masses are deconvoluted and the identity of the peaks are supported by the assessed MS/MS result (Fig. 4).

## 4 Notes

1. The glycoprotein amide column used in this experiment requires a specific conditioning, including injection of a glycan mix. Follow the column instructions provided by the supplier prior first-time use.
2. The KrudKatcher Ultra HPLC In-Line filter is used as an alternative to a guard column to protect the column and prolong the lifetime.

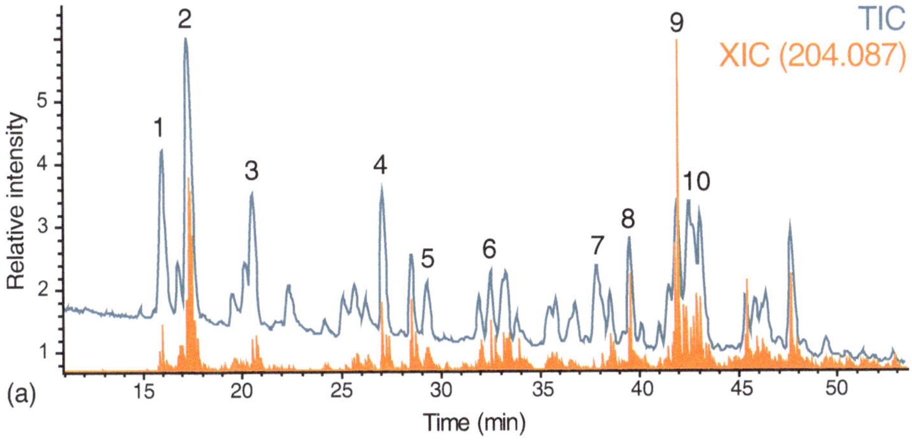


| No. | P. Range | Mr Exp. | Mr Theo. | Δ ppm | G. Composition |
|---|---|---|---|---|---|
| 1 | 48 - 62 | 2082.1140 | 2082.1058 | - 3.96 | Hex1HexNAc1-pep |
| 2 | 48 - 63 | 2185.1364 | 2185.1327 | - 1.67 | Hex1HexNAc1-pep |
| 3 | 64 - 70 | 1084.4672 | 1084.4659 | - 1.19 | Hex1HexNAc1-pep |
| 4 | 47 - 63 | 2635.3183 | 2635.3177 | - 0.22 | Hex2HexNAc2-pep |
| 5 | 76 - 82 | 1113.5303 | 1113.5289 | - 1.31 | Hex1HexNAc1-pep |
| 6 | 113 - 117 | 1258.5152 | 1258.5188 | + 2.80 | Hex2HexNAc2-pep |
| 7 | 31 - 46 | 2079.9450 | 2079.9444 | - 0.29 | Hex1HexNAc1-pep |
| 8 | 106 - 112 | 1869.7764 | 1869.7725 | - 2.04 | Hex3HexNAc3-pep |
| 9 | 27 - 46 | 2442.0887 | 2442.0882 | - 0.21 | Hex1HexNAc1-pep |
| 10 | 27 - 47 | 2543.1353 | 2543.1358 | + 0.22 | Hex1HexNAc1-pep |

(b)

**Fig. 4** HILIC-MS analysis of peptides generated by OpeRATOR digestion of the C1-inhibitor protein. (**a**) The overlaid TIC and XIC traces of HexNAc oxonium ion (204.087 *m/z*) confirms that most of the peaks are *O*-glycopeptides. (**b**) A selection of the identified peptide sequences and glycoforms verified by the corresponding MS/MS data are presented in the table

3. Make sure to use a PNGaseF preparation that is free from detergents, chelating agents, or other substances that may inhibit the enzymatic activities of the other enzymes. Detergents will in addition interfere with the following graphite clean up protocol.
4. The *O*-glycans of the C1 inhibitor studied here, are all situated at the N-terminal part of the protein. In this heavily *O*-glycosylated region from amino acid number 27 to 119 there are no disulfide bonds. The experimental setup may need adjustment depending on the nature of the specific protein to be studied. The amino acid sequence in combination with the potential/or known *O*-glycan sites, must be evaluated to find out if additional proteolytic enzymes are required to get *O*-glycopeptides of suitable sizes. If there are no *N*-glycans, PNGaseF can be

excluded. Disulfide bonds can usually be reduced with 4–6 M urea and up to 100 mM DTT or TCEP in a 1 h reaction at 37 °C. If this step is performed prior enzymatic digestion the sample need to be adjusted to <1 M urea before addition of the enzymes.

5. As OpeRATOR is the only protease used in this experiment, the most C-terminal *O*-glycan sites can only be inferred from the digestion pattern and not directly detected as a glycopeptide.
6. If another protease like trypsin is used, the graphite step will efficiently remove the more hydrophobic peptides that otherwise would interfere with the analysis, as nonglycosylated peptides generally show higher ionization efficiency in ESI-MS.
7. In the graphite spin protocol elution step, TFA is exchanged for FA as TFA can cause signal suppression in ESI-MS analysis.
8. If option 1 described below, vacuum centrifugation, is used to reach the right organic concentration, the *O*-glycopeptide recovery can be increased by applying a fresh volume of 0.1% FA in 50% acetonitrile for each of the repeated elution steps. The resulting elution fractions are pooled prior to evaporation in the vacuum centrifuge.
9. It is preferable that the samples are prepared in the operating mobile phase or (as in this case) in an organic concentration high enough for the peptides to be retained on the column. Visually inspect the samples prior injection on the LC as too high organic phase concentration might precipitate some peptides.
10. When evaporating acetonitrile–water mixtures, solvent bumping can occur and result in product loss or cross-contamination. By employing a controlled vacuum ramp this phenomenon can be avoided. For this mixture, an initial step at 450 mbar for 60 min was used, followed by a second (1 min) ramp down to 20 mbar where the samples were evaporated to dryness. Both steps were kept at 40 °C.
11. If the *O*-glycopeptide samples are evaporated to dryness, reconstitution is facilitated by the initial addition of a small volume of MQ, followed by the organic solution.
12. A higher injection amount can compensate to some degree for the general tendency of glycopeptides to ionize less well. If the injection volume is increased, the composition of the preparation buffer is crucial and needs to match the starting mobile phase conditions to ensure retention of the glycopeptides on the HILIC column.

13. The stepping energies and times applied here are according to Glycopeptide instant expertise™ acquisition method from Bruker [12].

14. Carbohydrate fragmentation via tandem mass spectrometry can be achieved using standard collision induced dissociation (CID) conditions, but to obtain optimal fragmentation data of the peptide moieties as well, especially with higher $m/z$ ratios, higher collision energies and increased time is required.

15. The amino acid numbering applied for human C1 inhibitor analysed in this study is based on the P05155 UniProtKB entry.

## References


1. Varki A (2017) Biological roles of glycans. Glycobiology 27:3–49. https://doi.org/10.1093/glycob/cww086
2. Houel S, Hilliard M, Yu YQ et al (2014) N- and O-glycosylation analysis of etanercept using liquid chromatography and quadrupole time-of-flight mass spectrometry equipped with electron-transfer dissociation functionality. Anal Chem 86:576–584. https://doi.org/10.1021/ac402726h
3. Jensen PH, Kolarich D, Packer NH (2010) Mucin-type O-glycosylation--putting the pieces together. FEBS J 277:81–94. https://doi.org/10.1111/j.1742-4658.2009.07429.x
4. Darula Z, Medzihradszky KF (2018) Analysis of mammalian *O*-glycopeptides-we have made a good start, but there is a long way to go. Mol Cell Proteomics 17:2–17. https://doi.org/10.1074/mcp.MR117.000126
5. Wada Y, Dell A, Haslam SM et al (2010) Comparison of methods for profiling *O*-glycosylation: human proteome organisation human disease Glycomics/proteome initiative multi-institutional study of IgA1. Mol Cell Proteomics 9:719–727. https://doi.org/10.1074/mcp.M900450-MCP200
6. Reiding KR, Bondt A, Franc V, Heck AJR (2018) The benefits of hybrid fragmentation methods for glycoproteomics. TrAC Trends Anal Chem 108:260–268. https://doi.org/10.1016/j.trac.2018.09.007
7. Ko BJ, Brodbelt JS (2015) Comparison of glycopeptide fragmentation by collision induced dissociation and ultraviolet photodissociation. Int J Mass Spectrom 377:385–392. https://doi.org/10.1016/j.ijms.2014.07.032
8. Yang S, Onigman P, Wu WW et al (2018) Deciphering protein O-glycosylation: solid-phase Chemoenzymatic cleavage and enrichment. Anal Chem 90:8261–8269. https://doi.org/10.1021/acs.analchem.8b01834
9. Yang W, Ao M, Hu Y et al (2018) Mapping the O-glycoproteome using site-specific extraction of O-linked glycopeptides (EXoO). Mol Syst Biol 14:e8486. https://doi.org/10.15252/msb.20188486
10. Feussner A, Kalina U, Hofmann P et al (2014) Biochemical comparison of four commercially available C1 esterase inhibitor concentrates for treatment of hereditary angioedema. Transfusion 54:2566–2573. https://doi.org/10.1111/trf.12678
11. Hinneburg H, Stavenhagen K, Schweiger-Hufnagel U et al (2016) The art of destruction: optimizing collision energies in quadrupole-time of flight (Q-TOF) instruments for glycopeptide-based glycoproteomics. J Am Soc Mass Spectrom 27:507–519. https://doi.org/10.1007/s13361-015-1308-6
12. Marx K, Pengelley S New Classification Workflow for N-Glycopeptides using Bruker oTOF Instruments and the Glycopeptide Instant Expertise™ Acquisition Method. Bruker Application Note TN-46

# Chapter 12

# Site-Specific O-Glycosylation Analysis by Liquid Chromatography–Mass Spectrometry with Electron-Transfer/Higher-Energy Collisional Dissociation

**Noritaka Hashii and Junya Suzuki**

## Abstract

O-glycosylation is a major post-translational modification of proteins. Accurate and detailed analysis to reveal O-glycosylation patterns at each site (site-specific O-glycosylation analysis) is essential to deeply understand glycoprotein function. Recent reports also demonstrated that unintended O-glycosylation occurs on therapeutic fusion glycoproteins; therefore, it is increasingly important to perform detailed and exhaustive O-glycosylation analysis during the development of therapeutic glycoproteins. Here, we describe a method of in-depth site-specific O-glycosylation analysis by liquid chromatography–mass spectrometry using electron-transfer/higher-energy collisional dissociation (EThcD) and database analysis.

**Key words** O-Glycosylation, Electron-transfer/higher-energy collisional dissociation, Liquid chromatography–mass spectrometry, Database search

## 1 Introduction

Glycosylation is one of the most common post-translational modifications in living organisms, including mammals, plants, and fungi [1]. Protein glycosylation is involved in several biological events such as cell differentiation and inflammation and in diseases such as cancer and congenital disorders [2, 3]. Therefore, glycosylation analysis is important to deeply understand the biological functions of glycoproteins and pathogenic glycosylation. In addition, recent reports demonstrated that unexpected O-glycosylation occurred on therapeutic fusion glycoproteins [4–7]. Detailed and exhaustive O-glycosylation analysis has become increasingly important in the development of therapeutic glycoprotein products.

Glycans are classified roughly into two kinds: N-linked glycans and O-linked glycans. Identification of N-glycosylation sites is easy because N-lined glycans are attached to asparagine (Asn) residues in

Arnaud Delobel (ed.), *Mass Spectrometry of Glycoproteins: Methods and Protocols*, Methods in Molecular Biology, vol. 2271, https://doi.org/10.1007/978-1-0716-1241-5_12, 

a consensus sequence (Asn-X-Ser/Thr). On the other hand, although O-linked glycans are attached at serine (Ser) or threonine (Thr) residues, identification of O-glycosylation sites is difficult because the consensus sequence for O-glycosylation is unknown; there are many kinds of O-glycosylation, such as O-*N*-acetylgalactosamine (O-GalNAc), O-*N*-acetylglucosamine (O-GlcNAc), O-mannose (O-Man), and O-fucose (O-Fuc) glycans; and heavy O-glycosylation often occurs at sequential multiple Ser/Thr residues in glycoproteins, including mucins [8, 9].

A new mass spectrometric fragmentation method called electron-transfer/higher-energy collisional dissociation (EThcD) has been used for site-specific glycosylation analysis in glycoproteomics [10]. An EThcD spectrum includes *b*-, *y*-, *c*-, and *z*-ions. As a result, EThcD enables a more complete fragmentation of a peptide moiety than higher-energy collisional dissociation (HCD) or electron-transfer dissociation (ETD) alone. Previously, we demonstrated the unexpected O-glycosylation of a commercially available therapeutic human glucagon-like peptide-1 (GLP-1)-Fc fusion protein with a glycine-serine $(G4S)_3$ linker peptide by liquid chromatography–mass spectrometry (LC-MS) with EThcD and a database search [11]. Actually, the glycoprotein sample was tryptically digested after denaturation using guanidine hydrochloride and carboxymethylation using dithiothreitol and monoiodoacetate. The resulting tryptic digests were separated by nano-flow LC with a reversed-phase column. MS/MS spectra of glycopeptides were obtained by two acquisition modes: data-dependent HCD-MS/MS and EThcD-MS/MS. The EThcD-MS/MS data acquisition was triggered by the presence of glycan oxonium ions in HCD-MS/MS. Finally, the microheterogeneity of each O-glycosylation site of glycopeptides was determined by a database search analysis using Bionic™ software as a search engine. Here, we show our detailed experimental sample preparation, LC-MS conditions and database search analysis for site-specific O-glycosylation analysis of glycoproteins.

## 2 Materials

### 2.1 Sample Preparation

1. Reaction vial and sample vial: PROTEOSAVE™ SS 1.5 mL micro tube (Cat. MS-4215 M, Sumitomo Bakelite, Tokyo, Japan) and 0.3 mL ultra-low adsorption sample vial (Cat. PSVial100, AMR, Tokyo Japan), respectively.
2. Denaturing buffer (0.5 M Tris–HCl (pH 8.6) containing 7 M guanidine hydrochloride and 5 mM EDTA): Dissolve 668.7 g of guanidine hydrochloride in ultrapure water and fill up to 800 mL. Add 60.57 g of Tris–HCl (Trizma Base) and 1.86 g of

EDTA. Then, adjust to pH 8.6 with 1 M hydrochloric acid and fill up to 1 L with ultrapure water. Store at room temperature.

3. Reducing buffer (1 M dithiothreitol): Dissolve 7.7 mg of dithiothreitol in 50 μL of ultrapure water. Freshly prepared.
4. Alkylation buffer (1 M sodium monoiodoacetate): Dissolve 10 mg of sodium monoiodoacetate in 48 μL of ultrapure water. Freshly prepared (*see* **Note 1**).
5. Stopping solution (1 M dithiothreitol solution): Reducing buffer was used.
6. Desalting solution: Ultrapure water.
7. Digestion buffer (50 mM Tris–HCl buffer, pH 8.3): Dissolve 605.7 mg of Tris–HCl in 80 mL of ultrapure water, then adjust to pH 8.3 with 1 M sodium hydroxide solution and fill up to 100 mL. Store at 4 °C.
8. Desalting column (PD MiniTrap G-25 column®, Cat. 28-9180-07, GE Healthcare Bio-Sciences, Buckinghamshire, England): Use the column according to the instruction manual.
9. Desalting buffer: Ultrapure water. Use in equilibration and elution processes.
10. PNGase F solution: Dissolve 250 units of PNGase F (Cat. 11-365-193-001, *N*-glycosidase F®, Roche Diagnostics, Mannheim, Germany) in 250 μL of ultrapure water.
11. Trypsin/Lys-C solution: Dissolve 20 μg of Trypsin/Lys-C Mix, Mass Spec Grade (Cat. V5073, Promega) in 20 μL of 50 mM acetic acid.
12. Analytical buffer: 0.1% (v/v) TFA (trifluoroacetic acid).

### 2.2 Liquid Chromatography–Mass Spectrometry and Database Search

1. Liquid chromatography: UltiMate 3000 RSLCnano LC system (Thermo Fisher Scientific).
2. Mass spectrometer: Hybrid ion trap-Orbitrap mass spectrometer (Orbitrap Fusion Lumos mass spectrometer, Thermo Fisher Scientific).
3. Database search software: Proteome Discoverer 2.0 software (Thermo Fisher Scientific).

## 3 Methods

The presented method is composed of three steps: sample preparation, LC-MS acquisition, and database searching (*see* **Note 2**). The detailed procedures of each step are described below. The buffers, solutions, vials and columns that are represented by italic letters are defined in Subheading 2.1.

### 3.1 Sample Preparation: Preparation of Tryptic Digests

1. Denaturation: Place a volume of a sample equivalent to 10 μg of a glycoprotein into a *reaction vial*, add 50 μL of *denaturing buffer* and mix (*see* **Note 3**).
2. Carboxymethylation: Add 2.0 μL of *reducing buffer* after denaturation and incubate at room temperature for 30 min. To the solution, add 4.8 μL of *alkylation buffer*, incubate at room temperature for 40 min in the dark, and then add 1.2 μL of *stopping solution* to stop the reaction. Remove reagents in the reaction mixture using a *desalting column* with *desalting solution* according to the instruction manual and then freeze-dry in a new *reaction vial.*
3. Add 20 μL of *digestion buffer* after freeze-drying and dissolve the carboxymethylated protein completely. Add 10 units of *PNGase F solution* and then incubate at 37 °C for 16 h (*see* **Note 4**).
4. Add 5 μL of *trypsin/Lys-C solution* (*see* **Note 5**) and then incubate at 37 °C for 16 h (*see* **Note 6**).
5. Dry by a centrifugal concentrator (Savant SpeedVac SPD1030, ThermoFisher Scientific) at room temperature, and then add 50 μL of *analytical buffer* and use as a sample for LC-MS analysis.

### 3.2 Liquid Chromatography–Mass Spectrometry

1. Analytical conditions of LC.

   In this LC system, a sample solution is loaded into a trap column using a micro-flow pump via a sample loop (Fig. 1). Adjust the loading time considering the flow rate, sample volume, and line volume from sample loop to trap column.

   (a) Analytical column: Reversed-phase column (Cat. ES800A, Easy-Spray LC column, C18, 3 μm, 0.075 × 150 mm, Thermo Fisher Scientific).

   (b) Trap column: Reversed-phase column (Cat. 752450, L-column 2 ODS, 5 μm, 0.3 × 5 mm, Chemicals Evaluation and Research Institute, Tokyo, Japan).

   (c) Mobile phase A: 0.1 (vol%) formic acid.

   (d) Mobile phase B: 0.1% (vol%) formic acid in acetonitrile.

   (e) Loading buffer: 0.1 (vol%) formic acid.

   (f) Mobile phase flow: Control the gradient by mixing mobile phases A and B as directed in Table 1.

   (g) Flow rate of nano-flow pump: 300 nL/min.

   (h) Flow rate of loading pump: 30 μL/min.

   (i) Injection volume: 5 μL.
2. Operating conditions for MS.

   In this MS operation, spectra are acquired from full-mass scan, data-dependent HCD-MS/MS using an Orbitrap and

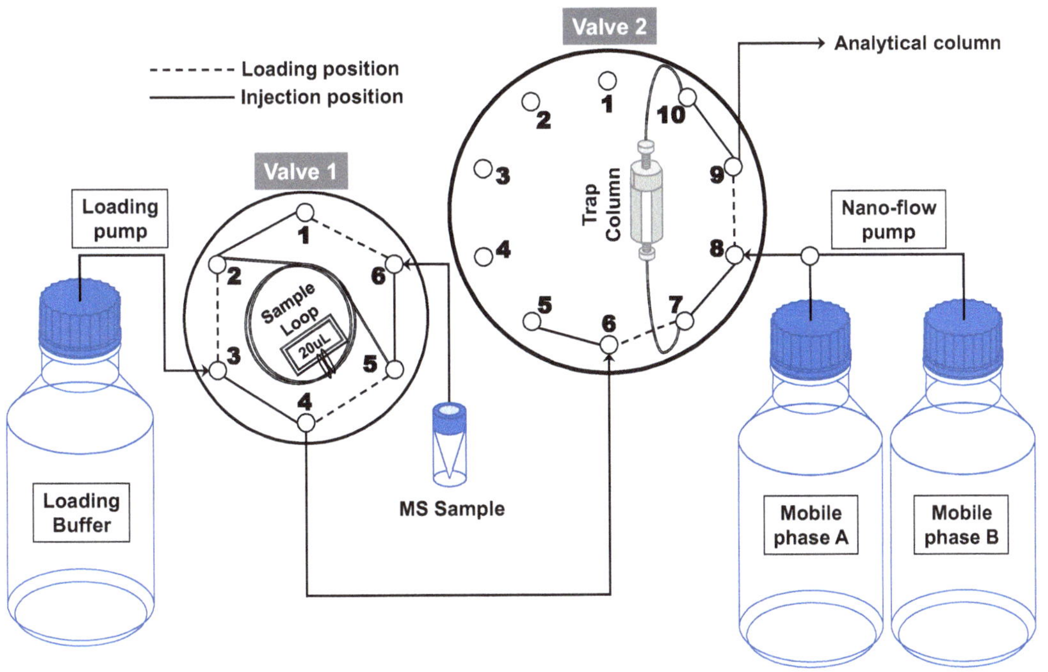


**Fig. 1** Illustration of the nano-flow liquid chromatography system. The sample solution is loaded into the trap column using a loading pump after being loaded in a sample loop

**Table 1**
**Gradient conditions of nano-flow liquid chromatography**

| Time after injection (min) | Mobile phase A (vol%) | Mobile phase B (vol%) |
|---|---|---|
| 0–3 | 98 | 2 |
| 3–90 | 98 → 60 | 2 → 40 |
| 90–91 | 60 → 10 | 40 → 90 |
| 91–95 | 10 | 90 |
| 95–96 | 10 → 98 | 90 → 2 |
| 96–125 | 98 | 2 |

triggered-EThcD-MS/MS using an ion trap. EThcD-MS/MS is performed only when glycan oxonium ions from HCD-MS/MS are detected (Fig. 2). The detailed operating conditions are as follows:

(a) Electrospray voltage: 2.0 kV in positive ion mode.

(b) Capillary temperature: 250 °C.

(c) Full mass resolution: 60,000.

(d) Full mass range: *m/z* 700–2000.

(e) Intensity threshold for precursor ions: $2.0 \times 10^4$.

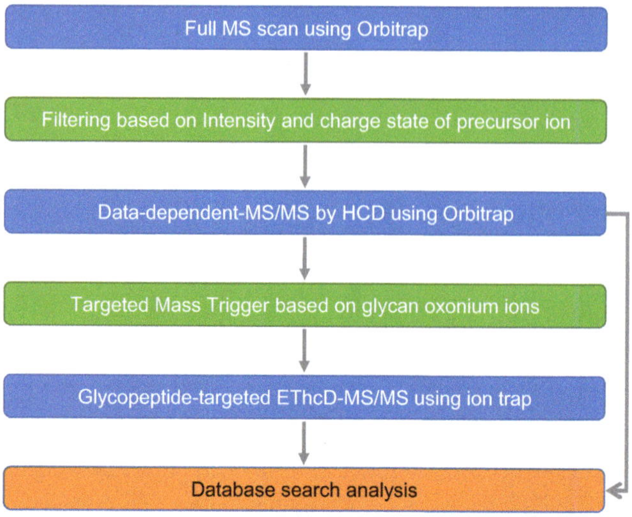


**Fig. 2** Flowchart of triggered electron-transfer/higher-energy collisional dissociation (EThcD) using glycan oxonium ions as the diagnostic ions of glycopeptides. Orbitrap, Kingdon trap; HCD, higher-energy collisional dissociation

(f) Range of charge state of precursor ions: 3–8.

(g) Dynamic exclusion: 30 s (mass tolerance: ±10 ppm).

(h) Conditions for HCD-MS/MS using Orbitrap.
- Resolution: 7500.
- Isolation window: *m/z* 3.0.
- Collision energy: 30%.
- Maximum injection time: 60 ms.

(i) Targeted mass trigger.
- Glycan oxonium ions: *m/z* 138.0545 (HexNAc fragmentation), 204.0867 (HexNAc), and 274.0921 (NeuNAc).
- Mass tolerance: ±15 ppm.

(j) Conditions for EThcD-MS/MS.
- Supplemental activation collision energy: 23% (*see* **Note 7**).
- Isolation window: *m/z* 3.0.
- Maximum injection time: 250 ms.

### 3.3 Database Search Analysis

1. Search engine in Proteome Discoverer: Bionic software (ProteinMetrics, CA, USA).
2. Database: In-house FASTA file for target protein(s).
3. Missed cleavages: One.

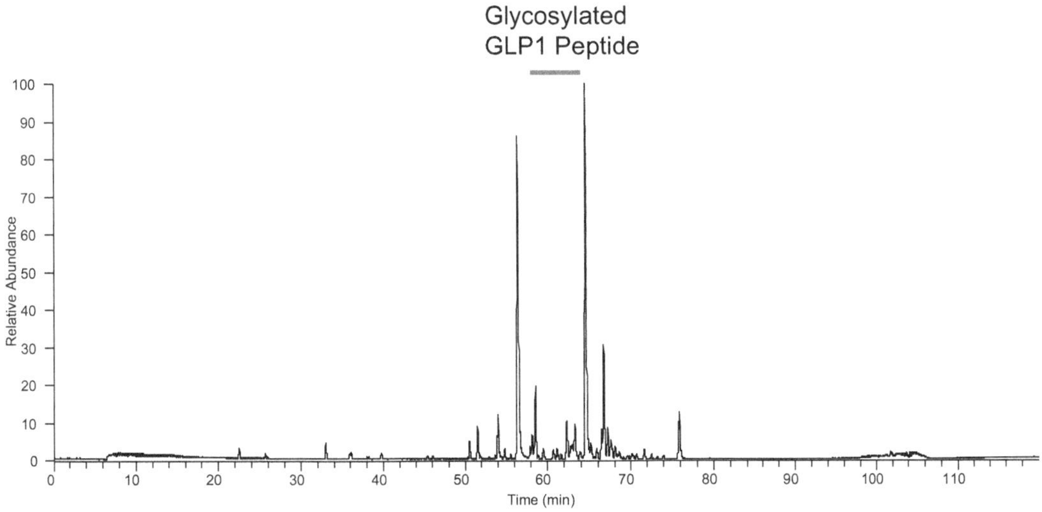


**Fig. 3** Base peak chromatogram (*m/z* 700–2000) acquired by LC-MS of tryptic digests of GLP1-Fc fusion protein

4. Precursor ion mass tolerance: 5 ppm.
5. Mass tolerance for fragment ions in HCD-MS/MS spectra: 20 ppm.
6. Mass tolerance for fragment ions in EThcD-MS/MS spectra: 0.4 Da.
7. Static modification: Carboxymethylation (58.005 Da) for Cys.
8. Dynamic glycosylation modifications: $[HexNAc]_1$ (203.079 Da), $[HexNAc]_2$ (406.385 Da), [Hex]1$[HexNAc]_1$ (365.132 Da), $[Hex]_1[HexNAc]_2$ (568.212 Da), $[Hex]_1[$-$HexNAc]_1[NeuAc]_1$ (656.228 Da), $[Hex]_1[HexNAc]_1[$-$NeuAc]_2$ (947.323 Da), [pentose ([Pent])$]_1$ (132.042 Da), $[Pent]_1[Hex]_1$ (294.095 Da), and $[Pent]_1[Hex]_1[NeuAc]_1$ (585.190 Da) for Ser and Thr.
9. Other dynamic modifications: Deamidation (0.984 Da) of Gln and Asn and oxidation (15.995 Da) of Met and Trp (*see* **Note 8**).
10. False discovery rate (FDR) for extraction of targeted peptide spectrum matches: <0.05 at the spectrum level.
11. Site probability threshold of peptide group modifications: 75.

Figure 3 shows a typical base peak chromatogram of trypsin/Lys-C digests of therapeutic GLP1-Fc fusion protein. This glycoprotein possesses a small amount of mucin-type O-glycans, such as on the GLP1 moiety [11]. Precursor ions from an O-glycosylated GLP1 peptide (*HGEGTFTSDVSSY-LEEQAAK*) were eluted at approximately 59 min. As shown in Fig. 4a, the precursor ion (monoisotopic *m/z* value,

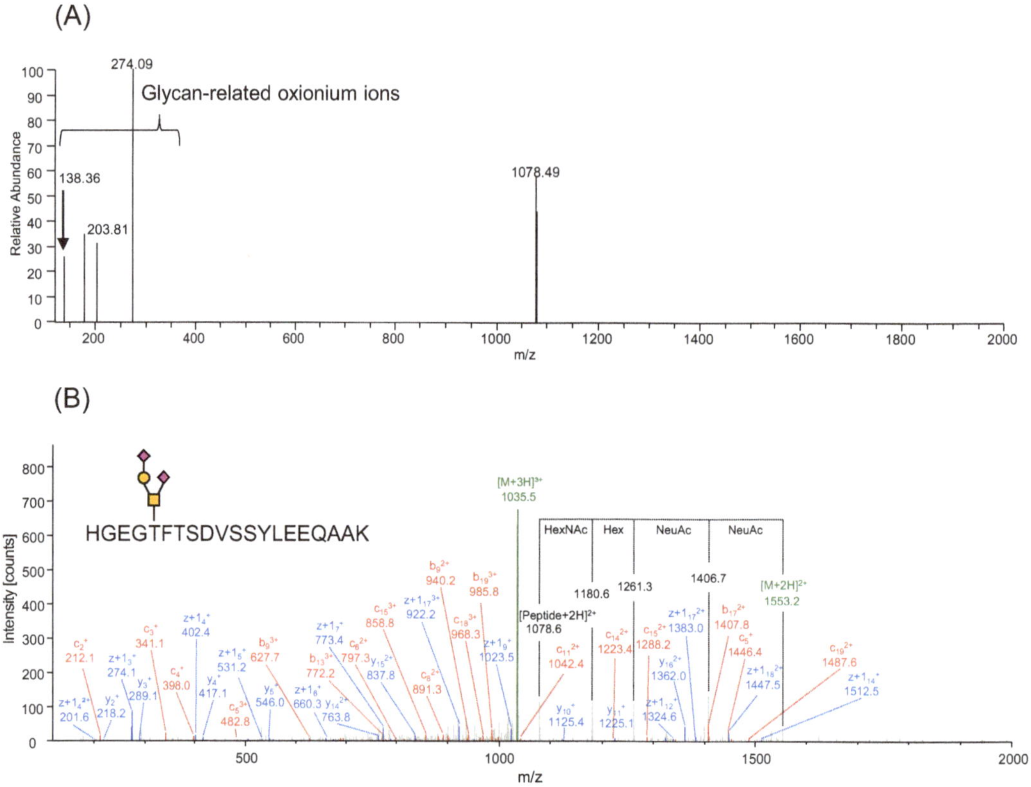


**Fig. 4** (**a**) Higher-energy collisional dissociation (HCD)-MS/MS and (**b**) triggered electron-transfer/higher-energy collisional dissociation (EThcD)-MS/MS spectra of the glycopeptide at *m/z* 1035.106 (3+). The *m/z* value of the calculated monoisotopic triply charged precursor ion of the glycopeptide (HGEGTFTSDVSSYLEE QAAK+$[Hex]_1[HexNAc]_1[NeuAc]_2$) is 1035.108. [Hex], hexose; [HexNAc], *N*-acetylhexosamine; [NeuAc], *N*-acetylneuraminic acid

1035.106; triply charged ion) of a peptide that eluted at 62 min yielded glycan-related oxonium ions by HCD-MS/MS. The triggered EThcD-MS/MS spectrum of the glycopeptide was identified to represent GLP1 bearing the glycan $[Hex]_1[HexNAc]_1[NeuNAc]_2$ by database analysis (Fig. 4b). The corresponding *c*-type ion and radical *z*-type ion (z+1) proved that the glycosylation site is Thr-5 of this glycopeptide. In addition, the glycan structure was deduced as NeuAc-Hex-(NeuAc)HexNAc based on a ladder of glycan-related fragment ions: $[peptide + 2H]^{2+}$ at *m/z* 1078, $[peptide + HexNAc + 2H]^{2+}$ at *m/z* 1180, $[peptide + HexNAc + Hex + 2H]^{2+}$ at *m/z* 1261, $[peptide + HexNAc + NeuAc + 2H]^{2+}$ at *m/z* 1406, and $[peptide + HexNAc + Hex + 2NeuAc + 2H]^{2+}$ ($[M + 2H]^{2+}$) at *m/z* 1553.

## 4 Notes

1. Store sodium monoiodoacetate at −20 °C in the dark.
2. Samples must be analyzed using an optimized LC-MS system. For example, when 100 fmoles of bovine serum albumin is analyzed, (1) typical precursor ions, such as those appearing at *m*/*z* 582.32, 723.38, and 997.59, are observed in the full mass spectrum, (2) the coverage rate (%) of the amino acid sequence is not less than 50%, and (3) multiple individual analyses show similar peaks at the same retention time. If the major peaks are asymmetric, check the fittings of the trap column and analytical column.
3. Change the amount of sample in accordance with experimental purposes. For example, in the case of an analysis of glycosylation with high-site occupancy, the amount of protein can be reduced to approximately 0.1 μg or even less.
4. If you analyze N-linked and O-linked glycopeptides at the same time, skip this step. However, the presented method is not optimized for N-glycosylation analysis. It must be noted that the analysis of glycopeptides having both N- and O-glycosylation sites could be more difficult due to the complication of MS/MS spectra caused by predominant fragmentation of N-glycans.
5. It is recommended that the protease is selected considering the amino acid sequence of the target glycoprotein. More importantly, short peptide fragments are not applicable to the EThcD-MS/MS analysis, which generally requires triply or more highly charged precursor ions to obtain *c*-/*z*-series fragment ions.
6. Change the incubation time because the digestion efficacy is different for each protein.
7. The supplemental activation collision energy of EThcD is the most important parameter to optimize for efficiently generating *c*-/*z*-series ions in the EThcD-MS/MS spectrum. Generally, the range of collision energy could range from 15 V to 25 V. Optimize the supplemental activation collision energy in advance.
8. Long-term digestion often leads to unintended chemical modifications, such as deamidation. In particular, Asn residues next to Gly are prone to be deamidated even if conventional digestion buffers are used. A database search considering deamidation in addition to O-glycosylation is recommended.

## Acknowledgments

This work was supported in part by Japan Agency for Medical Research and Development (AMED) under Grant Number [JP20ae0101022].

### References

1. Varki A, Cummings RD, Freeze HH, Stanley P, Bertozzi CR, Hart GW, Etzler ME (2008) Essentials of glycobiology, 2nd edn. Cold Spring Harbor Laboratory Press, Cold Spring Harbor New York
2. Reily C, Stewart TJ, Renfrow MB, Novak J (2019) Glycosylation in health and disease. Nat Rev Nephrol 15(6):346–366. https://doi.org/10.1038/s41581-019-0129-4
3. Kanagawa M, Kobayashi K, Tajiri M, Manya H, Kuga A, Yamaguchi Y, Akasaka-Manya K, Furukawa JI, Mizuno M, Kawakami H, Shinohara Y, Wada Y, Endo T, Toda T (2016) Identification of a post-translational modification with ribitol-phosphate and its defect in muscular dystrophy. Cell Rep 14 (9):2209–2223. https://doi.org/10.1016/j.celrep.2016.02.017
4. Wen D, Foley SF, Hronowski XL, Gu S, Meier W (2013) Discovery and investigation of *O*-xylosylation in engineered proteins containing a (GGGGS)n linker. Anal Chem 85 (9):4805–4812. https://doi.org/10.1021/ac400596g
5. Spencer D, Novarra S, Zhu L, Mugabe S, Thisted T, Baca M, Depaz R, Barton C (2013) O-xylosylation in a recombinant protein is directed at a common motif on glycine-serine linkers. J Pharm Sci 102 (11):3920–3924. https://doi.org/10.1002/jps.23733
6. Spahr C, Shi SD, Lu HS (2014) O-glycosylation of glycine-serine linkers in recombinant Fc-fusion proteins: attachment of glycosaminoglycans and other intermediates with phosphorylation at the xylose sugar subunit. MAbs 6(4):904–914. https://doi.org/10.4161/mabs.28763
7. Zhong X, Kieras E, Sousa E, D'Antona A, Baber JC, He T, Desharnais J, Wood L, Luxenberg D, Stahl M, Kriz R, Lin L, Somers W, Fitz LJ, Wright JF (2013) Pyroglutamate and O-linked glycan determine functional production of anti-IL17A and anti-IL22 peptide-antibody bispecific genetic fusions. J Biol Chem 288(2):1409–1419. https://doi.org/10.1074/jbc.M112.417717
8. Brockhausen I (2006) Mucin-type O-glycans in human colon and breast cancer: glycodynamics and functions. EMBO Rep 7 (6):599–604. https://doi.org/10.1038/sj.embor.7400705
9. Houel S, Hilliard M, Yu YQ, McLoughlin N, Martin SM, Rudd PM, Williams JP, Chen W (2014) N- and O-glycosylation analysis of etanercept using liquid chromatography and quadrupole time-of-flight mass spectrometry equipped with electron-transfer dissociation functionality. Anal Chem 86(1):576–584. https://doi.org/10.1021/ac402726h
10. Frese CK, Altelaar AF, van den Toorn H, Nolting D, Griep-Raming J, Heck AJ, Mohammed S (2012) Toward full peptide sequence coverage by dual fragmentation combining electron-transfer and higher-energy collision dissociation tandem mass spectrometry. Anal Chem 84(22):9668–9673. https://doi.org/10.1021/ac3025366
11. Hashii N, Suzuki J, Hanamatsu H, Furukawa JI, Ishii-Watabe A (2019) In-depth site-specific O-glycosylation analysis of therapeutic Fc-fusion protein by electron-transfer/higher-energy collisional dissociation mass spectrometry. Biologicals 58:35–43. https://doi.org/10.1016/j.biologicals.2019.01.005

# Chapter 13

# Profiling of *N*-Linked Oligosaccharides of a Glycoprotein by UPLC-FLR-ESI-MS After Derivatization with Fluorescent Anthranilamide

**Claire I. Butré, Eric Largy, and Arnaud Delobel**


## Abstract

*N*-glycans are described to have a large influence on the properties of therapeutic proteins, including safety and efficacy. For this reason, the extent and type of glycosylation is a characterization parameter for the analysis of antibodies and other therapeutic proteins. The method described here is a fast and high-throughput method for identification and semiquantification of *N*-glycans by HILIC-FLR-ESI-MS. Sample preparation has been optimized and simultaneous preparation of a large number of samples can be achieved within a day. The use of MS coupled to fluorescence detection is an additional tool for identifying the *N*-glycan type.


**Key words** HILIC, *N*-glycan, Monoclonal antibodies, Mass spectrometry

## 1 Introduction

Glycosylation is a common posttranslational modification of therapeutic proteins, including antibodies and fusion proteins. This linkage of a carbohydrate to a protein is known to influence the properties of the proteins such as stability, pharmacokinetics, and safety [1]. For instance, glycosylation is reported to protect the proteins from in vivo enzymatic degradations. This has been showed in vitro by comparing the degradation rate of three IgG, which were either glycosylated or deglycosylated [2]. For the three examples, faster degradation of the intact protein by papain was observed for the deglycosylated IgG compared to the glycosylated ones. Glycosylation is also reported as an influencing parameter of the cytotoxicity of the proteins. It was for instance showed that the removal of fucose on Fc-fusion proteins enhances the potent cytotoxicity toward target cells [3]. Pharmacokinetics is described to be related to the mannose content, where high-mannose glycans are found to increase the clearance rate of proteins [4]. From these

Arnaud Delobel (ed.), *Mass Spectrometry of Glycoproteins: Methods and Protocols*, Methods in Molecular Biology, vol. 2271, https://doi.org/10.1007/978-1-0716-1241-5_13, 

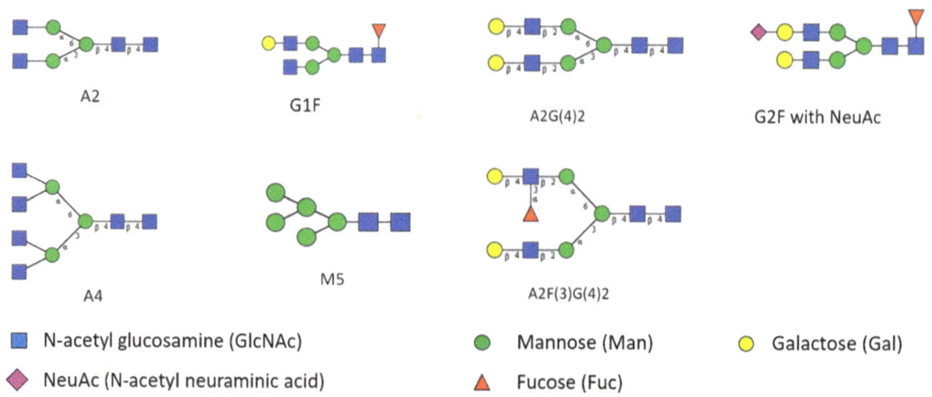


**Fig. 1** Examples of monosaccharide, depicted with symbols according to the CFG (consortium for functional glycomics)

examples, it is clear that the presence and the type of glycans are important properties of therapeutic proteins. Consequently, adequate methods for proper identification and quantification of *N*-glycans are a prerequisite for antibodies and recombinant proteins characterization.

Different types of *N*-glycosylation are found in therapeutic proteins depending on the host cell. The common core of *N*-glycans is made of two *N*-acetylglucosamine and three mannose residues (Fig. 1). From this, based on the monosaccharides they contain, the *N*-glycans can be defined as high mannose (with only mannose residues), complex with any number of other type of monosaccharides, and finally hybrids, with mannose residues on one branch and other saccharides on the other branch. Subclasses are defined based on the presence of a fucose residue (fucosylated glycans) or a sialic acid (charged glycans) and the number of antenna (tri/tetra antennary glycans).

For complete identification of all types of *N*-glycans, in the method presented here, the *N*-glycans are first cleaved from the protein using the endoglycosidase PNGase F. The cleaved *N*-glycans are then derivatized with the chromophore anthranilamide (2-aminobenzamide), leading to the formation of a Schiff's base, which is reduced into a secondary amine as the final reaction product (Fig. 2). The use of a chromophore allows optical detection, by fluorescence in this case. With this, the sensitivity of the analytical method is increased and semiquantification can be performed. If the approach is quite common, the originality of the method presented here is its high-throughput. Complete sample preparation including release and derivatization of the *N*-glycan can be done within a day, while previous methods would describe 3-day protocols [5]. In addition, besides the short time needed for deglycosylation, the use of a 96-well plate for purification instead of a column is a large gain in terms of time thanks to the large number

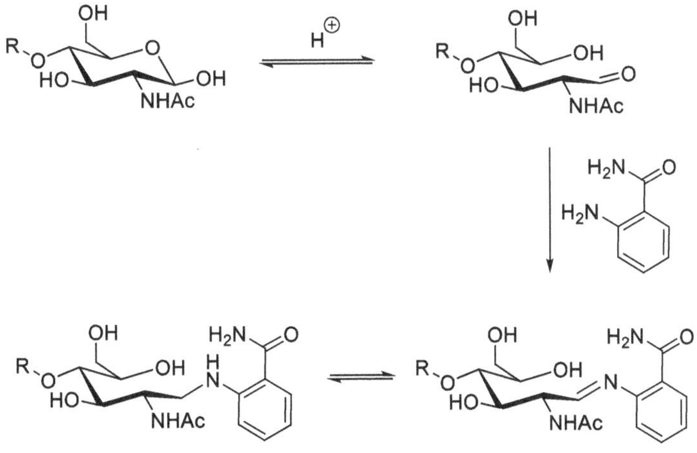


**Fig. 2** Derivatization reaction of *N*-glycan by anthranilamide

of samples that can be purified simultaneously. Finally, the method presented here by coupling the fluorescence detection with a mass spectrometric detection enables *N*-glycan identification.

## 2 Materials

All solutions are prepared using ultrapure water (prepared by purifying deionized water to attain a sensitivity of 18 MΩ·cm at 25 °C). Whenever possible, the use of mass spectrometry grade reagents is recommended.

### 2.1 Apparatus

1. UPLC system (autosampler and quaternary solvent manager, Acquity system, Waters).
2. Q-TOF mass spectrometer equipped with an electrospray ion source (XEVO G2-XS QTOF, Waters).
3. UPLC BEH Glycan (length 150 mm ID 2.1 mm, particle size 1.7 μm, Waters).
4. Fluorimetric FLR detector for Acquity (Waters).
5. Manifold equipped with a pressure regulation system and a pump.

### 2.2 UPLC Preparation

1. Mobile phase A: 100 mM ammonium formate pH 4.5. Weigh 6.31 g of ammonium formate in a 1.0 L volumetric flask and bring to volume with purified water. Adjust the pH to 4.50 ± 0.05 with formic acid.
2. Mobile phase B: acetonitrile.
3. Wash solvent: water–acetonitrile 50:50 v/v.
4. Purge solvent: water–acetonitrile 50:50 v/v.
5. Seal wash solvent: water–methanol 90:10.

**2.3 Sample Preparation**

1. 50 mg/mL RapiGest: Dissolve a 1 mg RapiGest vial (Waters, article no. 186001861) with 20 μL of Rapid PNGase F buffer 5× (Bioke, article no. P0710S) and homogenize with vortex.
2. Derivatization buffer. Weigh 4.0 g of sodium acetate trihydrate and 2.0 g of boric acid in a 100.0 mL volumetric flask. Dissolve and bring to volume with methanol.
3. Derivatization reagent. Weigh 300 mg of anthranilamide and 300 mg of cyanoborohydride in an amber 10.0 mL volumetric flask. Dissolve and bring to volume with the derivatization buffer. Store away from light (*see* **Note 1**).
4. 95% acetonitrile: acetonitrile–$H_2O$ (95:5 v/v).
5. 20% acetonitrile: acetonitrile–$H_2O$ (20:80 v/v).
6. 0.5 mL Amicon Ultra desalting and concentration devices with molecular weight cutoff of 3 kDa (Merck Chemicals, article no. UFC500324) or 10 kDa (Merck Chemicals, article no. UFC501024).
7. Oasis HLB μelution 30 μm purification plate (Waters, article no. 186001828BA).

**2.4 Standard**

1. Dissolve the dextran ladder (Waters, article no. 186006841) by adding 60 μL of water and 140 μL of acetonitrile to reach a final concentration of 1 μg/μL.
2. System performance standard. Dissolve the Glyko 2-AB (Human IgG *N*-linked glycan library, Prozyme Europe, article no. GKSB-005) by adding 15 μL of water and 35 μL of acetonitrile.

## 3 Methods

### 3.1 Sample Preparation (see Note 2)

Digestion can be done via a one-step protocol or a two-step protocol (*see* **Note 3**). Both protocols are described below.

#### 3.1.1 N-Glycan Release—One-Step Protocol

1. Desalt and/or concentrate samples using 0.5 mL Amicon Ultra centrifugal device with a molecular weight cutoff of 3 kDa or 10 kDa (*see* **Note 4**).
2. To 50–250 μg of protein, add 1 μL of Rapid PNGase F per 25 μg of protein and the volume of Rapid PNGase F buffer 5× necessary to reach a final 1× concentration (i.e., for 10 μL of a 5 mg/mL protein solution, add 3 μL of Rapid PNGase F buffer 5× and 2 μL of Rapid PNGase F).
3. Vortex and centrifuge.
4. Incubate at 50 °C for 30 min.

*3.1.2 N-Glycan Release—Two-Step Protocol*

1. Desalt and/or concentrate samples using 0.5 mL Amicon Ultra centrifugal device with a molecular weight cutoff of 3 kDa (*see* **Note 3**).
2. To 50–250 μg of protein, add the proper volume of 50 mg/mL RapiGest to reach a final concentration of 10 mg/mL (i.e., for 10 μL of a 5 mg/mL protein solution, add 3 μL of RapiGest in Rapid PNGase F buffer).
3. Vortex and centrifuge.
4. Incubate at 90 °C for 3 min.
5. Cool down for 3 min at room temperature.
6. Add 1 μL of Rapid PNGase F per 25 μg of protein; aspirate and dispense to mix.
7. Incubate at 50 °C for 5–30 min.

*3.1.3 Derivatization (see* **Note 5***)*

1. After cooling down of the samples to room temperature, add 2 μL of glacial acetic acid.
2. Vortex and centrifuge (*see* **Note 6**).
3. Add 100 μL of derivatization reagent, vortex and centrifuge.
4. Incubate at 80 °C for 1 h.
5. After cooling down of the samples to room temperature, centrifuge the samples 1 min at 12,000 × *g*.
6. Add 1.0 mL of 95% acetonitrile and vortex thoroughly.

*3.1.4 Purification*

1. Purification is done using Oasis HLB μelution 30 μm purification plate Vacuum of 2.5–4.0 in Hg should allow a slow and steady elution.
2. Condition the wells with 95% acetonitrile (5 × 500 μL).
3. Load the samples (2× 600 μL).
4. Centrifuge samples prior to second addition to ensure complete recovery.
5. Wash the wells with 95% acetonitrile (2 × 200 μL).
6. Elute the *N*-glycans with 20% acetonitrile (2 × 50 μL) (*see* **Note 7**).
7. Dry the samples using a Speed-Vac (*see* **Notes 8–10**).
8. Dissolve the samples in water–ACN 30:70 v/v (*see* **Note 11**).

### 3.2 UHPLC Separation

1. Inject 4–8 μL (*see* **Notes 12** and **13**).
2. The detector parameters are set to $\lambda_{excitation} = 330$ nm and $\lambda_{emission} = 420$ nm with a gain of 10 (*see* **Note 14**).
3. Samples are kept at 10 °C.
4. Column temperature is set to 40 °C.

5. The gradient is 0–2 min isocratic on 70% B at 0.4 mL/min, 2–34.8 min; linear gradient from 70% to 53% B at a flow rate of 0.4 mL/min, followed by 34.8–36 min a linear gradient from 53% to 20% B with flow rate decreasing to 0.25 mL/min; 36–39 min, isocratic on 20% B at 0.25 mL/min; 39–40 min from 20% to 70% B and flow rate up to 0.4 mL/min; and 40–45 min isocratic on 70% B at 0.4 mL/min.

### 3.3 MS Conditions

1. Samples are analyzed with an hyphenated Q-TOF MS spectrometer, in positive sensitivity ESI mode. Parameters listed below are provided for a Xevo G2-XS QTOF (Waters) and should be adapted for other MS systems.
2. The cone voltage is set to 80 V and the capillary voltage to 2.75 kV.
3. The source temperature is 120 °C and the cone gas flow is 100 L/h.
4. The desolvation gas temperature is 500 °C with a desolvation gas flow of 800 L/h.
5. Acquisition is done between *m/z* 100 and 2500 with a 1 s scan time. Acquisition is done in $MS^E$ mode (as defined by Waters, allowing both MS and "MS/MS-like" acquisition without prior knowledge of the *m/z* to be fragmented) with a low energy of 6 V and high energy ramp from 20 to 30 V (*see* **Note 15**).
6. Identification of the *N*-glycan obtained after derivatization is performed based on a glucose unit (GU) calibration (*see* **Note 16**) (dextran ladder, Fig. 3)) and confirmed with *m/z* determined by mass spectrometry (using a database of the software UNIFI, Waters for instance) (*see* **Note 17**).

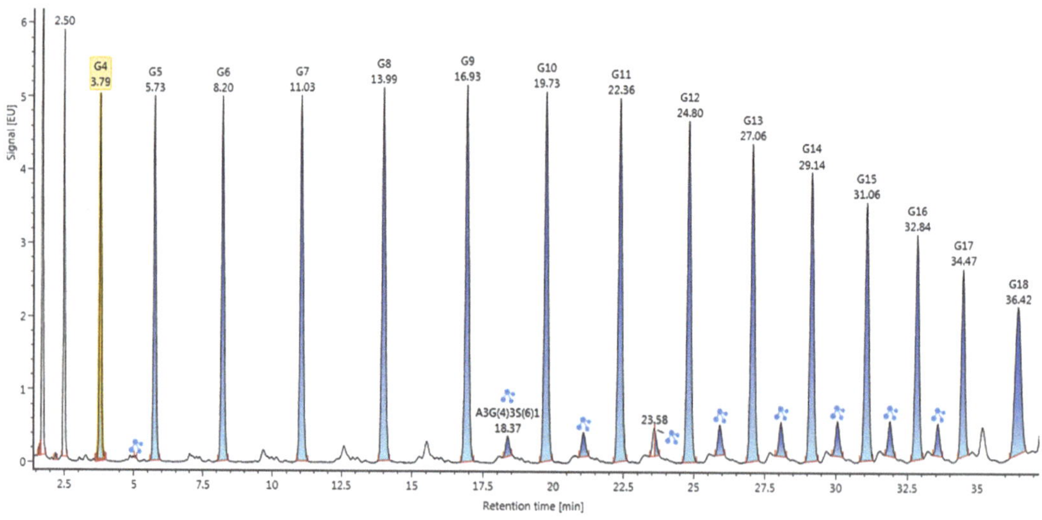


**Fig. 3** Fluorescence chromatogram of the dextran calibration ladder

7. Further confirmation can be obtained from the high energy MS data (obtained with the $MS^E$ mode). Relative abundance of each *N*-glycans is reported as relative fluorescence % peak area. Additionally, abundance can be reported by type of *N*-glycans (antennary number, sialylated, etc.).
8. Examples of chromatograms are provided in Figs. 4, 5, 6, 7.

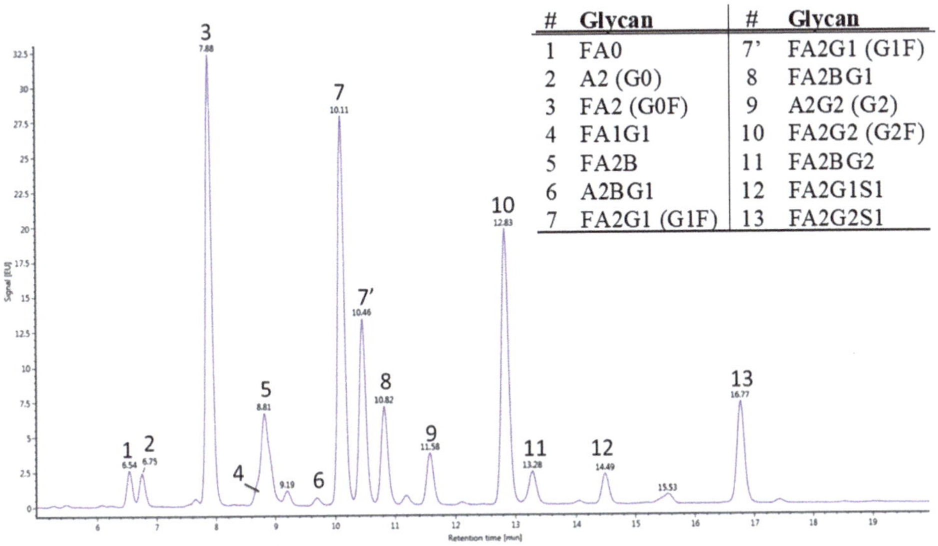


| # | Glycan | # | Glycan |
|---|---|---|---|
| 1 | FA0 | 7' | FA2G1 (G1F) |
| 2 | A2 (G0) | 8 | FA2BG1 |
| 3 | FA2 (G0F) | 9 | A2G2 (G2) |
| 4 | FA1G1 | 10 | FA2G2 (G2F) |
| 5 | FA2B | 11 | FA2BG2 |
| 6 | A2BG1 | 12 | FA2G1S1 |
| 7 | FA2G1 (G1F) | 13 | FA2G2S1 |

**Fig. 4** Fluorescence profile of the Glyko-2-AB-standard and peak annotation

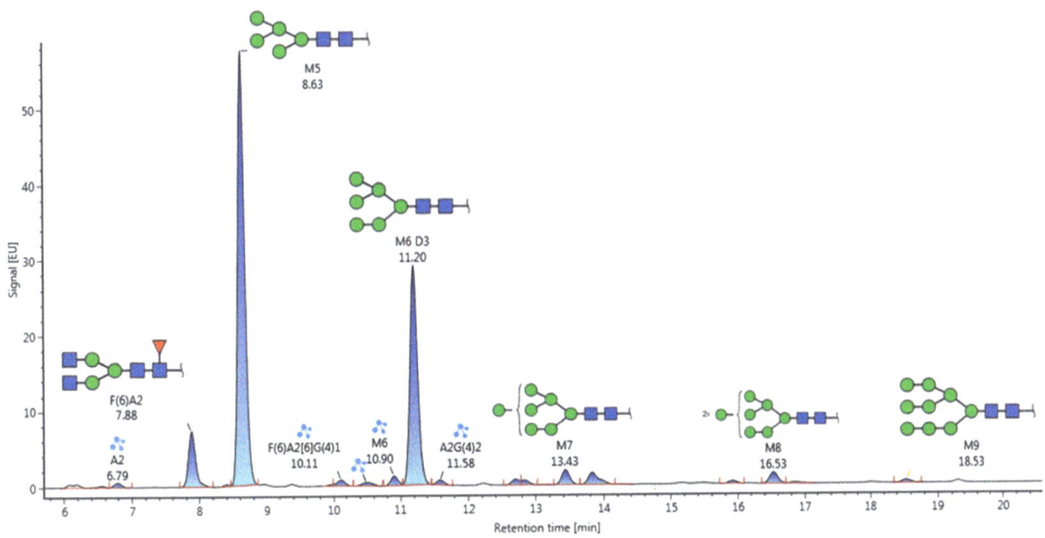


**Fig. 5** Fluorescence profile of Ribonuclease B (rich in mannose) with peak assignment

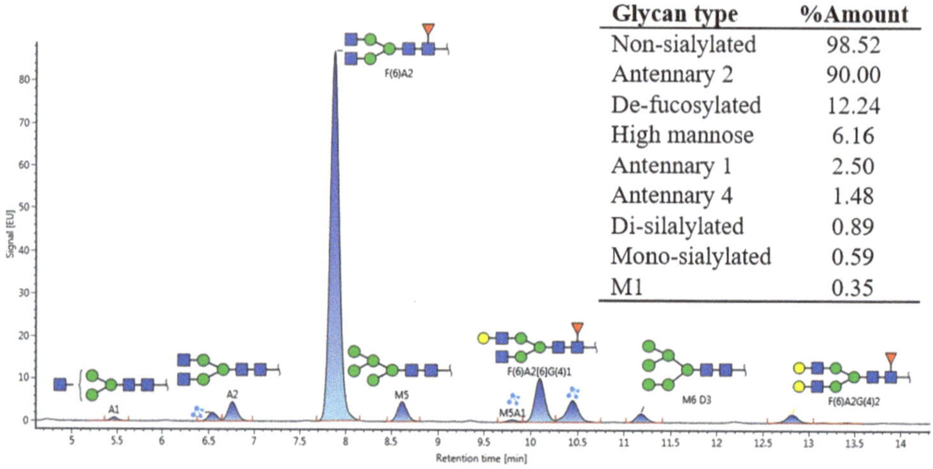


| Glycan type | %Amount |
|---|---|
| Non-sialylated | 98.52 |
| Antennary 2 | 90.00 |
| De-fucosylated | 12.24 |
| High mannose | 6.16 |
| Antennary 1 | 2.50 |
| Antennary 4 | 1.48 |
| Di-silalylated | 0.89 |
| Mono-sialylated | 0.59 |
| M1 | 0.35 |

**Fig. 6** Fluorescence profile of Adalimumab with peak assignment and summary by glycan type

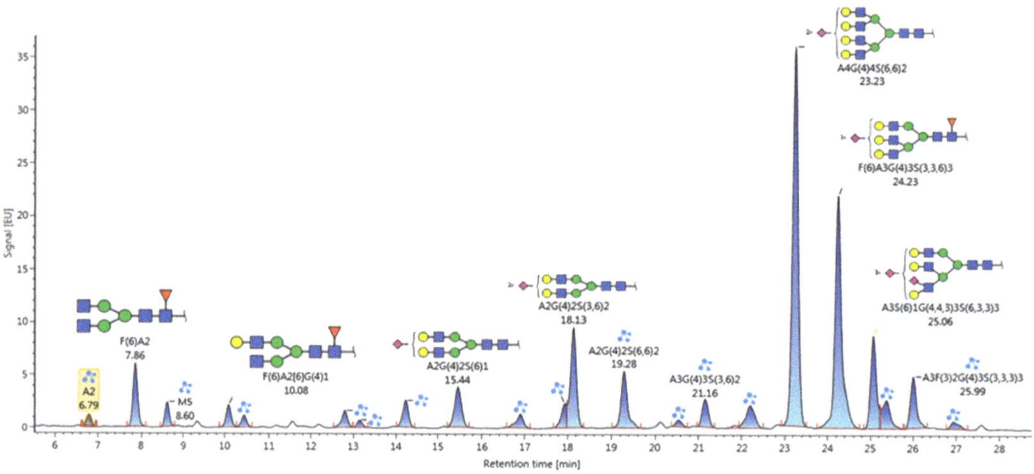


**Fig. 7** Fluorescence profile of Fetuin with peak assignment

## 4 Notes

1. Derivatization reagent cannot be stored more than a week and should be stored away from light.
2. Optimal protein amounts are 50–250 μg and protein concentrations of 1–10 mg/mL. Typical preparation is performed with 50 μg of protein. Protein amount can be scaled up to 250 μg when identification of species in low abundance is needed.
3. In some cases, the one step protocol is not fully efficient to remove some of the glycans, for instance *N*-glycans found on

the Fab of some mAb. The additional step of denaturation of the mAb in the presence of Rapigest as surfactant in the two-step protocol allows for a complete deglycosylation of the mAb. If no denaturant is added, mAb might precipitate.

4. The efficacy of the *N*-glycan release might be affected by molecules present in the buffer. SDS is known to inhibit the PNGase F for instance.
5. From the derivatization step onward, the sample should be protected from light.
6. After addition of acetic acid, a white precipitate may be visible.
7. Elution may be performed with larger volumes ($2 \times 100\ \mu L$) if the expected signal is low (e.g., for low amounts or very heterogeneous profile).
8. A drying step is necessary as the solvent used for elution is not compatible with the starting conditions of HILIC separation. Samples are then dried and resolubilized.
9. Partial desialylation might occur if the temperature in the Speed-vac rises above 28 °C.
10. As an alternative to speed-vac, samples can be freeze-dried.
11. Samples should be directly injected after resolubilization. Dry derivatized glycans can be stored for longer periods at −20 °C.
12. Column should be perfectly equilibrated with at least 20 column volumes as *N*-glycan identification relies on retention times.
13. Before injecting the samples, the gradient should be run three times (no injection) to ensure proper column equilibration.
14. Gain should be adapted to increase signal-to-noise ratio.
15. The calibration of the QTOF is performed by infusion of for instance a solution of NaI (2 mg/mL)/CsI at 2 mg/mL in 50:50 $H_2O$–isopropanol (v/v) on $m/z$ acquisition range 100 to 2500. External multipoint calibration is based on singly charged ions.
16. The dextran ladder (mixture of glucose polymers at different lengths) should be injected twice at the beginning of the sequence and twice at the end of the sequence. It can additionally be injected every 6 injections in duplicate.
17. Before detailed data analysis, the following points should be checked. There should not be significant offset between the fluorescence and MS detection. There should also not be significant retention time drift along the sample set as identification relies on retention time.

## References


1. Reusch D, Tejada ML (2015) Fc glycans of therapeutic antibodies as critical quality attributes. Glycobiology 25:1325–1334. https://doi.org/10.1093/glycob/cwv065
2. Zheng K, Bantog C, Bayer R (2011) The impact of glycosylation on monoclonal antibody conformation and stability. MAbs 3:568–576. https://doi.org/10.4161/mabs.3.6.17922
3. Shoji-Hosaka E, Kobayashi Y, Wakitani M et al (2006) Enhanced Fc-dependent cellular cytotoxicity of Fc fusion proteins derived from TNF receptor II and LFA-3 by fucose removal from asn-linked oligosaccharides. J Biochem 140:777–783. https://doi.org/10.1093/jb/mvj207
4. Goetze AM, Liu YD, Zhang Z et al (2011) High-mannose glycans on the Fc region of therapeutic IgG antibodies increase serum clearance in humans. Glycobiology 21:949–959. https://doi.org/10.1093/glycob/cwr027
5. Delobel A, Cantais F, Catrain A et al (2013) Therapeutic antibody glycosylation analysis: a contract research organization perspective in the frame of batch release or comparability support. Methods Mol Biol 988:115–143

# Chapter 14

# Evaluating *N*-Glycosylation of a Therapeutic Monoclonal Antibody Using UHPLC-FLR-MS with *Rapi*Fluor-MS Labeling

Rosie Upton, James Duffy, Sam Clawson, and David Firth

## Abstract

Released *N*-glycan analysis using the fluorescent label 2-AB (2-aminobenzamide) has been the "gold standard" method for released glycan analysis for several years. The more recent *Rapi*Fluor-MS™ labeling technique, however, offers enhanced mass spectrometric detection of released *N*-glycans, improving the sensitivity and detection limits of the method. The optimized multidimensional detection offers increased confidence in glycan identification which can be further supported by an exoglycosidase digestion array (optional). Here we describe the PNGase F release of *N*-glycans from a typical IgG1 monoclonal antibody (mAb) with subsequent labeling with *Rapi*Fluor-MS™ for detection by HILIC-FLR-MS. The method output quantifies the relative proportion of each glycan species including core afucosylation, sialylation, and high-mannose content, and has a limit of detection (LOD) of 0.01% relative abundance.

**Key words** Antibody, *Rapi*Fluor-MS™ labeling, LC-MS, Glycan profiling

## 1 Introduction

All IgG monoclonal antibodies (mAbs) [1] contain the prerequisite sequon for *N*-glycosylation (-Asn$^{297}$-X-Ser/ Thr-), where X can be any amino acid except proline) [2]. Typically this sequon occurs within the $CH_2$ domain of the Fc region [3]. The dominant glycoforms observed in IgGs expressed from CHO cells are G0F, G1F, and G2F, each of which contain the same trimannosyl core [4]. In biantennary complex oligosaccharides, two GlcNAc residues link this core to any galactose units, denoted G. The F denotes the presence of core fucosylation and for high-mannose species the integer refers to the total number of mannose subunits connected to the chitobiose core (GlcNAc-GlcNAc).

The degree and variety of glycosylation can significantly affect structural conformation, binding affinities and in some cases the mechanism of action, which, in turn, can impact the efficacy and toxicity of the therapeutic, ultimately affecting patient safety. In particular, high-mannose species have been linked to increased

Arnaud Delobel (ed.), *Mass Spectrometry of Glycoproteins: Methods and Protocols*, Methods in Molecular Biology, vol. 2271, https://doi.org/10.1007/978-1-0716-1241-5_14, 

serum clearance rates [5], NGNA sialic acids and α-galactose species can have a negative impact upon immunogenicity [6] and afucosylation (lack of core fucose) has been shown to positively affect binding to immune receptor cells therefore enhancing certain mechanisms of action [7, 8]. These glycan species are typically present in low abundances in mAbs yet can cause inordinately large changes toward effector function activities [9]. N-linked glycosylation is therefore classed as a critical quality attribute (CQA) by pharmaceutical manufacturers and governing bodies who require comprehensive characterization and monitoring throughout both the development and batch release testing stages of any mAb product [10].

The mAb glycan profile can be determined in situ either connected to the intact mAb or to Fc/2 fragments (e.g., following IdeS digestion). Analysis of a glycosylated protein however, can add complexity to the data and reduce the mass accuracy of the experiment, yielding quantitative information only for the most abundant glycan species. A more precise and homogeneous approach is to detach the glycans from the mAb protein, label them to aid detection, and then purify the sample to isolate the labeled released *N*-glycans. One of the most frequently used labels in recent years has been 2-aminobenzamide (2-AB) which binds to the glycan via reductive amination in a 1:1 ratio allowing for simple relative quantification calculations based on fluorescence detection [11]. The main limitation of 2-AB, however, is its poor ionization efficiency which can lead to nonreproducible and low-quality MS data, suppressing the benefit of MS detection. The specifically designed Waters *Rapi*Fluor-MS™ glycan label also binds in a 1:1 ratio and contains a quinoline fluorophore to facilitate fluorescence (FLR) detection and incorporates a tertiary amine group for enhanced ESI+ MS ionization (*see* Fig. 1). The *Rapi*Fluor-MS™ label has therefore been optimized for two-dimensional detection to allow for quantitative FLR and reproducible qualitative MS outputs [12].

Incorporating sensitive MS detection offers significantly improved limits of detection (0.01%), separation and relative

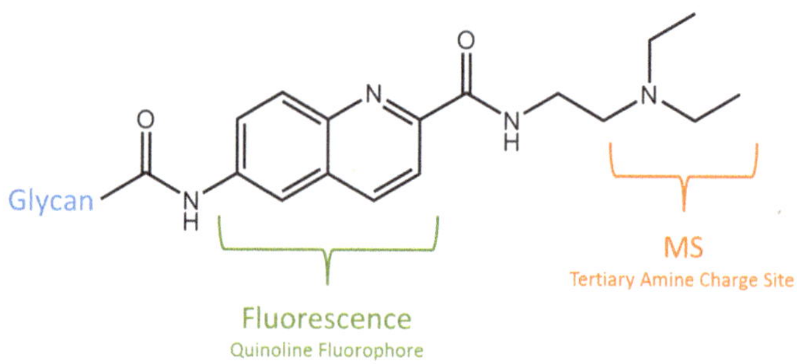


**Fig. 1** Released *N*-glycan structure following *Rapi*Fluor-MS™ derivatization

quantification of fluorescently coeluting species and increased confidence in low-abundance glycan assignments, equivalent to a more sensitive and robust method.

This chapter presents the Waters UHPLC-FLR-MS method utilizing the *Rapi*Fluor-MS™ released N-glycan kit but incorporates manual identification and interpretation of the data. The method involves the enzymatic removal of *N*-glycans from the mAb protein, labeling ready for fluorescent and MS detection and purification to homogenize the sample. Hydrophilic interaction liquid chromatography (HILIC) is used to separate the hydrophilic glycan moieties prior to FLR detection and use of a Dextran Ladder (*see* **Note 1**) provides FLR retention time calibration, specific to glycan analysis, for accurate identification. Detection by MS then offers increased experimental sensitivity and assignment confidence. Due to the presence of isobaric species within the glycan population, mass alone is not always sufficient to confirm identity. This method therefore incorporates an optional exoglycosidase digestion array to sequentially cleave the glycan residues which then offers structural information to aid confidence in certain glycan assignments. This additional step has greater benefit for more complex samples such as Fc-fusion proteins where larger, hybrid glycan species are more prevalent; however, the principal aspects are demonstrated for a typical IgG1 as part of this chapter.

## 2 Materials

Prepare all solutions using ultrapure Milli-Q water (18 MΩ-cm) and LC-MS grade reagents. All reagents should be stored at room temperature and used on the day of preparation (unless indicated otherwise).

### 2.1 Sample Preparation

1. *Rapi*Gest™ surfactant: 5% (w/v) *Rapi*Gest™ SF surfactant. Dissolve a 3 mg vial of *Rapi*Gest™ in 60 μL GlycoWorks Rapid™ buffer. Gently swirl to mix.
2. Deglycosylation enzyme: GlycoWorks™ Rapid PNGase F.
3. *Rapi*Fluor-MS™ labeling solution: Dissolve a 9 mg vial of *Rapi*Fluor™ labeling reagent in 131 μL dimethylformamide (DMF). Vortex briefly.
4. HILIC μElution SPE plate and vacuum manifold block.
5. SPE equilibration solution: 85% acetonitrile in water. Use within 1 month of preparation.
6. SPE sample wash solution: 90% acetonitrile + 0.1% formic acid in water. Use within 1 month of preparation.
7. Elution buffer: 95% ammonium acetate, 5% acetonitrile.
8. Sample diluent: 68% acetonitrile, 32% DMF.

9. Dextran ladder. Dissolve a 50 μg vial of *Rapi*Fluor-MS™ Dextran Ladder in 100 μL Milli-Q water. Subdivide into 15 μL aliquots and store at −80 °C. Use all aliquots within 12 months of preparation.

#### 2.1.1 Exoglycosidase Digestion (Optional)

1. QA-Bio CarboSeq™ N Kit: includes 5× Reaction Buffer, pH 5; α-(1-2,3,6,8,9) sialadase; β-(1-4)-galactosidase; and β-glucosaminidase.
2. Elution buffer and sample diluent (as above).

### 2.2 Mass Spectrometer Setup

1. Analytical column: Acquity UPLC BEH Amide, 130 Å, 1.7 μm, 2.1 × 150 mm.
2. Time-of-flight calibration lockspray (*see* **Note 2**): 200 fmol/μL [$Glu^1$]-Fibrinopeptide B. Firstly prepare a stock Glu-Fib solution (32 pmol/μL). Weigh 0.1 mg Glu-Fib and dissolve in 2 mL of 0.1% acetic acid in 50% $methanol_{(aq)}$. Subdivide into 625 μL aliquots and store at −80 °C. Use within 12 months of preparation. For the 200 fmol/μL final solution, take a 625 μL aliquot and add to 100 mL 0.1% formic acid in 50% $acetonitrile_{(aq)}$. Use within 6 months of preparation.
3. Mobile phase A and wash: 50 mM ammonium formate, pH 4.4. Add 10 mL Waters 5050 mM ammonium formate solution to 1 L of LC-MS grade water (*see* **Note 3**). Mix well and use within 1 month of preparation.
4. Mobile phase B and purge wash: Acetonitrile. Use manufacturer's expiry date.
5. Seal wash: 50% acetonitrile in water. Use within 3 months of preparation.

## 3 Methods

Carry out all procedures at room temperature unless otherwise specified.

### 3.1 Sample Denaturation and Deglycosylation

1. Set a heat block to at least 90 °C and a second heat block to 50 ± 2 °C.
2. Dilute the mAb sample to 2 mg/mL with Milli-Q water and vortex-mix.
3. In an Eppendorf tube, add 15.3 μL of Milli-Q water and 6 μL of 5% *Rapi*Gest™ to 7.5 μL (15 μg) of sample. (For a control blank combine 22.8 μL of Milli-Q water with 6 μL of 5% *Rapi*Gest™ and prepare in the same way as the protein samples).

4. Vortex-mix and heat at a minimum of 90 °C for 3 min in the heating block.
5. Allow the samples to cool to room temperature for 5 min and then centrifuge briefly.
6. Add 1.2 μL Rapid™ PNGase F and vortex-mix.
7. Heat the samples at 50 ± 2 °C for 5 min in a heating block.
8. Allow samples to cool to room temperature for 5 min and then centrifuge briefly.

### 3.2 Glycan Labeling

1. To the deglycosylated sample add 12 μL of the *Rapi*Fluor-MS™ labeling solution and vortex-mix.
2. Allow labeling to occur at room temperature for 5 min.
3. Dilute the labeled samples with 358 μL acetonitrile and vortex-mix.

### 3.3 SPE Cleanup

1. Place the HILIC μElution plate over a waste collection plate and condition the wells with 200 μL Milli-Q water. Place onto the vacuum block and pass through using minimal pressure (*see* **Note 4**).
2. Equilibrate the wells with 200 μL of 85% acetonitrile. Place onto the vacuum block and pass through using minimal pressure.
3. Load the labeled samples (~400 μL) into the conditioned wells and pass through using minimal pressure.
4. Wash each well with 2 × 600 μL 90% acetonitrile + 1% formic acid and pass through using minimal pressure.
5. Swap the waste collection plate for a 1 mL sample collection plate and elute the labeled glycans using 3 × 30 μL Elution Buffer using minimal pressure.
6. Dilute the eluted glycans with 310 μL Sample Diluent and mix by aspiration. (For optional exoglycosidase digestion do not perform this step and move to Subheading 3.4).
7. Transfer 200 μL of each sample to a UPLC vial and place in the UPLC autosampler (set to 10 °C) ready for injection. The labeled glycans can be stored for 72 h at 2–15 °C before expiry.

### 3.4 Exoglycosidase Digestion (Optional)

For the following enzymatic digestions an incubator will need to be set to 37 ± 2 °C.

1. Following Subheading 3.3, **step 5**, centrifuge and split the labeled glycan sample into 4 × 15 μL aliquots.
2. To one of the four aliquots add 52 μL Sample Diluent, vortex-mix, transfer to a UPLC vial and submit to the autosampler ready for injection.

3. To the remaining three aliquots, add 2 μL Reaction Buffer and 2 μL Sialidase. Cap one of the three aliquots, vortex-mix and incubate at 37 ± 2 °C overnight (*see* **Note 5**).
4. To the remaining two aliquots add 2 μL β-Galactosidase. Cap one aliquot, vortex-mix and incubate at 37 ± 2 °C overnight (*see* **Note 5**).
5. To the remaining aliquot add 2 μL β-glucosaminidase. Cap, vortex-mix, and incubate at 37 ± 2 °C overnight (*see* **Note 5**).
6. After overnight incubation, evaporate the exoglycosidase digested glycan samples to dryness using a centrifugal evaporator (lamp off, aqueous setting) for approximately 90 min.
7. Reconstitute the evaporated samples with 15 μL Elution Buffer and gently vortex until the glycan sample (any visible pellet) is dissolved.
8. Dilute the dissolved glycans with 52 μL sample diluent, vortex-mix briefly, and transfer to the UPLC vial before placing the samples in the UPLC autosampler. Analysis of samples on the day of preparation is recommended (*see* **Note 6**).

### 3.5 LC-FLR-MS Analysis

The method utilizes a Waters Vion Mass Spectrometer coupled to an Acquity H-Class Bio UPLC system with Fluorescence detection. UNIFI™ v1.8 controls the system, the acquisition of raw data, and the processing of acquired data.

1. Connect the solvent lines to the corresponding mobile phases and reagents (*see* Table 1). Purge all lines for a minimum of 5 min.
2. Connect the analytical column detailed in Subheading 2.2, **item 1**. Flush the column with a mobile phase ratio of 25:75 (A:B) at a flow rate of 0.4 mL/min for a minimum of 10 min.
3. Set up a Glycan analysis method within UNIFI™ to include the critical parameters detailed in Table 2. All other parameters were left as default settings.

**Table 1**
**Summary of the LC-MS mobile phases and wash solutions**

| Solvent line | Composition |
|---|---|
| Mobile phase A | 50 mM ammonium formate, pH 4.4 |
| Mobile phase B | Acetonitrile |
| Wash | 50 mM ammonium formate, pH 4.4 |
| Purge wash | Acetonitrile |
| Seal wash | 50% $acetonitrile_{(aq)}$ |

**Table 2**
**Critical parameters required for the LC-MS analysis of released *N*-glycans**

<table>
<tr><td>HILIC Gradient</td><td>
<table>
<tr><th>Time (min)</th><th>Flow Rate (mL/ min)</th><th>Composition A (%)</th><th>Composition B (%)</th></tr>
<tr><td>0.00</td><td>0.400</td><td>25.0</td><td>75.0</td></tr>
<tr><td>35.00</td><td>0.400</td><td>46.0</td><td>54.0</td></tr>
<tr><td>36.50</td><td>0.200</td><td>100.0</td><td>0.0</td></tr>
<tr><td>39.50</td><td>0.200</td><td>100.0</td><td>0.0</td></tr>
<tr><td>43.10</td><td>0.200</td><td>25.0</td><td>75.0</td></tr>
<tr><td>47.60</td><td>0.400</td><td>25.0</td><td>75.0</td></tr>
<tr><td>60.00</td><td>0.400</td><td>25.0</td><td>75.0</td></tr>
</table>
</td></tr>
<tr><td>Autosampler Temperature</td><td>10°C</td></tr>
<tr><td>Column Temperature</td><td>60°C</td></tr>
<tr><td>FLR Excitation Wavelength</td><td>265 nm</td></tr>
<tr><td>FLR Emission Wavelength</td><td>425 nm</td></tr>
<tr><td>Mass Analyser Mode</td><td>Sensitivity</td></tr>
<tr><td>Capillary Voltage</td><td>3 kV</td></tr>
<tr><td>Cone Voltage</td><td>80 V</td></tr>
<tr><td>Source Temperature</td><td>120°C</td></tr>
<tr><td>Desolvation Gas Temperature</td><td>350°C</td></tr>
<tr><td>Analysis Run Time</td><td>60 minutes</td></tr>
<tr><td>MS Scan Range</td><td>m/z 500-2500</td></tr>
<tr><td>Lock Mass</td><td>m/z 785.8421</td></tr>
</table>

4. Set up a sample set within UNIFI™. Include a column conditioning and blank injection (*see* **Notes 7** and **8**, respectively) first and then inject 1 μL of Dextran Ladder reference standard (*see* **Note 1**) before and after the released *N*-glycan samples. Define the dextran ladder as a standard in the sample set. Blank and sample injections are 30 μL.
5. Post analysis, flush the column with 50:50 (A:B) at 0.2 mL/ min for a minimum of 10 min prior to storage.

### *3.6 Released N-Glycan Analysis*

The experimental output from this method consists of a total ion chromatogram (TIC), a base peak intensity (BPI) trace, and a fluorescence (FLR) spectrum per sample.

**Table 3**
**Typical integration parameters (find 2D peak setting in UNIFI)**

| Parameter | Value |
|---|---|
| Integration window | 7–30 min |
| Peak width | 0.14 |
| Lift off | 0% |
| Touchdown | 0% |
| Peak rejection | Min area = 10,000 |

1. Make a note of the dextran ladder retention times from the FLR trace for each peak and edit the method to include the experimentally determined values (*see* **Note 9**).
2. Integrate the chromatograms using the parameters outlined in Table 3 (*see* **Note 10**).
3. UNIFI will generate a table which includes glycan assignment, peak area, relative percentages, GU values, and retention times. Some assignments will have mass information too based upon the glycan library; however, the standard approach within UNIFI is to assign glycans based on GU values-not mass (*see* **Note 11**).
4. Export the UNIFI results table into Excel and add a column for observed $m/z$.
5. Extract the mass spectrum from each of the TIC peaks identified by UNIFI and make a note of all species observed at that retention time in the observed $m/z$ column (*see* **Note 12**).
6. Using the Excel version of the Waters library assign the detected masses to expected glycan species. For glycan species not present in the Waters library a manual assignment will need to be calculated (*see* **Note 13**).
7. The final Excel table consists of all identified glycan assignments with their respective retention time, FLR peak area, relative FLR peak area (%) (*see* **Note 14**), GU value, $m/z$, and mass. Figure 2 shows an annotated FLR trace based upon the glycan assignments.

### 3.7 Optional Exoglycosidase Digestion

The main benefit from performing the additional exoglycosidase array is an increased confidence in *N*-glycan assignments. Due to the high number of structural variants, particularly for the larger glycans, mass alone is often not sufficient to confidently define the specific glycan species. The use of enzymes to sequentially cleave residues from the *Rapi*Fluor-MS™ labeled oligosaccharides can provide a visual representation of how the glycans are structured.

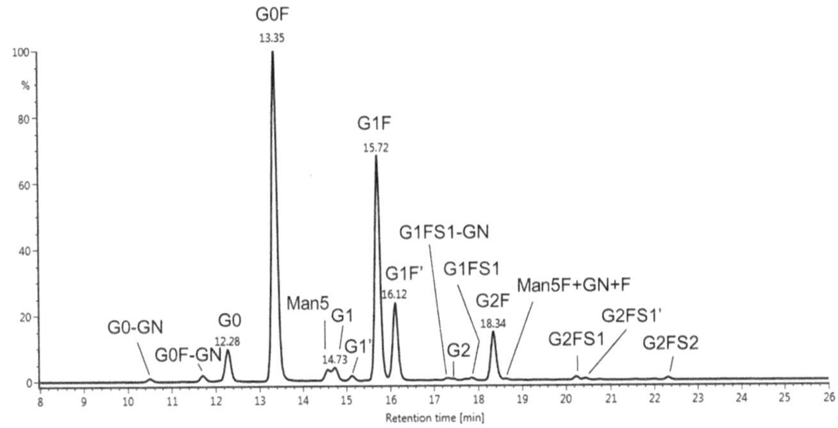


**Fig. 2** HILIC-FLR trace representing the *Rapi*Fluor-MS™ labeled *N*-glycans released from a typical IgG1 mAb

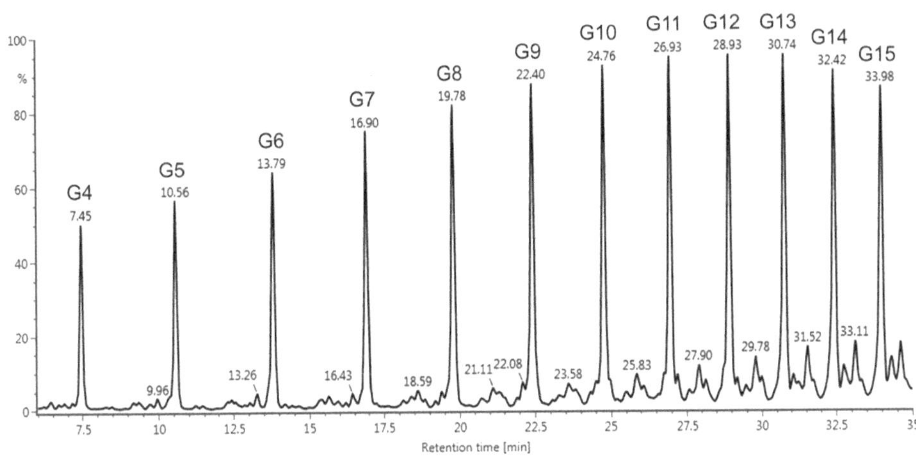


**Fig. 3** An example *Rapi*Fluor-MS™ dextran ladder HILIC-FLR chromatogram

1. Process the data in the same way as detailed in **steps 1–7**.
2. Compare the results from each digestion (*see* **Note 15**).

## 4 Notes

1. The dextran ladder consists of increasing glucose residues connected to one another (G1, G2, G3, etc.) which are fluorescently labeled. It is designed to be run under a specific HILIC gradient which gives sequential separation (*see* Fig. 3) according to the number of glucose units; producing a retention time calibration curve. This serves to compensate for inherent interinjection retention time drift and variability

between HILIC assays and LC systems. The Dextran Ladder is coupled to a glycan library supplied by Waters whereby specific GU (Glucose Unit) values are assigned to specific glycans which can help in glycan identification. However, if an alternative HILIC gradient is required to improve separation (for example) then the Dextran Ladder library is not compatible.

2. The mass correcting lockspray contains the peptide [Glu′]-fibrinopeptide B ($m/z$ 785.8421, 2+) which is infused simultaneously alongside the sample. At systematic intervals the baffle position in the source region of the mass spectrometer switches to block sample infusion and enable Glu-Fib infusion; such that both can be recorded within a single analysis. The lock mass acts as an internal calibrant to correct for any time-of-flight drift, resulting in improved mass accuracy.
3. If preparing the ammonium formate solution from scratch, it is advisable to avoid adjusting the pH with hydrochloric acid as this can cause chlorine adducts, which affect the sensitivity of the MS dimension of the assay; formic acid is a better alternative modifier for MS reagents.
4. If the samples are passed through the solid phase too quickly then the recovery rate could be affected, but applying no pressure and relying on gravity would significantly increase the analysis time. Applying a small amount of pressure therefore provides sufficient interaction with the extraction phase and keeps sample preparation time down. Do not cover any wells of the SPE plate that are not being used as this will cause a pressure build up and will draw the sample through too quickly.
5. Overnight digestion is recommended by the supplier. Shorter incubation times may be successful but need to be optimized on a case-by-case basis.
6. Evaporation of the organic sample diluent has been observed for samples stored in the LC autosampler for more than a day which can affect chromatographic peak shape and separation.
7. Conditioning the UHPLC system and HILIC column prior to the analytical run is best achieved by injection of a *Rapi*Fluor-MS™ labeled glycan sample. From experience, an equilibration injection of a control blank is less effective.
8. For released *N*-glycan analysis, the MS range covers $m/z$ 500–2500. Analysis of a blank is essential to allow for the comparison of instrument/setup related background noise with peaks of interest. A background subtraction can be performed within UNIFI which can clarify the data.
9. Retention time shifts can affect the GU values so updating the expected retention times with the experimentally determined values during each run ensures that the GU retention time

calibration is accurate. Typically, values for G4-G13 cover the retention time range of the eluting glycans.

10. Although standard software parameters integrate the majority of the chromatographic peaks for mAbs, it is worth checking that the automatic processing is correct. In particular, shoulders and smaller peaks can be missed and raised baselines can result in misplaced dropdowns. The peak rejection parameter may need to be altered depending upon the overall response of the data or depending on specific experimental thresholds. Selected parameters and any amendments to the automatic processing need to be kept consistent across repeat analyses to ensure a true experimental comparison. The data will also need to be reprocessed to account for any manual changes—ensure "keep manual changes" is selected when reprocessing.
11. A current limitation of relying on the glycan library within UNIFI is that only a single assignment can be made per glycan. To overcome this and to utilize the extensive MS data available, use the Investigate mode within UNIFI to manually search for and identify the glycans associated with each FLR/TIC peak based on mass.
12. Some glycan species coelute under a single FLR peak. For example, G1 and Man5 regularly coelute under the standard gradient conditions used; however, they are easily separated by mass ($m/z$ 895.86 and 773.81, respectively). The intensities of the MS signals for coeluting glycan species can be used to estimate the relative proportion of each, with respect to the overall area of the FLR peak. Ammonium, sodium, and potassium adducts are frequently observed alongside individual glycan species, particularly the larger ones. These help support the original glycan assignment but should not be mistaken for additional glycan species, even if the adduct variant is more abundant than the protonated glycan-as can be the case for Man8, for example. Figure 4 illustrates the most common adduct species observed via this method. In-source fragment ions may also be detected. G1F, for example, might be present as both a source ion of G1F+GN (~0.2%) as well as an abundant chromatographically resolved glycan species (~30–40%). The GU value will help differentiate these, given that a smaller glycan will typically elute from the column at an earlier retention time, whereas an in-source fragment ion will be detected at the same time as the precursor ion (*see* Fig. 5).
13. For detected signals that do not match components of the Waters library, a GU value will not be available. A good starting point for manual assignment is to highlight any patterns within the positively assigned glycans (e.g., frequent loss of GN or

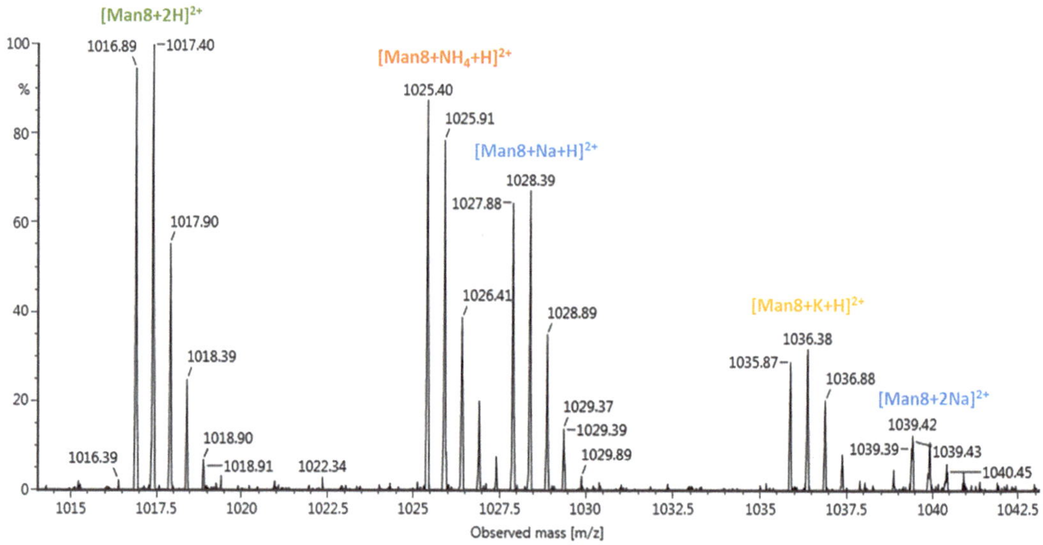


**Fig. 4** A typical mass spectrum for the Man8 glycan species highlighting the most commonly observed method specific adducts

sequential high mannose species) and to perform systematic additions or subtractions of different monosaccharide subunits until the mass matches that of the unknown. Occasionally a detected species will remain unidentified; however, an exoglycosidase digestion array may help with assignment. The Waters library is not exhaustive but can be added to manually over time based on acquired data.

14. Once all assignments (if possible) have been made, use Excel to calculate the % relative abundance of glycans within the population by dividing each response by the sum of all responses. Typically the most confident glycan assignments are made for species at >0.05% relative abundance. Anything below this threshold, that has not been identified based on mass, might be removed from the spreadsheet; Excel will update the calculations.

15. Following sialidase digestion, all sialylated *N*-glycans should be converted to their nonsialylated equivalents, provided that the sialic acid residue is directly connected to a terminal galactose residue; for example, G1FS1-GN and G2FS2 become G1F-GN and G2F, respectively. Similar changes are observed following combined sialidase, galactosidase, and glucosaminase digestions until only high mannose species remain undigested. An example of where this is particularly helpful is the case of Man4F+2GN and G1F', which are isobaric and coelute. Following galactosidase treatment G1F' would be converted to

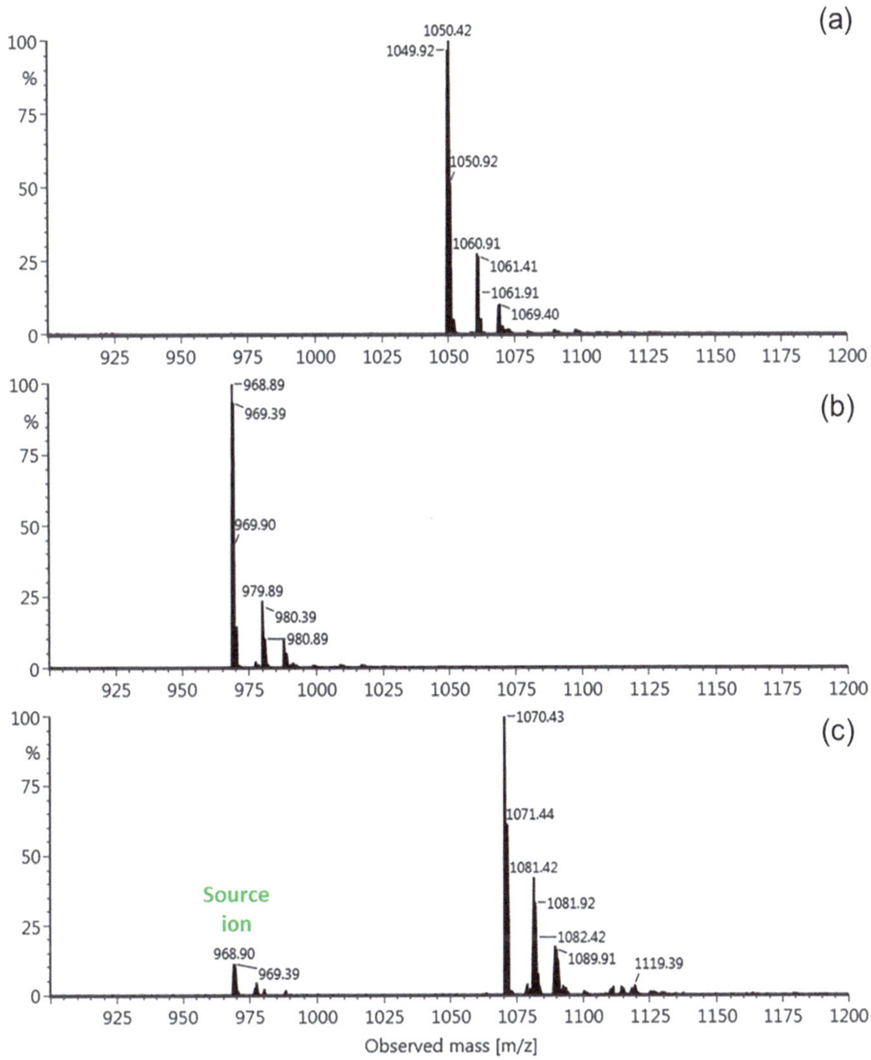


**Fig. 5** Extracted mass spectra for (**a**) G2F $[1049.92]^{2+}$, (**b**) G1F $[968.89]^{2+}$, and (**c**) G1F+GN $[1070.39]^{2+}$. The signal consistent with G1F observed in mass spectrum (**c**) is a source ion due to the same retention time as G1F+GN and results from the loss of GN. Each species has associated adducts to the right side of the main peak, as detailed in Fig. 4

G0F with Man4F+2GN remaining unaffected. Subsequent glucosaminase digestion would then convert G0F to Man3F, along with the majority of other glycan species, whereas Man4F+2GN would increase the observed signal for Man4F. Figure 6 summarizes an example exoglycosidase data set which demonstrates the structural conversions following each sequential cleavage.

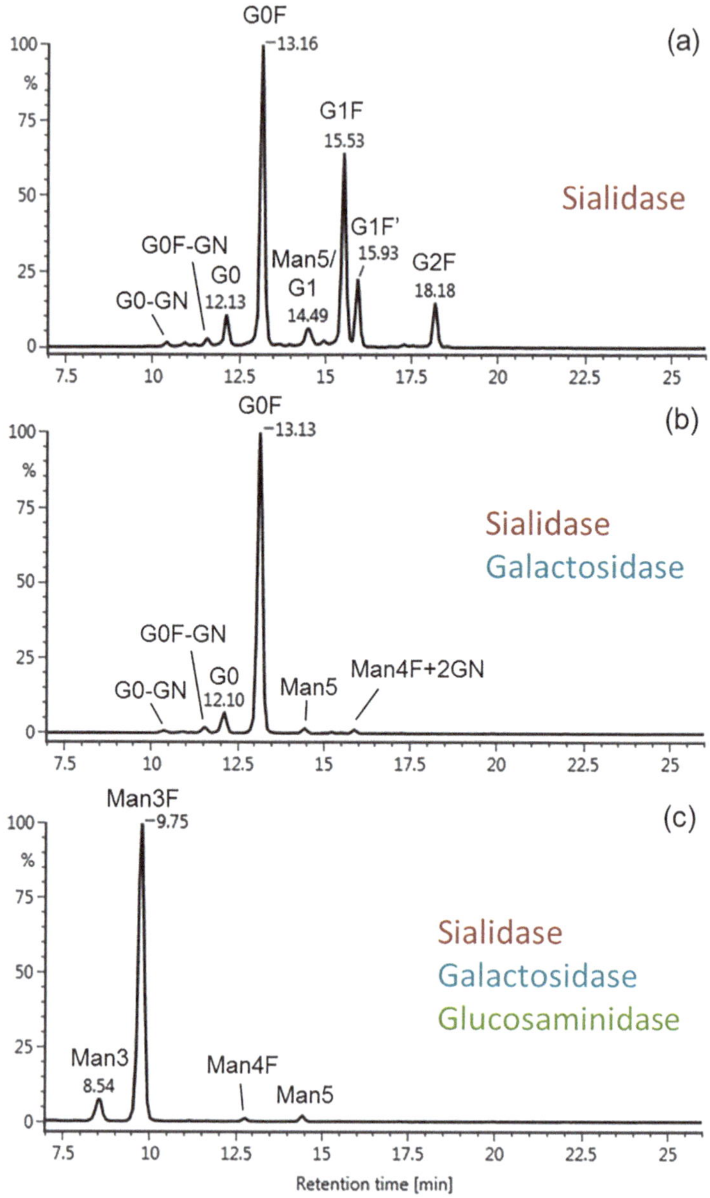


**Fig. 6** HILIC-FLR chromatograms representing the exoglycosidase array digestion of *Rapi*Fluor-MS™ labeled *N*-glycans released from a typical IgG1 mAb

## References

1. Leymarie N, Zaia J (2012) Effective use of mass spectrometry for glycan and glycopeptide structural analysis. Anal Chem 84:3040–3048. https://doi.org/10.1021/ac3000573
2. Yamane-Ohnuki N, Satoh M (2009) Production of therapeutic antibodies with controlled fucosylation. MAbs 1:230–236. https://doi.org/10.4161/mabs.1.3.8328
3. Jefferis R (2005) Glycosylation of recombinant antibody therapeutics. Biotechnol Prog 21:11–16. https://doi.org/10.1021/bp040016j

4. Beck A, Wagner-rousset E, Ayoub D et al (2013) Characterization of therapeutic antibodies and related products. Anal Chem 85:715–736
5. Goetze AM, Liu YD, Zhang Z et al (2011) High-mannose glycans on the Fc region of therapeutic IgG antibodies increase serum clearance in humans. Glycobiology 21:949–959. https://doi.org/10.1093/glycob/cwr027
6. Reusch D, Tejada ML (2015) Fc glycans of therapeutic antibodies as critical quality attributes. Glycobiology 25:1325–1334. https://doi.org/10.1093/glycob/cwv065
7. Okazaki A, Shoji-Hosaka E, Nakamura K et al (2004) Fucose depletion from human IgG1 oligosaccharide enhances binding enthalpy and association rate between IgG1 and FcγRIIIa. J Mol Biol 336:1239–1249. https://doi.org/10.1016/j.jmb.2004.01.007
8. Shields RL, Lai J, Keck R et al (2002) Lack of fucose on human IgG1 N-linked oligosaccharide improves binding to human FcGammaR-III and antibody-dependent cellular toxicity. J Biol Chem 277:26733–26740. https://doi.org/10.1074/jbc.M202069200
9. Fernandes D (2005) Demonstrating comparability of antibody glycosylation during biomanufacturing. Eur Biopharm Rev:106–110
10. Zhang P, Woen S, Wang T et al (2016) Challenges of glycosylation analysis and control: an integrated approach to producing optimal and consistent therapeutic drugs. Drug Discov Today 21:740–765. https://doi.org/10.1016/j.drudis.2016.01.006
11. Keser T, Pavic T, Lauc G, Gornik O (2018) Comparison of 2-aminobenzamide, procainamide and RapiFluor-MS as derivatizing agents for high-throughput HILIC-UPLC-FLR-MS N-glycan analysis. Front Chem 6:1–12. https://doi.org/10.3389/fchem.2018.00324
12. Lauber MA, Yu Y-Q, Brousmiche DW et al (2015) Rapid preparation of released N-glycans for HILIC analysis using a labeling reagent that facilitates sensitive fluorescence and ESI-MS detection. Anal Chem 87:5401–5409. https://doi.org/10.1021/acs.analchem.5b00758

# Chapter 15

# Routine Analysis of *N*-Glycans Using Liquid Chromatography Coupled to Routine Mass Detection

Ximo Zhang, Vithiya Vimalraj, and Mitul Patel

## Abstract

Analysis of *N*-glycans are commonly conducted via enzymatic release, labeling, and liquid chromatography (LC) separation and fluorescent detection. Mass spectrometry (MS) has been increasingly used as an orthogonal detection method to provide additional structural information and increase the confidence of *N*-glycan analysis. In this chapter, we describe a method to perform routine analysis of *N*-glycans including the sample preparation with a signal-enhancement label, LC-MS data generation, and data analysis. Using this method, up to 24 *N*-glycan samples can be prepared at one time and analyzed by LC-MS. With the addition of automation platform, up to 96 *N*-glycan samples can be prepared and analyzed in a high-throughput manner.

**Key words** LC-MS, *N*-glycan analysis, Routine mass detection, High-throughput analysis, Automation, Critical quality attribute, Cell line development, Process development, Analytical characterization

## 1 Introduction

*N*-glycosylation is a complex posttranslational modification of proteins and plays a critical role in protein folding, solubility, stability, pharmacokinetics, pharmacodynamics, and efficacy [1, 2]. Glycan profile differences in galactosylation, fucosylation and sialylation are now well studied as these specific glycans can influence the differential IgG functions which range from activating complement and triggering antibody-dependent cellular cytotoxicity to inhibitory or anti-inflammatory properties [3]. In addition, variation in glycoform can control the potency and clearance of therapeutic antibodies and stimulate immunogenic responses [4, 5]. Therefore, glycosylation is one of the critical quality attributes and is of major interest in both academic research and in biotherapeutic process development [6]. Regulatory authorities strongly encourage innovators in monitoring the type of glycan species, position, and level of sialylation [7]. They are also increasing the demands placed upon manufacturers to demonstrate how process can affect the glycan

Arnaud Delobel (ed.), *Mass Spectrometry of Glycoproteins: Methods and Protocols*, Methods in Molecular Biology, vol. 2271, https://doi.org/10.1007/978-1-0716-1241-5_15, 

composition and mandate human-friendly and consistent glycosylation to be retained within stringent limits to ensure safety and efficacy [1, 8].

Comprehensive *N*-glycan analysis is extensively performed in the biopharmaceutical industry to aid the final cell line selection, process development, and selection of the manufacturing process. There are several state-of-the-art analytical technologies for the identification and relative quantification of *N*-glycans. The most commonly used analytical approaches for the structural elucidation of carbohydrates derived from glycoproteins include liquid chromatography (LC), capillary electrophoresis (CE), nuclear magnetic resonance spectroscopy (NMR) and mass spectrometry (MS) [9]. Structural elucidation of *N*-glycans is demanding due to its inherent high diversity and complexity and usually a blend of multiple methods is used to provide the required information [10]. Released glycans are derivatized in most of the approaches to accelerate their separation, detection (fluorescence or MS) and improve sensitivity. There are various derivatization strategies that are applied and reductive amination, hydrazide labeling and permethylation are some of the widely employed labeling steps [11] Labeling compounds that are routinely used for reductive amination reaction are 2-aminobenzamide (2-AB), 2-aminobenzoic acid (2-AA) and 1-aminopyrene-3,6,8-trisulfonic acid (APTS) [11]. Normal Phase-HPLC has previously been used for relative quantification of 2-AB or 2-AA-labeled released *N*-glycans as a golden standard method. In this chapter, the methodology and application of a robust and sensitive *N*-glycan workflow that contains important attributes such as rapid tagging of reactive group with efficient fluorophore and enhanced ionization MS detection is described. Both manual and high-throughput aspects of this *N*-glycan method which employs hydrophilic interaction chromatography (HILIC) separation are covered.

## 2 Material

This section summarizes the materials needed for preparing 8 *N*-glycan samples released from monoclonal antibody. For 16 or 24 samples, scale up the volume of reagents and buffers accordingly.

### 2.1 Material for Sample Preparation of Released N-Glycan

1. Heat blocks: two separate modular heat blocks for 1 mL tubes with ability to heat up to more than 90 °C (*see* **Note 1**).
2. SPE equipment: a 96-well SPE extraction vacuum manifold, a SPE extraction pump, and a vacuum manifold shims (*see* **Note 2**).

3. Pipettes (1–10μL, 10–100μL, 20–200μL, and 100–1000μL capacity) and tips (1000, 20, 10μL).
4. Vortex.
5. SpeedVac (optional).
6. The GlycoWorks *Rapi*Fluor-MS *N*-Glycan kit for 24 samples: the kit contains the reagents and most materials needed for releasing and labeling *N*-glycans, including 3 vials of 3 mg/vial *Rapi*Gest surfactant, 250μL GlycoWorks rapid buffer (250 mM HEPES buffer, pH = 7.9), 1 vial of 35μL PNGase F, 3 vials of 9 mg/vial *Rapi*Fluor-MS labeling reagent powder, 3 vials of 1 mL anhydrous dimethylformamide (DMF), a 96-well HILIC μElution plate, waste tray, a 96-well sample collection plate, SPE elution buffer (200 mM ammonium formate in 5% acetonitrile) sample diluent (10:21 (v/v) DMF–acetonitrile), 1 mL reaction tubes, 600μL tapered bottom tubes, and caps (*see* **Note 3**).
7. Glycoprotein samples: 1.5μg/μL monoclonal antibody. Use 10μL (15μg) for each reaction (*see* **Note 4**).

### 2.2 LC-MS of Released N-Glycans

1. LC-MS system: HPLC or UPLC system with a fluorescence detector (FLR) and a mass detector (*see* **Note 5**).
2. Analytical column: Waters ACQUITY Glycan BEH Amide column, 2.1 × 150 mm, 1.7μm, 130 Å (*see* **Note 6**). For high throughput analysis: Waters ACQUITY Glycan BEH Amide column, 2.1 × 50 mm, 1.7μm, 130 Å.
3. UPLC vials and caps (*see* **Note 7**).
4. Solvent container: 4 1-L borosilicate lab-approved solvent containers (*see* **Note 8**).
5. Computer with informatic platform installed for instrument control and data analysis (*see* **Note 9**).
6. Mobile phase A: 50 mM ammonium formate in LC-MS grade water (pH = 4.4). Add 10 mL of Waters ammonium formate concentrate in the solvent container and add water to 1 L. Store this solution at room temperature and use within 1 week (*see* **Note 10**).
7. Mobile phase B: LC-MS grade acetonitrile. Add 1 L acetonitrile in the solvent container. Store this solution at room temperature and use within 6 months (*see* **Note 10**).
8. LC-MS grade 30:70 (v/v) water–acetonitrile (seal wash, strong and weak sample manager wash). Add 700 mL acetonitrile in the solvent container, then add water to 1 L. Sonicate the solution for 10 min to completely mix the solvents. Store this solution and use within 8 weeks (*see* **Note 10**).

## 3 Method

### 3.1 Sample Preparation of Released N-Glycans

*N*-glycans can be released from the glycoprotein and labeled with signal-enhancement tag within 1 h by following the procedures in the GlycoWorks *N*-Glycan kit protocol [20]. As shown in Fig. 1, sample preparation includes 4 steps: (1) Denaturation of glycoprotein; (2) Deglycosylation using PNGase F; (3) *N*-glycans labeling with *Rapi*Fluor-MS reagent for enhanced fluorescence and MS signal; (4) purification using a SPE μElution plate in HILIC mode.

1. Prepare RapiGest buffer solution: reconstitute the 1 vial (3 mg) of RapiGest surfactant in 60μL of GlycoWorks Rapid buffer and 40μL of 18.2 MΩ water. Vortex the content to mix (*see* **Note 11**).
2. Add 10μL 1.5μg/μL mAb sample to 1 mL reaction tube, then add 10μL GlycoWorks RapiGest buffer solution. Aspirate and

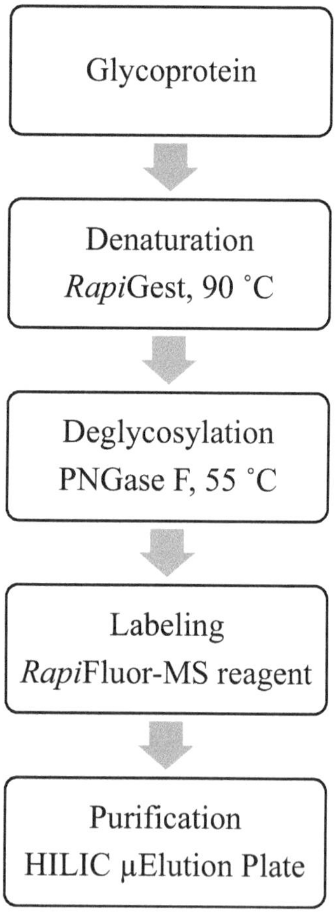


**Fig. 1** Workflow for *N*-glycan releasing and labeling using the *Rapi*Fluor-MS *N*-glycan kit

dispense well. Heat the mixture for 3 min using a heat block at 90 °C for protein denaturation (*see* **Note 12**).

3. Remove the 1 mL tube from heat block and cool for 3 min.
4. Prepare PNGase F solution: add 220μL 18.2 MΩ water to the 1 vial (35 mL) of PNGase F.
5. Add 10μL PNGase F solution to the 1 mL reaction tube, aspirate and dispense to mix. Incubate this mixture at 50 °C for 5 min. Then remove the 1 mL tube from the heat block and cool at room temperature for 3 min.
6. Dissolve 9 mg of *Rapi*Fluor-MS reagent in 110μL of anhydrous DMF. Aspirate and dispense the solution or vortex to mix. Add 10μL of *Rapi*Fluor-MS solution to the deglycosylated mixture in the 1 mL tube. Aspirate and dispense to mix.
7. Allow the labeling reaction to proceed at room temperature for 5 min. Then dilute the reaction with 360μL acetonitrile (*see* **Note 13**).
8. Set up a GlycoWorks HILIC μElution Plate on a vacuum manifold. Place a waste tray in the manifold to collect the flow through.
9. Condition and equilibrate the reaction wells to be used on the μElution plate with 200μL of 18.2 MΩ water and 200μL of 15:85 water–acetonitrile, respectively.
10. Load the 400μL acetonitrile diluted samples. Wash the well twice with 600μL of 1:9:90 (v/v/v) formic acid–water–acetonitrile each time.
11. Empty the waste tray. Place a 96-well collection plate with 600μL tapered bottom tubes in the vacuum manifold to collect the flow through glycans.
12. Elute glycans with 30μL of GlycoWorks SPE elution buffer three times (*see* **Note 14**).
13. Dilute the 90μL eluate with 310μL sample diluent in the GlycoWorks *Rapi*Fluor-MS *N*-glycan kit (*see* **Note 15**). Cap the tapered bottom tube using the provided cap mats, store at 4 °C and use within 1 week (*see* **Note 16**).

### 3.2 Routine LC-MS Analysis of Released N-Glycans

#### 3.2.1 LC-FLR-MS Data Generation

1. Connect the UPLC lines to the mobiles phases A and B, and to seal wash, strong and weak sample manager wash (*see* **Note 17**). Prime mobile phase A and B line for 5 min, respectively. Prime seal wash for 5 min. Wash sample manager syringe and needle for 1 min and 15 cycles (*see* **Note 18**).
2. Connect the column to the UPLC instrument. At 60 °C, wash the column at 0.2 mL/min flow rate using 50:50 mobile phase A–B for 30 min (*see* **Note 19**).
3. Configure instrument and set instrument method based on Table 1

4. Set up gradient elution method according to Table 2 (*see* **Note 20**).
5. Set up QDa and fluorescence detectors with the parameters in Table 3 (*see* **Note 21**).

**Table 1**
**Analytical conditions**

| | |
|---|---|
| Sample loop volume | 10μL |
| Syringe volume | 100μL |
| Injection volume | 10μL |
| Column temperature | 60 °C |
| Sample temperature | 6 °C |

**Table 2**
**Chromatographic gradient**

| Time (min) | Flow rate (min) | % A | % B |
|---|---|---|---|
| Initial | 0.400 | 25.0 | 75.0 |
| 35.00 | 0.400 | 46.0 | 54.0 |
| 36.50 | 0.200 | 80.0 | 20.0 |
| 39.50 | 0.200 | 80.0 | 20.0 |
| 43.10 | 0.200 | 25.0 | 75.0 |
| 47.60 | 0.400 | 25.0 | 75.0 |
| 55.00 | 0.400 | 25.0 | 75.0 |

**Table 3**
**Fluorescence and QDa detection parameters**

| | | |
|---|---|---|
| FLR detector | Excitation wavelength | 265 nm |
| | Emission wavelength | 425 nm |
| QDa mass detector | Mass range | 500–1250 *m/z* |
| | Mode | ESI positive |
| | Mass range | 500–1250 *m/z* |
| | Collection mode | Centroid |
| | Sample rate | 2 points/s |
| | Cone voltage | 15 V |
| | Probe temp | 400 °C |
| | Capillary voltage | 1.5 kV |

6. Prepare samples. Transfer the glycan sample from 600μL tube to a UPLC sample vial. Add 500μL 18.2 MΩ water in a separate sample vial and use as the "blank" sample (*see* **Note 22**). Place all sample vials in the sample manager.
7. Equilibrate the column for at least 15 min at the initial condition of the instrument methods: 0.4 mL/min, 25% mobile phase A at 60 °C (*see* **Note 23**).
8. Set up a sample set containing the sequence of injections: 10μL injection of water followed by 10μL injection of each glycan sample, then another 10μL injection of water.
9. After the pressure ripple is less than 10 psi in 1 min, start the injection sequence to collect data.
10. After the injections are completed, set the flow rate at 0.05 mL/min with 50/50 A/B and reduce the column temperature to 30 °C, or run a shutdown method with the same set of parameters (*see* **Note 24**).

*3.2.2 Data Analysis*

1. Open and review the FLR and MS data of the blank injection. A typical chromatographic profile is shown in Fig. 2. No peaks should present in the range between 10 and 35 min.
2. Open and review the FLR and MS data of the glycan samples. Most glycans should elute between 10 and 35 min. Figure 3

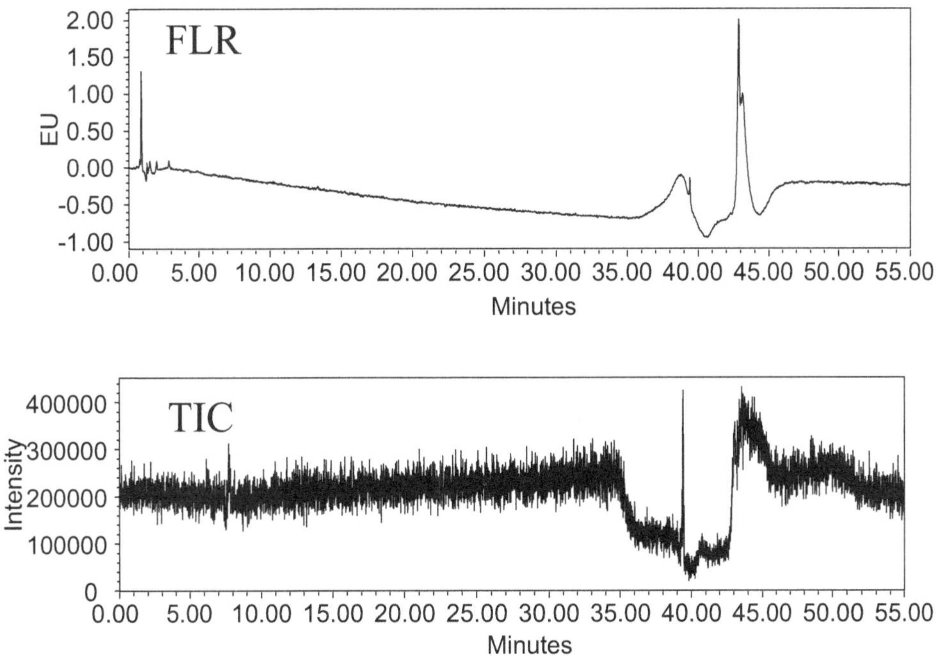


**Fig. 2** A typical set of FLR (top) and TIC (bottom) chromatogram for blank injection in routine LC-MS analysis. No peak should be eluted between 10 and 35 min

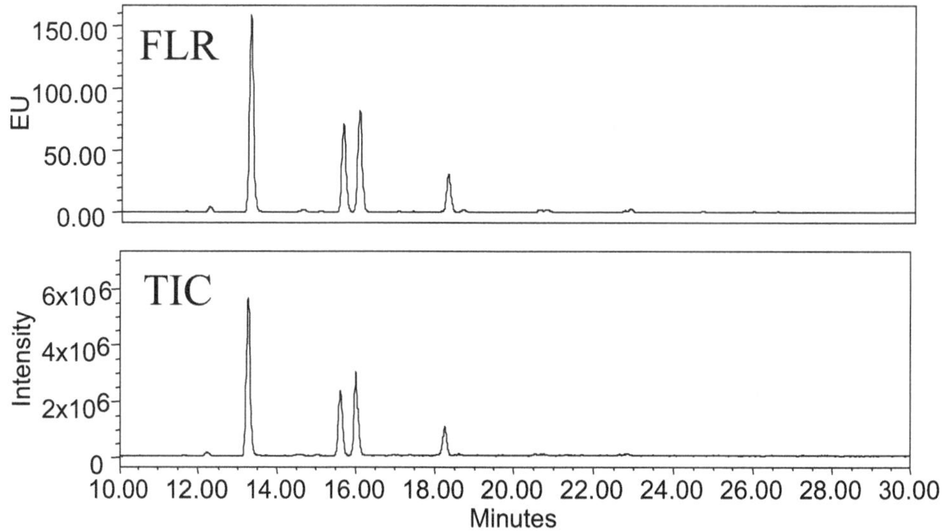


**Fig. 3** A typical set of FLR (top) and TIC (bottom) chromatogram of released glycans from mAbs in routine LC-MS analysis

shows the chromatographic profile of released glycans from NIST mAb (*see* **Note 25**).

3. Create processing method for FLR and MS data. Review data and set up a retention time window and a threshold of peak area/height for integration (*see* **Note 26**).
4. In the processing method, enter the name and expected peak information such as retention time for the known glycans in the process method (*see* **Note 27**).
5. Set up acceptance criteria for critical glycan species in the processing method (*see* **Note 28**).
6. Save the processing method and process the sample set (*see* **Note 29**).
7. Review the processed results. All peaks above the defined threshold should be integrated within the set retention time window. A table of the detailed peak integration results should be provided, including the peak name, retention time, % amount, base peak, etc. (*See* **Note 30**) If acceptance criteria were entered, out-of-spec results will be labeled in the peak result table. An example of the processed results is shown in Fig. 4.

### 3.3 High-Throughput Release N-Glycan Analysis

*N*-glycan data are routinely used to guide process development and hence a largely automated, high-throughput approach is required to monitor the levels of *N*-glycans. Manual sample preparation is laborious and time-consuming and hence would not be able to support the analysis of large number of samples. In order to address

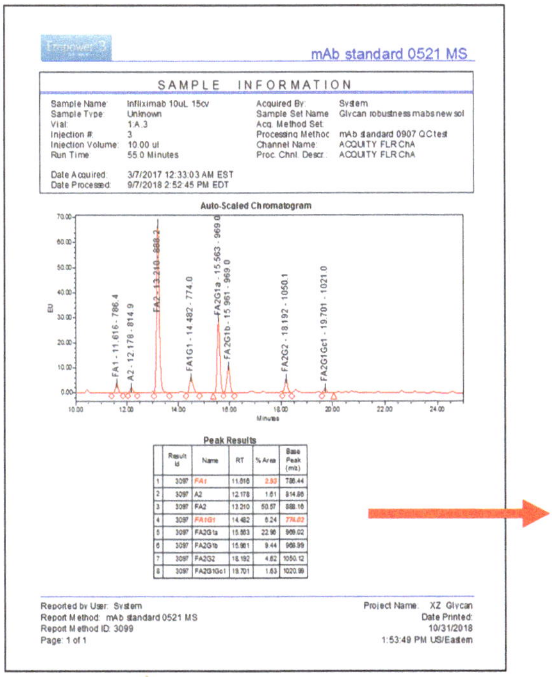


**Peak Results**

| | Result Id | Name | RT | % Area | Base Peak (m/z) |
|---|---|---|---|---|---|
| 1 | 3097 | FA1 | 11.616 | 2.93 | 786.44 |
| 2 | 3097 | A2 | 12.178 | 1.61 | 814.86 |
| 3 | 3097 | FA2 | 13.210 | 50.57 | 888.16 |
| 4 | 3097 | FA1G1 | 14.482 | 6.24 | 774.02 |
| 5 | 3097 | FA2G1a | 15.563 | 22.96 | 969.02 |
| 6 | 3097 | FA2G1b | 15.961 | 9.44 | 968.99 |
| 7 | 3097 | FA2G2 | 18.192 | 4.62 | 1050.12 |
| 8 | 3097 | FA2G1Gc1 | 19.701 | 1.63 | 1020.99 |

**Fig. 4** An example of the report for routine analysis of released glycans, showing the analysis information, FLR chromatogram, and peak result table. Out-of-spec results are flagged in red

these shortcomings, a robust high-throughput workflow from sample preparation to data analysis with minimal hands-on time and user expertise has been developed. Sample preparation has been completely automated using Tecan Freedom EVO 100® liquid handling robot employing 96-well plates [19].

### 3.4 Applications of RapiFluor-MS N-Glycan Method

Recombinant monoclonal antibodies have emerged as an important class of therapeutics for treating a wide range of diseases including cancer and autoimmune diseases. All approved monoclonal antibodies (mAbs) for therapeutic treatments belong to the IgG class and are manufactured in mammalian cell lines such as Chinese hamster ovary (CHO), mouse NS0 or SP2/0 myeloma cells. These expression systems allow the generation of full-length monoclonal antibodies with a *N*-glycosylation profiles which is human like. However, the glycan profile can be influenced by multiple factors such as production conditions, media composition, and a clonal variability [12]. This poses a significant challenge to the production of biotherapeutics with consistent product quality [13]. Many biopharmaceutical companies are keen to understand and control the microheterogeneity of IgG antibody glycoforms and dealing and resolving this challenge has led to the new opportunities for tailoring antibody therapeutics to maximize their efficacy and safety profile [14]. One of the difficulties in achieving this target is the availability of the rapid analytical methods for the in-depth analysis of *N*-glycans. *Rapi*Fluor-MS *N*-glycan method provides an ideal high-throughput platform for the detailed characterization of *N*-glycan species.

#### 3.4.1 Cell Line and Process Development

There has been continuous improvement in generating cell lines with high titer, these include expression vector optimisation, high-throughput clonal selection methods and improved media formulation [8]. These recent improvements in the cell lines and cell culture conditions means that there is now an increased industry wide focus on product quality rather than titer improvement. Since glycosylation plays such a pivotal role in influencing molecule's biophysical and biological activities, *N*-glycan analysis is being performed routinely to assess the product quality. Typically, hundreds to thousands of clones are screened through several stages of evaluation to identify the cell lines with high titer, long-term stability, and acceptable product quality [8]. High-throughput *N*-glycan analysis provides detailed product characterization for many cell lines at early stages of cell line development to identify the most favorable condition [15]. Modifications in the cell culture media, process, and vessel types means that glycosylation behavior of a cell line at an initial stage might not be reflective of that at a late stage [8]. Hence, it is pivotal to monitor and characterize the glycan profile from early cell line selection to a late stage manufacturing process.

*N*-glycan testing is also extensively performed to aid the process development activities. *N*-glycan analysis is required not only to ensure candidate cell lines meet the desired glycan profiles but also to identify critical process parameters (CPPs) that affect glycan profiles and to assess the impact of these variables during process development [16]. Furthermore, the introduction of high-throughput automated mini-bioreactor systems has significantly reduced the process development timelines for early stage biopharm programs. These automated cell culture platforms such as Ambr® 15 and 250 have demonstrated the ability to provide reliable estimates of process performance and product quality from small scale to large pilot scales [17]. Besides being able to evaluate many conditions such as media and feed screening in mini-bioreactor systems for an initial clone, a high-throughput glycan-profiling workflow is also required to rapidly assess the impact of process variables on the glycan profiles [18]. The *Rapi*Fluor-MS *N*-glycan method provides a sensitive, robust and high-throughput *N*-glycan workflow to support the product quality monitoring for samples from an Ambr® platform. This in turn helps to finely tune the process parameters to achieve the desired glycosylation profile [6]. Upstream process development is often accelerated by the implementation of a Design of Experiment (DOE) approach to generate the desired product quality [18].

*3.4.2 Analytical Characterization*

The product critical quality attributes are identified and assessed at the initial stage of the project and need to be confirmed at each stage of a product development cycle. This strategy allows the development of high quality and potent biotherapeutics with minimal compromise to the estimated time and budget [18]. Confirmation of batch-to-batch consistency to ensure product safety and efficacy by routine glycan profiling is essential to establish that glycan structures are maintained within specific ranges [14].

## 4 Notes

1. Alternatively, a thermocycler can also be used for incubation. The 200μL PCR tubes being used for most thermocyclers is not enough for the volume of 400μL upon ACN dilution. To complete the reaction without sample transferring, scale down the liquid volume to 2× concentration.
2. A positive pressure manifold with a spacer can be used instead.
3. This kit includes three sets of reagents and buffers. Each set can be used to prepare eight released glycan samples. Use one set of materials in this protocol.
4. Dilute with water if the sample is from a stock solution with higher concentration. if sample is in complex matrices, perform a buffer exchange step with pH 7.9 HEPES buffer.
5. In this protocol, a Waters Acquity UPLC H-Class Bio system was used for illustration with a Waters fluorescence detector (FLR) and a QDa mass detector for in-line fluorescence and MS detection via electrospray ionization. Other instrument systems can also be used if meeting the following criteria. The LC system should be configured with a solvent pump to deliver gradient flow, a sample manager to automatically inject up to 10μL sample, and a column manager with the ability to heat up the column to more than 60 °C. The pressure limit of the LC system needs to be at least 8000 psi. A 10 mm flow cell is recommended for the fluorescence detector to provide high sensitivity. If other type of mass detector is being used, the experimental parameters should be adjusted accordingly for the best performance of MS detection.
6. Other analytical columns that are designed for glycan separation can also be used but will provide different profiles of glycan separation. Data analysis parameters need to be optimized accordingly.
7. Use high recovery sample vials to maximize the sample recovery.
8. Contamination in glassware can lead to noises and adducts in MS spectra. It is recommended to use manufacturer certified

clean bottles to avoid contamination. Otherwise follow the cleaning protocol for standard borosilicate glassware [21]: sonicate in LC-MS grade 50/50 isopropanol/water with 0.1% formic acid for 20 min. Rinse with LC-MS grade acetonitrile and dry before use.

9. Software programs from any instrument vendors can be used for routine analysis if all the required functionalities are provided. Specific software programs from each vendor can be used if compliance is required. Empower is used in this protocol for illustration.
10. Use LC-MS grade solvent to perform LC-MS analysis of released glycans. LC-MS grade solvents contain less metal contaminants, which can reduce the adduct level and provide higher intensity of glycan signals in MS spectra.
11. The reconstituted *Rapi*Gest surfactant is stable for 1 week at 2–8 °C. Long-term storage of frozen aliquots is not recommended due to possible solubilization issues.
12. This step is critical. Heat the 1 mL tube to a measured heat block surface temperature of 105–110 °C. This is to ensure the glycoprotein has been heat denatured before PNGase F digestion. For challenging samples, a higher temperature and possibly even near-boiling conditions can be used to achieve complete denaturation. Longer denaturing time such as 10–30 min can also be considered.
13. The *Rapi*Fluor-MS labeling is a self-quenching reaction. If the reaction proceeds at room temperature for more than 5 min, a quenching step is not needed.
14. The recovery at this step is about 75%. If higher recovery is desired, add another 30μL volumes of GlycoWorks SPE elution buffer and concentrate the obtained glycans via SpeedVac.
15. The sample diluent provided in the GlycoWorks *Rapi*Fluor-MS *N*-glycan kit contains DMF and acetonitrile at a 10:21 ratio.
16. The prepared glycan sample can be stored at −80 °C for long term storage.
17. If the instrument has A2/B2 lines (binary pump) or C/D lines (quaternary pump), place line A2 (or C) in 100% water and line B2 (or D) in 100% acetonitrile.
18. The priming time in this protocol is set for the Waters UPLC H-Class Bio system. Depending on the LC systems, the priming time can be different. If the LC instrument does not have seal wash lines, skip the seal wash priming step.

19. If this is a new column, follow the column Care and Use manual to condition the column as a new column needs longer time and more mobile phases to equilibrate.

20. This is the generic gradient for released *N*-glycan separation and has been proven as useful for most applications. It should be noted that the % A at the washing step (36.5–39.5 min) is at 80%. Avoid using 100% aqueous solvent at washing step for increased the column life.

21. Due to the high proton affinity of *Rapi*Fluor-MS label, the derivatized glycans preferentially adopt high charge states during positive ion mode electrospray ionization. The predominant charge state for small neutral glycan is $M+2H^+$. [21] MS source conditions may require optimization to achieve high signal intensity, limited in-source fragmentation, and minimized adduct formation.

22. If the LC sample manager is configured with a 96-sample plate format, the released glycans can be injected directly from the 600μL tube.

23. When starting the flow, increase the flow rate gradually, for example, 0.1→0.2→0.3→0.4, to prevent pressure surge on the column and detector flow cell.

24. It is recommended to set up a shutdown method as it can save resources such as solvent, optical detector lamp hours and column life.

25. Retention times can have slight shift due to the variation in each component in the system, including column, dwell volume, and instrument models.

26. For routine analysis of *N*-glycans, it is highly recommended to create one method to process both FLR and MS data, and use the FLR data for integration, relative quantification, and result review. The threshold is defined by users, which can be determined based on either detection limit of the method or different laboratory requirements.

27. The identity of glycans in the sample should be obtained prior to the routine analysis.

28. Acceptance criteria are dependent on the samples used in the analysis and should be determined by the user based on prior knowledge, such as the correlation between glycan abundance and PK/PD, and performance of the instrument. Acceptance criteria can be set up for retention time (or relative retention time), % amount (relative abundance), and base peak (most abundant $m/z$) of the critical species.

29. Other processing parameters can also impact the quality of results, such as noise reduction and smoothing. User should

consider optimizing these parameters for different analyses when reviewing data and designing reporting template.

30. The % amount of each peak is calculated automatically upon integration using the following equation:

$$\%\text{Amount of the glycan} = \frac{\text{Peak area}}{\text{Total peak area}} \times 100$$

## Acknowledgments

The authors would like to thank Mike Molloy, Steve Warr, and Gary Finka from GlaxoSmithKline, Stevenage; Dr. Robert Birdsall, Steven Calciano, and Nick Pittman from Waters Corporation for content review and expertise suggestion on the manuscript.

### References

1. Jefferis R (2008) Glycosylation of recombinant antibody therapeutics. Biotechnol Prog 21 (1):11–16
2. Reusch D, Haberger M, Maier B, Maier M, Kloseck R, Zimmermann B, Hook M, Szabo Z, Tep S, Wegstein J, Alt N, Bulau P, Wuhrer M (2014) Comparison of methods for the analysis of therapeutic immunoglobulin G fc-glycosylation profiles—part 1: separation-based methods. MAbs 7(1):167–179
3. Cobb B (2019) The history of IgG glycosylation and where we are now. Glycobiology 30 (4):202–213
4. Read E, Park J, Brorson K (2011) Industry and regulatory experience of the glycosylation of monoclonal antibodies. Biotechnol Appl Biochem 58(4):213–219
5. Lalonde M, Durocher Y (2017) Therapeutic glycoprotein production in mammalian cells. J Biotechnol 251:128–140
6. Reusch D, Tejada M (2015) Fc glycans of therapeutic antibodies as critical quality attributes. Glycobiology 25(12):1325–1334
7. Zwebb H (2019). Glycosylation main approval issue with biosimilars/Conferences/Home—GaBI Online—Generics and Biosimilars Initiative. Gabionline.net. http://gabionline.net/Conferences/Glycosylation-main-approval-issue-with-biosimilars. Accessed 9 Sep 2019
8. Zhang P, Woen S, Wang T, Liau B, Zhao S, Chen C, Yang Y, Song Z, Wormald M, Yu C, Rudd P (2016) Challenges of glycosylation analysis and control: an integrated approach to producing optimal and consistent therapeutic drugs. Drug Discov Today 21(5):740–765
9. Szigeti M, Guttman A (2017) Automated *N*-glycosylation sequencing of biopharmaceuticals byy capillary electrophoresis. Sci Rep 7 (1):11663
10. Mittermayr S, Bones J, Guttman A (2013) Unraveling the Glyco-puzzle: glycan structure identification by capillary electrophoresis. Anal Chem 85(9):4228–4238
11. Ruhaak L, Zauner G, Huhn C, Bruggink C, Deelder A, Wuhrer M (2010) Glycan labeling strategies and their use in identification and quantification. Anal Bioanal Chem 397 (8):3457–3481
12. Simonov V, Ivanov S, Smolov M, Abbasova S, Piskunov A, Poteryaev D (2019) Control of therapeutic IgG antibodies galactosylation during cultivation process and its impact on IgG1/FcγR interaction and ADCC activity. Biologicals 58:16–21
13. Saldova R, Kilcoyne M, Stöckmann H, Millán Martín S, Lewis A, Tuite C, Gerlach J, Le Berre M, Borys M, Li Z, Abu-Absi N, Leister K, Joshi L, Rudd P (2017) Advances in analytical methodologies to guide bioprocess engineering for bio-therapeutics. Methods 116:63–83
14. Jefferis R (2009) Glycosylation as a strategy to improve antibody-based therapeutics. Nat Rev Drug Discov 8(3):226–234
15. Doherty M, Bones J, McLoughlin N, Telford J, Harmon B, DeFelippis M, Rudd P (2013) An automated robotic platform for rapid profiling oligosaccharide analysis of monoclonal antibodies directly from cell culture. Anal Biochem 442(1):10–18

16. Sha S, Agarabi C, Brorson K, Lee D, Yoon S (2016) N-glycosylation design and control of therapeutic monoclonal antibodies. Trends Biotechnol 34(10):835–846
17. Manahan M, Nelson M, Cacciatore J, Weng J, Xu S, Pollard J (2019) Scale-down model qualification of ambr® 250 high-throughput mini-bioreactor system for two commercial-scale mAb processes. Biotechnol Prog 35(6):e2870
18. Gupta S, Shukla P (2018) Glycosylation control technologies for recombinant therapeutic proteins. Appl Microbiol Biotechnol 102 (24):10457–10468
19. Waters.com. (2019). GlycoWorks RapiFluor-MS N-Glycan Kit – Automation Care and Use Manual: Waters. https://www.waters.com/webassets/cms/support/docs/715005359en.pdf
20. Waters.com (2015). GlycoWorks RapiFluor-MS N-Glycan Kit—24 Samples. https://www.waters.com/webassets/cms/support/docs/715004903.pdf
21. Waters.com. (2016). Best practices in the analysis of RapiFluor-MS labeled glycans using the ACQUITY QDa detector (performance model). https://www.waters.com/webassets/cms/library/docs/720005655en.pdf

# Chapter 16

# Online 2D-LC for Complex *N*-Glycan Analysis from Biopharmaceuticals


## Sonja Schneider, Edgar Naegele, and Sonja Krieger



## Abstract

EPO has a complex glycosylation pattern with differently branched and charged glycans. A combination of hydrophilic interaction chromatography (HILIC) with weak anion exchange chromatography (WAX) enables highly orthogonal separation. Comprehensive 2D-LC analysis with HILIC in the first and WAX in the second dimension provides high resolution 2D chromatography together with simultaneous charge profiling. Meanwhile, multiple heart-cutting 2D-LC analysis combining WAX and HILIC separation provides a flexible alternative whereby the user can select multiple peaks to be analyzed in the second dimension and, moreover, run longer gradients in the second dimension.



**Key words** Biopharmaceutical, Glycan analysis, Glycosylation, 2D-LC, EPO, Comprehensive 2D-LC, Multiple heart-cutting 2D-LC, HILIC, WAX, High resolution


## 1 Introduction

Erythropoietin (EPO) is a 30,400 Dalton (Da) glycoprotein hormone that regulates the production of red blood cells (erythropoiesis) [1]. The molecule consists of a 165 amino acid single polypeptide chain and a complex carbohydrate addition that amounts to 40% of the total molecular weight [2]. Due to the flexible molecular structure of the glycans, they cover probably most of the surface of EPO [2]. The glycosylation of EPO is highly variable because it contains multiple glycosylation sites, each of which can have a wide variety of glycan structures [3]. This results in a huge complexity of glycan structures that is referred to as microheterogeneity [3]. The glycosylation portion of EPO consists of three *N*-linked glycosylation sites at Asn 24, 38, and 83, and one *O*-linked glycosylation site at Ser 126 [2, 3], *see* Fig. 1. In this scheme, as an example, four differently branched and charged glycans are displayed, that typically occur in EPO [1, 3]. Each of the three *N*-linked glycans is likely to contain up to four sialic acids (*N*-acetylneuraminic acid, NeuAc) [4].


Arnaud Delobel (ed.), *Mass Spectrometry of Glycoproteins: Methods and Protocols*, Methods in Molecular Biology, vol. 2271, https://doi.org/10.1007/978-1-0716-1241-5_16, © Springer Science+Business Media, LLC, part of Springer Nature 2021

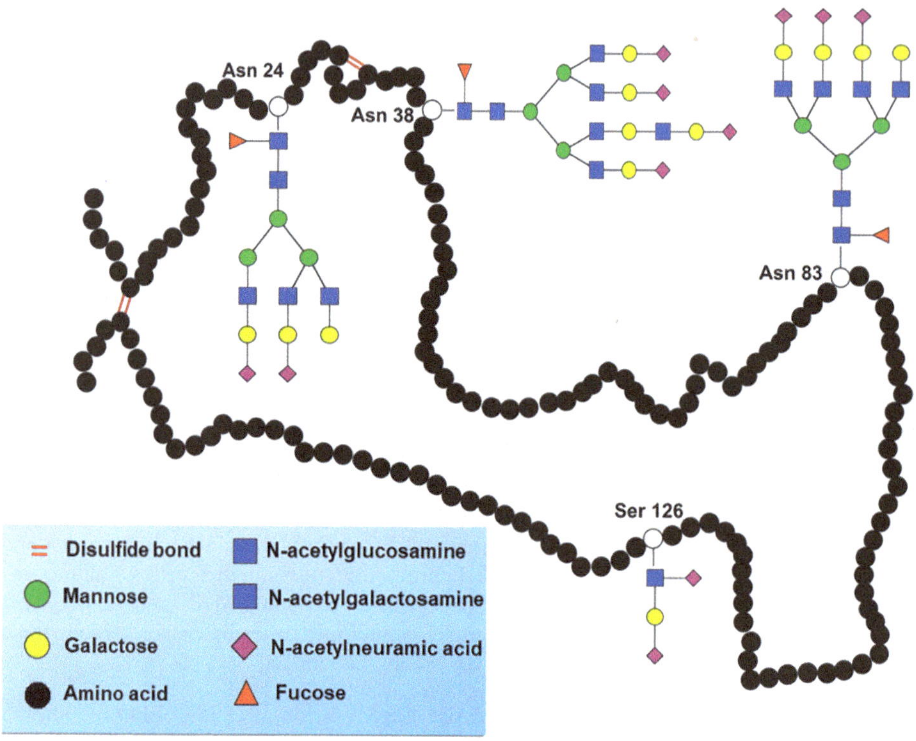


**Fig. 1** EPO structure with four examples of *N*-glycans typically occurring in EPO

Recombinant human EPO (rhEPO) has proven to be highly efficient in the treatment of different diseases such as anemias associated with cancer, chronic renal failure, and HIV infection. Furthermore, rhEPO plays an important role in surgical settings [4]. Detailed characterization of the glycan profile of biopharmaceuticals is a regulatory requirement as differences in glycosylation can affect both the pharmacodynamics and pharmacokinetic behavior in the human body. Therefore, it is necessary to develop advanced analytical technologies for efficient and detailed glycosylation analysis.

The complex mixture of highly branched glycans on EPO must first be enzymatically released from the protein before they can be studied in detail. Even after the glycans have been isolated from the protein backbone, their analysis still poses a huge analytical challenge. The method of choice is typically hydrophilic interaction chromatography (HILIC) after labelling with 2-aminobenzamide (2AB) for sensitive fluorescence detection [5]. While HILIC efficiently separates glycans according to hydrodynamic radius, it is insufficient to fully resolve the complex mixture of branched glycan structures that are present in samples such as EPO or fetuin [6]. Fortunately, weak/strong anion exchange chromatography (WAX/SAX) provides a highly orthogonal separation that depends on the number and arrangement of acidic monosaccharides in the glycan. In the case of EPO and fetuin, these are typically sialic acids known as *N*-acetylneuraminic acid or NeuAc [7, 8].

A combination of WAX/SAX and HILIC has a huge potential to enhance peak capacity in two-dimensional liquid chromatography (2D-LC) due to the highly orthogonal nature of these two separation techniques. Bones et al. showed an offline 2D analysis with a combination of WAX separation in the first dimension with fraction collection of ten peaks, followed by HILIC analysis in the second dimension [6]. The complete offline 2D analysis of ten EPO *N*-glycan peaks resulted in a total cycle time of over 4 h with additional hands on time for moving the samples from the fraction collector into a sample injector for reinjection [6]. This type of 2D workflow typically also requires additional time to dry down the first dimension fractions in a vacuum centrifuge prior to resuspending them in a suitable matrix and volume for the second dimension analysis [6].

Online 2D-LC workflows for either comprehensive or (multiple) heart-cutting analysis enable the user to run highly automated workflows without manual interference. Comprehensive 2D-LC analysis, using two sample loops within a 2-position/-4-port-duo valve, leaves no peak missed from the first dimension [9]. If higher resolution is desired in the second dimension, the 1290 Infinity Multiple Heart-Cutting 2D-LC Solution enables more flexibility, for example longer cycle times or columns. This solution is comprised of two external valve drives with 6-position/-14-port valves, each with six preinstalled 40 μL loops, resulting in twelve loops [10].

Online 2D-LC concepts are shown for the *N*-glycan analysis of therapeutic EPO using a combination of HILIC/WAX for comprehensive analysis and WAX/HILIC for detailed multiple heart-cutting analysis [11].

## 2 Materials

Use only LC grade solvents. Prepare all solvents and samples with fresh ultrapure water (e.g., from a Milli-Q Integral system equipped with a 0.22 μm membrane point-of-use cartridge) (Millipak). Prepare all samples at room temperature (unless indicated otherwise). Diligently follow all waste disposal regulations when disposing of waste materials.

### 2.1 Sample Preparation

1. 50 mM sodium phosphate buffer, pH 7.5.
2. Denaturation solution: 0.2% SDS –100 mM 2-mercaptoethanol in water.
3. 15% octylglucoside.

4. GlycoProfile 2-AB Labeling Kit (Sigma-Aldrich) or equivalent individual reagents.
5. DMSO–acetic acid solution: add 150 μL of acetic acid glacial to the entire ampoule of DMSO (350 μL). Note that DMSO forms a semisolid at cooler temperatures, hence warming may be required.
6. Glycan Clean-up Cartridges (Sigma-Aldrich).
7. 30% acetic acid solution: Mix 6 volumes of ~50% acetic acid solution (part of Glycan Clean-Up kit) with 4 volumes of HPLC Grade Water. A minimum volume of 5 mL is required per sample.
8. HPLC grade water. A minimum volume of 3 mL is required per sample.
9. 96% acetonitrile solution: Prepare a 96% (v/v) acetonitrile solution by mixing 96 volumes of HPLC grade acetonitrile with 4 volumes of HPLC Grade Water.
10. Centrifugal evaporator.
11. Heating block.

## 2.2 (2D)-LC/MS Analysis

### 2.2.1 Comprehensive 2D-LC HILIC/WAX Workflow

1. 1D—Mobile phase A: 1 L acetonitrile.
2. 1D—Mobile phase B: Prepare 10 mM ammonium formate by weighing 0.63 g of ammonium formate, and fill it up with 1 L of water, adjust to pH 4.5 using formic acid.
3. 2D—Mobile phase A (40% acetonitrile): Add 400 mL of acetonitrile to 600 mL water.
4. 2D—Mobile phase B: Prepare 250 mM ammonium formate by weighing 15.764 g of ammonium formate and fill it up with 1 L of water, adjust to pH 4.5 using formic acid. Add 400 mL of acetonitrile to 600 mL of 250 mM ammonium formate solution.

### 2.2.2 Multiple Heart-Cutting 2D-LC WAX/HILIC Workflow

1. 1D—Mobile phase A (40% acetonitrile): Add 400 mL of acetonitrile to 600 mL water.
2. 1D—Mobile phase B: Prepare 250 mM ammonium formate by weighing 15.764 g of ammonium and fill it up with 1 L of water, adjust to pH 4.5 using formic acid. Add 400 mL of acetonitrile to 600 mL of 250 mM ammonium formate solution.
3. 2D—Mobile Phase A: 1 L acetonitrile.
4. 2D—Mobile phase B: Prepare 50 mM ammonium formate by weighing 3.15 g of ammonium formate, and fill it up with 1 L of water, adjust to pH 4.5 using formic acid.

*2.2.3 LC Columns*

1. Agilent AdvanceBio Glycan Mapping Column, 2.1 × 150 mm, 1.8 μm.
2. Agilent AdvanceBio Glycan Mapping Column, 4.6 × 50 mm, 2.7 μm.
3. Agilent Bio WAX Column, 2.1 × 250 mm, 5 μm.
4. Agilent Bio WAX Column, 2.1 × 50 mm, 5 μm.

*2.2.4 Apparatus and Software*

1. Agilent 1260 Infinity Bio-inert Quaternary Pump in the first dimension.
2. Agilent 1290 Infinity Binary Pump in the second dimension.
3. Agilent 1290 Infinity Autosampler.
4. Agilent 1290 Infinity Thermostated Column Compartment.
5. Agilent 1290 Infinity Valve Drive with 2-Position/-4-Port-Duo valve equipped with either two 40-μL loops or.
6. Agilent Multiple Heart-Cutting Valve Upgrade Kit.
7. Agilent 1260 Infinity Fluorescence Detector.
8. Agilent 6530 Accurate Mass QTOF LC/MS system.
9. Agilent OpenLAB CDS ChemStation Edition for LC & LC/MS Systems.
10. Agilent MassHunter Workstation Software.
11. Glycan structures were created with GlycoWorkbench.

## 3 Methods

### 3.1 Protein Deglycosylation

1. Prepare a 500 units/mL PNGase F solution by adding 100 μL of high purity water to a vial of 50 units (*see* **Note 1**). This results in an enzyme solution containing ~5 mM potassium phosphate buffer, pH 7.5. The vial containing PNGase F should be centrifuged briefly before opening to ensure the lyophilized material is at the bottom of the tube.
2. Prepare a 2 mg/mL solution of denatured glycoprotein (Fetuin and ß-EPO) by adding 100 μg of lyophilized glycoprotein to 45 μL of 50 mM sodium phosphate buffer, pH 7.5, and then adding 5 μL of denaturation solution. Heat the solution to 100 °C for 10 min to denature the glycoprotein.
3. Allow the solution to cool. Add 5 μL of 15% octyl glucoside (*see* **Note 2**).
4. Add 20 μL of PNGase F enzyme solution (500 units/mL, *see* **Note 3**) to the reaction mixture and incubate at 37 °C for 3 h.
5. Stop the reaction by heating to 100 °C for 5 min.
6. Dry the glycan-containing samples completely using a centrifugal evaporator (Speed-vac). Be sure to keep the temperature

during evaporation below 28° to avoid desialylation of the EPO Glycans. Also, excess moisture will have negative effects on labeling and stability.

### 3.2 Glycan Labeling (GlycoProfile 2-AB Labeling Kit)

1. Prepare the Labeling Solution immediately prior to labeling. Allow the unopened vial to reach RT and ensure that most of the material is at the bottom of the vial by gently tapping on a solid surface. Gently remove the cap, reconstitute with the desired volume of water or buffer, replace the cap, and mix thoroughly.
2. Tap or briefly spin down component vials to avoid loss of reagent in the cap or on the walls of ampules. To open ampules, hold both the body and the top of the ampoule, then gently, but firmly, snap open at the colored break-ring, directing the break away from your body.
3. Add 100 μL of the DMSO–acetic acid mixture to the entire vial of 2-AB (5 mg). Mix or vortex until completely dissolved.
4. Add the entire volume of 2-AB solution (**step 3**) to the vial of reductant (6 mg of sodium cyanoborohydride). Mix until completely dissolved. If insoluble particulates remain, the solution may be heated to 65 °C for up to 3 min. Any remaining particulates can be dissolved by adding 10 mL of water (HPLC grade or better). This is the Labeling Solution and should be protected from light. It is stable for 1 h.
5. Add 5 mL of the Labeling Solution (**step 4**) to each dried glycan sample. Cap and mix thoroughly. Tap or spin down to collect the dissolved sample in the bottom of each vial.
6. Incubate glycan samples for 3 h at 65 °C in a heating block. Avoid moist heat. Note: Incubation periods of between 2 and 4 h will not significantly affect results.
7. If insoluble particulates are present, vortex the preheated samples from **step 5** for 30 min.
8. Following incubation, briefly spin the vials to recollect each sample at the bottom of the reaction vial. Allow the samples to cool to ambient temperature prior to proceeding with postlabeling analysis.

### 3.3 Glycan Purification

Sample cleanup is highly recommended to remove excess dye and labeling reagents, *see* also **Note 4**.

#### 3.3.1 Glycan Adsorption

1. Wash each cartridge with 1 mL of HPLC Grade Water.
2. Wash each cartridge with 5 mL of 30% Acetic Acid Solution.
3. Allow the cartridge to drain completely and then wash with 1 mL of 96% Acetonitrile Solution.

4. While the membrane is still wet (following the acetonitrile wash), spot a sample (maximum volume of 10 μL) onto each cartridge membrane, ensuring that the sample is spread over the entire membrane surface. If the cartridge membrane has dried, apply 0.5 mL of 96% Acetonitrile Solution to moisten.
5. The sample vials may be rinsed with 100 mL of 96% Acetonitrile Solution. Each rinse may then be applied to the appropriate membrane for maximum recovery.
6. Allow each cartridge to stand for 15 min to ensure complete adsorption of the glycans to the membrane.
7. Wash each cartridge with six consecutive 1 mL volumes of 96% Acetonitrile Solution. Allow the cartridge to drain between each 1 mL application. Discard the organic solvent wash appropriately.

#### 3.3.2 Glycan Elution

1. Place each cartridge over a collection vessel sufficient for collecting 1.5 mL of water. If aseptic filtration is required, place the cartridge over a 5 mL syringe fitted with a 0.45 mm PTFE filter.
2. Elute the glycans by washing with three consecutive 0.5 mL volumes of HPLC Grade Water. Allow each volume to completely drain between elutions.
3. Filter the sample as appropriate. Dry the purified glycan samples using a centrifugal evaporator (Speed Vac).
4. Redissolve the purified glycan samples in a desired volume of solvent or water as appropriate for additional analysis.
5. Store samples at −20 °C protected from light in preparation for further downstream analysis (*see* **Notes 5** and **6**).

### 3.4 2D-LC/MS Analysis

The separation within a one-dimensional HILIC setup is not sufficient to resolve all *N*-glycans of EPO or fetuin, as shown in Fig. 2a (EPO) and Fig. 2b (fetuin) (*see* **Note 7**).

#### 3.4.1 Comprehensive 2D-LC Analysis HILIC/WAX

1. Valve & Loop configuration: 2pos/4port duo 2 loops (cocurrent)—Loop size 40 μL.
2. Use LC parameters described in Table 1.
3. Use MS parameters described in Table 2.
4. With parallel MS QTOF analysis, the peaks can be assigned to the corresponding charge (which for most peaks is equal to the number of sialic acids contained in the glycan). The detected parent ion masses are entered into the GlycoMod tool from Expasy to find the related glycans structures (http://web.expasy.org/glycomod/).

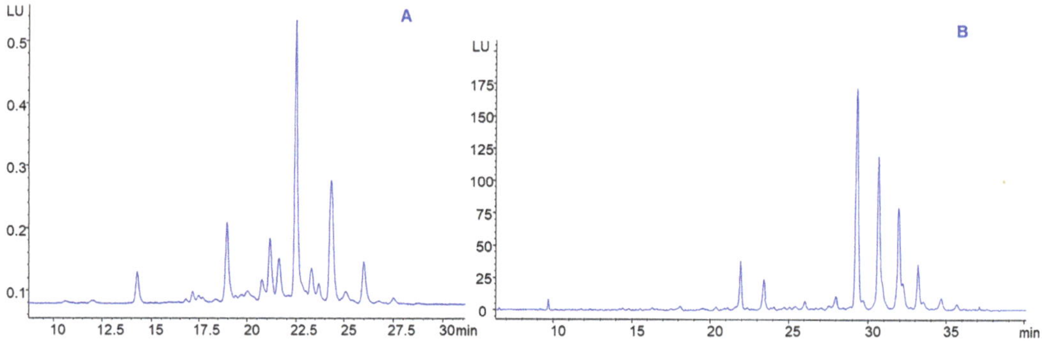


**Fig. 2** 1D HILIC separation on an Agilent AdvanceBio Glycan Mapping Column, 2.1 × 150 mm, 1.8 μm. (**a**) EPO, (**b**) fetuin

**Table 1**
**Comprehensive 2D-LC parameters**

| | EPO | Fetuin |
|---|---|---|
| 1D flow rate | 0.05 mL/min | |
| 1D gradient | 0 min—28% B<br>110 min—47% B<br>110 min—100% B | 0 min—30% B<br>110 min—44% B<br>110 min—100% B |
| 1D stop time | 120 min | |
| 1D post time | 45 min | |
| 2D parameter mode | Comprehensive | |
| 2D gradient stoptime | 0.36 min | |
| Modulation time | 0.50 min | |
| Flow | 1.50 mL/min | |
| Idle flow | 0.50 mL/min | |
| 2D gradient | 0.00 min—5% B<br>0.35 min—95% B<br>0.36 min—5% B | |
| Injection volume | 20 μL | 3 μL |
| Thermostat autosampler | 5 °C | |
| 1D column temperature | 40 °C | |
| 2D column temperature | 50 °C | |
| FLD | Ex. 260 nm, Em. 430 nm | |
| Peak width | >0.0031 min (0.063 s resp. time) (37.04 Hz) | |

**Table 2**
**MS parameters given for an Agilent 6530 Accurate Mass QTOF LC/MS system**

| | |
|---|---|
| Gas temp | 250 °C |
| Sheath gas temp | 250 °C |
| Gas flow | 8 l/min |
| Sheath gas flow | 8 l/min |
| Nebulizer | 25 psi |
| Vcap | 3500 V |
| Nozzle | 1000 V |
| Skimmer | 45 V |
| Oct 1 RF Vpp | 550 |
| Mode | MS |

5. As expected, the second dimension separation groups glycans according to their charge (*see* Fig. 3). The neutral glycans, which elute immediately with the injection peak, are shortly followed by the singly charged glycans. More clearly separated, the double, triple, quadruple, and few quintuple (fetuin) charged glycans elute with increasing salt gradient in the second dimension. Therefore, in addition to increase the peak capacity, the WAX separation assists peak assignment and furthermore provides the glycan charge profile that is required in the analysis of EPO glycosylation (*see* **Note 8**).

*3.4.2 Multiple Heart-Cutting 2D-LC Analysis WAX/HILIC*

1. Valve & Loop configuration: 2pos/4port duo 2 × 6 loops (cocurrent)—Loop size 40 μL (*see* **Note 9**).
2. Use LC parameters described in Table 3.
3. The chromatogram of the first dimension separation can be loaded into the 2D-LC acquisition setup to facilitate method development, *see* Fig. 4. The peaks that were chosen to be reanalyzed, can be selected either by peak triggering using a first dimension detector or using time segments with certain loop fill times. This enables the reduction of solvent from the first dimension (*see* **Note 10**).
4. Figure 4 shows the 2D pump setup with the loaded first dimension chromatogram with eleven peaks to be reanalyzed (yellow marks), also represented by eleven 2D time segments (red circle). Although only ten loops are available for storing the peaks, while two loops are always in the flow path, this setup enables the analysis of more than ten peaks. After the first peak

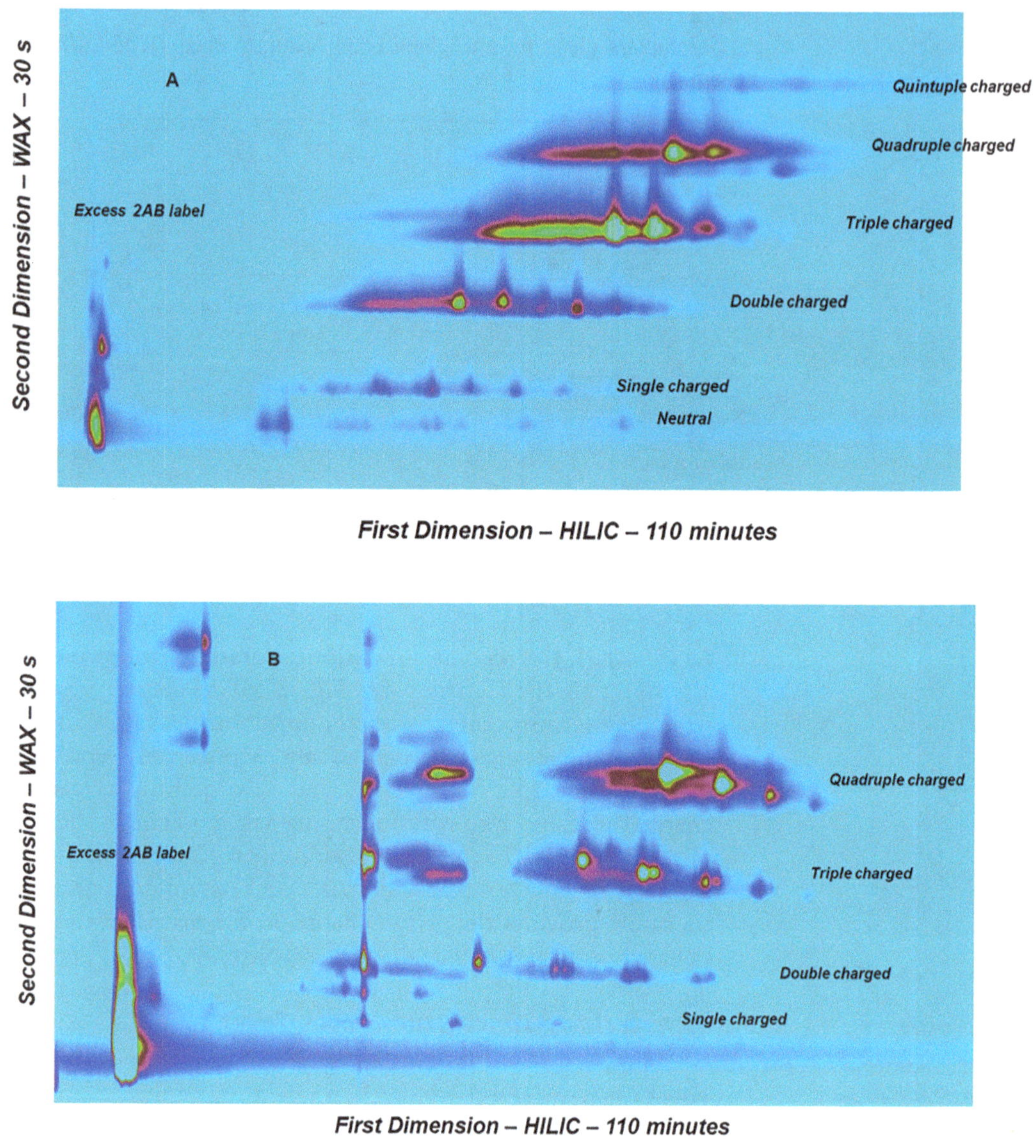


**Fig. 3** Comprehensive HILIC/WAX 2D-LC separation of fetuin (**a**) and EPO (**b**), showing highly orthogonal separation. The ion exchange chromatography in the second dimension reveals the charge pattern of the glycans

is collected in the first loop of the first 6-position/-14-port valve, it is immediately injected onto the second dimension column by switching the central 2D-LC Valve. After the 2D-LC Valve has switched, the loops of the second 6-position/-14-port valve can be filled with up to five peaks. As soon as the first 2D gradient has finished, the 2D-LC Valve switches back so that the loops of the first 6-position/-14-port

**Table 3**
**Multiple heart-cutting LC parameters**

| | |
|---|---|
| 1D flow rate | 0.25 mL/min |
| 1D gradient | 0 min—5% B<br>20 min—60% B<br>21 min—5% B |
| 1D stop time | 72 min |
| 2D parameter mode | Multiple heart-cutting |
| 2D gradient stoptime | 3.5 min |
| 2D cycle time | 4.9 min |
| Flow | 2 mL/min |
| Idle flow | 0.50 mL/min |
| 2D gradient | 0.00 min—65% B<br>3.5 min—57% B |
| 2D-time segments (→heart cutting time points)<br>(time based, loop fill time 0.1 min) | 6.85 min<br>7.45 min<br>7.85 min<br>11.15 min<br>11.45 min<br>11.75 min<br>12.30 min<br>14.60 min<br>14.85 min<br>15.25 min<br>15.70 min |
| Injection volume | 20 μL |
| Thermostat autosampler | 5 °C |
| 1D column temperature | 25 °C |
| 2D column temperature | 40 °C |
| FLD | Ex. 260 nm, Em. 430 nm |
| Peak width | 0.025 min (0.5 s resp. time) (18.52 Hz) |

valve can be filled. This, however, requires that the 2D analysis of the first peak is finished. Therefore, for method development, the adjustment of cycle time in the second dimension is critical for flexible peak selection.

5. A gradient time of 3.5 min is used with a reequilibration time of 1.4 min (*see* **Note 11**). Six examples are shown in Fig. 5 to demonstrate the resolving power of the HILIC separation within the multiple heart-cutting setup (peaks 1, 4, 5, 8,

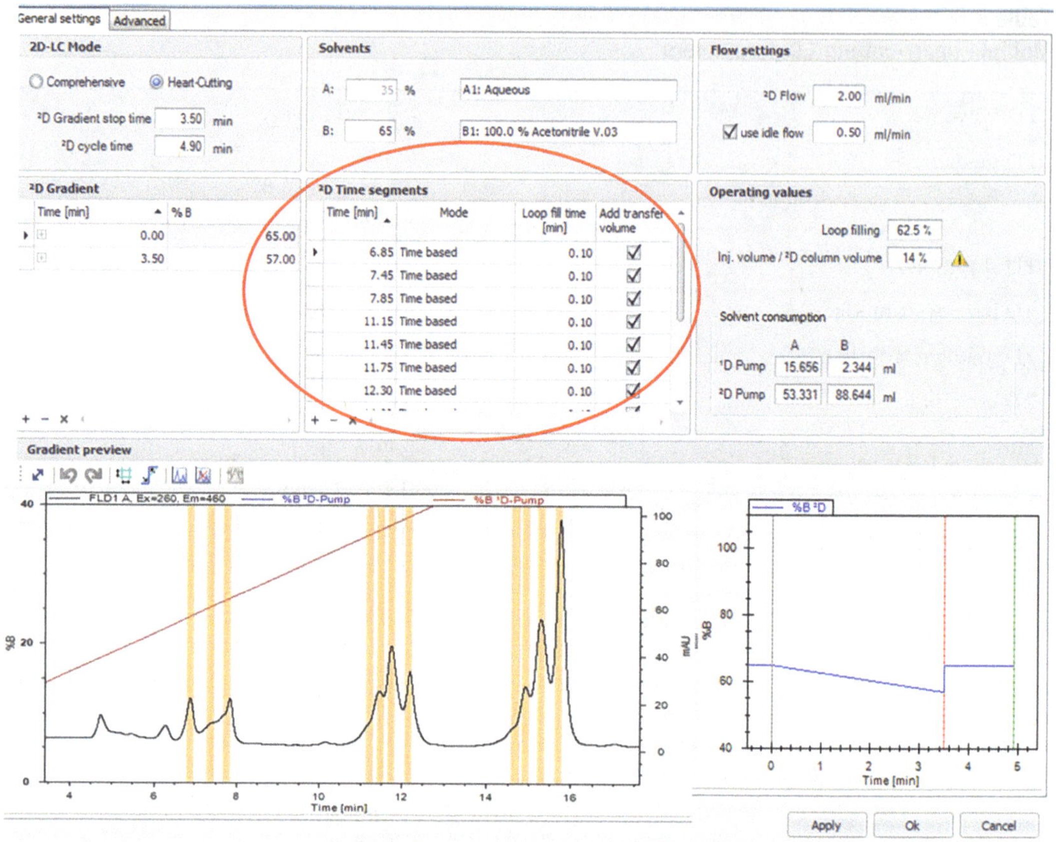


**Fig. 4** Setup of 2D Pump, loaded WAX chromatogram from the first dimension with eleven peaks—chosen for reanalysis in the second dimension using HILIC: eleven 2D time segments represent the chosen peak areas (red circle)

9, and 10). Areas that are only visible as shoulders in the first dimension (e.g., peak 8) revealed at least eight peaks in the second dimension. Under most of the peaks, which are only showing one major peak in the first dimension, several underlying peaks are detected and resolved.

## 4 Notes

1. Activity of PNGase F: One unit of PNGase F will catalyze the release of *N*-linked oligosaccharides from one nanomole of denatured Ribonuclease B in 1 min at pH 7.5 at 37 °C. PNGase F is active in the pH range of 6–10 with an optimal pH of 8.6.

   One Sigma unit of PNGase F activity is equal to 1 IUB milliunit. PNGase F Cleaves *N*-linked oligosaccharides between innermost GlcNAc and asparagines from *N*-linked

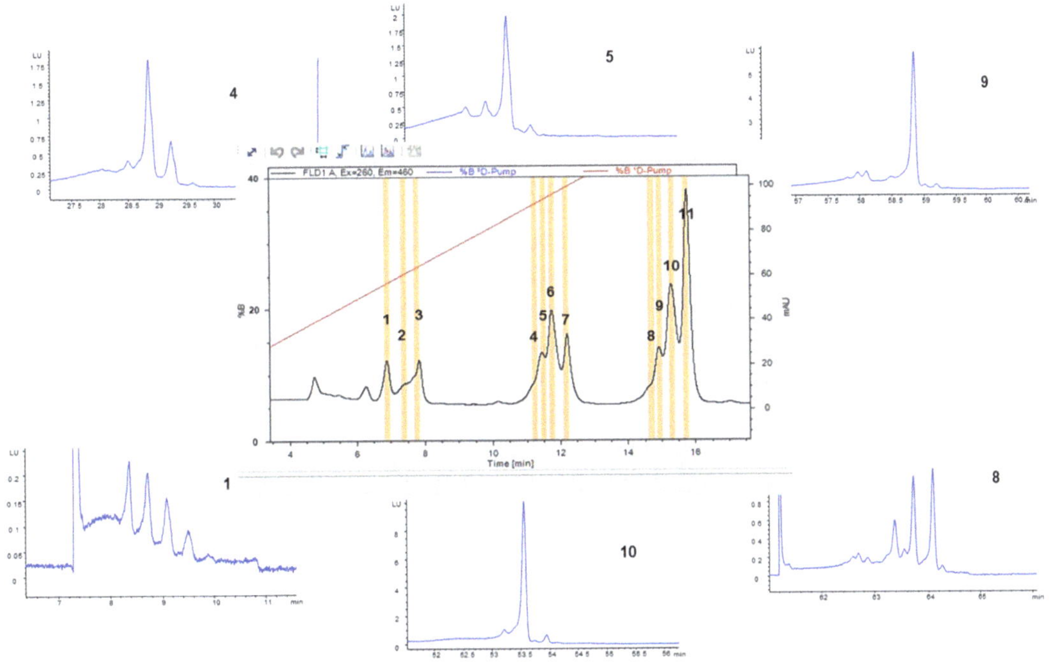


**Fig. 5** Six examples for the high resolving power of the multiple heart-cutting setup, resolving up to eight peaks under the area that is marked with number 8

glycoproteins. EPO is a 166 amino acids protein containing three *N*-glycosylation sites (Asn-24, Asn-38, and Asn-83) and 1 *O*-glycosylation site (Ser-126). As PNGase F cleaves only *N*-linked oligosaccharides, the *O*-linked glycans are not considered in this protocol.

As EPO as well as fetuin contains three glycosylation sites, the amount of PNGase F needs to be tripled to ensure complete cleavage of all attached *N*-Glycans.

2. The rate of cleavage can be increased up to 100 times with the usage of detergents and heat denaturation, so both is used in this protocol.

3. Calculation example for EPO: 100 μg of EPO is used with a molecular weight cutoff of 30,400 Da.

   n = m/M.

   n = 3.3 nmol EPO.

As one Unit PNGase F is recommended to cleave the *N*-linked Glycans from one glycosylation site from Ribonuclease B, three Units were used to cleave the *N*-glycans from three glycosylation sites of 1 nmol EPO.

In total, 10 Units (20 μL of the PNGase F solution) were used for cleavage of the EPO glycans.

4. Sample Clean up: The amount of attached glycans is reported to be about 40% of the total weight of the protein for EPO [2]. 100 μg EPO is used per sample, so about 40 μg of glycans is expected in the sample. As the recommendation is that the maximum sample size is 10 μL of glycan containing solution and/or 20 μg of glycan, split the sample into two for the clean-up.

   Note: The acetonitrile concentration is critical. Acetonitrile solutions of <96% will result in sample loss by the inappropriate elution of glycans, particularly low molecular weight glycans. A minimum volume of 5 mL is required per sample.

   During the wash procedure the solvent flow may be obstructed by air gaps. Should this occur, apply a slight pressure to the top of the cartridge in order to clear the trapped air and resume normal flow. The liquid samples should be at or below room temperature before spotting onto the cartridge membrane.

   The acetonitrile concentration is critical. Acetonitrile solutions of <96% will result in sample loss by the inappropriate elution of glycans, particularly low molecular weight glycans. A minimum volume of 5 mL is required per sample.

5. Extended periods of time between incubation and analysis may result in desialylation of labeled glycans and consequently should be avoided.

6. Consider dilution with acetonitrile for optimal peak shape in the HILIC chromatogram.

7. To improve resolution and to enhance peak capacity, a combination of WAX and HILIC separation was used for a highly orthogonal separation. First, a comprehensive WAX/HILIC 2D-LC setup was tested using fetuin. A 110-min WAX gradient was used in the first dimension, followed by a 30 s second dimension comprehensive HILIC run using a 4.6 × 50 mm HILIC column. Although 40% ACN was used in the first dimension solvents, the glycans were not retained on the HILIC column (data not shown). Typically, HILIC columns require a longer reequilibration time than other types of columns. Presumably, the 30 s cycle times in the second dimension are not compatible with HILIC separations of glycans. Moreover, the high amount of water (60%) in the first dimension effluent, together with a relatively high loop volume of 40 μL and the super short gradient, did not allow for good glycan retention on the short 50 mm HILIC column.

8. EPO isoforms are classified according to their net charge (epoetin alpha, beta, etc.). This setup enables simultaneous charge profiling in combination with a well resolved glycan peak pattern.

9. Due to the multiple heart-cutting setup using twelve 40 μL loops in two 6-position/-14-port valves, it is possible to "park" peaks from the first dimension, enabling longer second dimension gradients.
10. The HILIC gradient in the second dimension starts with a total amount of 35% water. If too much water is injected into the second dimension column, glycan retention can be reduced. The amount of water from the first dimension eluent can be reduced, if the 40 μL loops are only partly filled with the peaks from the first dimension. Here, the loops are filled with 62.5% with solvent from the first dimension, lowering the total amount of water that is injected into the second dimension column.
11. In contrast to the comprehensive 2D-LC solution, the multiple heart-cutting solution allows analysis by HILIC in the second dimension. This is, because the multiple heart-cutting approach allows for the use of longer gradient and reequilibration times in the second dimension.

## References


1. Mocini D et al (2007) Structure, production and function of erythropoietin: implications for therapeutical use in cardiovascular disease. Curr Med Chem 14(21):2278–2287
2. Egrie JC et al (1986) Characterization and biological effects of recombinant human erythropoietin. Immunobiology 172:213–224
3. Rush RS et al (1995) Microheterogeneity of erythropoietin carbohydrate structure. Anal Chem 67:1442–1452
4. Egrie JC, Browne JK (2001) Development and characterization of novel erythropoiesis stimulating protein (NESP). Br J Cancer 84(Supplement 1):3–10
5. Melmer M et al (2010) HILIC analysis of fluorescence-labeled N-glycans from recombinant biopharmaceuticals. Anal Bioanal Chem 398:905–914
6. Bones C et al (2011) 2D-LC analysis of BRP 3 erythropoietin N-glycosylation using anion exchange fractionation and hydrophilic interaction UPLC reveals long poly-N-acetyl lactosamine extensions. Anal Chem 83 (11):4154–4162
7. Jelkmann W (1992) Erythropoietin: structure, control of production, and function. Physiol Rev 72(2):449–489
8. Nwosu CC et al (2011) Simultaneous and extensive site-specific N- and O-glycosylation analysis in protein mixtures. J Proteome Res 10 (5):2612–2624
9. Buckenmaier S (2015) Agilent 1290 Infinity 2D-LC Solution for Multiple Heart-Cutting. Agilent Technologies Application Note, Publication Number 5991-5615EN. Agilent Technologies, Inc., Waldbronn, Germany
10. Schneider S (2016) N-Glycan Analysis of mAbs and Other Glycoproteins with UHPLC and Fluorescence Detection. Agilent Technologies Application Note, publication number 5990-9774EN. Agilent Technologies, Inc., Waldbronn, Germany
11. Schneider S, Naegele E, Krieger S (2017) Online 2D-LC Analysis of Complex N-Glycans in Biopharmaceuticals Using the Agilent 1290 Infinity 2D-LC Solution. Agilent Technologies Application Note, publication number 5991-5349EN. Agilent Technologies, Inc., Waldbronn, Germany

# Chapter 17

# Profiling, Relative Quantification, and Identification of Sialylated *N*-Linked Oligosaccharides by UPLC-FLR-ESI/MS After Derivatization with Fluorescent Anthranilamide

Claire I. Butré, Eric Largy, Fabrice Cantais, and Arnaud Delobel

## Abstract

The presence of sialic acids is one characteristic of glycosylated therapeutic proteins. The presence of these charged monosaccharides is critical for the immunogenicity properties and structural properties of the proteins. Profiling of the *N*-glycans and their charge state is a requisite for complete protein characterization. Two analytical methods developed on released *N*-glycans are described in this chapter, allowing for the determination of the sialoglycosylation with different levels of details. In the first method (AEX-HILIC/FLR), *N*-glycans are separated based on their charge and the average charge state can be determined from the fluorescence profile. In the second method (AEX-RP-FLR-MS), *N*-glycans are also separated based on their charge and the sialylation level is determined based on the fluorescence signal. In addition, in this method, the *N*-glycans are also separated by type and identified with the hyphenated MS. For both methods, an optimized protocol with fast and high-throughput sample preparation and purification is presented.

**Key words** *N*-glycans, Sialic acids, AXH, AXR, Mass spectrometry

## 1 Introduction

Glycosylation is a common posttranslational modification of therapeutic proteins (including monoclonal antibodies and fusion proteins) and corresponds to the linkage of a sugar moiety on a protein. There are two distinct types of glycosylation: *N*-glycans linked to asparagine (in Asn-X-Ser and Asn-X-Thr consensus sequences, where X is any amino acid except proline) and *O*-glycans linked on Ser or Thr. The common core of *N*-glycans is composed of two *N*-acetylglucosamine and three mannose residues, and from this, different types of *N*-glycans are found depending on the expression system. These can be high-mannose *N*-glycans if only mannose residues are present or complex *N*-glycans if any type of monosaccharide is linked to the core. One type of monosaccharide found in glycoproteins is sialic acids, which are charged

Arnaud Delobel (ed.), *Mass Spectrometry of Glycoproteins: Methods and Protocols*, Methods in Molecular Biology, vol. 2271, https://doi.org/10.1007/978-1-0716-1241-5_17, 

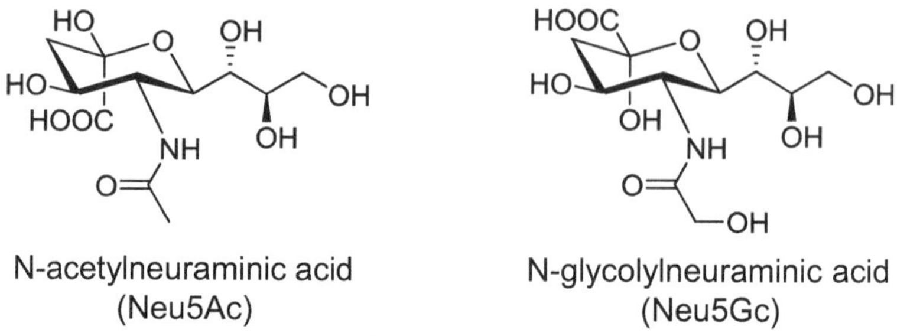


**Fig. 1** Structure of two sialic acids, *N*-acetylneuraminic acid and *N*-glycolylneuraminic acid

monosaccharides. Sialic acids are usually present on the terminal galactose of the *N*-glycan. The two main types of sialic acids found in therapeutic proteins are *N*-acetyl-neuraminic acid (Neu5Ac) and to a lower extent *N*-glycolylneuranimic acid (Neu5Gc) (Fig. 1). Neu5Ac is considered as normal human-type sialylation while Neu5Gc is found in nonhuman glycoproteins.

This type of sugar is critical as it is directly related to immunogenicity of proteins [1]. It was for instance showed that the removal of the entire *N*-Glycan or the removal of the sialic acid of intravenous gamma globulin would both results in a loss of anti-inflammatory activity [2]. This indicated that the anti-inflammatory activity was only present when sialic acids were attached to the immunoglobulins. Sialic acids also play a role on the structure and conformational stability of the antibodies. In one example, it was shown that the presence of sialylated glycans destabilizes the $CH_2$ domain in Chinese hamster ovary-expressed IgG1 [3]. Finally, it was shown for some therapeutic proteins that the sialylation level could influence the pharmacokinetic properties [4–6] of the protein or the binding to receptors [7, 8].

The determination of the extent of sialylation is critical as it can have a clear influence on the properties of the proteins, making it a critical quality attribute (CQA) for therapeutic glycoproteins.

The content of sialic acid can be for instance quantified after cleavage from the oligosaccharide and after derivatization using an external calibration of the free acids. For this, the free sialic acids are separated by reversed phase chromatography and quantified by fluorescence detection [9]. The study of the sialylation at the protein or peptide level is also possible with current MS techniques, but is still laborious work due to the complexity of the saccharides [10]. Finally, characterization of the sialylation on the *N*-glycans after release from the protein is also possible. The release is commonly done with PNGase but the protocol is time-consuming. The method for sample preparation presented here allows a fast and high throughput sample preparation. For this, the use of the Rapid PNGase allows for a release within 30 min. Samples are derivatized

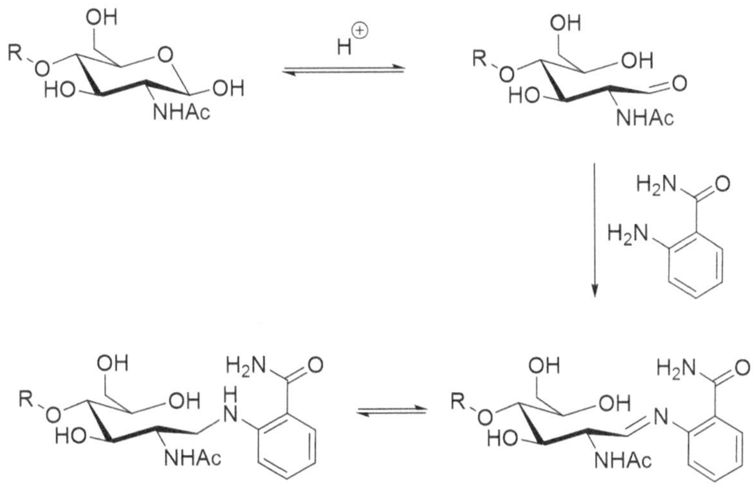


**Fig. 2** Derivatization reaction of *N*-glycan by anthranilamide

using anthranilamide and further purified on 96-well plate, which eases in terms of time the purification of a large number of samples. Anthranilamide is a fluorophore used here for optical detection of *N*-glycans. The derivatization reaction is first the reaction of anthranilamide with the oligosaccharide, leading to the formation of a Schiff's base. This base is then reduced into a secondary amine (Fig. 2), as the final reaction product.

With this sample preparation, two profiling methods with different levels of information were set up. The first approach is to use an AEX-HILIC mixed-mode column which separates the glycan in order of sialylation level, independently of the rest of the structure of the *N*-glycan. Based on the fluorescence level, the percentage of charged glycan is determined. The second approach increases the level of details, by using a different mixed-mode column but also by coupling the system with a MS system. This second mixed mode column is an AEX-RP column, in which the AEX separates here as well the glycan by order of sialylation level and, the RP mode allows the separation of the *N*-glycans within each sialylation level. The percentage of charged glycan can be determined by the fluorescence level and sialylated *N*-glycans are identified by the MS.

## 2 Materials

All solutions are prepared using ultrapure water (prepared by purifying deionized water to attain a sensitivity of 18 MΩ cm at 25 °C). Whenever possible, the use of mass spectrometry grade reagents is recommended.

### 2.1 Relative Quantification of the Sialylation of N-Glycan by UPLC-FLR

#### 2.1.1 Apparatus

1. UPLC system (autosampler and quaternary solvent manager, Acquity system, Waters).
2. Fluorimetric FLR detector for Acquity (Waters).
3. Manifold equipped with a pressure regulation system and a pump.
4. GlycanPac AXH-1 column (length 250 mm ID 2.1 mm, particle size 1.9 μm, Thermo Scientific).

#### 2.1.2 UPLC Preparation

1. Mobile phase A: water.
2. Mobile phase B: acetonitrile.
3. Mobile phase C: 100 mM ammonium formate pH 4.5. Weigh 6.31 g of ammonium formate in a 1.0 L volumetric flask and bring to volume with purified water. Adjust the pH to 4.50 ± 0.05 with formic acid.
4. Wash solvent: acetonitrile–water (50:50 v/v).
5. Purge solvent: acetonitrile–water (50:50 v/v).

#### 2.1.3 Sample Preparation

1. 50 mg/mL RapiGest: Dissolve a 1 mg RapiGest vial (Waters, article no. 186001861) with 20 μL of Rapid PNGase F buffer 5× (Bioke, article no. P0710S) and homogenize with vortex.
2. Derivatization buffer. Weigh 4.0 g of sodium acetate trihydrate and 2.0 g of boric acid in a 100.0 mL volumetric flask. Dissolve and bring to volume with methanol.
3. Derivatization reagent. Weigh 300 mg of anthranilamide and 300 mg of cyanoborohydride in an amber 10.0 mL volumetric flask. Dissolve and bring to volume with the derivatization buffer (*see* **Note 1**).
4. 95% acetonitrile: acetonitrile–$H_2O$ (95:5 v/v).
5. 20% acetonitrile: acetonitrile–$H_2O$ (20:80 v/v).

### 2.2 Charge-Based Profiling of N-Glycans by UPLC-FLR-$MS^E$

#### 2.2.1 Apparatus

1. UPLC system (autosampler and quaternary solvent manager, Acquity system, Waters).
2. Q-TOF mass spectrometer equipped with an electrospray ion source (XEVO G2XS QTOF, Waters).
3. Fluorimetric FLR detector for Acquity (Waters).
4. Manifold equipped with a pressure regulation system and a pump.
5. GlycanPac AXR-1 (length 150 mm ID 2.1 mm, particle size 1.9 μm, Thermo Scientific).

#### 2.2.2 UPLC Preparation

1. Mobile phase A: water.
2. Mobile phase B: acetonitrile.

3. Mobile phase C: 100 mM ammonium formate pH 4.5. Weigh 6.31 g of ammonium formate in a 1.0 L volumetric flask and bring to volume with purified water. Adjust the pH to 4.50 ± 0.05 with formic acid.
4. Wash solvent: water.
5. Purge solvent: water.

#### 2.2.3 Sample Preparation

1. 50 mg/mL RapiGest: Dissolve a 1 mg RapiGest vial (Waters, article no. 186001861) with 20 μL of Rapid PNGase F buffer 5× (Bioke, article no. P0710S) and homogenize with vortex.
2. Derivatization buffer. Weigh 4.0 g of sodium acetate trihydrate and 2.0 g of boric acid in a 100.0 mL volumetric flask. Dissolve and bring to volume with methanol.
3. Derivatization reagent. Weigh 300 mg of anthranilamide and 300 mg of cyanoborohydride in an amber 10.0 mL volumetric flask. Dissolve and bring to volume with the derivatization buffer (*see* **Note 1**).
4. 95% acetonitrile: acetonitrile–$H_2O$ (95:5 v/v).
5. 20% acetonitrile: acetonitrile–$H_2O$ (20:80 v/v).
6. 0.5 mL Amicon Ultra desalting and concentration devices with molecular weight cutoff of 3 kDa (Merck Chemicals, article no. UFC500324) or 10 kDa (Merck Chemicals, article no. UFC501024).

## 3 Methods

### 3.1 Relative Quantification of the Sialylation of N-Glycan by UPLC-FLR

Digestion can be performed using a one-step or two-step protocol (*see* **Note 2**). Both protocols are described below.

#### 3.1.1 Sample Preparation (One-Step Digestion)

1. Desalt and/or concentrate samples using 0.5 mL Amicon Ultra centrifugal device with a molecular weight cutoff of 3 kDa or 10 kDa. The final volume should be in the 30–40 μL range (*see* **Note 3**).
2. Glycan release: To 250–500 μg of protein (*see* **Note 4**), add 1 μL of Rapid PNGase F per 25 μg of protein and the volume of Rapid PNGase F buffer 5× necessary to reach a final 1× concentration (i.e., for 50 μL of a 5 mg/mL protein solution, add 15 μL of Rapid PNGase F buffer 5× and 10 μL of Rapid PNGase F).
3. Vortex and centrifuge.
4. Incubate at 50 °C for 30 min.

5. Derivatization (*see* **Note 5**): After cooling down of the samples to room temperature, add 2 μL of glacial acetic acid.
6. Vortex and centrifuge (*see* **Note 6**).
7. Add 100 μL of derivatization reagent, vortex and centrifuge.
8. Incubate at 80 °C for 1 h.
9. After cooling down of the samples to room temperature, centrifuge the samples 1 min at 12,000 × *g*.
10. Add 1.0 mL of acetonitrile 95% and vortex thoroughly.
11. Purification: Purification is done using Oasis HLB μelution 30 μm purification plate (Waters, article no. 186001828BA). Vacuum of 2.5–4.0 inHg should allow a slow and steady elution.
12. Condition the wells with acetonitrile 95% (5 × 500 μL).
13. Load the samples (2× 600 μL).
14. Samples are centrifuged prior to the second addition to ensure complete recovery. Wash the wells with 95% acetonitrile (2 × 200 μL).
15. Elute the *N*-glycans with 20% acetonitrile (2 × 100 μL).
16. Dry the samples using a Speed-Vac (*see* **Notes** 7 and **8**).
17. Redissolve the samples in 10 μL water and then 90 μL acetonitrile (*see* **Note 9**).

#### 3.1.2 Sample Preparation (Two-Step Digestion)

1. Desalt and/or concentrate samples using 0.5 mL Amicon Ultra centrifugal device with a molecular weight cutoff of 3 kDa or 10 kDa. The final volume should be in the 30–40 μL range (*see* **Note 3**).
2. Glycan release: To 250–500 μg of protein (*see* **Note 4**), add the proper volume of 50 mg/mL RapiGest in 5× Rapid PNGase F buffer to reach a final concentration of 10 mg/mL (i.e., for 50 μL of a 5 mg/mL protein solution, add 15 μL of RapiGest in Rapid PNGase F buffer).
3. Vortex and centrifuge.
4. Incubate at 90 °C for 3 min.
5. Cool down for 3 min at room temperature.
6. Add 1 μL of Rapid PNGase F per 25 μg or protein (i.e., for 50 μL of a 5 mg/mL protein solution, add 10 μL of Rapid PNGase F), aspirate and dispense to mix.
7. Incubate at 50 °C for 5–30 min.
8. Derivatization (*see* **Note 5**): After cooling down of the samples to room temperature, add 2 μL of glacial acetic acid.
9. Vortex and centrifuge (*see* **Note 6**).
10. Add 100 μL of derivatization reagent, vortex, and centrifuge.

11. Incubate at 80 °C for 1 h.
12. After cooling down of the samples to room temperature, centrifuge the samples 1 min at 12,000 × *g*.
13. Add 1.0 mL of acetonitrile 95% and vortex thoroughly.
14. Purification: Purification is done using Oasis HLB μelution 30 μm purification plate (Waters, article no. 186001828BA). Vacuum of 2.5–4.0 inHg should allow a slow and steady elution.
15. Condition the wells with acetonitrile 95% (5× 500 μL).
16. Load the samples (2× 600 μL).
17. Samples are centrifuged prior to the second addition to ensure complete recovery. Wash the wells with 95% acetonitrile (2× 200 μL).
18. Elute the *N*-glycans with 20% acetonitrile (2× 100 μL).
19. Dry the samples using a Speed-Vac (*see* **Notes** 7 and **8**).
20. Redissolve the samples in 10 μL water and then 90 μL acetonitrile (*see* **Note 9**).

#### 3.1.3 UHPLC Separation

1. Inject 1–2 μL (*see* **Notes 10–12**).
2. The detector parameters are set to $\lambda_{excitation}$ = 330 nm and $\lambda_{emission}$ = 420 nm with a gain of 10 (*see* **Note 13**).
3. Column temperature is set to 30 °C and samples are kept at 5 °C.
4. The gradient is 0–4 min isocratic on 90% B, 10% C, 4–8 min; linear gradient to 40% A, 30% B, 30% C; and isocratic in these conditions to 13 min, linear gradient from 13–15 min to 90% B, 10% C; and isocratic for equilibration in these conditions for 10 min. The flow rate is 0.25 mL/min.

#### 3.1.4 Data Analysis

1. Peaks are identified by the sialylation level: neutral, mono-, di-, tri-, . . ., sialylated.
2. Report the fluorescence % area as a semiquantification parameter of the sialylation level (Figs. 3 and 4).

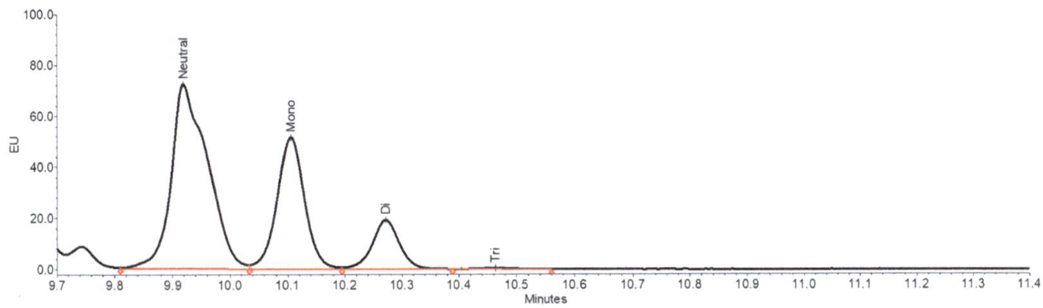


**Fig. 3** FLR profile of etanercept *N*-glycans and sialylation levels

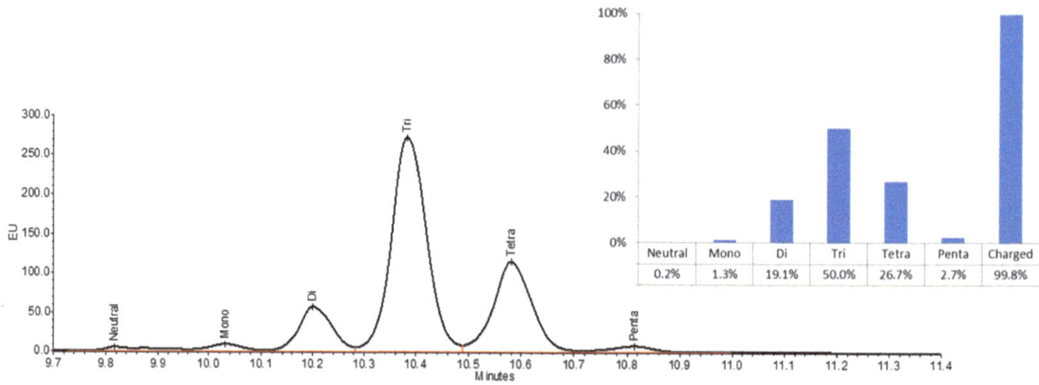


**Fig. 4** FLR profile of a fetuin *N*-glycan and inset to show the calculated sialylation levels

### 3.2 Charge-Based Profiling of N-Glycans by UPLC-FLR-MS$^E$

#### 3.2.1 Sample Preparation

The sample preparation (*N*-glycan release, derivatization, and purification) is identical to the sample preparation for relative quantification of the sialylation of *N*-glycan by UPLC-FLR. The samples are dried and, prior to injection, are redissolved in 50 μL water (*see* **Note 9**).

#### 3.2.2 UHPLC Separation

1. Inject 1–10 μL (*see* **Note 11**).
2. The detector parameters are set to $\lambda_{\text{excitation}} = 350$ nm and $\lambda_{\text{emission}} = 420$ nm with a gain of 10 (*see* **Note 13**).
3. Column temperature is set to 50 °C and samples are kept at 10 °C (*see* **Note 14**).
4. The gradient is 0–60 min linear gradient from 93% A, 7% C to 50% A, 15% B, 35% C and 60–60.1 min; linear gradient to 100% C; 60.1–65 min isocratic on 100% C and 65–70 min, linear gradient to starting conditions 93% A and 7% C; and equilibration for 10 min in these conditions.
5. Flow rate is 0.4 mL/min.

#### 3.2.3 MS Conditions

1. Analyze samples with an hyphenated Q-TOF MS spectrometer, in positive ESI mode (sensitivity).
2. The cone voltage is set to 20 V and the capillary voltage to 2.75 kV.
3. The source temperature is 100 °C and the cone gas flow is 100 L/h.
4. The desolvation gas temperature is 425 °C with a desolvation gas flow of 800 L/h.
5. Acquisition is done between $m/z$ 100 and 2500 with a 1 s scan time.

6. Acquisition is done in $MS^E$ mode (as defined by Waters, allowing both MS and "MS/MS-like" acquisition without prior knowledge of the *m/z* to be fragmented) with a low energy of 6 V and high energy ramp from 20 to 30 V (*see* **Note 15**).

### 3.2.4 Data Analysis

1. Properly integrate all desired peaks.
2. Report the percentage of sialylation by summing the %area of all sialylated peaks.
3. Relative abundance of each *N*-glycans can be given by their fluorescence % area. Additionally, relative abundance can be reported by type (antennary number, fucosylated, etc.). *N*-glycans can be identified from their *m/z* and identification can be confirmed using the high energy MS data, which yields characteristic fragments.
4. Examples of fluorescence chromatograms for etanercept and fetuin are presented in Figs. 5 and 6, respectively.

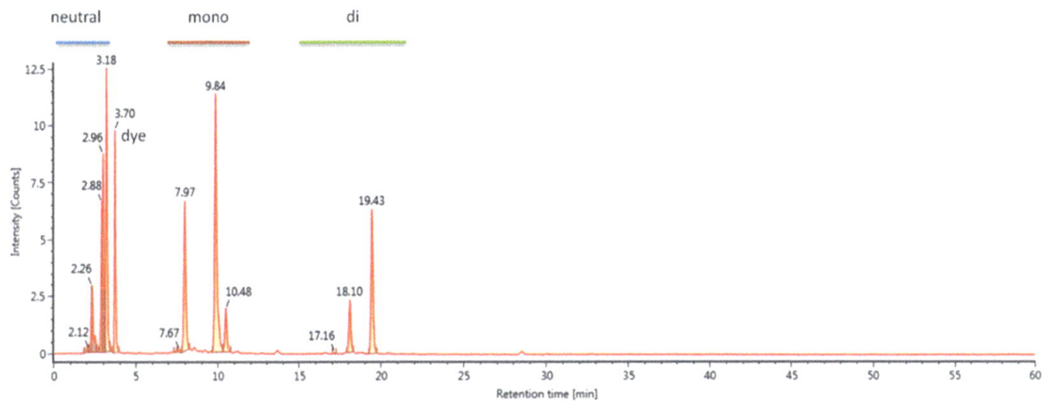


**Fig. 5** FLR profile of etanercept *N*-glycans and sialylation levels

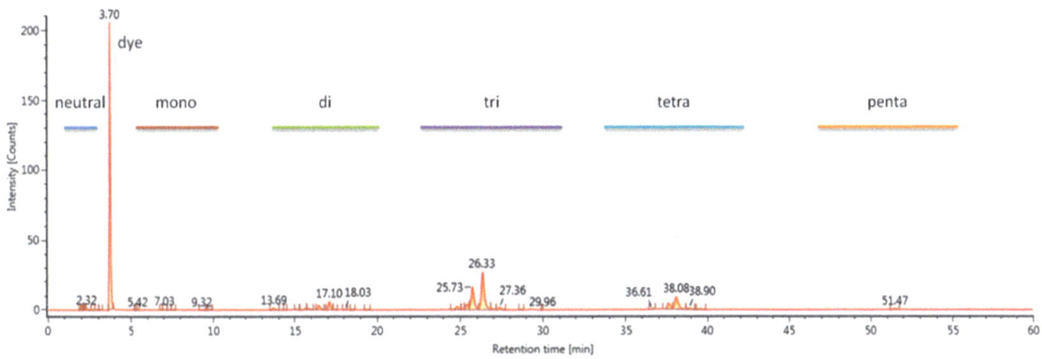


**Fig. 6** FLR profile of fetuin *N*-glycans and sialylation levels

## 4 Notes

1. Derivatization reagent should be stored away from light and cannot be stored more than a week.
2. For most proteins, the one step protocol can be performed but to ensure a complete deglycosylation, a preheating step is required for some proteins (e.g., Fab *N*-glycans).
3. The efficacy of the *N*-glycan release might be affected by molecules present in the buffer. SDS is known to inhibit the PNGase F for instance.
4. Typical preparation is performed with 250 μg of protein.
5. From the derivatization step onward, the sample should be protected from light.
6. After addition of acetic acid, a white precipitate may be visible.
7. Partial desialylation might occur if the temperature in the Speed-vac rises above 28 °C.
8. As an alternative to speed-vac, samples can be freeze-dried.
9. The injected solutions are stable only 78 h at low temperature (e.g., in the injector tray of the UPLC system). After this period, peaks may exhibit altered shapes or splitting.
10. A column cleaning gradient must be performed before the analysis. For that, an additional solution is prepared (50 mM sodium pyrophosphate in 100 mM ammonium formate pH 4.5, set on line D of the quaternary solvent manager). The gradient is as follows: 0–8 min, linear gradient from 50% B and 50% C to 20% B and 80% C. 8–16 min, to 20% B and 80% D. Isocratic on 20% B, 80% D from 16 to 56 min. From 56 to 57.6 min, linear gradient from 20% B, 80% D to 20% B, 80% C and kept isocratic to 20% B–80% C to 80 min. From 80 to 96 min, linear gradient to 80% B, 20% C and linear gradient to 97.6 min, 20% A, 78% B and 2% C and kept in these conditions to 112 min. From 112 min to 130, linear gradient from 20% A, 78% B and 2% C to 90% B, 10% C. The flow rate is 0.25 mL/min. This gradient can also be used in case of performance loss.
11. Before injecting the samples, the gradient should be run three times (no injection) to ensure proper column equilibration.
12. The injection volume is limited to 2 μL; greater volumes will lead to peak distortion.
13. Gain should be adapted to increase signal-to-noise ratio.
14. The temperature of the column (50 °C) does not lead to desialylation of the *N*-glycans, based on comparison performed for five column temperatures from 30 to 50 °C.

15. The calibration of the QTOF is performed by infusion of for instance a solution of NaI (2 mg/mL)/CsI at 2 mg/mL in 50/50 $H_2O$/isopropanol (v/v) on acquisition range *m/z* 100 to 2500. External multipoint calibration is based on singly charged ions.

## References


1. Ghaderi D, Taylor RE, Padler-Karavani V et al (2010) Implications of the presence of N-glycolylneuraminic acid in recombinant therapeutic glycoproteins. Nat Biotechnol 28:863–867. https://doi.org/10.1038/nbt.1651
2. Kaneko Y, Nimmerjahn F, Ravetch JV (2006) Anti-inflammatory activity of immunoglobulin G resulting from Fc sialylation. Science 313:670–673. https://doi.org/10.1126/science.1129594
3. Zhang Z, Shah B, Richardson J (2019) Impact of Fc *N*-glycan sialylation on IgG structure. MAbs 11:1381–1390. https://doi.org/10.1080/19420862.2019.1655377
4. Jelkmann W (2002) Pharmacology, pharmacokinetics and safety of recombinant human erythropoietin (rhEPO). Recomb Hum Erythrop Clin Oncol:203–221. https://doi.org/10.1007/978-3-7091-7658-0_11
5. Teare JM, Kates DS, Shah A, Garger S (2019) Increased branching and sialylation of n-linked glycans correlate with an improved pharmacokinetic profile for BAY 81–8973 compared with other full-length rFVIII products. Drug Des Devel Ther 13:941–948. https://doi.org/10.2147/DDDT.S188171
6. Morimoto K, Tsuda E, Said AA et al (1996) Biological and physicochemical characterization of recombinant human erythropoietins fractionated by mono Q column chromatography and their modification with sialyltransferase. Glycoconj J 13:1013–1020. https://doi.org/10.1007/BF01053197
7. Wada R, Matsui M, Kawasaki N (2019) Influence of N-glycosylation on effector functions and thermal stability of glycoengineered IgG1 monoclonal antibody with homogeneous glycoforms. MAbs 11:350–372. https://doi.org/10.1080/19420862.2018.1551044
8. Darling RJ, Kuchibhotla U, Glaesner W et al (2002) Glycosylation of erythropoietin affects receptor binding kinetics: role of electrostatic interactions. Biochemistry 41:14524–14531. https://doi.org/10.1021/bi0265022
9. Delobel A, Cantais F, Catrain A et al (2013) Therapeutic antibody glycosylation analysis: a contract research organization perspective in the frame of batch release or comparability support. Methods Mol Bio 988:115–143
10. Zhang Q, Li Z, Wang Y et al (2018) Mass spectrometry for protein sialoglycosylation. Mass Spectrom Rev 37:652–680. https://doi.org/10.1002/mas.21555

# Chapter 18

# Linkage Analysis of Oligosaccharides and Polysaccharides: A Tutorial

Ian Black, Christian Heiss, Russell W. Carlson, and Parastoo Azadi

## Abstract

Polysaccharides and oligosaccharides are a diverse group of natural polymers with important biological functions. The diversity of carbohydrate polymers is vast, ranging from small oligosaccharides of defined composition decorating proteins, to large, complex heteropolymers comprising integral cell wall components of plants, fungi and bacteria. An important step in the elucidation of unknown carbohydrate structures in a sample is the assessment of the various linkages present. This is accomplished by performing linkage analysis of the sample. The analysis proceeds as a successive series of chemical steps in which unlinked carbohydrate hydroxyls are marked with methyl groups, the sample is hydrolyzed into monosaccharides and reduced to alditols, and finally free hydroxyls are acetylated. Gas chromatography-mass spectrometry (GC-MS) analysis is employed to analyze the resultant partially methylated alditol acetates (PMAAs). The following paper reviews the major literature pertaining to the specific protocol for linkage analysis of carbohydrates outlined herein. The review details additional steps necessary for the completion of uronic acid linkage analysis, as well as analysis of chitin containing polymers. It also gives chromatographic examples of common erroneous results which the first time practitioner will want to be aware of. Our hope is that this protocol will serve as a definitive guide, allowing novice researchers to perform linkage analysis of carbohydrates in their own lab.

**Key words** PMAAs, Partially methylated alditol acetates, Carbohydrate, Linkage analysis, Polysaccharide, GC-MS, Gas chromatography

## 1 Introduction

Carbohydrate polymers are a diverse class of biological compounds. They comprise the major structural component of plant and fungal cell walls. As glycans, they are also attached to proteins, where they play essential functional and regulatory roles in all multicellular organisms. In addition, they are integral components of the outer surface membranes of many bacteria, where they can mediate host pathogen response, as well as survival in the environment.

Understanding the structure of carbohydrate polymers is an important step toward understanding their functional significance. There are a host of analytical procedures available for this structural

Arnaud Delobel (ed.), *Mass Spectrometry of Glycoproteins: Methods and Protocols*, Methods in Molecular Biology, vol. 2271, https://doi.org/10.1007/978-1-0716-1241-5_18, 

elucidation. Composition analysis gives quantitative information on which monosaccharides are present in a sample. Composition-based methods for determining anomeric configuration are also available. Liquid chromatographic procedures are often used for the molecular weight analysis of large polysaccharides. Furthermore, full structural analysis of polysaccharides and oligosaccharides can be obtained by NMR or mass spectrometry, respectively.

Linkage analysis is the analytical method for determining which hydroxyl groups of each monosaccharide in a polysaccharide are linked to other substituents. To accomplish this, free hydroxyl groups, that is, those not substituted by another monosaccharide or involved in ring formation, are marked with methyl groups via a Williamson ether synthesis. The substituted hydroxyl groups are then exposed via hydrolysis of the carbohydrate sample to partially methylated monosaccharides, which are then reduced to form partially methylated alditols (PMAs). Lastly, the free hydroxyls of the PMAs (that were the substituted positions and ring position in the original carbohydrate) are marked with acetyl groups by acetylation (Fig. 1). The end result of the analysis is a complex mixture of partially methylated alditol acetates (PMAAs). While neutral sugars can be analyzed via the simplified scheme outlined above, some carbohydrates have functional groups (e.g., the 6-position of a uronic acids is a carboxylic group, amino sugars have amino groups in place of hydroxyls) that require modifications in order to be properly analyzed. Other sugars, such sialic acid, require more highly specialized procedures not covered here.

Although many variations of the different steps have been published, proper linkage analysis without artifactual peaks requires that the methylation reaction proceed to a high degree of completion. Therefore, most research (and a large percentage of the background presented here) has focused on the methylation step, as methylation has been the step in the linkage analysis workflow that is most difficult to achieve with quantitative yield. A significant amount of literature has been dedicated to finding optimal reagents and reaction conditions necessary to give complete methylation of different samples. The hydrolysis, reduction and acetylation reactions were detailed satisfactorily early on with only slight modifications over the years.

The analysis of PMAAs utilizes gas chromatography–mass spectrometry (GC-MS), a powerful tool for the structural elucidation of the multiple similar epimers of a given PMAA generated during the procedure. The superior resolution of gas chromatography allows for the separation of dozens of chemically similar carbohydrate derivatives. However, the number of peaks present in complex carbohydrate samples can still lead to significant overlap, especially for low abundance monosaccharides. Differences in retention times distinguish between different glycose (i.e., glucose, mannose, galactose, etc.) PMAAs, and the mass spectra allow for

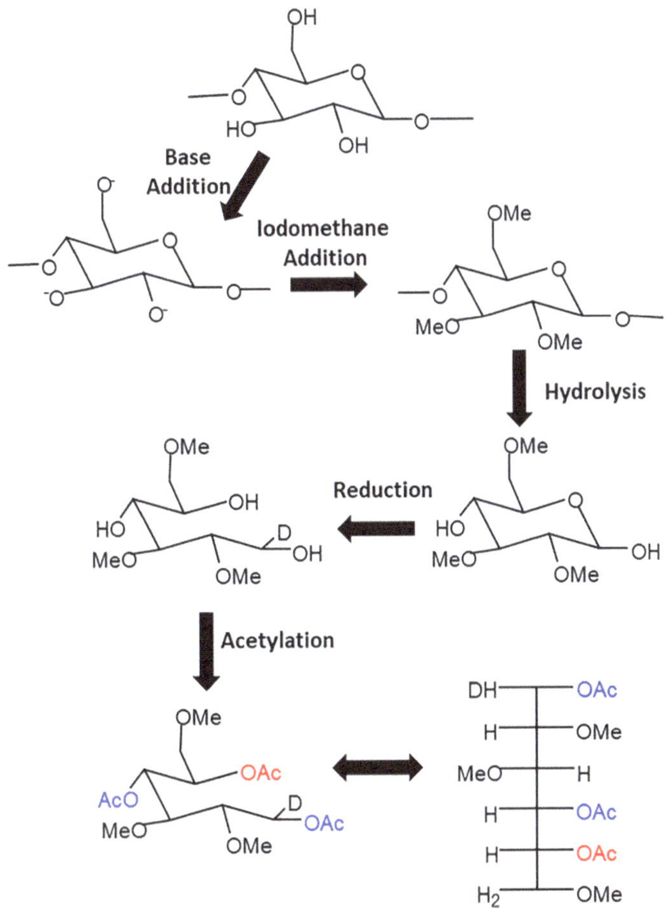


**Fig. 1** The process of linkage analysis of a 4-linked hexose polymer. Permethylation is seen as a two-step reaction; first a conversion of each free hydroxyl to an alkoxide using a strong base, then a substitution reaction using methyl iodide adds a methyl group to the alkoxide. The methylated polymer is then hydrolyzed in acid and reduced using sodium borodeuteride. The deuterated version of this reducing agent is used to label the original reducing end since the reduction can lead to symmetrical alditols where C-1 and C-6 become indistinguishable. Finally, acetylation then marks the newly generated hydroxyls with acetyl groups. The analyst is left with a linear PMAA structure containing unique acetylation and methylation patterns that can be interpreted to ascertain the linkage positions

unambiguous assignment of the linkage positions of each PMAA. Further, interpretation of mass spectral fragmentation allows for the assignment of monosaccharides whose exact structure is unknown, which is important given that nature generates a tremendous diversity of unusual and rare monosaccharides.

The following section details a review of the literature pertaining to the analysis of carbohydrate linkages via the production of PMAAs. We do not attempt to be exhaustive, but instead focus on papers whose results informed the protocols described below. Our lab uses and teaches these protocols extensively and has found them to be the most effective method for generating linkage data. We also describe detailed modifications to these protocols required for samples that contain amino and acidic sugar residues. We also give examples of common mistakes in the analysis to demonstrate the specific results of problematic analyses with a view of helping the analyst troubleshoot the workflow that is being used. Thus, this tutorial is intended as a straightforward entry point for nonexperts into the linkage analysis method and also as a practical resource for those that perform the linkage analysis on a regular basis.

### 1.1 Chemistry of the Linkage Analysis

Early in the twentieth century, different procedures for the methylation of carbohydrates were described in the literature [1, 2]. These reactions employed silver oxide and methyl sulfate, but the methods involved long reaction times and frequently resulted in poor methylation efficiencies. As described by Hakomori [3], "the ability to methylate complex carbohydrates was a painstaking procedure requiring a minimum of 10 g of highly purified sample and several months of work to yield full methylation." Even then, incomplete methylation was common. In 1964, Hakomori published a method that allowed for the methylation of a glycolipid sample using a simplified procedure requiring 30 min of reaction time and generated completely methylated glycolipids with minimal contamination due to undermethylation [4]. The key to the simplified methylation was use of sodium hydride in DMSO, which created a methylsulfinyl carbanion (dimsyl anion), a strong base able to ionize carbohydrate hydroxyl groups, which could then react with methyl iodide to form stable methyl ethers. While the procedure was relatively fast for small oligosaccharides or glycolipids, Hakomori found it necessary to extend the reaction times overnight or even perform two methylations in succession for the larger polymer glycogen, and still the methylation was not 100% complete. This indicated that methylation of large polysaccharides is less efficient than small oligosaccharides. Nevertheless, the procedure proved to be much quicker and simpler than most of the previously established methods. Further work by Sandford and Conrad [5] expanded on Hakomori's procedure to allow linkage analysis of uronic acid-containing polysaccharides from bacteria. Uronic acids cannot directly be analyzed by methylation, because the initially formed uronic acid methyl ester is cleaved during the subsequent hydrolysis of the methylated polysaccharide, reforming a negatively charged carboxyl group. This makes the methyl derivatives of uronic acids nonvolatile, preventing passage through the GC column. However, uronic acids were derivatized by reducing the uronic acid methyl ester intermediate to the primary alcohol. After a second round of methylation to derivatize the newly formed free hydroxyl, the authors hydrolyzed the samples, allowing for analysis of the partially methylated monosaccharides. Chemical characterization of the monosaccharides by GC-MS allowed for identification of the individual residues and their linkages. Thus, the authors put forth a viable procedure of methylation, reduction and hydrolysis for analyzing the linkages of carbohydrates containing uronic acids in a relatively fast and reproducible manner.

Until the 1960s, the methylation analysis consisted of permethylation followed by hydrolysis and chromatographic separation of the resulting partially methylated monosaccharides. The widely varying volatility of these derivatives made them difficult to analyze. They were routinely separated by paper chromatography, a tedious, slow, and low-resolution method. In 1963, Biemann [6] addressed this shortcoming by introducing *O*-acetylation of the free hydroxyls in monosaccharides, increasing the volatility of the derivatized sugars and making it possible to analyze them by mass spectrometry. Notably, the authors extolled the small sample quantity needed and ease with which acetylated monosaccharides could be prepared. Later work by Oades showed the utility of reduction followed by acetylation for quantitative analysis of monosaccharides via the generation of alditol acetates [9].

Another shortcoming in the separation of partially methylated monosaccharides was the equilibration of sugar anomers and ring forms during hydrolysis leading to multiple peaks for each monosaccharide residue constituting the polysaccharide. Lindberg [8] outlined a method employing a reduction step to the analysis, converting the partially methylated monosaccharides to the corresponding alditols before the final acetylation and gas liquid chromatography analysis of the resulting PMAAs.

These improvements were combined with Hakomori's method and became the widely adopted method for the permethylation of carbohydrates [8]. Phillips and Fraser [7] further added to Hakomori's work by showing that the dimsyl base could be more quickly produced using potassium hydride instead of sodium hydride. Furthermore, it was noted that the potassium iodide side product of the reaction is much less soluble in chloroform than the sodium iodide counterpart, and thus the extraction of methylated carbohydrates produced by the use of dimsyl potassium yielded cleaner product than that obtained by dimsyl sodium.

Despite its utility, the dimsyl base has drawbacks. Obtaining pure sodium or potassium hydride is difficult, and the presence of impurities resulting from side reactions can result in noncarbohydrate peaks being present in GC chromatograms. Furthermore, the reagents can be challenging to work with for novice researchers. Also, the base is sensitive to moisture and carbon dioxide, making long-term storage problematic. Following up on earlier work showing that the addition of potassium tert-butoxide to DMSO would generate dimsyl anions in equilibrium with butoxide base

[10], Finne et al. argued that the use of this reagent for base preparation was preferable over the metal hydrides as it gave equivalent methylation while also being more stable in storage [11].

Many modifications intended to improve the efficiency and ease of the linkage analysis protocol have been suggested. Waeghe et al. demonstrated that by minimizing the reagents used and being careful to not over dry samples and thus evaporating the derivatized carbohydrate, linkage analysis could be performed on less than 50 μg of sample [12]. Thus, the multistep reaction procedure could be performed on small sample volumes as long as it is carried out with care. This would prove important for glycoprotein analysis as yields of glycans from proteins are generally quite low. The authors also reported the use of sodium borodeuteride for prior reduction of the reducing end to prevent peeling, a degradation reaction that can destroy oligomers under basic conditions [13]. Though this step would later be shown to generate artificial methyl glycosides by York et al. [14], Harris et al. reported a complete review of each step in the linkage analysis, optimizing them for speed and recovery [15]. These authors demonstrated that the methylation of pullulan was 100% using potassium hydride derived dimsyl anion but only 50% using tert-butoxide derived dimsyl anion. They also noted that increasing the time of methylation from 10 min to 1 h did not increase the methylation efficiency, indicating that the methylation reaction happens relatively quickly for the sample they analyzed, after which extending the reaction time did not improve results.

The preferred choice of base for methylation analysis changed drastically with the work of Ciucanu and Kerek [16] who argued that the effective base in the tert-butoxide procedure was hydroxide anions rather than dimsyl base, whose presence was due to small amounts of water in the starting tert-butyl alcohol. They even pointed out that use of dry tert-butyl alcohol resulted in very little methylation. The authors thus reasoned it would be more efficient to use alkali metal hydroxides in DMSO to ionize carbohydrate hydroxyls. Accordingly, they showed that the addition of finely ground sodium hydroxide and iodomethane to a sample dissolved in DMSO resulted in efficient methylation. These results were obtained relatively rapidly and eliminated the need for more exotic base preparations using metal hydrides or potassium tert-butoxide. As such, the new method rapidly gained favor as it was more facile and did not require specialized equipment or advanced technical expertise. This work was expanded upon by Anumula and Taylor who noted that the same results could be achieved using a base prepared from a 50% solution of sodium hydroxide [17]. In this instance, 50% sodium hydroxide is washed repeatedly with DMSO to remove the water (*see* Subheading 3 below). The purpose is to generate an extremely fine suspension of sodium hydroxide in DMSO which can then be added to the sample that is dissolved in

DMSO. This proved more convenient than the method of Ciucanu and Kerek, which requires grinding solid sodium hydroxide into a fine and uniform powder, a task Anumula and Taylor describe as "difficult and hazardous." Some drawbacks to methylation using sodium hydroxide were noted. York et al. published results showing that linkage analysis of oligosaccharides of known structures resulted in the detection of anomalous linkages [14]. The authors showed that methylation with sodium hydroxide resulted in oxidation of oligoglycosyl alditols derived from plant cell-wall xyloglucans, thus giving small but significant false linkage peaks in the chromatogram when alditols are present in the samples. They further showed that oxidation products from terminal, nonreducing glucopyranosyl residues were not detected, indicating terminal residues were being degraded, which resulted in incorrect terminal to linked ratios between monosaccharides. They concluded that the results of oxidative degradation during linkage analysis might limit its utility in the structural elucidation of highly purified products where even small oxidative losses/conversions could result in erroneous conclusions. Work by Needs and Selvendran [18, 19] demonstrated the significance of the order of reagent addition to the success of the methylation reactions. It was shown that oxidation is avoided if the base is added for a period of time prior to the methylating agent. This order of addition had been followed by Hakomori, but Ciucanu and Kerek added base and methyl iodide simultaneously. Ciucanu and Costello later reported that along with the order of base and iodomethane addition, the anhydrous nature of the reaction was partially responsible for oxidative degradation [20]. The authors found that in the presence of trace amounts of water, the oxidation reaction reported by York et al. was eliminated, although it is important to avoid too much water, which can result in undermethylation. While the authors continued to use powdered sodium hydroxide, which acts as a strong desiccant, their results explain why oxidation has not been observed in the alternative preparation of sodium hydroxide developed by Anumula and Taylor. Starting with a 50% solution of sodium hydroxide ensures that the final base preparation will continue to have trace amounts of water present.

## 2 Materials

### 2.1 Preacetylation

1. Pyridine.
2. Acetic anhydride.
3. Borosilicate screw top test tubes with Teflon-lined caps.

### 2.2 Methylation

1. 50% NaOH.
2. Anhydrous methanol.

3. DMSO.
4. Potassium hydride.
5. Hexane.
6. Ethanol.
7. Iodomethane.
8. Dichloromethane (DCM).
9. Acetonitrile.
10. Borosilicate screw top test tubes with Teflon-lined caps.
11. C18 disposable purification columns.
12. Plastic pipets.
13. Glass pipets.
14. Desktop centrifuge.
15. Three-necked flask.
16. Vortex mixer.
17. Magnetic stirrer and stir bars.
18. Ice bath.
19. Argon or nitrogen gas.

### 2.3 Reduction of Uronic Acid Methyl Ethers

1. 20 mg/mL solution of lithium borodeuteride in 90% THF.
2. Acetic acid.
3. Methanol/acetic acid solution (9:1, v/v).
4. NaOH/DMSO base (*see* Subheading 3.2.1).
5. Dionex OnGuard II cartridges.
6. Nitrogen gas.

### 2.4 Hydrolysis

1. 2 M trifluoroacetic acid (TFA).
2. Isopropanol.
3. Nitrogen gas.

### 2.5 Reduction

1. 10 mg/mL solution of sodium borodeuteride in 20 mM ammonium acetate.
2. Acetic acid.
3. Methanol–acetic acid solution (9:1, v/v).
4. Nitrogen gas.

### 2.6 Acetylation

1. Acetic anhydride.
2. Pyridine.
3. Trifluoroacetic acid (TFA).
4. Isopropanol.

5. Dichloromethane (DCM).
6. GC vial insert.
7. Nitrogen gas.

### 2.7 GC-MS Analysis

1. GC-MS system.
2. SP-2331 bonded phase fused silica capillary column (Supelco).
3. Equity-1 fused silica capillary column (Supelco).

## 3 Methods

What follows below is a detailed method for the generation of PMAAs. It is slightly different from previously published procedures [21] in that it does not focus on a single type of sample but can be used to analyze many different types of polysaccharides. As such, some steps presented below may be unnecessary depending on the type of sample being analyzed. The method is equally applicable to polysaccharides and oligosaccharides, as long as the methylation step is suitably adapted. The procedure includes steps for working with samples containing neutral, amino, and acidic glycosyl residues. It is important to choose specific steps based on the type of sample to be analyzed. Preacetylation is required only if the sample contains chitin or chitosan. Likewise, prereduction is not necessary for samples containing only neutral sugar residues.

### 3.1 Preacetylation

For some samples containing amino sugars that are sparingly soluble, a preacetylation step may be employed to help with solubilization. Chitin-containing samples are one such example. Because acetylation is a labile modification, no further additional steps are required, and the procedure does not alter the end result, as any added acetyl groups are labile under the basic conditions of the permethylation.

1. Place 100–300 μg of sample in a clean test tube.
2. Add 150 μL pyridine to the sample and sonicate 30 s.
3. Add 150 μL acetic anhydride to the sample.
4. Heat the sample to 100 °C for 1 h (expect a brown/red color to form in the sample solution. This color will remain throughout the analysis) (*see* **Note 1**).
5. Dry the cooled sample under nitrogen.
6. Once completely dry, proceed to the appropriate methylation step.

### 3.2 Methylation

Proper methylation of samples is one of the most important steps of linkage analysis as mistakes in the methylation procedure inevitably lead to erroneous results [25]. The type of procedure used is dependent on the types of carbohydrates expected in the sample. For uronic acid containing samples, the expectation of β-elimination requires the use of dimsyl anion. For neutral or amino containing linkage analysis use of sodium hydroxide is preferred.

#### *3.2.1 NaOH Base Preparation*

1. Add 100 μL of 50% NaOH to a new borosilicate test tube using a plastic pipette.
2. To this, add 200 μL anhydrous methanol and vortex for 10 s (the solution should become transparent).
3. Add 2 mL DMSO and vortex for 5 s. The resultant solution should be cloudy.
4. Centrifuge at low speed. A desktop centrifuge is sufficient. The result should be a precipitate on the walls and at the top of a clear DMSO layer with a gel–like substance (the base) at the bottom of the tube.
5. Remove the top DMSO layer, including washing the white precipitate off the tube walls with the DMSO before removing. Be careful not to disturb or remove the gelatinous base at the bottom of the tube.
6. Add 2 mL of fresh DMSO and repeat the washing step five times until the tube no longer contains white precipitate at the top. (Air bubbles in the solution can make it difficult to see the precipitate, but a brief sonication will remove the air bubbles making it easier.)
7. The final base should be dissolved in 2 mL of DMSO for use.

#### *3.2.2 Potassium Dimsyl Base Preparation*

The preparation of dimsyl from potassium hydride requires some minimal inert-gas equipment as the reaction requires anhydrous conditions. To this effect, it is important to dry the glassware in an oven prior to use (cool the glassware under an inert-gas flow prior to start of the preparation). Also, the DMSO should be dried with molecular sieves for 48 h prior to use (*see* **Note 2**).

1. Weigh out 4.8 g of potassium hydride dissolved in mineral oil in a clean three-necked flask under an argon atmosphere and containing a stir bar.
2. Add 10 mL hexane to the flask and stir for 15 s to mix. Turn off the stirring and allow the potassium hydride to collect at the bottom of the flask.
3. Remove the hexane layer, being cautious to not remove any of the potassium hydride in the process. Discard in a beaker containing ethanol.
4. Repeat **steps 3** and **4** five times to fully remove the mineral oil.
5. Allow the purging argon to evaporate the hexane while continuing moderate stirring (*see* **Note 3**).

6. Once the hexane has been completely evaporated, slowly add 10 mL of dry DMSO to the potassium hydride dropwise while continuing to stir at moderate speed. The products of this reaction, in addition to dimsyl base, are hydrogen gas and heat; therefore, adding DMSO too quickly can cause the reaction to overheat and ignite the hydrogen gas resulting in a fire. (Expect vigorous bubbling as the hydride reacts with the DMSO and hydrogen is released. Potassium hydride is very reactive and extreme care must be taken to ensure it is not exposed to moisture).
7. After bubbling subsides, the resulting DMSO solution/suspension of dimsyl base can be used directly or aliquoted into glass vials for storage (*see* **Note 4**).

#### 3.2.3 Methylation of the Sample with NaOH Base

1. Add 50–400 μg of purified sample to a test tube. (Use more sample if you suspect it is not pure).
2. Add 200 μL dry DMSO to the sample.
3. Add a clean, dry stir bar to the sample.
4. Allow the sample to stir for 5 min. If the sample is difficult to solubilize, it can be heated to 80 °C for 5 min, making certain that it is allowed to cool before base addition. It can also be stirred overnight or for several days prior to analysis.
5. Add 200 μL of the NaOH/DMSO base slurry (prepared as described in Subheading 3.2.1) to the sample and allow the mixture to stir for 10 min. Alternatively, if the sample is hard to solubilize, it can be sonicated for 10 min.
6. Add 100 μL iodomethane to the sample and allow the sample to stir for 30 min. (For oligosaccharides, only 10–15 min reaction time is needed for this step).
7. For polysaccharides, repetition of **steps 5** and **6** can be employed to ensure complete methylation of the polymer.
8. Add 2 mL of water to the sample. The iodomethane will no longer be soluble and the solution will appear cloudy.
9. Add 2 mL of dichloromethane to the samples.
10. Vortex the sample for 20 s, then use low speed centrifugation to separate the aqueous and organic layers.
11. Remove the aqueous (top) layer with a glass pipet, being careful to not remove any of the organic layer.
12. Add a fresh portion of 2 mL of water and repeat **steps 10** and **11**.
13. After washing a total of five times, carefully remove the bottom organic layer and transfer to a new tube (*see* **Note 5**).
14. Dry the organic layer under nitrogen.

*3.2.4 Methylation of the Sample with Dimsyl Base*

1. Add 50–400 μg of purified sample to a test tube. (Use more sample if you suspect it is not pure).
2. Add 200 μL dry DMSO to the sample.
3. Add a clean, dry stir bar to the sample.
4. Allow the sample to stir for 5 min. If the sample is difficult to solubilize, it can be heated to 80 °C for 5 min, making certain that it is allowed to cool before base addition. It can also be stirred overnight or for several days prior to analysis.
5. Add 80 μL of the DMSO dimsyl base (prepared as described in Subheading 3.2.2) to the sample and allow the mixture to stir for 10 min. Alternatively, if the sample is hard to solubilize, it can be sonicated for 10 min.
6. Place the sample in an ice bath with a stir plate, allowing the sample to stir while chilling. (The sample should be cold but not frozen at this step. If it begins to freeze, start iodomethane addition. The iodomethane addition will prevent or reverse freezing at this point).
7. Slowly add 100 μL of ice cold iodomethane dropwise from a glass syringe to the sample. Let the sample stir for 15 min in ice water.
8. Add 2 mL of water to the sample.
9. Insert a pipet into the sample solution and slowly purge with nitrogen to evaporate the iodomethane. The cloudy sample will become clear as the iodomethane evaporates.
10. Load the sample solution onto a preconditioned, disposable C18 column.
11. Wash the C18 column with 4 mL water.
12. Elute the sample from the C18 column with 2 mL acetonitrile.
13. Evaporate the acetonitrile under nitrogen.

**3.3 Reduction of Uronic Acid Methyl Ethers**

Samples containing uronic acids should be methylated using dimsyl base (as described in Subheading 3.2.4), and the resulting carboxylic acid methyl ester reduced. After reduction, the sample is then methylated again using NaOH base methylation as a single methylation step of polysaccharides almost always results in incomplete methylation. Also, the reduction of the uronic acid residues converts the carboxyl groups to primary alcohols resulting in free hydroxyls that require the second methylation step. Lithium borodeuteride is a suitable reducing agent for uronic acid methyl esters as it is stronger and more effective than $NaBD_4$. The deuterated reducing agent is used so that the uronic acid will be converted to its hexose (–COOH is reduced to $–CD_2OH$) derivative, and the incorporation of deuterium will also distinguish the former uronic acid (now a $-CD_2OH$–containing hexose from a normal $-CH_2OH$–containing hexose, which may also be present in the sample. The use of lithium borodeuteride is an important distinction from the use of sodium borodeuteride during the reduction of the carbonyl called for later in the procedure. The methyl ester of the carboxylic acid is less reactive than the carbonyl, and therefore reduction using sodium borohydride would be incomplete. Therefore, we use lithium borodeuteride, as it is a more powerful reducing agent. Lithium aluminum deuteride is an even more powerful reducing agent and can be used as well. However, this reagent reacts violently with water and must be used and stored under strict exclusion of moisture. In addition, $LiAlD_4$ tends to form emulsions that are difficult to separate due to the production of aluminum hydroxide during the reaction. Therefore we use lithium borodeuteride as an alternative. Neutralization of the reducing agent leaves a lot of salt in the sample which must be removed in order to get complete methylation of the sample during the second methylation step. Failure to remove the salt prior to the second methylation results in lower recovery of the sample. The uronic acid reduction procedure is as follows.

1. To the dry methylated sample, add 400 μL of a 20-mg/mL solution of lithium borodeuteride in 90% THF.
2. Heat the sample at 100 °C for 4 h.
3. Neutralize the reaction by adding five drops acetic acid.
4. Dry the sample under nitrogen.
5. Add 500 μL water which will dissolve the salt and pass through Dionex OnGuard II Cartridge.
6. To minimize sample loss, collect the eluent from the cartridge in the original sample tube.
7. Dry the sample under nitrogen to remove the THF. Excess water can be removed via lyophilization after the THF has been removed.
8. In order to ensure complete removal of the borate (which can interfere with subsequent methylation), dry twice with 200 μL of a methanol: acetic acid (9:1) solution.
9. Methylate the sample using the NaOH/DMSO base as detailed above in Subheading 3.2.3 (only one round of methylation is necessary).

### 3.4 Hydrolysis

For hydrolysis, organic acids as opposed to mineral acids are generally employed as they are easily removed via evaporation. Care must be taken to remove all the acid after the hydrolysis as any residual acid left in the test tube will consume reducing reagent, possibly resulting in incomplete reduction of the samples.

1. To the dried sample, add 200 μL of 2 M trifluoracetic acid (TFA).
2. Heat at 120 °C for 2 h.
3. Dry the cooled sample under nitrogen. To aid evaporation, several drops of isopropanol can be added to the sample (*see* **Note 6**).
4. Once the sample is dry, 100 μL of isopropanol is added to the sample, and the mixture is vortexed for 5 s.
5. The sample is dried again.
6. **Steps 4** and **5** are repeated twice more to ensure all the TFA is evaporated from the sample.

### 3.5 Reduction

The reduction of the partially methylated monosaccharides serves to reduce the complexity of the sample chromatogram. Without reduction, each monosaccharide linkage would appear as 2 to 4 peaks due to α- and β-anomer mixtures as well as possible furanose and pyranose ring forms. By reducing the sample, both anomeric and ring form complexity is eliminated. Furthermore, we use sodium borodeuteride instead of borohydride. The deuteride labels the anomeric carbon (e.g., conversion of –CHO to –CHDOH). On GC/MS analysis of the final PMAAs, this allows one to distinguish mass spectrum fragments arising from C1-end from those arising from the C6-end of each PMAA.

1. To the dry sample, add 300 μL of a 10 mg/mL solution of sodium borodeuteride in 20 mM ammonium acetate.
2. Allow the sample to sit at room temperature for a minimum of 3 h to overnight.
3. Neutralize the sample by adding three drops of acetic acid.
4. Dry the samples under nitrogen.
5. To the dry samples, add 5 drops 9:1 methanol–acetic acid. Vortex for 5 s and dry.
6. Repeat **step 5** two more times. A dry, white, salty crust should appear in the tube.

### 3.6 Acetylation

Acetylation of the partially methylated and reduced monosaccharides can be accomplished by using acetic anhydride and either a base or acid catalyst. For samples containing neutral sugars or reduced uronic acids, we use TFA as a catalyst, as the reaction is efficient and relatively quick. For samples containing amino sugars, we employ pyridine as the catalyst. A longer reaction time and higher temperature are necessary for complete acetylation of both the free hydroxyls and the primary amines of the monosaccharides. If the contents of the sample are unknown, then it is best to use the pyridine acetylation to ensure everything is accounted for.

#### 3.6.1 For Neutral Partially Methylated Alditols

1. Add 200 μL acetic anhydride and 200 μL concentrated TFA to the samples.
2. Heat the samples to 35 °C for 15 min.
3. Add 1 mL isopropanol to the sample and allow to dry under nitrogen (the sample contains a lot of salt left over from the reduction step and will therefore only dry down to a syrup).
4. Once dried to a syrup, add 2 mL of water to the sample, followed by 2 mL dichloromethane.
5. Vortex the sample for 20 s, then use low speed centrifugation to separate the aqueous and organic layers.
6. Remove the aqueous layer via aspiration, being careful to not remove any of the organic layer.
7. Add a fresh 2 mL of water and repeat **steps 5** and **6**.
8. After washing a total of five times, carefully remove the bottom organic layer and transfer to a new tube.
9. Dry the dichloromethane layer until only ~50 μL remains.
10. Pipet the dichloromethane into a GC vial insert.

#### 3.6.2 For Amino Containing Sugars

1. Add 150 μL acetic anhydride and 150 μL pyridine to the samples.
2. Heat the samples at 100 °C for 40 min.
3. Dry the cooled samples under nitrogen.
4. Once dry, add 2 mL of water to the sample followed by 2 mL dichloromethane.
5. Vortex the sample for 20 s, then use low speed centrifugation to separate the aqueous and organic layers.
6. Remove the aqueous layer via aspiration, being careful to not remove any of the organic layer.
7. Add a fresh 2 mL of water and repeat **steps 5** and **6**.
8. After washing a total of five times, carefully remove the bottom organic layer and transfer to a new tube.
9. Dry the dichloromethane layer until only ~50 μL remains.
10. Pipet the dichloromethane into a GC vial insert.

### 3.7 GC-MS Analysis

Analysis of the neutral PMAAs is accomplished using a 30-m Supelco SP-2331 bonded phase fused silica capillary column. For analysis of amino containing PMAAs, a Supelco Equity-1 fused silica capillary column is required. Tables 1 and 2 provide oven conditions for analysis of PMAAs using both columns.

### 3.8 Interpretation of Spectra

A full review of the mass spec analysis of the PMAAs generated during linkage analysis is beyond the scope of this tutorial. However, papers with tables assigning linkages based on mass

**Table 1**
**Oven ramp conditions for SP2330 column used for neutral PMAAs analysis**

| Oven Ramp (°C/min) | End Temp. (°C) | Hold Time (min) |
|---|---|---|
| | 60 | 1 |
| 27.5 | 170 | 0 |
| 4 | 235 | 2 |
| 3 | 240 | 12 |

**Table 2**
**Oven ramp conditions for EC-1 column used for amino containing PMAA analysis**

| Oven Ramp (°C/min) | End Temp. (°C) | Hold Time (min) |
|---|---|---|
| | 80 | 1 |
| 4 | 220 | 0 |
| 20 | 260 | 25 |

fragmentation have been published [22], along with a publicly accessible database of common PMAA spectra available at the Complex Carbohydrate Research Center website [24]. Protocols for making standards to aid in identifying linkages [23] have also been published. Use of linkage standards is especially important for distinguishing monosaccharides whose fragmentation is similar. For example, mannose, glucose, and galactose are all hexose residue and therefore all produce the same fragmentation for a given linkage. In order to identify the hexose linkages from each other, it is necessary to know the elution order (4–linked mannose comes out before 4–linked galactose, followed by 4–linked glucose on a particular GC column). This is accomplished by making and running standards using the individual sugars present in the sample of interest.

### 3.9 Example Chromatograms

Below are several example chromatograms. The purpose of this section is to demonstrate the results of linkage analysis from several different samples. Here, we also include results of failed analyses in order to help the reader troubleshoot any problems in their analyses.

Figure 2 shows the effect of preacetylation on the analysis of chitin. Both samples give similar chromatograms, with the expected 4-linked *N*-acetylglucosamine being the predominant peak arising from the linkage analysis. Smaller linkages are also present in the sample, along with a medium sized terminal *N*-acetylglucosamine peak. At first glance, both samples give similar amounts of 4-linked

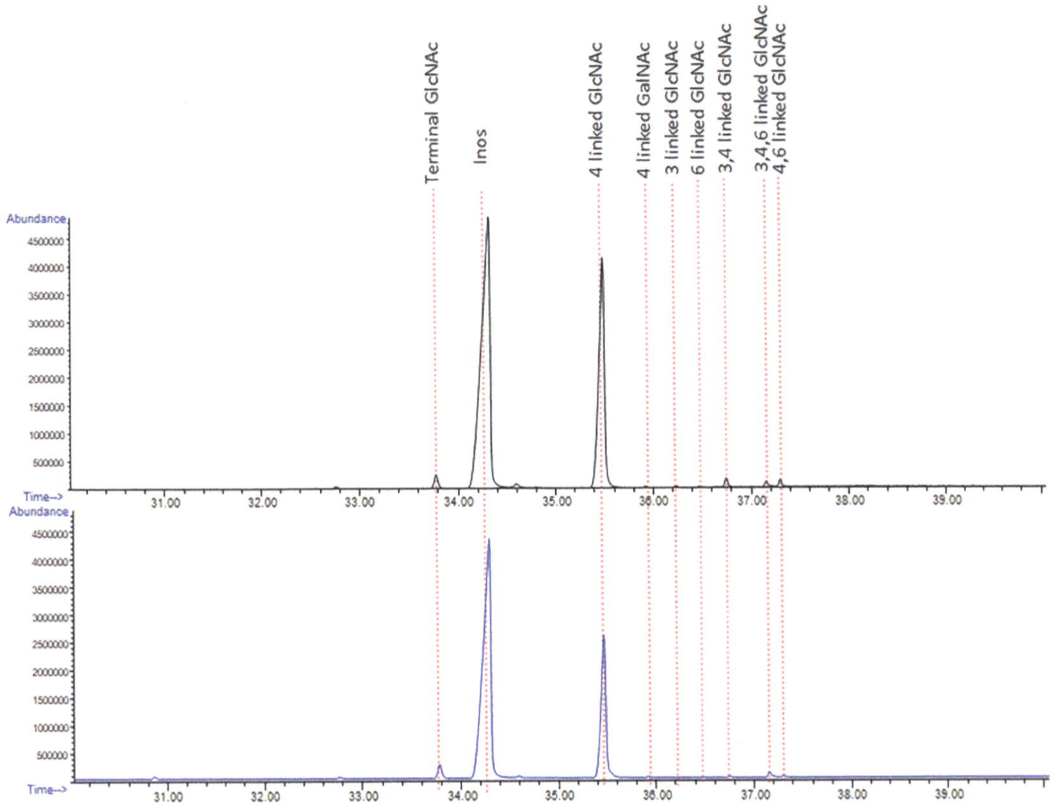


**Fig. 2** Linkage analysis of chitin (Sigma, from shrimp shell) with (top, black) and without (bottom, blue) the use of a preacetylation step. 150 μg chitin was used for the analysis with preacetylation. 250 μg chitin was used for the analysis without acetylation. 30 μg inositol was added to both samples as an internal standard

*N*-acetylglucosamine, with the ratio of the nonacetylated peak being only slightly smaller in intensity than its inositol peak. However, the nonacetylated chromatogram represents a starting weight of 250 μg chitin, whereas the acetylated sample is only 150 μg chitin. Thus, despite having 40% less starting material, the preacetylated chitin still has a slightly higher amount of 4-linked *N*-acetylglucosamine detected compared to internal standard. The reason is that native chitin is not soluble, even in the presence of strong sodium hydroxide base. This poor solubility results in some of the chitin not being methylated and thus being lost during the extraction steps after methylation. It is therefore advisable to preacetylate chitin or chitosan samples.

Figure 3 shows the results of incomplete methylation and poor recovery. Because the samples are comprised of polysaccharides, it is necessary to extend the methylation time in order to fully methylate them. The top chromatogram in Fig. 3 shows the results of not extending the methylation time. In this example, the methylation time was 20 min (5 min base treatment, 15 min after iodomethane addition), and only one round of methylation, rather than two, was

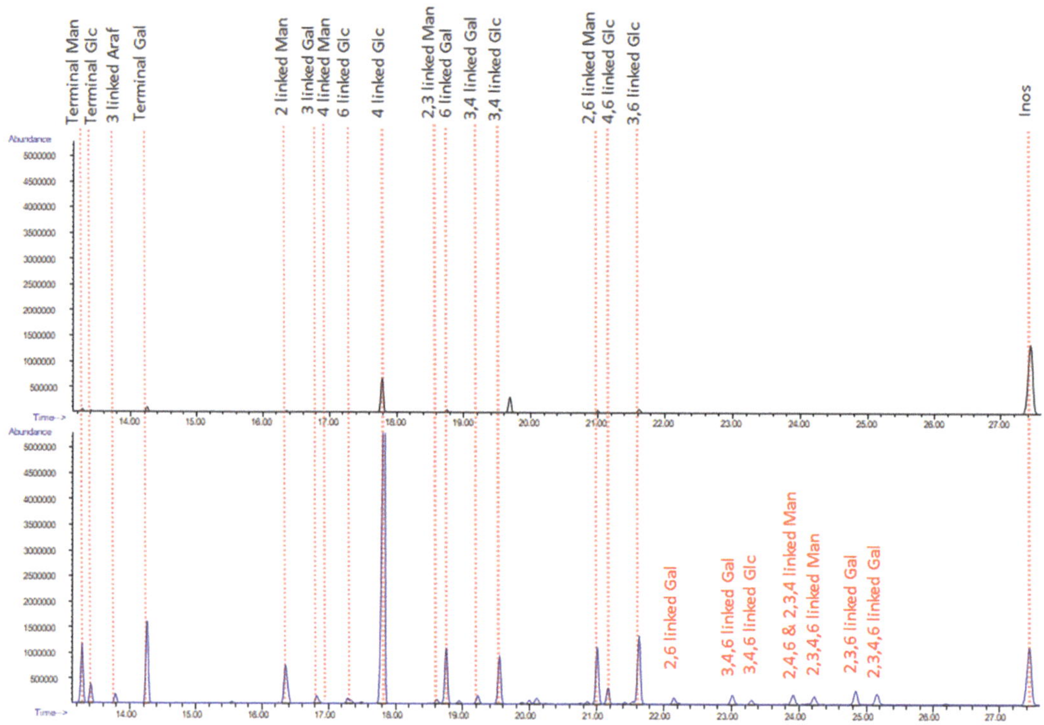


**Fig. 3** Analysis of the same sample containing multiple neutral polysaccharides. For the analysis, 100 μg commercially sourced mannan, arabinogalactan, and starch were mixed to use as a standard. 10 μg inositol was used as internal standard. Poor recovery or undermethylation are shown, caused by abbreviated methylation time (top), and by addition of 20μL water to the sample (bottom) respectively

performed. Comparison of the inositol peaks to the carbohydrate linkages in the top chromatogram shows that most of the sample was not methylated and as a result was lost during the sample cleanup. This explains why, despite having the same amount of starting material, the carbohydrate peaks are of lower intensity compared to the undermethylated sample peaks seen in the bottom chromatogram. The bottom chromatogram shows the results when the samples were not completely dried. The sample gives intense peaks, and the correct linkages are visible, but there are also artifacts of "multiply linked" peaks (labeled in red) that are the result of undermethylation. Undermethylation commonly occurs in linkage analysis. It can be a result of samples that have not been fully dried, of not using enough base, or from analysis of a poorly soluble sample. In each case the results are that numerous "multiply linked" peaks appear in the chromatogram. In our example these undermethylated peaks are not very intense, so it is tempting to simply ignore them. However, many times undermethylation can result in large "multiply linked" peaks. As a result, distinguishing actual linkages from those due to undermethylation can become a

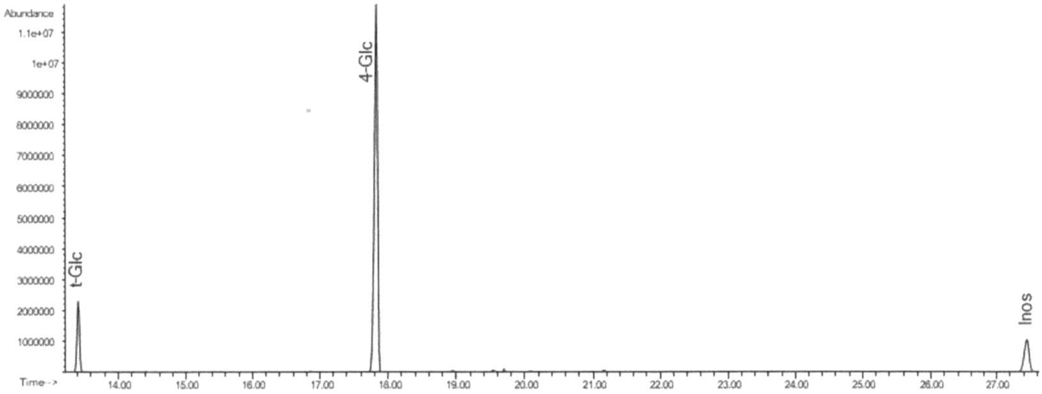


**Fig. 4** Analysis of 100 μg maltoheptose using the same abbreviated methylation conditions as was used to obtain the top chromatogram of Fig. 2. 10 μg inositol was used as an internal standard. The sample was run alongside the samples shown in Fig. 3

challenge. Sims et al. list several examples of published results in which the samples showing clear signs of undermethylation are misinterpreted as legitimate linkage peaks.[25]

Figure 4 shows that oligosaccharides can be fully methylated with shorter reaction times than polysaccharides. Full methylation of maltoheptaose was achieved in 20 min. The 4-linked glucose peak is much larger than the inositol, indicating good sensitivity compared to the results of Fig. 3 (top). Thus, the requirements of the methylation are different depending on the sample being analyzed. Working with polysaccharides requires extending the methylation time and doing two rounds to ensure complete methylation. But when working with small oligosaccharides, a single round with shorter methylation time is sufficient.

Figure 5 shows the importance of using dimsyl base for methylation of acidic residues. The results of the linkage analysis using dimsyl base shown in the top chromatogram are as expected. The major residue in the sample is a mixture of 4-linked galactose and 4-linked 6-$D_2$-galactose (i.e., galacturonic acid), with the galacturonic acid being the major contributor to the peak area. This contrasts with the bottom chromatogram, which was obtained after permethylation with NaOH and shows several neutral carbohydrate peaks that are present in both chromatograms but only a small peak of 4-linked galactose and no indication of galacturonic acid. This is the result of performing the methylation with NaOH base, which causes degradation of the 4-linked galacturonic acid via β-elimination, leaving only the nonacidic sugars behind in the analysis. Hence, Fig. 5 shows the necessity of methylating uronic acid–containing samples with dimsyl base rather than with sodium hydroxide.

Linkage analysis of glycans can be difficult as the amount of sample is usually limited. One must also deal with the added difficulty of other noncarbohydrate components in the sample if the

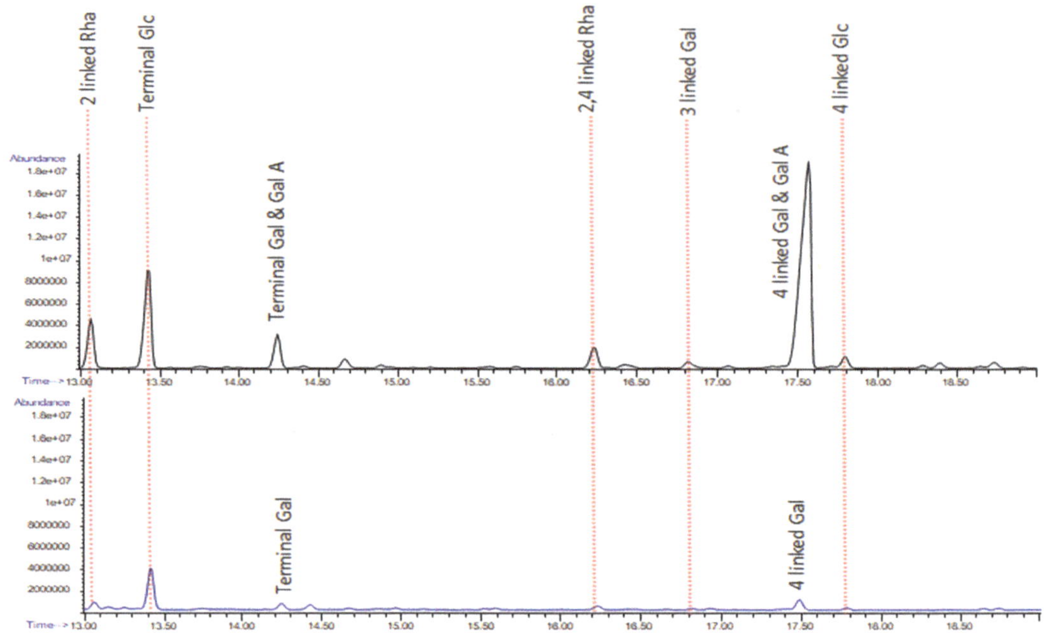


**Fig. 5** Analysis of 4-linked polygalacturonic acid (Megazyme) containing sample. Sample methylated using dimsyl anion (top) and sodium hydroxide base (bottom). 1.5 mg of each sample was used, and the samples were run in parallel

glycans are not purified beforehand. Thus, one important consideration is the background noise caused by noncarbohydrate components that may be in the sample. Because the abundance of carbohydrate may be low, it is important to reduce background noise and contamination as much as possible. One way to achieve this is to thoroughly wash the sample after the methylation step and again after the acetylation step. Figure 6 shows the results of linkage analysis on 100 μg fetuin. Here, the sample glycans are not purified in any way prior to analysis. The top chromatogram shows both wash steps repeated five times. The chromatogram shows several PMAAs due to the glycosyl linkages present for the major glycans on fetuin. Although the signal is not as strong as those seen above using polysaccharides, it is easy to make out the major linkages as well as some noncarbohydrate peaks. This is contrasted with the bottom chromatogram in Fig. 6 showing the results from a sample that was washed only two, rather than five times. The baseline peaks are noticeably more intense although the amount of fetuin was equal in both experiments. Also, there are more noncarbohydrate peaks (denoted with asterisks) present at higher intensities. Figure 6 shows that the extra washes, while time-consuming, are important for enhancing the signal-to-noise ratio of the samples and result in fewer noncarbohydrate peaks cluttering the chromatogram. And this is especially important when starting sample amount is limited.

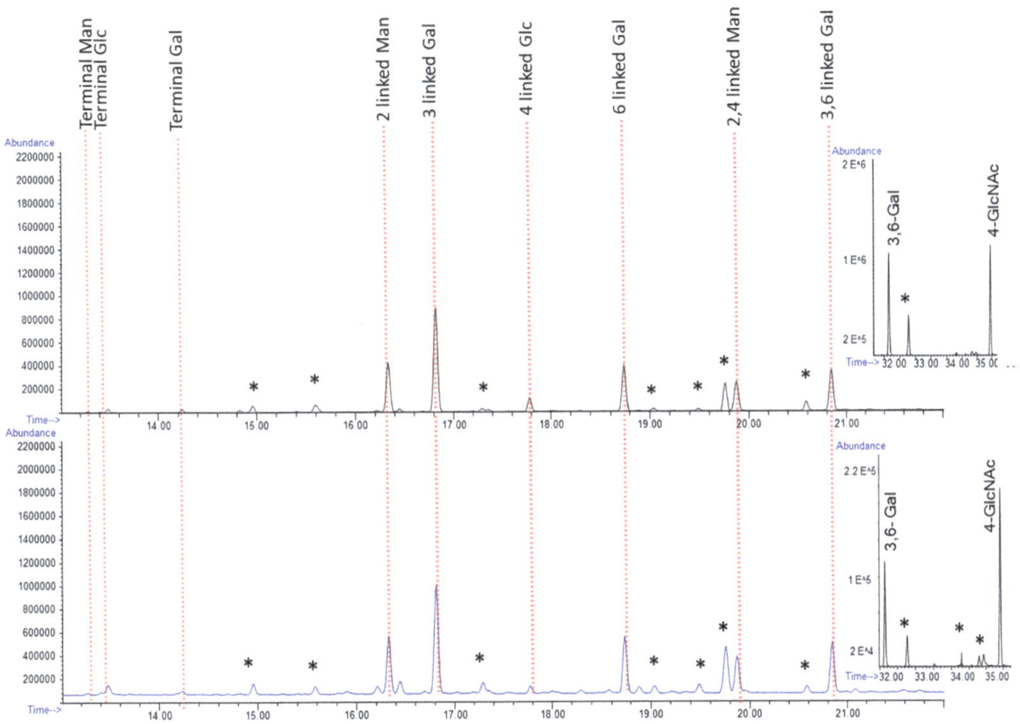


**Fig. 6** Analysis of 100 μg fetuin. Top chromatogram—sample washed five times after methylation and again after the acetylation. Bottom chromatogram—sample washed only two times after methylation and acetylation. Inset chromatogram—samples run on a separate GC column used to analyze amino sugars

### *3.10 Conclusion*

Although the linkage analysis has been improved dramatically over the past century and has been used with great success in the structural characterization of numerous polysaccharides and oligosaccharides, it continues to be a challenging procedure fraught with many pitfalls. As a result, only a small number of expert labs routinely perform this analysis. It has been our goal in this chapter to contribute to lowering the barrier of entry into the field of carbohydrate analysis and to making the linkage analysis protocol accessible to a greater number of nonexperts. For this reason, we included detailed procedures including tips and tricks that are often omitted in the methods sections of most publications. We also included the results of mistakes that are commonly made by those who lack the experience required to optimize the reaction conditions to obtain accurate results. We hope that this tutorial will serve as a practical and useful guide to those who are starting out in the challenging area of carbohydrate analysis, as well as those who are well versed in it but are still looking to improve their skills.

## 4 Notes

1. While performing the preacetylation for amino containing samples, if the sample clumps together, several brief repeated sonication steps can be performed during the heating to help dissolve it.
2. The reaction described for the dimsyl preparation can be miniaturized to only prepare enough base for a single set of samples. Reducing the amount of potassium hydride used will allow for a quicker and safer base preparation.
3. During the dimsyl base preparation, the dried potassium hydride should be the consistency of talcum powder. If it does not reach this consistency, then mineral oil is still present. Further extractions with hexane are then necessary to completely remove the mineral oil.
4. Since the dimsyl base reacts with air and moisture, storage should be done under argon and in a freezer. However, even under these conditions, the base will develop a noticeable black film after several months, indicating it has expired.
5. While performing the dichloromethane wash after the methylation using sodium hydroxide, an alternative to transferring the dichloromethane layer to a new tube is to remove the top aqueous layer. The dichloromethane layer can then be dried under nitrogen in the original tube. However, some water will invariably remain with the dichloromethane, and the sample will need to be lyophilized to fully remove it. While effective for polysaccharides, this procedure could lead to evaporative losses in the case of small oligosaccharides.
6. During the drying down after hydrolysis, some methods call for the addition of methanol instead of isopropanol to aid in drying, as methanol has a lower boiling point and would therefore allow for faster drying. However, any residual methanol consumes the reducing agent faster, similar to acid, so its use is avoided in this procedure.

## Acknowledgments

This work was supported by the U.S. Department of Energy, Office of Science, Basic Energy Sciences, under Award # DE-SC0015662 (DOE Center for Plant and Microbial Complex Carbohydrates at the Complex Carbohydrate Research Center).

## References


1. Haworth WN, Machemer H (1932) Polysaccharides. Part XIII. The chainlength of methylated cellodextrins. J Chem Soc 0:2372–2374
2. Purdie T, Irvine J (1903) The alkylation of sugars. J Chem Soc 83:1021–1037
3. Hakomori S (1980) Citation classic—a rapid permethylation of glycolipid, and polysaccharide catalyzed by methylsulfinyl carbanion in dimethyl sulfoxide current content. Life Sci 23:12
4. Hakomori S (1964) A rapid permethylation of glycolipid, and polysaccharide catalyzed by methylsulfinyl carbanion in dimethyl sulfoxide. J Biochem 55:205–208
5. Sandford PA, Conrad HE (1966) The structure of the Aerobacter aerogenes A3 (S1) polysaccharide. I. a reexamination using improved procedures for methylation analysis. Biochemistry 5(5):1508–1517
6. Biemann K, DeJongh DC, Schnoes K (1963) Application of mass spectrometry to structure problems. Acetates of Pentoses and hexoses. J Am Chem Soc 85(12):1763–1771
7. Phillips LR, Fraser BA (1980) Methylation of carbohydrates with dimsyl potassium in dimethyl sulfoxide. Carbohydr Res 90 (1):149–152
8. Lindberg B (1972) Methylation analysis of polysaccharides. Methods Enzymol 28:178–195
9. Oades JM (1967) Gas-liquid chromatography of alditol acetates and its application to the analysis of sugars in complex hydrolysates. J Chromatogr 28(2):246–252
10. Brauman JJ, Bryson JA, Kahl DC, Nelson NJ (1970) Equilibrium acidities in dimethyl sulfoxide. J Am Chem Soc 92(22):6679–6680
11. Finne J, Krusius T, Rauvala H (1980) Use of potassium tert-butoxide in the methylation of carbohydrates. Carbohydr Res 80(2):336–339
12. Waeghe TJ, Darvill AG, McNeil M, Albersheim P (1983) Determination, by methylation analysis, of the glycosyl-linkage compositions of microgram quantities of complex carbohydrates. Carbohydr Res 123(2):281–304
13. Barker SA, Law AR, Somers PJ, Stacey M (1967) Linkage analysis of carbohydrates by using saccharinic acid formation. Carbohydr Res 3(4):435–444
14. York WS, Kiefer LL, Albersheim P, Darvill AG (1990) Oxidation of oligoglycosyl alditols during methylation catalyzed by sodium hydroxide and iodomethane in methyl sulfoxide. Carbohydr Res 208:175–182
15. Harris PJ, Henry RJ, Blakeney AB, Stone BA (1984) An improved procedure for the methylation analysis of oligosaccharides and polysaccharides. Carbohydr Res 127:59–73
16. Ciucanu I, Kerek F (1984) A simple and rapid method for the permethylation of carbohydrates. Carbohydr Res 131:209–217
17. Anumula KR, Taylor PB (1992) A comprehensive procedure for preparation of partially methylated alditol acetates from glycoprotein carbohydrates. Anal Biochem 203(1):101–108
18. Needs PW, Selvendran RR (1993) An improved methylation procedure for the analysis of complex polysaccharides including resistant starch and a critique of the factors which lead to undermethylation. Phytochem Anal 4:210–216
19. Needs PW, Selvendran RR (1993) Avoiding oxidative degradation during sodium hydroxide/methyl iodide-mediated carbohydrate methylation in dimethyl sulfoxide. Carbohydr Res 245(1):1–10
20. Ciucanu I, Costello CE (2003) Elimination of oxidative degradation during the per-O-methylation of carbohydrates. J Am Chem Soc 125 (52):16213–16219
21. Pettolino F, Walsh C, Fincher GB, Bacic A (2012) Determining the polysaccharide composition of plant cell walls. Nat Protocol 7:1590–1607
22. Björndal H, Lindberg B, Svensson S (1967) Mass spectrometry of partially methylated alditol acetates. Carbohydr Res 5(4):433–440
23. Doares SH, Albersheim P, Darvill AG (1991) An improved method for the preparation of standards for glycosyl-linkage analysis of complex carbohydrates. Carbohydr Res 210:311–317
24. https://www.ccrc.uga.edu/specdb/ms/pmaa/pframe.html
25. Sims I, Carnachan S, Bell T, Hinkly S (2018) Methylation analysis of polysaccharides: Technical advice. Carbohydr Polym 188:1–7

# Chapter 19

# Use of Exoglycosidases for the Structural Characterization of Glycans

Elizabeth McLeod, Paula Magnelli, and Xiaofeng Shi

## Abstract

The use of sequential exoglycosidase digestion of oligosaccharides followed by LC-FLD, LC-MS or CE analysis provides detailed carbohydrate structural information. Highly specific exoglycosidases cleave monosaccharides from the nonreducing end of an oligosaccharide and yield information about the linkage, stereochemistry and configuration of the anomeric carbon. Here we use combinations of exoglycosidases to precisely characterize glycans on the Fc domain of therapeutic antibodies and dimeric fusion proteins. The workflow described includes glycan release with Rapid™ PNGase F (NEB #P0710), direct labeling of released glycans with procainamide (PCA) or 2-aminobenzamide (2AB), cleanup of labeled glycans and a 3 h enzymatic digestion with exoglycosidases. This protocol is designed for completion within an 8 h time frame to allow for subsequent LC-FLD, LC-MS, or CE analysis overnight.

**Key words** Exoglycosidase, Glycan Structure, Rapid™ PNGase F, LC-FLD, LC-MS, CE

## 1 Introduction

Characterization of glycans on therapeutic IgGs is critical as the stability, half-life, and clinical efficacy are affected by the glycoforms present on the molecule. The inherent complexity of protein glycosylation poses a daunting analytical challenge. Multiple orthogonal methods are often used to elucidate structure, but even with techniques such as LC-MS, which has the advantage of an associated mass corresponding to each chromatographic peak, there can be ambiguities when assigning structures. There are often several possible glycan isoforms associated with an identical mass.

Highly specific exoglycosidases can be used to cleave monosaccharides from the nonreducing end of an oligosaccharide yielding information about the linkage, stereochemistry and configuration of the anomeric carbon. The use of exoglycosidase digestion panels is an experimentally straightforward approach that was first described in the 1990s by Pauline Rudd et al. [1]. For many years, the widespread application of this approach was

Arnaud Delobel (ed.), *Mass Spectrometry of Glycoproteins: Methods and Protocols*, Methods in Molecular Biology, vol. 2271, https://doi.org/10.1007/978-1-0716-1241-5_19, 

hindered by inconsistent supplies of highly pure enzymes with minimal contaminating activities. In addition, optimized enzyme formulations and reaction conditions that allow multiple enzymatic reactions in a single step were not available. In this chapter, we describe protocols that utilize high-quality enzymes and reagents from reliable commercial sources that are compatible with most commonly used labeling strategies and detection methods, allowing for accurate structural glycan assignment.

## 2 Materials

Prepare all solutions with HPLC grade solvents to avoid contamination during mass spectrometric analysis.

1. Rapid PNGase F (enzyme and buffer) (New England Biolabs).
2. Procainamide (PCA) stock solution: Dissolve 550 mg of PCA in 1 mL of DMSO. This solution is stable for several weeks at −20 °C and numerous freeze–thaw cycles.
3. 2-aminobenzamide (2-AB) stock solution: Dissolve 250 mg of 2-AB in 1 mL of DMSO. This solution is stable for several weeks at −20 °C and numerous freeze–thaw cycles.
4. 200 mg/mL sodium cyanoborohydride solution: Dissolve 200 mg of sodium cyanoborohydride in 1 mL of water. This solution is stable for several weeks at −20 °C and numerous freeze–thaw cycles.
5. 50 mM ammonium formate Buffer, pH 4.4: Add 7.66 mL of formic acid to 3.9 L of water. Add ammonium hydroxide to bring the solution to pH 4.4 (~11 mL of ammonium hydroxide).
6. 85% acetonitrile (ACN): mix 85 mL of acetonitrile with 15 mL of water.
7. 1% formic acid: to 99 mL of water, add 1 mL of formic acid.
8. SPE buffer: 200 mM ammonium acetate.
9. Glacial acetic acid.
10. Methanol.
11. Exoglycosidases (α2-3,6,8,9 neuraminidase A, α1-3,4,6 galactosidase, β1-4 Galactosidase S, β-*N*-acetylglucosaminidase S, α1-2,4,6 fucosidase O, α1-2,3,6 Mannosidase) and Glycobuffer 1, or *N*-glycan sequencing kit (New England Biolabs).
12. HILIC Plate, or HILIC Spin Column.
13. GlycoWorks *Rapi*Fluor-MS *N*-Glycan Kit (Waters).

## 3 Methods

Carry out all steps at room temperature, unless otherwise directed.

### 3.1 Protocol Using Procainamide (PCA) or 2-Aminobenzamide (2AB)

#### 3.1.1 Rapid Deglycosylation

1. Using PCR tubes (200 μL), add 30 μg of monoclonal antibody (*see* **Note 1**) to a final volume of 16 μL.
2. Add 4 μL of Rapid PNGase F Buffer and mix.
3. Incubate the mixture at 80 °C for 2 min and cool to room temperature.
4. Add 1 μL of Rapid PNGase F.
5. Incubate for 10 min at 50 °C in a thermocycler or heat block.

#### 3.1.2 Fluorescent Labeling with Procainamide (PCA) or 2-Aminobenzamide (2AB)

1. Prepare acidified PCA or 2-AB labeling solution by adding one volume of glacial acetic acid to eight volumes of PCA or 2-AB stock solution.
2. Add 18 μL of acidified PCA or 2-AB labeling solution and 24 μL of 200 mg/mL sodium cyanoborohydride solution to the deglycosylation reactions.
3. Incubate for 45 min at 65 °C in a thermocycler.
4. Cool reactions to room temperature. The labeled glycans can then be purified with a HILIC plate or column.

#### 3.1.3 Glycan Purification with a 96-Well HILIC Plate

1. Add 350 μL acetonitrile (ACN) to the labeled reactions to a final concentration of 85% ACN.
2. Set up a HILIC elution plate with shims or spacer and waste tray if necessary (*see* **Note 2**).
3. Condition well with 200 μL of $H_2O$.
4. Equilibrate well with 200 μL of 85% ACN.
5. Load PCA or 2-AB labeled samples diluted with ACN (~410 μL) onto the HILIC plate.
6. Wash wells with 3× 200 μL of 1% formic acid, 90% ACN.
7. Replace waste tray with a collection plate.
8. Elute glycans with 3× 30 μL of SPE buffer (*see* **Note 3**) into the collection plate.
9. Dry the 90 μL sample in a speed vac at 35 °C or lyophilize overnight (*see* **Note 4**).
10. Resuspend the sample in 30 μL of $H_2O$ for subsequent exoglycosidase reactions.

#### 3.1.4 Glycan Purification with a HILIC Spin Column

1. Add 350 μL of acetonitrile (ACN) to the labeled reactions for a final concentration of 85% ACN.

2. Using either a vacuum manifold or centrifuge adaptor (following manufacturer's instructions), condition a HILIC spin column with 350 μL of water.
3. Equilibrate the column with 350 μL of 85% ACN.
4. Load PCA labeled samples diluted with ACN onto the HILIC column and spin or vacuum-aspirate.
5. Wash column with 5× 300 μL of 1% formic acid, 90% ACN.
6. Elute glycans with 3× 30 μL of SPE Buffer into a collection tube for a final volume of 90 μL.
7. Dry the 90 μL sample in a speed vac at 35 °C or lyophilize overnight (*see* **Note 4**).
8. Resuspend the sample in 30 μL of water for subsequent exoglycosidase reactions.

#### 3.1.5 Digestion of PCA or 2AB Labeled Glycans with Exoglycosidases

Note: Exoglycosidases can be used in single digests or in combinations to elucidate information about the total glycan profile.

1. In PCR tubes (200 μL), mix 5 μL of PCA-labeled *N*-glycans (equivalent to 5 μg of starting material) from previous step with 2 μL of 10× Glycobuffer 1, the recommended volume of exoglycosidase (as presented in Table 1) and water to a final reaction volume of 20 μL (*see* **Note 5**).
2. Incubate reactions for 3 h at 37 °C.
3. Add 10 μL of 50 mM ammonium formate buffer, pH 4.4 and 90 μL acetonitrile to each 20 μL reaction for a final acetonitrile concentration of 70%.

**Table 1**
**Preparation of exoglycosidase solutions for enzymatic digestion of PCA or 2-AB labeled glycans**

| Component | RXN 1 | RXN 2 | RXN 3 | RXN 4 | RXN 5 | RXN 6 |
|---|---|---|---|---|---|---|
| PCA/2-AB-labeled *N*-glycans | 5 μL | 5 μL | 5 μL | 5 μL | 5 μL | 5 μL |
| 10× Glycobuffer 1 | 2 μL | 2 μL | 2 μL | 2 μL | 2 μL | 2 μL |
| $H_2O$ | 13 μL | 11 μL | 10 μL | 9 μL | 8 μL | 6 μL |
| α2-3,6,8,9 Neuraminidase A | | 2 μL | 2 μL | 2 μL | 2 μL | 2 μL |
| α1-3,4,6 Galactosidase | | | 1 μL | 1 μL | 1 μL | 1 μL |
| β1-4 Galactosidase S | | | | 1 μL | 1 μL | 1 μL |
| β-*N*-Acetylglucosaminidase S | | | | | 1 μL | 1 μL |
| α1-2,4,6 Fucosidase O | | | | | | 2 μL |
| **Total** | **20 μL** | **20 μL** | **20 μL** | **20 μL** | **20 μL** | **20 μL** |

4. Samples are now ready for LC or LC-MS analysis (*see* **Note 6**). A representative chromatogram obtained for infliximab using PCA labeling is presented in Fig. 1.

### 3.2 Protocol Using Waters GlycoWorks RapiFluor-MS N-Glycan Kit

There are a number of "Instant Label" products such as Instant 2AB, Instant Procainamide and *Rapi*Fluor-MS available that provide enhanced fluorescence response and MS sensitivity for glycan detection. Here we provide a protocol for labeling using Waters GlycoWorks *Rapi*Fluor-MS *N*-Glycan Kit and subsequent exoglycosidase digestion.

#### 3.2.1 Rapid Deglycosylation

1. Using PCR tubes (200 μL), add 15 μg of monoclonal antibody to a final volume of 22.8 μL (*see* **Note 1**).

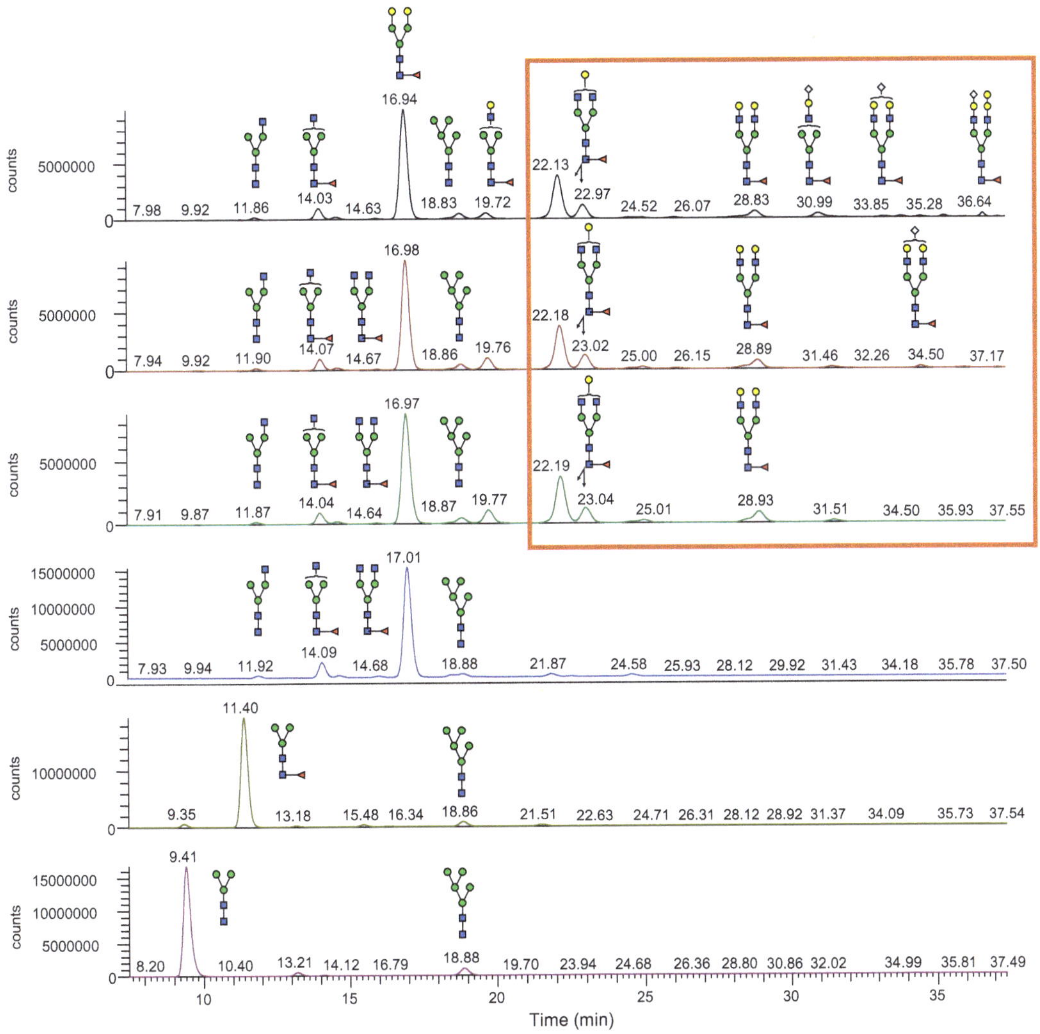


**Fig. 1** (**a**) *N*-glycans release from infliximab, labeled with PCA, and digested for 3 h with exoglycosidases (**b**) Expanded lower abundance profile of Infliximab glycans (*see* **Note 7**)

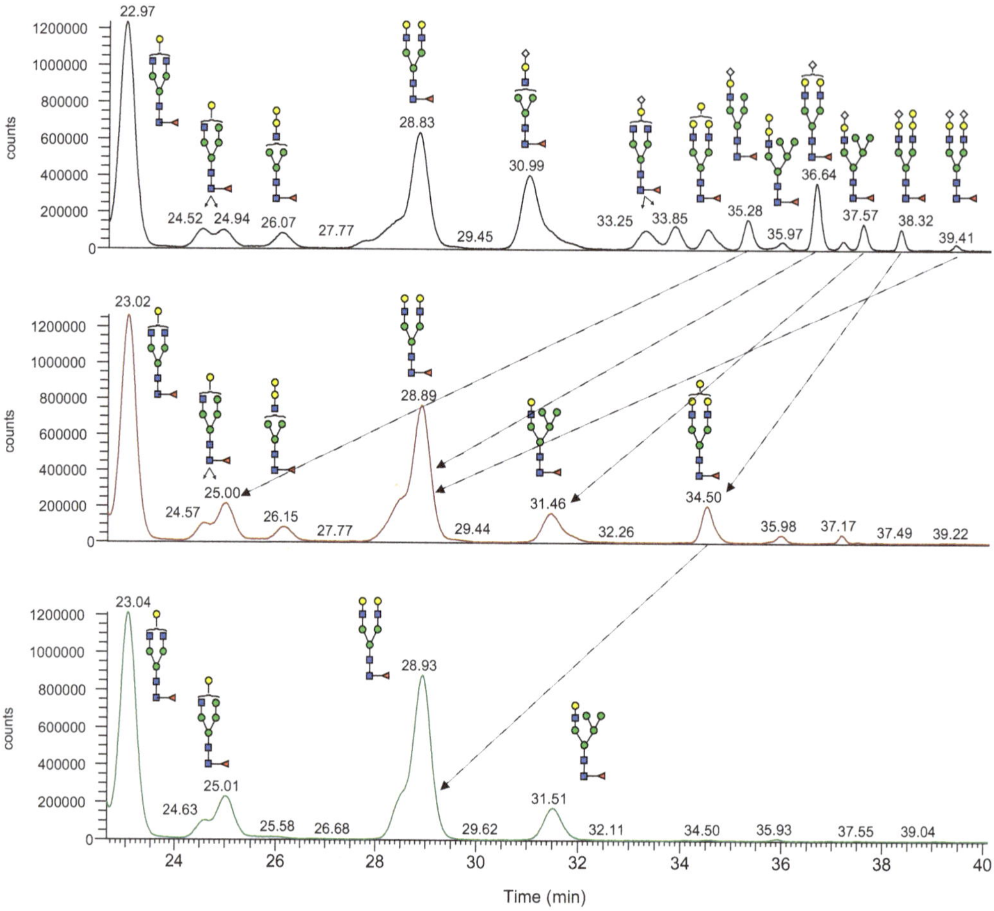


**Fig. 1** (continued)

2. Add 6 μL of 5% (w/v) Buffered RapiGest SF and mix.
3. Incubate mixture at 80 °C for 2 min and cool.
4. Add 1.2 μL of Rapid PNGase F for a final volume of 30 μL.
5. Incubate for 10 min at 50 °C in a thermocycler or heat block.

*3.2.2 Fluorescent Labeling of Released Glycans with RapiFluor*

1. Prepare reagent solution by dissolving one vial of 23 mg of the GlycoWorks *Rapi*Fluor-MS Reagent Powder in 335 μL of GlycoWorks Reagent Solvent (anhydrous DMF). Mix several times to ensure the reagent is dissolved.
2. Add 12 μL of the *Rapi*Fluor-MS Reagent Solution to the deglycosylation reactions and mix thoroughly.
3. Incubate the reaction to at room temperature for 5 min.
4. Add 358 μL of ACN to the reaction.

*3.2.3 Glycan Purification with a 96-Well HILIC Plate*

1. Set up a GlycoWorks HILIC μElution Plate with shims or spacer and waste tray if necessary (*see* **Note 2**).
2. Condition wells to be used on the μElution plate with 200 μL of water.
3. Equilibrate wells with 200 μL of 85% ACN.
4. Load the entire 400 μL acetonitrile diluted sample to each well.
5. Wash the well with 2× 600 μL volumes of 1:9:90 (v/v/v) formic acid–water–acetonitrile.
6. Replace the waste tray with a 96-well collection plate.
7. Elute glycans with 3× 30 μL volumes of GlycoWorks SPE Elution Buffer (*see* **Note 3**).
8. Dry the 90 μL sample in Speedvac at 35 °C for about 1.5 h (*see* **Note 4**).
9. Resuspend dried pellet in 25 μL of water.

*3.2.4 Digestion of RapiFluor-Labeled Glycans with Exoglycosidases*

Note: Exoglycosidases can be used in single digests or in combinations to elucidate information about the total glycan profile.

1. In PCR tubes (200 μL), mix 3 μL of *Rapi*Fluor-labeled *N*-glycans (equivalent to 5 μg of starting material) from previous step with 2 μL 10× Glycobuffer 1, the recommended volume of exoglycosidase (as presented in Table 2) and water to a final reaction volume of 20 μL (*see* **Note 5**).
2. Incubate reactions for 3 h at 37 °C.
3. Add 10 μL of 50 mM ammonium formate Buffer pH 4.4 and 90 μL acetonitrile to each 20 μL reaction for a final acetonitrile concentration of 70%.
4. Samples are now ready for analysis (*see* **Note** 7).

**Table 2**
**Preparation of exoglycosidase solutions of enzymatic digestion of *Rapi*Fluor-labeled glycans**

| Component | RXN 1 | RXN 2 | RXN 3 | RXN 4 | RXN 5 | RXN 6 |
|---|---|---|---|---|---|---|
| *Rapi*Fluor-labeled *N*-glycans | 3 μL | 3 μL | 3 μL | 3 μL | 3 μL | 3 μL |
| 10× Glycobuffer 1 | 2 μL | 2 μL | 2 μL | 2 μL | 2 μL | 2 μL |
| $H_2O$ | 15 μL | 13 μL | 12 μL | 11 μL | 10 μL | 8 μL |
| α2-3,6,8,9 Neuraminidase A | | 2 μL | 2 μL | 2 μL | 2 μL | 2 μL |
| α1-3,4,6 Galactosidase | | | 1 μL | 1 μL | 1 μL | 1 μL |
| β1-4 Galactosidase S | | | | 1 μL | 1 μL | 1 μL |
| β-*N*-Acetylglucosaminidase S | | | | | 1 μL | 1 μL |
| α1-2,4,6 Fucosidase O | | | | | | 2 μL |
| **Total** | **20 μL** | **20 μL** | **20 μL** | **20 μL** | **20 μL** | **20 μL** |

## 4 Notes

1. Glycans constitute approximately 2% of the molecular weight of an antibody. In general, 30 μg of antibody is a sufficient quantity of glycoprotein to generate a labeled glycan substrate with a good fluorescence and mass spectrometric signals with PCA label. 2-AB labeled glycan fluorescence and mass spectrometric signals are typically not as strong as PCA and may require more labeled substrate to get an adequate MS signal. Instant labels such as *Rapi*Fluor-MS that provide enhanced fluorescence response and MS sensitivity for glycan detection require less labeled substrate. Avoid buffers containing SDS, as it inhibits PNGase F. Common stabilizing reagents such as Tween, Triton X-100, NP-40, octyl glucoside, and nondetergent sulfobetaine, as well as traces of organic solvents, can prevent optimal rapid deglycosylation. Antibody substrate used here Infliximab was dialyzed in 20 mM Tris, 50 mM NaCl pH 7.5 to remove sucrose.
2. Positive or negative pressure manifold can be used to purify labeled glycans. Begin using low pressure and increase until flow is approximately 1 drop per second.
3. SPE Buffer: 200 mM Ammonium Acetate.
4. Alternatively, divide the 90 μL eluent sample in 8 aliquots of 11.25 μL each and dry in speed vac at 35 °C for 30 min to shorten drying step. After sample has completely dried add 3 μL of water to each tube.
5. Exoglycosidase digestions are prone to evaporation due to the extended reaction time and the small volume. This can result in incomplete digestion of the substrate. PCR machines are ideal for performing digests as evaporation is minimized.
6. In this particular experiment, *N*-glycan samples are separated using a XBridge™ BEH Amide column (Waters) on a Dionex UltiMate® LC equipped with fluorescent detection in line with an LTQ™ Orbitrap Velos™ Spectrometer equipped with a heated electrospray standard source (HESI-II probe).
7. There are several software programs available to correctly assign glycan structures, some based on GU values in conjunction with mass. The following is one example of a web based tool: https://glycananalyzer.neb.com/exoanalyze/main_page.action

### Reference

1. Rudd PM et al (1997) Oligosaccharide sequencing technology. Nature 388:205–207

# Chapter 20

# Determination of Isomeric Glycan Structures by Permethylation and Liquid Chromatography–Mass Spectrometry (LC-MS)

**Byeong Gwan Cho, Alireza Banazadeh, Wenjing Peng, Jingfu Zhao, Mona Goli, Sakshi Gautam, Ahmed Hussein, and Yehia Mechref**

## Abstract

The existence of glycans in isomeric forms is responsible for the multifariousness of their properties and biological functions. Their altered expression has been associated with various diseases and cancers. Analysis of native glycans is not very sensitive due to the low ionization efficiency of glycans. These facts necessitate their comprehensive structural studies and establishes a high demand for sensitive and reliable techniques. In this chapter, we discuss the strategies for effective separation and identification of permethylated isomeric glycans. The sample preparation for permethylated glycans derived from model glycoproteins and complex biological samples, analyzed using LC-MS/MS, is delineated. We introduce protein extraction and release of glycans, followed by strategies to purify the released glycans, which are reduced and permethylated to improve ionization efficiency and stabilize sialic acid residues. High-temperature LC-based separation on PGC (porous graphitized carbon) column is conducive to isomeric separation of glycans and allows their sensitive identification and quantification using MS/MS.

**Key words** Glycomics, Isomeric separation, LC-MS, Permethylation, Porous graphitized carbon

## 1 Introduction

Among the various posttranslational modifications (PTMs) that proteins undergo in living organisms, glycosylation is one of the most common modifications, especially on secreted and membrane proteins [1]. It has long been known that glycans are involved in a wide range of biological processes and play important biological roles [2]. Examples of such functions include directing the folding of nascent proteins [3], protecting proteins from degradation catalyzed by proteases [4], as well as intrinsic and extrinsic recognition of cells [5]. Also, the alteration of glycan profiles has been correlated to many mammalian diseases, such as Alzheimer's disease [6], immune deficiencies [7], and several types of cancers [8]. Studies

Arnaud Delobel (ed.), *Mass Spectrometry of Glycoproteins: Methods and Protocols*, Methods in Molecular Biology, vol. 2271,
https://doi.org/10.1007/978-1-0716-1241-5_20, 

indicating the relevance of isomeric glycans to cancers also highlighted the importance of characterizing isomeric glycan forms [9–11].

The two major types of protein glycosylation, *N*-linked and *O*-linked glycosylation, are the covalent attachment of glycans to asparagine residues and serine/threonine residues of a protein, respectively. The glycosylation of proteins occurs in the endoplasmic reticulum and Golgi apparatus. *N*-glycans are built from the nucleotide sugar precursors and involves numerous enzymes such as glycosyltransferases and exoglycosidases [12–14]. Therefore, the glycan profiles depend on the availability and expressions of these enzymes and sugar nucleotides. Although the number of monosaccharides involved in glycosylation is limited, the diversity of glycan structures originates from the fact that monosaccharides are assembled into glycans in ways that not only vary by sequence, composition, and length but also by linkages and branching.

Although mass spectrometry (MS) is considered the method of choice for glycan profiling due to the high sensitivity and capacity for structural elucidation, MS based glycomics is still challenging because of microheterogeneity of glycan structures and limited capability of MS in differentiating glycan isomers. Such limitation emphasizes the necessity of coupling MS with separation techniques, such as ion mobility (IM) [15–17], capillary electrophoresis (CE) [18, 19], and liquid chromatography (LC) [20]. However, IM and CE are less amenable for MS-based glycomics as compared to LC due to the low resolution of IM and inefficient interfacing of CE to MS.

Various LC separation techniques have been applied in glycomics, including hydrophilic interaction liquid chromatography (HILIC) [21, 22], reverse-phase (RP) LC [23], porous graphitized carbon (PGC) [11, 24–26], and high pH anion-exchange chromatography (HPAEC) [27]. For reducing end-labeled glycans, efficient isomeric separation can be attained using HILIC columns. However, the accurate glycan structure assignment and quantitation involving HILIC is hindered by sialic acid loss occurring in the electrospray ionization (ESI) source and fucose migration observed in $MS^2$ [28, 29].

To address these concerns, derivatization of glycans is often performed. Permethylation, one of the most utilized glycan derivatization techniques, is significantly beneficial for MS-based glycomics because of its ability to address issues of glycan stability and low ionization efficiency [29]. Moreover, it facilitates the separation of glycans using RP-LC by converting glycans into more hydrophobic structures. Separation of permethylated glycans using RP-LC, however, induces low isomeric resolution. Therefore, the isomeric separation of permethylated glycans on PGC column was recently developed in our laboratory which is capable of achieving baseline separation of both linkage and positional isomers by elevating the temperature of the separation column [25, 26]. The interaction between PGC materials and glycan is based on hydrophobicity and electrostatic interaction [30], thus making it structure selective and capable of discriminating isomers. The high-temperature PGC-LC-MS platform has been successfully employed to profile isomeric glycans in biological samples [9, 11, 31].

This chapter summarizes the different strategies needed to characterize glycan isomers through permethylation and PGC-LC-MS. Detailed sample preparation steps from complex sample matrix such as cell lines, tissues and human blood serum are described. Release, purification and permethylation of glycans procedures to ensure sensitive isomeric glycan analysis are also described. Finally, different methods to characterize glycan isomers, using glycan standards, exoglycosidases, $MS^n$, and 3D modeling, are described in detail.

## 2 Materials

### 2.1 Protein Extraction

1. Ammonium bicarbonate (ABC).
2. Sodium deoxycholate (SDC).
3. 2.0 mL microcentrifuge tubes, conical, with screw cap.
4. BeadBug microtube homogenizer (Benchmark Scientific, Edison, NJ).
5. 400 μm molecular biology grade zirconium beads (OPS Diagnostics, LLC, Lebanon, NJ).
6. Micro BCA protein assay kit (Thermo Fisher Scientific Inc., Rockford, IL).

### 2.2 N-Glycan Release

#### 2.2.1 In-Solution N-Glycan Release

1. Ammonium bicarbonate (ABC).
2. PNGase F (glycerol-free, 500,000 units/mL) (New England Biolabs, Ipswich, MA).

#### 2.2.2 Filter Aided N-Glycan Release

1. Ammonium bicarbonate (ABC).
2. Amicon ultra-0.5 mL centrifugal filter unit (Ultracel-10 K) (Sigma-Aldrich, St. Louis, MO).
3. PNGase F (glycerol-free, 500,000 units/mL) (New England Biolabs, Ipswich, MA).

### 2.3 O-Glycan Release

1. Pronase (Roche Diagnostics GmbH, Mannheim, Germany).
2. Borane-ammonia complex, 97%.
3. Ammonium hydroxide, 28%.

### 2.4 Glycan Purification

#### 2.4.1 Sodium Deoxycholate Removal

1. Formic Acid >98.0%.

#### 2.4.2 Protein Precipitation

1. ACS/USP grade ethanol.

#### 2.4.3 Dialysis

1. Biotech cellulose ester dialysis membrane (MWCO: 500–100 Da) (Spectrum Labs, Rancho Dominquez, CA).

#### 2.4.4 C18 Glycan Purification

1. HPLC grade methanol.
2. Acetic acid.
3. HyperSep C18 cartridge (100 mg bed weight, 1 mL column volume, Thermo Fisher Scientific, San Jose, CA).

### 2.5 Reduction of Reducing End of Glycans

1. Borane-ammonia complex, 97%.
2. HPLC grade methanol.

### 2.6 Solid-Phase Permethylation

1. Dimethyl sulfoxide, >99.9%.
2. Sodium hydroxide beads, 20–40 mesh, 97% (Sigma-Aldrich, St. Louis, MO).
3. Iodomethane containing copper as a stabilizer, 99.5%.
4. HPLC grade acetonitrile.
5. Empty micro spin column, 5 μm frit (Harvard Apparatus, Holliston, MA).

### 2.7 PGC-LC Separation

1. Loading solution (98% HPLC water, 2% acetonitrile and 0.1% formic acid).
2. Mobile phase A (98% HPLC water, 2% acetonitrile and 0.1% formic acid).
3. Mobile phase B (100% acetonitrile with 0.1% formic acid).
4. C18 precolumn (3 μm particle size, 100 Å pore size, 75 μm i. d., 2 cm length, Thermo Scientific, Sunnyvale, CA).
5. HyperCarb PGC column (5 μm particle size, 75 μm i.d., 100 mm length, Thermo Scientific, Sunnyvale, CA).
6. Ultimate 3000 Nano LC system (Thermo Scientific, Sunnyvale, CA).

### 2.8 Mass Spectrometry

1. LTQ Orbitrap Velos mass spectrometer (Thermo Fisher Scientific Inc., Waltham, MA).

## 3 Methods

The procedures below describe detailed steps from preparing permethylated *N*- and *O*-glycans from complex biological samples to LC-MS conditions and finally, data processing and determination of isomers using glycan standards, exoglycosidases, $MS^n$, and 3D modeling. Workflow of the methods is depicted in Fig. 1.

### 3.1 Protein Extraction

Before glycan release, protein extraction is often required, especially for tissues and cells to remove interfering molecules such as lipids and other small molecules. For other biological fluids such as serum, plasma, or cerebrospinal fluid, the protein extraction step may be omitted since PNGase F can release *N*-glycans directly in the sample matrix [9, 32]. Protein extraction steps are also unnecessary for model glycoproteins, such as bovine fetuin or ribonuclease B.

#### 3.1.1 Cell Line Protein Extraction

1. To an empty 2.0 mL microtube, add enough zirconium beads to cover 0.5 cm of the microtube.
2. To a cell pellet, add 100 μL of 50 mM ABC buffer (pH = 7.5) and mix by pipetting.
3. Carefully transfer the cell pellet + buffer solution to the 2.0 mL microtube containing zirconium beads.
4. Add 100-μL of 5% SDC (aq) solution to the microtube containing the cell pellet, then transfer to a 2.0 mL microtube to recover the remaining cells.
5. Homogenize using a BeadBug microtube homogenizer at 4,000 rpm (1,541 × *g*), 4 °C, for 90 s, three times with a 30 s break between each cycle.
6. To determine the protein concentration from the prepared cell lysate, use a bicinchoninic acid (BCA) protein assay kit (Thermo Scientific).
7. Protein extraction from tissue samples can be completed similarly (*see* **Note 1**).

### 3.2 N-Glycan Release

There are two methods of releasing *N*-glycans: in-solution and by filter-aided *N*-glycan separation (FANGS) [33]. The FANGS method is convenient because it incorporates protein extraction steps using a 10k MWCO filter prior to PNGase F digestion. However, in our previous study, a significant sample loss was associated with FANGS in comparison to in-solution release [32].

#### 3.2.1 In-Solution N-Glycan Release

1. Dilute the sample (biological fluid, cell lysate, or tissue protein extract) with 50 mM ABC buffer (pH = 7.5) by adding nine times the sample volume.
2. Denature the sample in a 90 °C water bath for 10 min.

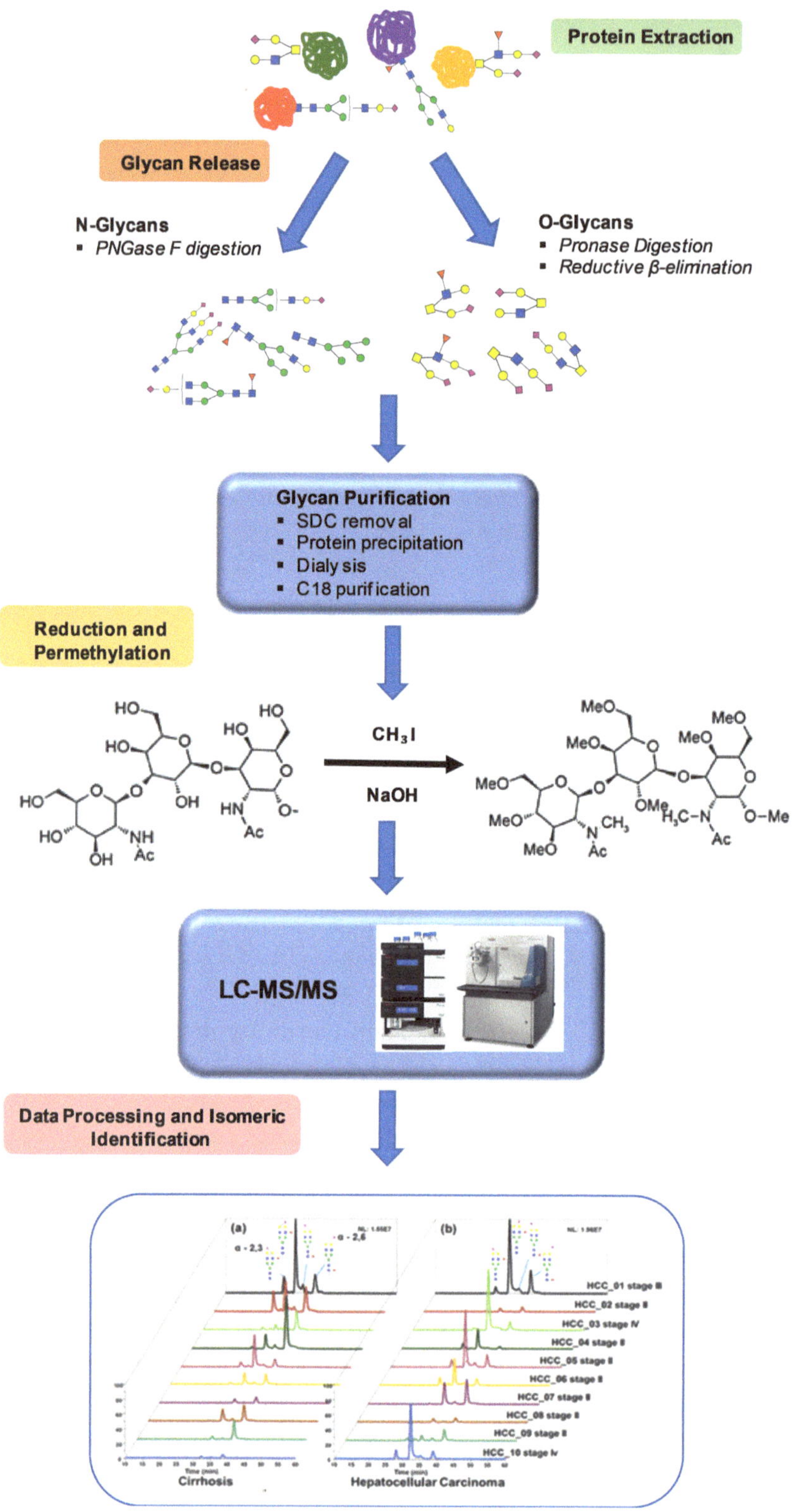


**Fig. 1** Typical workflow for the isomeric characterization of glycans using permethylation

3. Cool the sample down to room temperature.
4. Add a 1 μL aliquot of stock PNGase F solution (500 U).
5. Incubate the sample in a 37 °C water bath for 18 h.

#### 3.2.2 Filter-Aided N-Glycan Release

1. Dilute the sample (biological fluid, cell lysate, or tissue protein extract) with 50 mM ABC buffer (pH = 7.5) by adding nine times the sample volume.
2. Denature the sample in a 90 °C water bath for 10 min.
3. Cool the sample down to room temperature.
4. Add 400 μL of HPLC grade water to the 10k MWCO filter (Sigma).
5. Centrifuge the filter unit at 14,000 × *g* for 15 min.
6. Discard the flow-through.
7. Add the sample to the filter unit.
8. Centrifuge the filter unit at 14,000 × *g* for 15 min.
9. Discard the flow-through.
10. Add 400 μL of 50 mM ammonium bicarbonate (pH = 7.5).
11. Centrifuge the filter unit at 14,000 × *g* for 15 min.
12. Discard the flow-through.
13. Repeat **steps 10–12** two times.
14. Attach a new collection tube to the filter unit.
15. Add 100 μL of 50 mM ammonium bicarbonate (pH = 7.5).
16. Add a 1 μL aliquot of stock PNGase F solution (500 U) and mix well by pipetting.
17. Incubate the sample in a 37 °C water bath for 18 h.
18. Centrifuge the filter unit at 14,000 × *g* for 15 min.
19. Add 400 μL of water to the filter unit.
20. Centrifuge the filter unit at 14,000 × *g* for 15 min.
21. Repeat **steps 19** and **20** two more times.
22. Collect the flow-through and dry.

### 3.3 O-Glycan Release

Although enzymes are commercially available to liberate *O*-glycans from serine and threonine residues, enzymatic digestion is limited because these enzymes are only able to release *O*-glycans with specific core structures. To overcome this issue, the enzymatic/chemical release of *O*-glycans [34] is performed where proteins are digested by proteolytic enzymes before the release of *O*-glycans by either reductive β-elimination [35] or permethylation. It should be noted, however, that *O*-glycan release by permethylation produces free reducing end, resulting in more complex chromatography although it has been previously reported that enzymatic/chemical release of *O*-glycans using permethylation can significantly enhance sensitivity [34].

*3.3.1 Pronase Digestion*

1. Prepare 10 mg/mL pronase solution in HPLC grade water.
2. Adjust the protein concentration to 2 mg/mL and pronase concentrations 0.2 mg/mL. For example, if the protein concentration of your sample is 10 mg/mL and volume of your sample is 100 μL, add 390 μL of HPLC grade water and 10 μL of 10 mg/mL pronase solution.
3. Incubate at 55 °C for 48 h.
4. Dry in the Speed-Vac concentrator.

*3.3.2 Reductive β-Elimination*

1. Prepare 5 mg/mL of borane-ammonia complex in a 28% $NH_4OH$ solution.
2. Add the borane-ammonia solution to the sample (1 μL of the solution per 1 μg of protein).
3. Vortex and incubate at 45 °C for 18 h.
4. Dry in the Speed-Vac concentrator.

*3.3.3 Solid-Phase Permethylation*

1. To the dried pronase-digested sample, add 30 μL of dimethyl sulfoxide, 1.2 μL of water, and 20 μL of iodomethane.
2. Prepare sodium hydroxide beads in dimethyl sulfoxide.
3. Cut 0.5 cm off from a 1000 μL pipette tip.
4. Pipet the sodium hydroxide beads into a clean and empty microspin column. Make sure the amount of sodium hydroxide beads is enough to fill approximately 1.5 cm but does not exceed 2.0 cm.
5. Place the packed microspin column into a holder with an empty 2.0 mL Eppendorf tube.
6. Place the assembly in a microcentrifuge and centrifuge at 1800 rpm (300 × *g*) for 2 min.
7. Add 200 μL of dimethyl sulfoxide to the column and centrifuge again.
8. Discard the flow-through.
9. Replace the 2.0 mL Eppendorf tube with a new tube.
10. Apply the sample onto the column.
11. Incubate at room temperature for 25 min (*see* **Notes 2** and **3**).
12. Add another 20 μL aliquot of iodomethane onto the column and incubate for another 10 min (*see* **Notes 2** and **3**).
13. Centrifuge at 1800 rpm (300 × *g*) for 2 min.
14. Apply 50 μL of acetonitrile and centrifuge again.
15. Remove the microspin column and its holder from the tube and dry the sample in the Speed-Vac concentrator.

### 3.4 Glycan Purification

Prior to glycan derivatization, sample clean-up steps are essential, especially when detergent that is not mass spectrometry compatible, such as sodium deoxycholate, is used [32]. Some of the methods described below are optional based on the sample matrix or detergents used. It should be noted that although the C18 purification procedure can remove hydrophobic lipids, proteins, and other small molecules, it does not remove salts from a sample [6]. For desalting, either dialysis or activated charcoal microspin columns can be used.

#### 3.4.1 Sodium Deoxycholate Removal (For Cell Line and Tissue Samples)

1. Add enough formic acid to make a final concentration of 1% formic acid within the SDC containing the sample solution.
2. Vortex and centrifuge at 14,800 rpm (21,100 × *g*) for 10 min.
3. Collect and transfer the supernatant to a new, clean tube.
4. Dry in the Speed-Vac concentrator.

#### 3.4.2 Protein Precipitation

1. Add enough ice-cold ethanol to bring the final ethanol concentration to 90%.
2. Vortex and incubate the sample at −20 °C for 1 h.
3. Centrifuge the sample at 14,800 rpm (21,100 × *g*) at 4 °C for 10 min.
4. Collect and transfer the supernatant to a new, clean tube.
5. Dry in the Speed-Vac concentrator.

#### 3.4.3 Dialysis (See Note 4)

1. Cut the dialysis membrane (MWCO: 500–1000 Da) so that it is large enough to fit the dialysis device.
2. Ensure that deionized water is flowing through the lower chamber of the dialysis device.
3. Pipet the sample that has been resuspended in 50 μL of water onto the top of the membrane.
4. Dialyze overnight.
5. Collect the dialyzed sample by pipetting out from the dialysis well, washing the well with 100 μL of 0.1% formic acid (aq).
6. Dry in the Speed-Vac concentrator.

#### 3.4.4 C18 Glycan Purification (See Note 5)

1. Wash the C18 cartridge (100 mg, 1 mL) with 1 mL methanol three times.
2. Condition the C18 cartridge with 1 mL 5% acetic acid (aq) three times.
3. Apply the sample that has been resuspended in 0.25 mL 5% acetic acid (aq).
4. Collect the flow-through in a new, clean tube.

5. Wash the cartridge with 5% acetic acid (aq) three more times, collecting each fraction.
6. Dry in the Speed-Vac concentrator.

### 3.5 Reduction of the Reducing End of Glycans

Reducing the reducing end of glycans is critical because free end glycans can be separated by liquid chromatography, displaying α and β anomers. Anomeric peaks can complicate determination of isomeric glycans due to the overlapping of anomeric and isomeric peaks.

1. Prepare a 10 mg/mL borane-ammonia complex solution in HPLC grade water.
2. Add 10 μL of the borane-ammonia complex solution to the purified and dried glycans.
3. Mix well and incubate in the water bath at 60 °C for 1 h.
4. Add 0.5 mL of methanol.
5. Dry in the Speed-Vac concentrator.
6. Repeat **steps 4** and **5** until all white residues disappear.

### 3.6 Solid-Phase Permethylation

Permethylation is an ideal derivatization technique for glycans as it enhances ionization.

Moreover, sialic acid loss and fucose migration are prevented through permethylation [29, 35, 36]. Detailed procedure is described under *Solid-phase permethylation* in Subheading 3.3.

1. To dried reduced *N*- or *O*-glycan samples, add 30 μL of dimethyl sulfoxide, 1.2 μL of water and 20 μL of iodomethane.
2. Follow procedures starting from **step 2** of "*Solid-Phase Permethylation*" under Subheading 3.3.

### 3.7 PGC-LC Conditions

After permethylation, the increased hydrophobicity of the glycans leads to a stronger interaction between glycan isomers and PGC particles, thus prompting efficient separation of isomeric glycans. The isomeric separation is performed on a nano PGC column at high temperature to attain the high resolution and reduced the backpressure. Prior to the separation using the PGC column, C18 trap column was utilized to purify permethyated glycans (*see* **Note 6**).

1. Resuspend the reduced and permethylated glycan sample in 20% ACN, 80% water, and 0.1% formic acid. The injection amount depends on the sample type. Generally, for cell line samples, glycans released from equivalent of 50 μg proteins are injected; for human blood sera, glycans released from 1 μL of serum are injected; for model glycoproteins (such as RNase B and bovine fetuin), glycans released from 0.5 μg of proteins are injected; for glycan standards, 2 ng of glycan standards are injected.

2. Set the column oven temperature to 75 °C (*see* **Note 7**) and allow the column to condition at this temperature for at least 30 min to stabilize the temperature inside the column.
3. The multistage gradient for *N*-glycan isomeric analysis is as follows:

   The flow rate of the nano pump is set to 0.6 μL/min; the flow rate of the loading pump is set to 3 μL/min. Mobile phase A is 2% ACN in water with 0.1% formic acid. Mobile phase B is 100% ACN with 0.1% formic acid.

   0–10 min: 20% B; the position of the 10-port valve is set to the 1–2 position.

   10–20 min: 20–50% B; the position of the 10-port valve is switched to the 10-1 position at 10 min.

   20–40 min: 50–80% B.

   40–60 min: 80–95% B.

   60–87 min: 95% B.

   87–88 min: 95–20% B; the position of the 10-port valve is switched to the 1–2 position at 85 min.

   88–90 min: 20% B.
4. The multistage gradient for *O*-glycan isomeric analysis is as follows:

   0–10 min: 20% B; the position of the 10-port valve is set to the 1–2 position.

   10–11 min: 20–35% B; the position of the 10-port valve is switched to the 10-1 position at 10 min.

   11–20 min: 35–60% B.

   20–46 min: 60–95% B.

   46–84 min: 95% B.

   84–85 min: 95–20% B; the position of the 10-port valve is switched to the 1–2 position at 85 min.

   85–90 min: 20% B.
5. Wash the column between two runs using 95% ACN with 0.1% formic acid for 15 min (Use at least 10 times the column volume) and then condition for 5 min using the initial condition of % B in the next run.

### 3.8 MS Conditions

A nanoESI source is utilized for ionization (*see* **Note 8**). The nanoESI voltage is set to 1.6 kV. The temperature of the transfer tube is set to 275 °C. The duty cycle contains three evens. Event 1 is a full MS with a resolution of 100,000 using orbitrap. The scan range of the full MS is set to 700–2000 *m/z* for permethylated *N*-glycomics and 200–2000 *m/z* for permethylated *O*-glycomics, respectively. Event 2 is a collision-induced dissociation (CID) using a linear ion trap (IT). Data-dependent acquisition (DDA) is used to select the top six most intense ions from the full MS (event 1) to perform CID fragmentation. The dynamic exclusion parameters are set as follows to prevent repetition of ion selection: repeat count 2; repeat duration 10 s; exclusion list size 50; and exclusion duration 30 s. The CID threshold is 5000 counts; isolation width is 3.0 *m/z*; normalized collision energy is 30; activation Q is 0.25, and activation time is 15 ms. Ejected charge state 1 is checked for *N*-glycomics and unchecked for *O*-glycomics. Event 3 is a higher energy collision dissociation (HCD) using HCD cell. The ion signal threshold is 1000 counts; normalized collision energy is 45, and the activation time is 0.1 ms. HCD fragments are analyzed in FTMS at a resolution of 7500. Other parameters for HCD are the same as those for CID.

### 3.9 Data Processing

Glycomics data acquired from Orbitrap mass spectometers is initially processed by Multiglycan software [37] which is a rapid and reliable tool for automated glycan compositional identification. Software parameter used for identification of permethylated glycans is displayed in Fig. 2. Data processed by Multiglycan is then manually validated for glycan isomer identification which is initially performed based on $MS^1$ full scan. Then, isomers of each validated glycan structures are confirmed by their $MS^2$ scans. Confirmed glycan isomers are quantified by summing all peak areas including multiple adducts and charge states of the corresponding isomers.

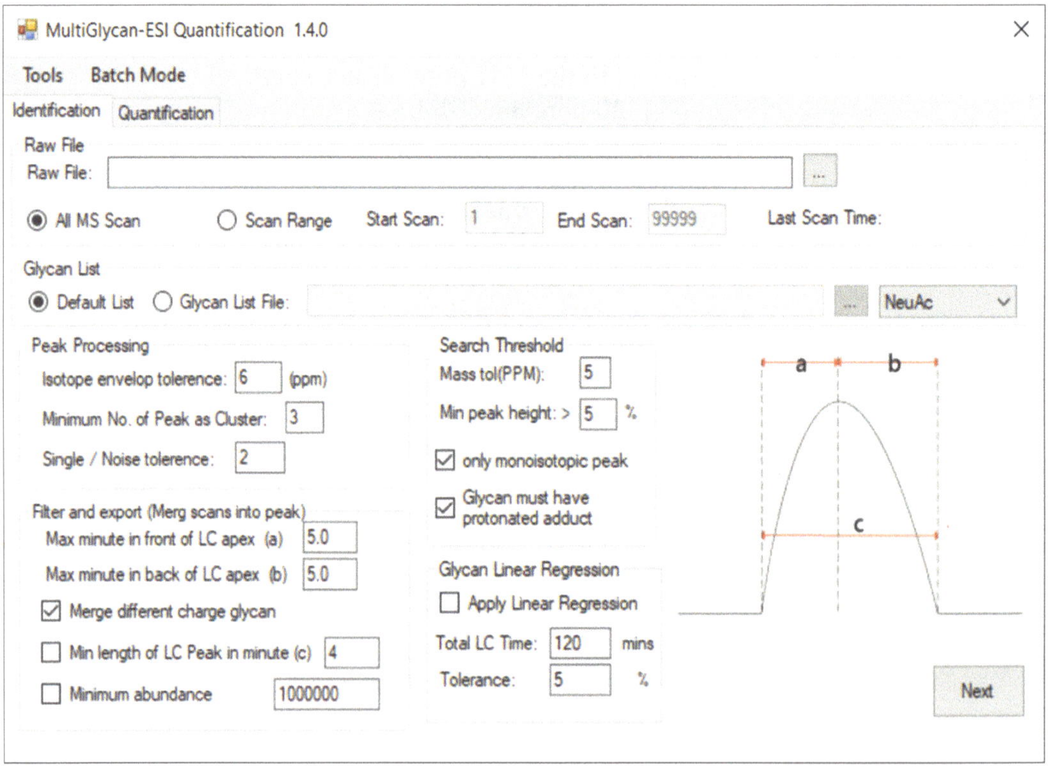


**Fig. 2** Interface and parameter settings in MultiGlycan. The parameters in this figure are utilized for glycan compositional identification

*3.9.1 Glycan Isomer Identification and Quantitation*

To eliminate false positives, manual peak inspection is performed after the initial screening of MultiGlycan. Although the MultiGlycan algorithm enables reliable identification, it may introduce false positives caused by incorrect isotopic envelopes. Therefore, manually inspecting chromatograms and mass spectra using Xcalibur Qual Browser is necessary to correctly identify glycan structures. The peaks containing ions that match theoretical $m/z$ values are extracted from raw files with a 10 ppm mass tolerance. Then, monoisotopic mass and isotopic distribution of target ions are evaluated. An ion that has a correct monoisotopic mass and isotopic distribution is considered to be a true glycan structure. Next, the glycan structure is further confirmed using CID $MS^2$ data. HCD $MS^2$ data can be complementary to CID when examining fragments whose $m/z$ is less than the low mass cutoff threshold of IT. An extracted ion chromatograph (EIC) is utilized to designate glycan isomers with a mass tolerance of 10 ppm. EIC may result in multiple peaks from the same $m/z$ value. Each peak is considered to be an isomer from the same glycan composition. The quantitation of glycan isomers is performed by adding up the peak areas of EICs of all charge states and adducts. (*see* **Note 9**) The resulting chromatogram depicting multiple permethylated glycan structures derived from human blood serum is shown in Fig. 3.

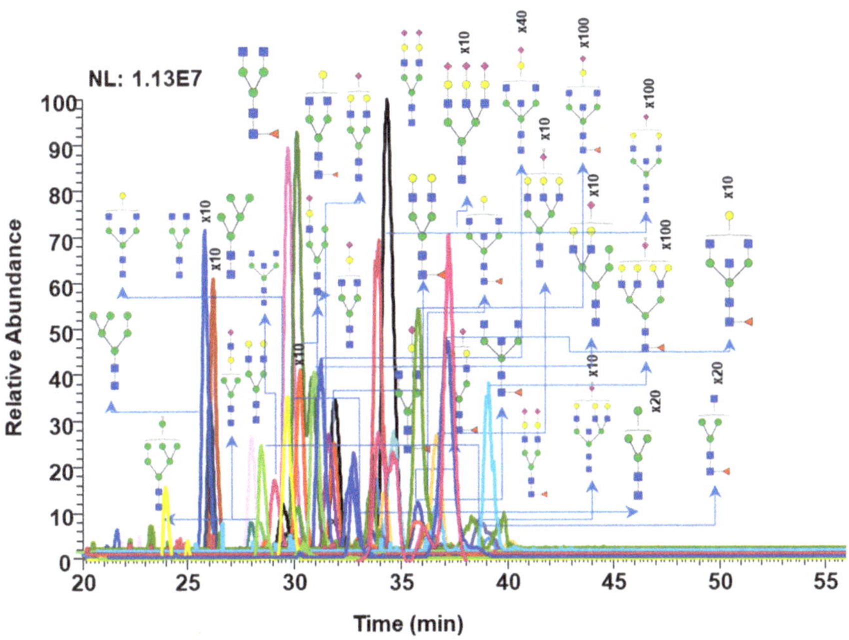


**Fig. 3** Representative chromatogram of permethylated glycans derived from human blood serum generated from PGC-LC-MS

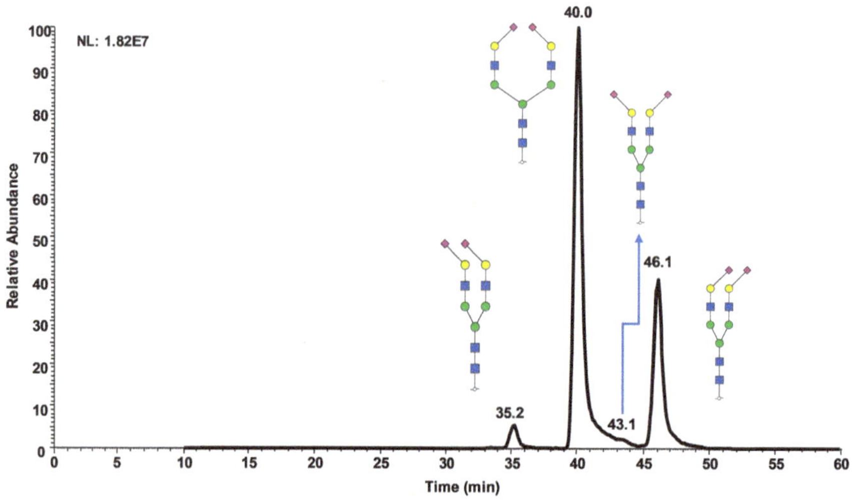


**Fig. 4** EIC of reduced and permethylated biantennary bisialylated glycans derived from bovine fetuin. (From ref. 26 with modifications)

### *3.10 Glycan Isomer Identification*

#### *3.10.1 Identification by Comparison to NMR Data*

Isomeric glycan structures can be assigned by comparing the peak areas of each isomer with previously reported quantitative NMR data [25, 26]. The EIC in Fig. 4 represents the biantennary bisialylated glycans, $HexNAc_4Hex_5NeuAc_2$, released from bovine fetuin. The four peaks at 35.2, 40.0, 43.1, and 46.1 min represent one structure with two $\alpha$-2,3 linked sialic acids, two structures with one $\alpha$-2,3 and one $\alpha$-2,6 linked sialic acid, and one structure with two $\alpha$-2,6 linked sialic acids, respectively. To identify the specific structures for the four peaks, the peak area ratios were compared to the previously reported NMR data for the structural study of *N*-glycans derived from fetuin. According to the published NMR data [26], the abundances of the structures with two $\alpha$-2,3 linked sialic acids, the two structures with one $\alpha$-2,3 and one $\alpha$-2,6 linked sialic acid on different branches, and the structure with two $\alpha$-2,6 linked sialic acids are 10.7%, 57.0%, less than 1.0%, and 32.2%, respectively. This result is comparable with peak areas of four isomeric peaks (Fig. 4) produced by PGC-LC-MS in high temperature. Slight differences between the obtained data and the reported NMR data may be attributed to the source of fetuin samples and the measurement uncertainties inherent to the methods utilized.

#### *3.10.2 Identification Using Exoglycosidase Digestion*

Isomeric characterization of glycans can be performed with an array of exoglycosidases, by using the sequential application of specific exoglycosidases to cleave terminal monosaccharides [25, 26]. To determine galactose linkage isomers from a glycan structure with terminal sialic acids, sialidase is initially applied. Then, the glycan sample is treated with $\beta$-1,3 galactosidase, as shown in Fig. 5a where galactose linkage isomers were determined from a $HexNAc_5Hex_6NeuAc_3$ glycan structure derived from fetuin. As shown in

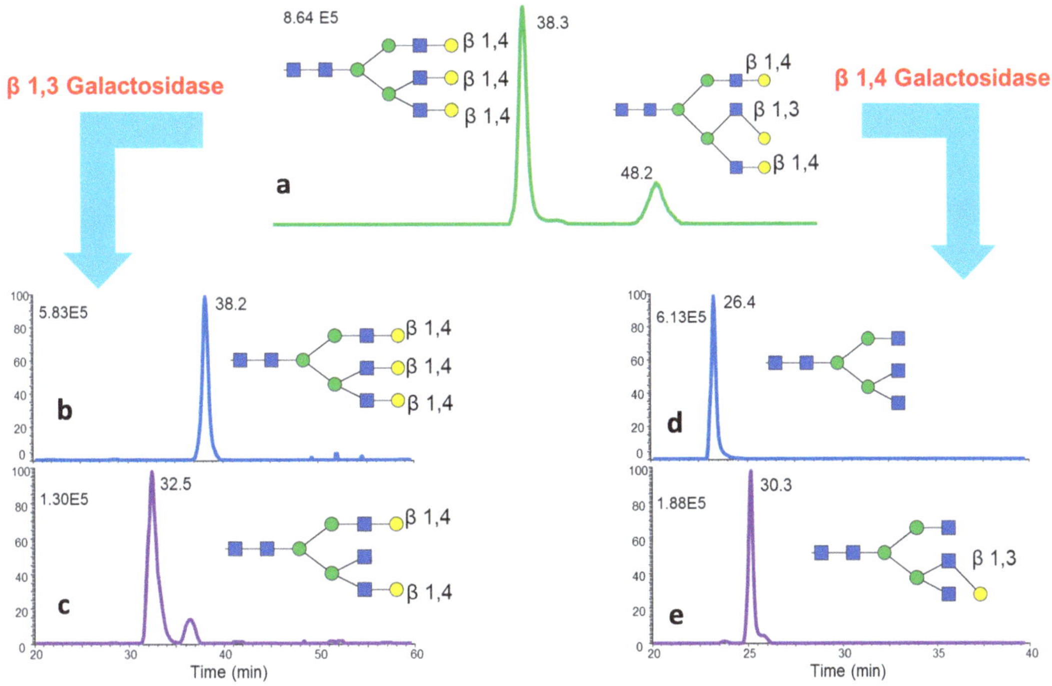


**Fig. 5** EIC of reduced and permethylated triantennary trisialylated glycans derived from bovine fetuin after sialidase (**a**), β-1.3 galactosidase (**b**, **c**), β-1.4 galactosidase (**d**, **e**). (From ref. 25 with modifications)

Fig. 5b, the first peak at 38.3 min remained intact with the same *m/z* value and retention time, confirming that the first isomer had no $\beta$-1,3 linked galactose. The second peak at 48.2 min disappeared, and the peak of $HexNAc_5Hex_5$ appeared at 32.5 min (Fig. 5c), which confirmed the second peak to be a glycan with one $\beta$-1,3 linked galactose. Next, the sample was treated with $\beta$-1,4 galactosidase. After the treatment, the peak at 38.3 min disappeared, and the peak of $HexNAc_5Hex_3$ appeared at 26.4 min (Fig. 5d). All galactose residues were cleaved from the first isomer, while the second isomer was converted to $HexNAc_5Hex_4$ at 30.3 min (Fig. 5e), indicating that the second isomer had two galactoses connected with a $\beta$-1,4 linkage.

#### *3.10.3 Identification Using MS$^n$ Data*

MS$^n$ data can provide useful information for the assignment of isomeric glycans [25, 26]. For instance, we assigned the above two isomers ($\beta$-1,3 and $\beta$-1,4 galactose linkage isomers of triantennary glycan derived from bovine fetuin), by using CID and HCD data [25]. Fig. 6 depicts the CID and HCD spectra for the two isomers. As can be seen in Fig. 6b and c, both structures showed similar fragments with similar intensity. However, there was a fragment ion, *m/z* 1161.52, in the lower trace derived by losing one galactose from the precursor ion. The data demonstrated that the later-eluting isomer contained one unique galactose that was different from the other galactoses, which is in agreement

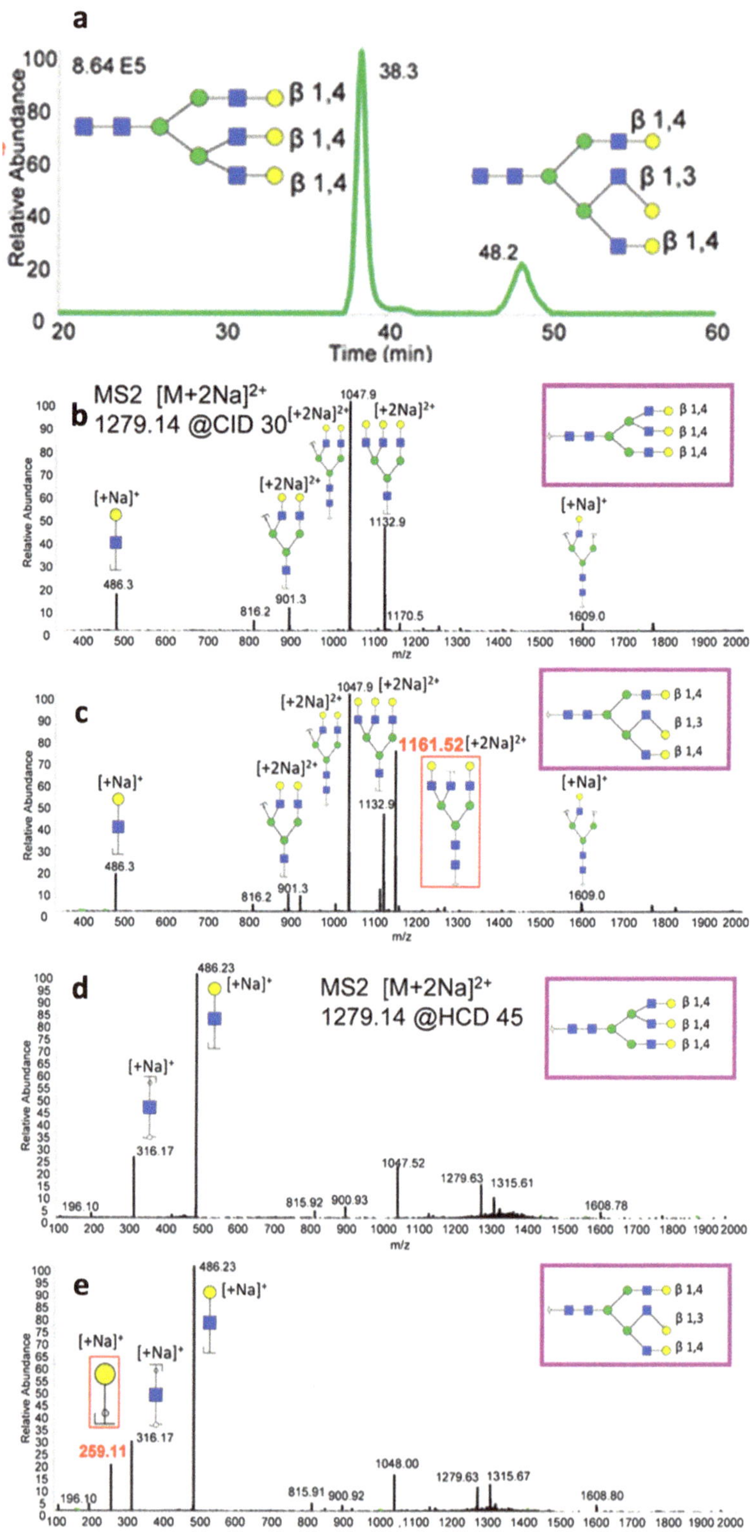


**Fig. 6** EIC of reduced and permethylated triantennary trisialylated glycans derived from bovine fetuin after sialidase treatment (**a**) and their CID (**b**, **c**) and HCD (**d**, **e**) spectra. (From ref. 25 with modifications)

with previously reported NMR data [26]. To further confirm this isomeric structure, an HCD fragmentation experiment was performed to observe the fragmented galactose ion, because ions with a value of less than 400 $m/z$ were not present in the CID spectra due to the limitations of the ion trap instrument. In the HCD spectra (Fig. 6d and e), most fragments were identical except for the unique fragment at $m/z$ 259.1 in the MS/MS spectra for the later-eluting isomer. The fragment ion of $m/z$ 259.1 could be assigned to a free galactose. These data demonstrate the advantages of using CID and HCD data for the assignment of isomeric glycan structures.

#### 3.10.4 Identification Using Standard Glycan Isomers

Standard glycan isomers can also be used for peak assignment by elution orders of each isomer [26]. Figure 7a shows the EIC for mixed standard permethylated glycans, which are core- and

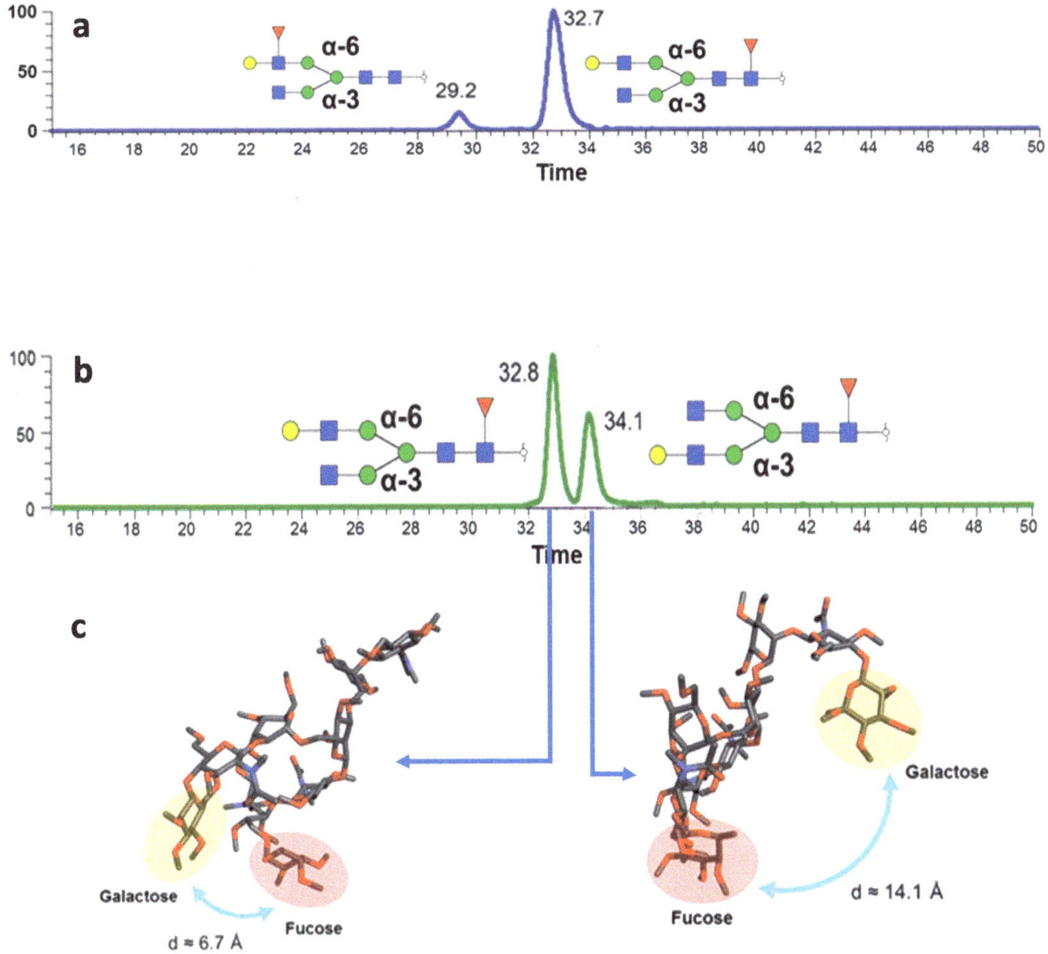


**Fig. 7** (**a**) EIC of reduced and permethylated core- and branched-fucosylated standard glycans. (**b**) EIC of reduced and permethylated with galactose positional isomers derived from human blood serum and (**c**) their 3D modeling structures. (From ref. 26 with modifications)

branch-fucosylated glycans. The peak at retention time 29.2 min represents a branch-fucosylated structure, while the peak at 32.7 min corresponds to a glycan structure with core fucosylation, which demonstrates the isomeric separation of permethylated fucosylated glycans using PGC-LC-MS at 75 °C. These data can be used for the assignment of glycans derived from biological samples. For example, Fig. 7b shows the EIC of a $HexNAc_4Hex_4Fuc_1$ glycan released from human blood serum with two peaks observed, at retention times 32.8 and 34.1 min, both representing core-fucosylated isomers confirmed by $MS^2$ spectra. These peaks therefore represent isomers that originated from galactose residues residing on different branches ($\alpha$-3 or $\alpha$-6 branches). Due to the lack of unique diagnostic fragment ions present in the $MS^2$ spectra, the retention times of standard glycans displayed in Fig. 7a were utilized; where a core-fucosylated glycan standard with an α-6 galactose branch displayed a retention time of 32.7 min. From this observation, it can be concluded that the retention time at 32.8 min appearing in Fig. 7b is a core-fucosylated glycan with $\alpha$-6 galactose branch, while the retention time at 34.1 min is a core-fucosylated glycan with an $\alpha$-3 galactose branch (*see* **Note 10**).

#### *3.10.5 Identification Using Modeling Data*

Separation on PGC columns is facilitated by the combination of RP behavior, based on the hydrophobicity of analytes, and by a polar retention effect, because of the high polarizability of graphitic carbon material. Furthermore, since the column material is planar, the 3D structure of the analyte also influences retention [38, 39]. Data shown in Fig. 7 represent how structural modeling aided in confirming the identification of permethylated glycan isomers [26]. In Fig. 7a, representing two isomers associated with galactose residue residing on different branches ($\alpha$-3 or $\alpha$-6 branches). The retention times of these structures on PGC column, were already confirmed by using standard glycans (*see* Subheading 3.10.4). Structural modeling data (Fig. 7c), using quantum chemical calculations based on density functional theory (DFT) [26], for the two isomers indicated that when the galactose residue is located on the α-6 arm, the fucose residue is closer to the galactose-containing antenna and the molecule is more compact, resulted in less exposed methyl groups to interact with PGC. Conversely, when the galactose is on the $\alpha$ -3 arm, the distance between the fucose and galactose residues is considerably greater ($\approx$14.1 compared to 6.7 Å), making the molecule less compact, thus increasing the nonpolar surface area. Consequently, the less nonpolar surface area of the compact structure prompted less hydrophobic interactions with PGC column, causing the early elution of this isomer, as the first peak shown in Fig. 7a. The consistence of the theoretical elution order estimated by the structural modeling and their real elution order acquired in the experiment demonstrated the reliable correlation of permethylated glycan isomer models to their retention times on PGC column.

## 4 Notes

1. For tissue samples, they must be washed 1–3 times with ice-cold 10 mM PBS buffer solution to remove salts and other contaminants.
2. It should be noted that a permethylation reaction should be performed in low humidity conditions due to the hygroscopic nature of sodium hydroxide beads.
3. It is recommended to permethylate no more than 12 samples per batch to ensure complete permethylation.
4. Dialysis with MWCO 500–1000 Da can be used to desalt released *N*-glycans. However, it does not remove other biomolecules such as proteins or lipids. This procedure is not applicable to *O*-glycans due to their molecular weight being lower than the dialysis membrane molecular weight cutoff.
5. C18 cartridge is a useful tool to remove hydrophobic molecules while it does not retain hydrophilic molecules such as glycans. However, it does not remove salts from the sample.
6. The 10-port valve has two positions, 1-2 and 10-1. The flow path of each position can be designed differently. In our case, at the 1-2 position, the inlet of the C18 trap column is connected to a loading pump while the outlet is connected to waste. The inlet of the PGC column is connected to nano pumps, and the outlet is connected to a nanoESI source. The 1-2 position is used for sample loading and online purification. At the 10-1 position, the inlet of the C18 trap column is connected to nano pumps, and the outlet is connected to the inlet of the PGC column. The outlet of the PGC column is connected to a nanoESI source. The 10-1 position is used for isomeric separation after sample loading and online purification.
7. 75 °C is the maximum temperature that can be achieved by Ultimate 3000 Nano LC system. However, temperature higher than 75 °C can be used to perform isomeric separation of permethylated glycans using a PGC column.
8. It is easy for the spray needle to be clogged, hindering the spray. Thus, it is important to continually monitor the needle tip. Ensure that the needle is not clogged and the spray is clear and stable before analyzing samples.

   The ejected charge state 1 setting of CID should be checked for *N*-glycomics and unchecked for *O*-glycomics. This is an important parameter to obtain a high-quality $MS^2$. This setting needs to be unchecked when analyzing *O*-glycans, because most *O*-glycans are smaller than *N*-glycans and are thus likely to be singly charged.

9. Other software, such as Skyline, may be used for automatic identification and quantitation of glycan isomers. However, manual inspection is still necessary to obtain better, more reliable identification and quantitation.
10. It is worth noting that the use of standard glycans as the only source of structural information is not recommended for two reasons: (1) there is a lack of broad glycan structure coverage, and (2) it is not cost-effective.

## Acknowledgments

This work was supported by grants from the National Institutes of Health, NIH (1R01GM112490-04 and 1R01GM130091-01).

### References

1. Xu C, Ng DTW (2015) Glycosylation-directed quality control of protein folding. Nat Rev Mol Cell Biol 16:742
2. Varki A, Gagneux P (2015) Biological functions of glycans. In: Essentials of glycobiology. Cold Spring Harbor Laboratory Press, Cold Spring Harbor (NY), pp 77–88
3. Roth J et al (2010) Protein N-glycosylation, protein folding, and protein quality control. Mol Cells 30(6):497–506
4. Sola RJ, Griebenow K (2009) Effects of glycosylation on the stability of protein pharmaceuticals. J Pharm Sci 98(4):1223–1245
5. Ohtsubo K, Marth JD (2006) Glycosylation in cellular mechanisms of health and disease. Cell 126(5):855–867
6. Cho BG, Veillon L, Mechref Y (2019) N-glycan profile of cerebrospinal fluids from Alzheimer's disease patients using liquid chromatography with mass spectrometry. J Proteome Res 18(10):3770–3779. https://doi.org/10.1021/acs.jproteome.9b00504
7. Back NK et al (1994) An N-glycan within the human immunodeficiency virus type 1 gp120 V3 loop affects virus neutralization. Virology 199(2):431–438
8. Christiansen MN et al (2014) Cell surface protein glycosylation in cancer. Proteomics 14 (4-5):525–546
9. Peng W et al (2019) Revealing the biological attributes of N-glycan isomers in breast cancer brain metastasis using porous graphitic carbon (PGC) liquid chromatography-tandem mass spectrometry (LC-MS/MS). J Proteome Res 18(10):3731–3740. https://doi.org/10.1021/acs.jproteome.9b00429
10. Alley WR Jr, Novotny MV (2010) Glycomic analysis of sialic acid linkages in glycans derived from blood serum glycoproteins. J Proteome Res 9(6):3062–3072
11. Huang Y et al (2017) LC-MS/MS isomeric profiling of permethylated N-glycans derived from serum haptoglobin of hepatocellular carcinoma (HCC) and cirrhotic patients. Electrophoresis 38(17):2160–2167
12. Kelleher DJ, Gilmore R (2006) An evolving view of the eukaryotic oligosaccharyltransferase. Glycobiology 16(4):47R–62R
13. Colley KJ, Varki A, Kinoshita T (2015) Cellular organization of glycosylation, in essentials of glycobiology. Cold Spring Harbor Laboratory Press, Cold Spring Harbor (NY), pp 41–49
14. Burda P, Aebi M (1999) The dolichol pathway of N-linked glycosylation. Biochim Biophys Acta 1426(2):239–257
15. Gaye MM et al (2012) Ion mobility-mass spectrometry analysis of serum N-linked glycans from esophageal adenocarcinoma phenotypes. J Proteome Res 11(12):6102–6110
16. Hofmann J, Pagel K (2017) Glycan analysis by ion mobility-mass spectrometry. Angew Chem Int Ed 56(29):8342–8349
17. Jin C et al (2019) Separation of isomeric O-glycans by ion mobility and liquid chromatography-mass spectrometry. Anal Chem 91(16):10604–10613. https://doi.org/10.1021/acs.analchem.9b01772
18. Mechref Y (2011) Analysis of glycans derived from glycoconjugates by capillary electrophoresis-mass spectrometry. Electrophoresis 32(24):3467–3481

19. Suzuki S (2013) Recent developments in liquid chromatography and capillary electrophoresis for the analysis of glycoprotein glycans. Anal Sci 29(12):1117–1128
20. Dong X et al (2018) Advances in mass spectrometry-based glycomics. Electrophoresis 39(24):3063–3081
21. Tousi F et al (2013) Differential chemical derivatization integrated with chromatographic separation for analysis of isomeric sialylated N-glycans: a nano-hydrophilic interaction liquid chromatography-MS platform. Anal Chem 85(17):8421–8428
22. Wuhrer M, de Boer AR, Deelder AM (2009) Structural glycomics using hydrophilic interaction chromatography (HILIC) with mass spectrometry. Mass Spectrom Rev 28(2):192–206
23. Ruhaak LR et al (2010) Glycan labeling strategies and their use in identification and quantification. Anal Bioanal Chem 397 (8):3457–3481
24. Ashwood C et al (2019) Standardization of PGC-LC-MS-based glycomics for sample specific glycotyping. Analyst 144(11):3601–3612
25. Zhou S et al (2017) LC-MS/MS analysis of permethylated N-glycans facilitating isomeric characterization. Anal Bioanal Chem 409 (2):453–466
26. Zhou S et al (2017) Isomeric separation of Permethylated Glycans by porous graphitic carbon (PGC)-LC-MS/MS at high temperatures. Anal Chem 89(12):6590–6597
27. Behan JL, Smith KD (2011) The analysis of glycosylation: a continued need for high pH anion exchange chromatography. Biomed Chromatogr 25(1-2):39–46
28. Wuhrer M et al (2006) Mass spectrometry of proton adducts of fucosylated N-glycans: fucose transfer between antennae gives rise to misleading fragments. Rapid Commun Mass Spectrom 20(11):1747–1754
29. Zhou S et al (2017) Direct comparison of derivatization strategies for LC-MS/MS analysis of N-glycans. Analyst 142(23):4446–4455
30. Bapiro TE, Richards FM, Jodrell DI (2016) Understanding the complexity of porous graphitic carbon (PGC) chromatography: modulation of Mobile-stationary phase interactions overcomes loss of retention and reduces variability. Anal Chem 88(12):6190–6194
31. Dong X et al (2018) LC-MS/MS glycomics of idiopathic rapid eye movement sleep behavior disorder. Electrophoresis 39(24):3096–3103
32. Zhu R et al (2018) Enhanced quantitative LC-MS/MS analysis of N-linked Glycans derived from glycoproteins using sodium Deoxycholate detergent. J Proteome Res 17 (8):2668–2678
33. Rahman SA et al (2014) Filter-aided N-glycan separation (FANGS): a convenient sample preparation method for mass spectrometric N-glycan profiling. J Proteome Res 13 (3):1167–1176
34. Goetz JA, Novotny MV, Mechref Y (2009) Enzymatic/chemical release of O-Glycans allowing MS analysis at high sensitivity. Anal Chem 81(23):9546–9552
35. Huang YP et al (2002) Matrix-assisted laser desorption/ionization mass spectrometry compatible beta-elimination of O-linked oligosaccharides. Rapid Commun Mass Spectrom 16(12):1199–1204
36. Desantos-Garcia JL et al (2011) Enhanced sensitivity of LC-MS analysis of permethylated N-glycans through online purification. Electrophoresis 32(24):3516–3525
37. Yu CY et al (2013) Automated annotation and quantification of glycans using liquid chromatography-mass spectrometry. Bioinformatics 29(13):1706–1707
38. Veillon L et al (2017) Characterization of isomeric glycan structures by LC-MS/MS. Electrophoresis 38(17):2100–2114
39. West C, Elfakir C, Lafosse M (2010) Porous graphitic carbon: a versatile stationary phase for liquid chromatography. J Chromatogr A 1217 (19):3201–3216

# Chapter 21

# Optimization of Multiple Glycosidase and Chemical Stabilization Strategies for *N*-Glycan Isomer Detection by Mass Spectrometry Imaging in Formalin-Fixed, Paraffin-Embedded Tissues

Connor A. West, Xiaowei Lu, Grace Grimsley, Kim Norris-Caneda, Anand S. Mehta, Peggi M. Angel, and Richard R. Drake

## Abstract

The analysis of *N*-glycan distributions in formalin-fixed, paraffin-embedded (FFPE) tissues by matrix-assisted laser desorption/ionization (MALDI) imaging mass spectrometry (IMS) is an effective approach for characterization of many disease states. As the workflow has matured and new technology emerged, approaches are needed to more efficiently characterize the isomeric structures of these *N*-glycans to expand on the specificity of their localization within tissue. Sialic acid chemical derivatization can be used to determine the isomeric linkage (α2,3 or α2,6) of sialic acids attached to *N*-glycans, while endoglycosidase F3 (Endo F3) can be enzymatically applied to preferentially release α1,6-linked core fucosylated glycans, further describing the linkage of fucose on *N*-glycans. Here we describe workflows where *N*-glycans are chemically derivatized to reveal sialic acid isomeric linkages, combined with a dual-enzymatic approach of endoglycosidase F3 and PNGase F to further elucidate fucosylation isomers on the same tissue section.

**Key words** Imaging mass spectrometry, Formalin-fixed, Paraffin-embedded tissue imaging, MALDI imaging mass spectrometry, Peptide N-glycosidase F, Chemical derivatization, Endoglycosidase F3, Sialic acid, Core fucose

## 1 Introduction

Over the past 7 years, robust MALDI imaging mass spectrometry approaches to spatially map the distribution of multiple *N*-linked glycans in frozen and formalin-fixed, paraffin-embedded (FFPE) tissues have been developed and continue to evolve. Our collective research group published the initial report in 2013 for the use of sprayed peptide *N*-glycosidase F (PNGaseF) to release *N*-glycans for detection by MALDI-IMS [1]. This was followed by adaptation of the methods to FFPE tissue blocks and tissue microarrays (TMA) [2, 3]. Following digestion and spraying of chemical matrix, the

Arnaud Delobel (ed.), *Mass Spectrometry of Glycoproteins: Methods and Protocols*, Methods in Molecular Biology, vol. 2271, https://doi.org/10.1007/978-1-0716-1241-5_21, 

released *N*-glycans are detected by MALDI-FT-ICR MS, MALDI-TOF MS, and more recently MALDI-Q-TOF MS instruments [3, 4]. The key to the approach is the spraying of a molecular layer of PNGaseF on the tissue of interest, making it an essentially solid-phase reaction. The use of solvent sprayers to apply the enzymes has made this approach very reproducible and portable [5, 6], and has facilitated multiple research groups to apply the *N*-glycan MALDI IMS approach to different tissues and diseases [7–15].

Several advantages of the approach have emerged, as well as continued challenges. The advantages include use of essentially unlimited numbers of FFPE tissues stored worldwide, signal detection that is dependent on PNGase F activity, no purification or enrichment of the glycans are required prior to analysis, and routinely 30 or more glycans can be detected depending on the MALDI instrument used [3, 4, 10, 12]. This in turn facilitates the generation of different glycan panels associated with specific histopathology features and tissue subregions useful for differentiation of disease from normal regions. There are still limitations associated with MALDI analysis of *N*-glycans dependent on instrumentation used, such as loss of sialic acid residues and differentiation of isomeric structures [3, 4]. New approaches linked with MALDI IMS analyses have been developed and are the subject of the chapter.

There is an innate ability of individual sugar residues to form glycosidic bonds in multiple combinations, allowing for biological diversity and the complex natures of glycoconjugates like *N*-linked glycans [16]. Key biologically relevant examples include the ability of fucose residues to be linked in α1,2; α1,3; α1,4; or α1,6 configurations, directed by specific fucosyltransferase genes. For sialic acids in terminal positions, these can be linked in either α2,3 or α2,6 configurations. There are many approaches to studying the isomeric structures, reviewed in Kolarich [16], and we describe workflows using chemical and enzymatic approaches specifically applied to MALDI IMS tissue analysis workflows using sialic acid stabilization strategies [7] and use of a core-fucose specific endoglycosidase F3 [17].

## 2 Materials

Prepare all solutions in HPLC grade water. Follow all safety and waste disposal regulations.

### 2.1 Solutions for MALDI Imaging Mass Spectrometry

1. Antigen retrieval solution/Low pH tissue clearing solution: Pour 50 mL water into a clean 50–100 mL bottle. Add 50 μL of citraconic acid anhydride buffer to the water. Add 4 μL of

12 M HCl and mix. Add water to the bottle for a total of 100 mL and mix. Check that pH is around 3.0 ± 0.5. Use the same day.

2. Preparation of 95% ethanol. Add 950 mL 200 proof ethanol to a clean bottle. Add 50 mL water and mix.
3. Preparation of 70% ethanol: Add 700 mL 200 proof ethanol to a clean bottle. Add 300 mL water and mix.
4. 25% trifluoroacetic acid (TFA): Add 3 mL water to a clean bottle. Carefully add 1 mL of trifluoroacetic acid to the water and mix.
5. Matrix solvent (50% acetonitrile (ACN), 0.1% TFA): Add 25 mL water to a clean 1 L bottle. Carefully add 400 μL of trifluoroacetic acid to the water and mix. Add 500 mL of acetonitrile and mix. Add water to 1 L and mix. Store for up to 2 months at room temperature.
6. MALDI matrix for *N*-glycan imaging (alpha-cyano-4-hydroxycinnamic acid (CHCA), 7 mg/mL in 50% acetonitrile–0.1% TFA): Weigh out 42.0 ± 0.1 mg CHCA. Add the solid CHCA to a clean 50 mL falcon tube. Bring to volume with 6 mL of matrix solvent. Vortex briefly and sonicate 5 min using a benchtop sonicator. Filter CHCA solution using a 13 mm 0.2 μm PTFE hydrophilic syringe filter graded for use with HPLC solvents.
7. Preparation of pH 4.5 Endoglycosidase F3 solution: Add 500 mL water to a clean bottle. Carefully add 15.8 μL of 1 M HCl. Carefully mix. Store at room temperature.

### 2.2 Tissue Clearing Solution

1. Preparation of high pH tissue clearing solution (10 mM Tris, pH 9). Add 500 mL water to a clean 1000 mL bottle. Add 1.21 ± 0.12 g Tris Base and mix. Add water to 900 mL and adjust pH to 9.0 ± 0.03 using 1 M HCl. Add water to a total volume of 1000 mL.

### 2.3 Enzyme and TM-Sprayer™ Solutions

1. Preparation of pH 4.5 Endoglycosidase F3 solution: Add 500 mL water to a clean bottle. Carefully add 15.8 μL of 1 M HCl. Carefully mix. Store at room temperature.
2. Preparation of 1 mL of 0.1 μg/μL endoglycosidase F3 solution. To 100 μg of endoglycosidase F3 [17], add 1 mL of prepared pH 4.5 solution and mix. Use same day. This prepares enough enzyme to cover 4 microscope slides (*see* **Note 1**).
3. Preparation of 1 mL of 0.1 μg/μL PNGase F PRIME™ solution. To 100 μg of PNGase F PRIME™ (from N-Zyme Scientifics) add 1000 μL of water and mix. Use same day. This prepares enough enzyme to cover 4 microscope slides.

4. TM Sprayer™ push solvent (50% methanol/water): Add 500 mL methanol to a clean bottle. Add 500 mL water and mix. Solvent may be kept at room temperature during the duration of use.

### 2.4 Amidation Reaction Solutions

1. Amidation Solvent 1: In a glass vial, add 22 μL of 1-(3-dimethylaminopropyl)-3-ethylcarbomiide (EDC) (*see* **Note 2**) and 15.8 μL of dimethylamine. Add 42.2 mg of 1-Hydroxybenzotriazole hydrate (HOBt) (*see* **Note 3**). Finally, add 0.5 mL of Dimethyl Sulfoxide (DMSO) and mix until dissolved. Solvent may be kept at room temperature for the duration of use. This prepares enough solution to cover 2 microscope slides.
2. Amidation Solvent 2: In a glass vial, add 150 μL of 28–30% ammonia in water to 350 μL of DMSO and mix. Solvent may be kept at room temperature for duration of use. This prepares enough solution to cover 2 microscope slides.
3. Carnoy's solution (30% chloroform, 60% ethanol, 10% acetic acid): Add 300 mL of chloroform to a clean bottle. Add 600 mL of 200 proof ethanol and 100 mL of glacial acetic acid and mix. Solvent may be kept at room temperature.
4. TFA–ethanol solution (0.1% TFA in ethanol): Add 99.9 mL of 200 proof ethanol to a clean bottle. Add 0.1 mL of trifluoroacetic acid and mix. Store at room temperature.

## 3 Methods

Carry out all procedures at room temperature, unless otherwise specified.

### 3.1 Heating and Dewaxing

1. Incubate the slides with tissue face up in a 60 °C oven for 1 h.
2. Prepare Coplin jars of solvent for dewaxing by pouring the following solutions into Coplin jars: Xylenes, two Coplin jars; 200 proof ethanol, USP grade, two Coplin jars; 95% ethanol, one Coplin jar; 70% ethanol, one Coplin jar; double distilled water, two Coplin jars. Solvent should be added to a level that will allow complete immersion of the tissue sections.
3. Dewax the slides in the following order with the specified times: xylenes, two times at 2 min each; 200 proof ethanol, two times at 1 min each; 95% ethanol solution, one time at 1 min; 70% ethanol solution, one time at 1 min; HPLC grade water, two times at 3 min each. For each step, immerse the slides completely in freshly poured solution for the stated length of time. At the end of the incubation time, agitate the slides briefly as they are removed from the Coplin jars.
4. Dry the slides in a vacuum desiccator for 5 min.

### 3.2 Slide Scanning

1. This step produces an optical image needed for selecting tissue regions for imaging analysis. Fiducials, or reference points, are needed to accurately "teach" the instrument where the tissue is located on the slide.
2. Use a reflective metallic marker to make a small circle at each corner of the microscope slide. Use a black marker to draw a cross or hash mark on top of each silver circle. The reflective marker provides a contrasting background for clear visualization of the black mark to use as a fiducial.
3. For images that will be acquired by mass spectrometry at ≥100 μm spatial resolution, scan the whole slide at a minimum of 1200 ppi resolution. For images that will be acquired with a ≤50 μm stepsize scan the slide at a minimum of 2400 ppi resolution. Save the images.
4. After scanning and prior to antigen retrieval, slides may be stored overnight in a desiccator. For longer times over 2 days, store the slides at −20 to −80 °C. It is preferable to proceed with the next step immediately.

### 3.3 Amidation–Amidation (AA) Reaction

This AA reaction described is adapted the original protocol of Holst et al. [7]. A schematic of the AA reactions is shown in Fig. 1. Representative spectra for two mono-sialylated biantennary *N*-glycans after AA treatment are shown in Fig. 2.

1. Add 200 microliters of Amidation Solvent 1 to the heated and dewaxed tissue slide.
2. Cover the tissue with a glass coverslip and ensure that all bubbles are removed underneath the coverslip.
3. Place slide in a sealable glass Pyrex lunch box. Dimensions are 10 cm length by 4 cm height × 6 cm width at the base of the vessel. Elevate the slides from the bottom of the chamber using

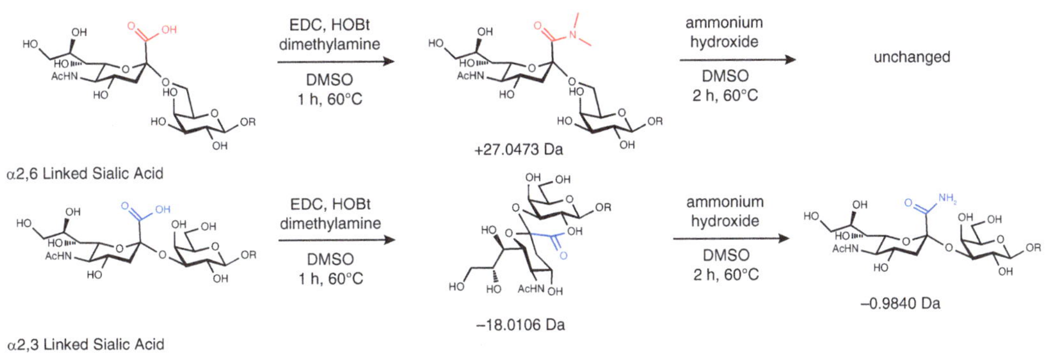


**Fig. 1** Reaction schematic for amidation–amidation chemical derivatization. Schematic of sialic acid derivatization via amidation–amidation reaction [7]. Top: reaction scheme for α2,6 linked sialic acids; Bottom: reaction scheme for α2,3 linked sialic acids. Changes in mass are shown following each step in reaction scheme

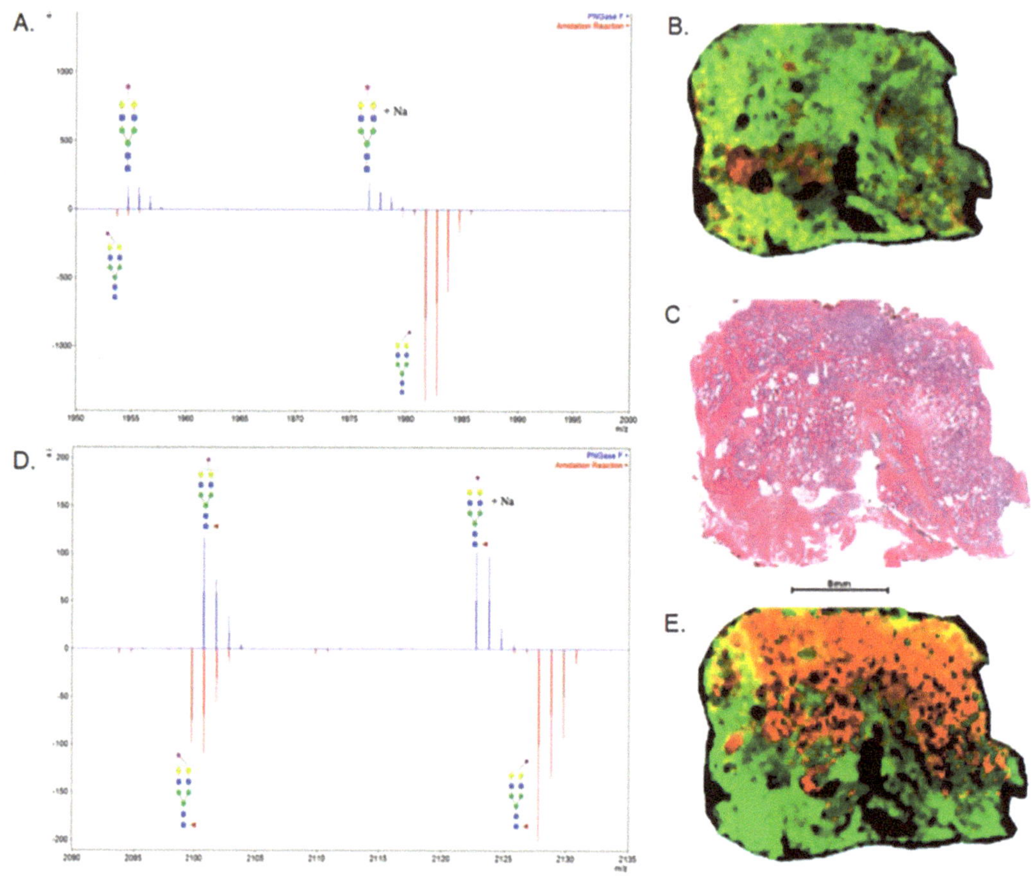


**Fig. 2** Example amidation–amidation chemical derivatization on tissue. The amidation–amidation reaction was performed on a FFPE prostate cancer tissue. Representative spectra and selected images of amidated sialic acid glycans are shown following N-glycan MALDI imaging mass spectrometry analysis. (**a**) Spectra for an Nglycan, Hex5HexNAc4, with and without amidation–amidation. The top spectra in blue represents the nonderivatized N-glycan specie, carrying one or two Na ions. The bottom spectra in red shows the mass shift resulting from the amidation derivatization. Sialic acids residues are shown as purple diamonds. Those angled to the left indicate α2,3 linkage while those angled to the right represents an α2,6 linkage. (**b**) The distribution of the two glycans shown in panel (**a**) in the whole tissue are shown. The resulting image overlay shows localization of the α2,3 linked N-glycan, m/z = 1953.741 (red, −1 m.u.), and α2,6 linked, 1981.782 (green, +27 m.u.). (**c**) Hematoxylin and eosin stain of the tissue section analyzed. (**d**) Spectra for the fucosylated N-glycan Hex5HexNAc4Fuc1 with and without the amidation–amidation reaction. Top spectra in blue is nonderivatized and the bottom spectra in red is showing the mass shift following derivatization. (**e**) Resulting image overlay showing localization of α2,3 linked N-glycan, m/z = 2099.810 (red, −1 m.u.) and α2,6 linked, m/z = 2127.832 (green, +27 m.u.)

glass 13 × 100 centrifuge tubes. Seal and place the box into a prewarmed 60 °C oven for 1 h. This size vessel allows two slides to be done per reaction. Alternate incubation chambers can be used (*see* **Note 4**).

4. After incubation, extract the slide and gently remove the coverslip, avoiding disfiguring the tissue (*see* **Note 5**).

5. Place slide perpendicular on a paper towel to remove residual reaction solvent. A lab tissue can be used to wick peripheral fluids.
6. Using a pipette tip, vacuum dry the slide for 10 s over the tissue. A standard in-house laboratory vacuum is recommended. Be careful to get close to the tissue, but do not touch the tissue (*see* **Note 6**).
7. Add 200 μL of DMSO to the tissue.
8. Repeat the vacuum drying (**step 6**).
9. Repeat **steps 8** and **9** two more times for a total of three DMSO washes.
10. Add 200 microliters of Amidation Solvent 2 to the tissue.
11. Cover the tissue with a glass coverslip and remove any bubbles below the coverslip.
12. Place slide in the same Pyrex lunch box set up and incubated in the oven at 60 °C for 2 h.
13. Remove slides and place perpendicular on paper towel to remove residual reaction solution.
14. Prepare a series of Coplin jars for solvent washing: 200 proof ethanol, USP grade, 2 Coplin jars; Carnoy's solution, 2 Coplin jars; double distilled water, 1 Coplin jar; 200 proof ethanol, USP grade, 2 Coplin jars. Solvent should be added to a level that will allow complete immersion of the tissue sections. A total of 4 Coplin jars for ethanol, 2 Coplin jars for water, and 2 Coplin jars for Carnoy's solution are needed.
15. Wash the slides in the following order with the specified times, using an individual Coplin jar for each step: Add to 200 proof ethanol for 2 min, repeat for 2 min in next ethanol Coplin jar; Add to Carnoy's solution for 10 min, repeat for 10 min in the next Carnoy's Coplin jar; Add to double distilled water for one wash for 1 min; Add to 200 proof ethanol for 2 min; repeat for 2 min in next ethanol Coplin jar. For each step, immerse the slides completely in freshly poured solution for the stated length of time. At the end of the incubation time, agitate the slides briefly as they are removed from the Coplin jars.
16. Measure out 10 mL of the TFA–ethanol solution. Carefully and slowly pour the solution over the slide for 30 s.
17. Do not allow the slides to dry, and proceed directly to the antigen retrieval step (Subheading 3.3) (*see* **Note 7**).

### 3.4 Antigen Retrieval

1. This step details antigen retrieval for imaging mass spectrometry using a vegetable steamer or decloaker apparatus (*see* **Note 8**).

2. Fill all parts of the vegetable steamer or decloaker to the marked water level. This includes the overall basin and the slide holder jars.
3. Add ~10 mL of the antigen retrieval solution to a plastic five slide mailer vessel (leave top open).
4. Place three slides into each five slide mailer with top opened. Slides in positions 1 and 5 are placed with tissue facing outward to the solution. Position 3 may face either way. This allows good solvent access to the tissue.
5. Fill the slide mailer the rest of the way with the antigen retrieval solution so that all tissue is completely covered.
6. Holes from an 18 gauge needle or equivalent can be punched in the slide mailer lid. If the mailer has no holes punched in the lid, snap close one corner of the mailer. This allows steam to exit. Place the mailer in the steamer or decloaker.
7. For use of the vegetable steamer, heat for 30 min. Temperature should reach 95 °C for a minimum of 20 min.
8. For use of the decloaker, options for two temperature and pressure settings are provided. The lower temperature option is a good default setting, and the higher temperature and pressure option can work well in tissues with higher fat content (*see* **Note 9**).
9. Decloaker Temperature Setting Option 1: Begin program of 95 °C for 30 min and no pressure setting. Temperature should reach 95 °C for a minimum of 30 min, although slides should be in the decloaker for the preheating as well.
10. Decloaker Temperature Setting Option 2: Begin program of 115 °C for 15 min and pressure of 5.4 psi. Temperature should reach 115 °C for a minimum of 15 min, although slides should be in the decloaker for the preheating as well.
11. Place the hot slide mailer in a container holding cool water (*see* **Note 10**). Cooling water should come midway up the side of the mailer. Cool for 5 min in the water bath.
12. Remove half the buffer from the mailer and replace with distilled water. Cool on the countertop for 5 min. Repeat removal of half the buffer two more times, each with 5 min of cooling. Complete by rinsing in 100% distilled water.
13. Dry the slides for 5 min in a desiccator.

### 3.5 Endo F3 Application by the TM-Sprayer™

1. A syringe pump with 0.05% accuracy in pumping 25 μL/min is used for enzyme application. A glass or plastic 1-mL syringe with Luer lock is used for loading enzyme into the sprayer.
2. Fill a glass or plastic 1-mL syringe with Luer lock syringe with prepared Endo F3 solution ensuring that there are no bubbles

in syringe (*see* **Note 11**). Fasten the syringe to the TM-Sprayer™ line used for enzyme spraying. Secure the syringe in the pump.

3. Set the pump to a flow rate of 25 μL/min with an inner diameter matching that of the used syringe. Do not turn on the pump at this time.
4. Place the microscope slides with tissue samples on the TM-Sprayer™ sample area, fastening them with lab tape.
5. Turn on the TM-Sprayer™ and then the controlling computer.
6. Open the nitrogen gas tank valve, setting the regulator to 10 psi.
7. In the TM-Sprayer™ software, set the temperature to 45 °C. Temperature will not adjust without the nitrogen gas flowing.
8. Program the TM-Sprayer™ to cover the appropriate number of slides, allowing a 5 mm additional edge distance for spray-head turn round.
9. Program the TM-Sprayer™ method for Endo F3 to use 15 passes, crisscross pattern, velocity of 1200, 3.0 mm track spacing, and a dry time of zero. The tip of the sprayhead should be 40 mm distance from the surface of the slide.
10. Start the syringe pump. Place a blank microscope slide under the nozzle of the spray head to check the TM-Sprayer™ to monitor the start of enzyme solution spraying. It generally takes about 1–3 min to start emitting solution.
11. Once moisture is detected on the blank slide, press "Start" in the TM-Sprayer™ software. The Endo F3 enzyme will be applied in a thin layer to the slide.

### 3.6 Incubation for On-Tissue Digestion

1. The same incubation approach is used for both Endo F3 and PNGase F PRIME™ digestion and is performed immediately after application of Endo F3 or PNGase F PRIME™.
2. Prepare an incubation chamber for enzyme digestion using a plastic 100 × 15 mm cell culture dish. Fit a single layer paper towel (Wypall ×60) on the bottom of the dish. Fold two 4 × 6 Kimwipes and place at opposite sides of the dish. Using a spray bottle of water, add water to saturate the paper towels and Kimwipes. Stop adding water when excess water accumulates in the dish, observed by tilting the dish to one side. For the specified towels and Kimwipes, this is about 5 mL of water.
3. The incubation chamber should be preheated for 15–30 min in an oven at 37.5 °C to produce a thin layer of condensation on the top of the incubation dish. The incubation dish is heated with lid in place; no lab tape or other sealant is required.

4. Place the slide with the tissue facing upward into the incubation chamber using the Kimwipes as supports. Gently push the slide down slightly so that when the cover is placed on, the tissue does not touch the incubation chamber cover.
5. Incubate for 2 h in the oven set at 37.5 ± 1.5 °C. Ensure that the internal oven temperature is at the correct temperature using a secondary thermometer.
6. After incubation, remove the slide slowly while holding it parallel with the countertop. Wipe off the condensation to prevent liquid rolling onto the tissue surface and delocalizing *N*-glycans (or peptides).
7. Store the slide in a 5 slide mailer to protect the released *N*-glycans. If matrix cannot be sprayed the same day, store briefly in a desiccator (6–12 h) or at −20 °C for long term (2–3 days).
8. It is recommended to immediately spray matrix onto the slide.

### 3.7 MALDI Matrix Application by the TM-Sprayer™

1. MALDI matrix application is performed after Endo F3 digestion and again after PNGase F PRIME™ digestion.
2. An isocratic pump with 0.05% accuracy in pumping 100 μL/min is used to for matrix application. A glass 5-mL syringe with Luer lock for loading matrix into the sprayer.
3. Ensure that the isocratic pump is set to pump 100 μL/min. Solvent may be degassed to limit flow variation.
4. Turn on the TM-Sprayer™ and controlling computer.
5. Open the nitrogen gas tank, setting the regulator to 10 psi.
6. In the TM-Sprayer™ software, set the temperature to 80 °C. Temperature will not adjust without the nitrogen gas flowing.
7. Program the TM-Sprayer™ method for CHCA matrix application to use 8 passes, crisscross pattern, velocity of 1300, 2.5 mm track spacing and zero dry time. The tip of the sprayhead should be 40 mm distance from the surface of the slide.
8. Place the samples on the TM-Sprayer™ platform, fastening them with lab tape.
9. Program the TM-Sprayer™ to cover the appropriate number of slides, allowing a 5 mm additional edge distance for sprayhead turn round.
10. Fill a glass 5-mL syringe with the filtered MALDI matrix solution, ensuring that there are no bubbles in the syringe.
11. Fasten the syringe to the TM-Sprayer™ line going to the 6-port valve. With the valve switch in "Load" position, inject the MALDI matrix solution into the 5 mL loop.
12. Ensure that the pump is flowing at 100 μL/min and that appropriate pump pressure readouts are stable.

13. Move the six-port valve switch to "Spray."
14. Use a blank microscope slide to check the TM-Sprayer™ nozzle for spraying of solution. Once matrix is detected as an opaque film on the dummy slide, press "Start" in the TM-Sprayer™ software.
15. CHCA solution will be applied in a thin layer onto target tissues. When finished, matrix coated slides may be imaged immediately by mass spectrometry or stored in a desiccator.
16. The reaction scheme for Endo F3 and representative spectra from the reaction is provided in Fig. 3. Examples of the MALDI imaging data using Endo F3 is shown in Fig. 4.

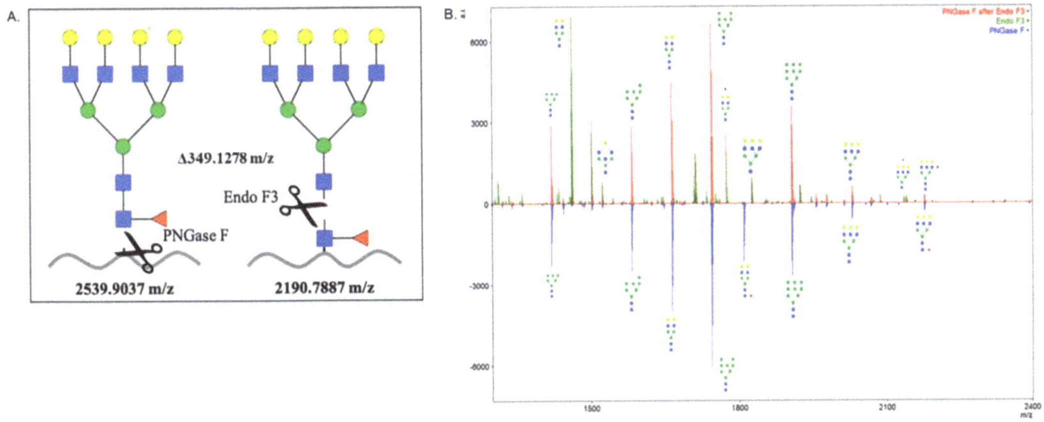


**Fig. 3** Endoglycosidase F3 cleavage and resulting mass shift. (**a**) Example of differential cleavage of Endoglycosidase F3 and PNGase F. Endoglycosidase F3 will leave a fucose and *N*-acetylglucosamine residue on the protein, resulting in a −349.1278 *m/z* mass shift from the parent *N*-glycan mass. (**b**) Representative spectra of Endoglycosidase F3 cleaved glycans versus PNGase F cleaved glycans

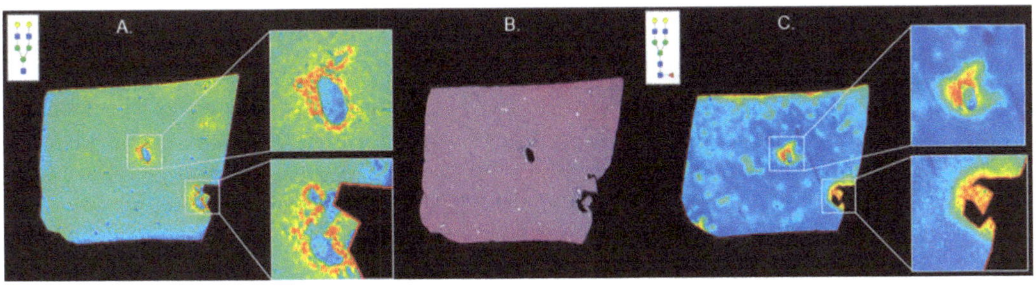


**Fig. 4** MALDI imaging mass spectrometry using endoglycosidase F3. Resulting image for *N*-glycan *m/z* = 1809.6393 from Endoglycosidase F3 digestion (**a**, *m/z* = 1460.5266 *m/z*) and separate PNGase F digestion in normal liver FFPE tissue slices (**c**). Zoom in areas indicate areas of interest for core fucosylation localization. (**b**) Hematoxylin and eosin stain of tissue used for Endoglycosidase F3 digestion

#### 3.8 Tissue Clearing of Matrix and Residual Endo F3 Cleaved N-Glycans

1. The purpose of tissue clearing is to remove matrix, Endo F3 and *N*-glycans prior to PNGase F PRIME™ digestion.
2. Remove matrix by incubation the slide in 200 proof/100% ethanol for 1 min.
3. Incubate the slide in 95% ethanol solution for 1 min followed by 70% ethanol for 1 min.
4. Remove hydrophilic *N*-glycans by incubating the slide in water for 1 min, high pH solution for 1 min, water for 1 min, low pH solution (citraconic buffer) 1 min, and water 1 min. For each step, agitate the slides 3–5 times at the end of the incubation.
5. Wipe excess water off the back of the slide and dry for 5 min in desiccator.
6. Application of PNGaseF PRIME™ is done by TM Sprayer as described for endo F3 (Subheading 3.5). The same amount of enzyme is used.
7. Perform PNGase F PRIME™ digestion following Subheading 3.6, Incubation for On-tissue Digestion.
8. After On-tissue Digestion, complete Subheading 3.7 MALDI Matrix Application by the M3 TM-Sprayer™.
9. Upon completion, matrix coated slides may be imaged immediately by mass spectrometry or stored in a desiccator until imaging experiments are completed (*see* **Note 12**).

## 4 Notes

1. The indicated amount is appropriate for four slides, which is the capacity of the M3 TM-Sprayer™. This amount can be adjusted for additional slides if an M5 TM-Sprayer™ is being used.
2. EDC is moisture sensitive and decomposes with trace amount of water. Once opened, store the EDC in a dry, desiccated chamber. EDC should be prepared fresh immediately prior to use.
3. The HOBt stock used herein contains 20% water. If anhydrous HOBt is used, the amount to be weighed should be adjusted. Using too much HOBt will decrease the selectivity of the reaction between the 2,3 and 2,6 linked sialic acids.
4. A lidded glass Petri dish can be used, but must be sealed with Parafilm. This seal is critical to retain dimethyl amine and ammonia concentrations during the reaction. Two slides can be done in a 100 $cm^2$ dish.
5. Observe the liquid tension between the coverslip and slide surface. In some cases, the coverslip can be gently slid off the edge of the slide. In other cases, it is easier to flip a corner of the coverslip upward to break this liquid seal.

6. Vacuum dry long enough to remove any obvious liquid present, and then proceed immediately to **step** 7.
7. Be prepared to start the antigen retrieval step by preparing the indicated buffers. We do not recommend allowing the slide to dry or stop at this step for storage.
8. Antigen retrieval is necessary to reverse the formalin crosslinks of the proteins in the tissues. If used, this step is done after the chemical modification steps for sialic acid stabilization, and is also done before application of the endoF3 or PNGaseF PRIME™ enzymes.
9. The high temperature and pressure setting will cause terminal sialic acids residues to cleave off, thus the chemical stabilization must be done for these tissues and settings if detection of sialic acid glycans are desired. However, if stabilization is not done, use of the 95 °C setting for 30 min and no pressure will minimize sialic acid loss during this step.
10. Use hot pads or heat resistant gloves to remove the slide mailer, as it is full of near boiling liquid and could be a potential burn hazard to skin and eyes.
11. To remove any bubbles within the syringe barrel following loading of the Endo F3 solution, pull a small volume of air into the syringe. Holding the syringe with the tip upward, gently dispense the syringe until the air bubble is gone.
12. Prepared slides can be stored in the desiccator for 3 months.

## Acknowledgments

This work was supported in part by the National Institutes of Health/National Cancer Institute grant U01CA242096 (RRD, ASM, and PMA), R01 CA120206 (ASM), U01 CA168856 (ASM), U54 MD010706 (RRD and PMA), and the Hollings Cancer Center Predoctoral Fellowship (CAW).

### References

1. Powers TW, Jones EE, Betesh LR, Romano PR, Gao P, Copland JA, Mehta AS, Drake RR (2013) Matrix assisted laser desorption ionization imaging mass spectrometry workflow for spatial profiling analysis of N-linked glycan expression in tissues. Anal Chem 85 (20):9799–9806
2. Powers TW, Neely BA, Shao Y, Tang H, Troyer DA, Mehta AS, Haab BB (2014) Drake RR. MALDI imaging mass spectrometry profiling of N-glycans in formalin-fixed paraffin embedded clinical tissue blocks and tissue microarrays. PLoS One 9(9):e106255
3. Drake RR, Powers TW, Jones EE, Bruner E, Mehta AS, Angel PM (2017) MALDI mass spectrometry imaging of N-linked Glycans in cancer tissues. Adv Cancer Res 134:85–116
4. Drake RR, West CA, Mehta AS, Angel PM (2018) MALDI mass spectrometry imaging of N-linked Glycans in tissues. Adv Exp Med Biol 1104:59–76

5. Angel PM, Mehta A, Norris-Caneda K, Drake RR (2018) MALDI imaging mass spectrometry of N-glycans and tryptic peptides from the same formalin-fixed, paraffin-embedded tissue section. Methods Mol Biol 1788:225–241
6. Drake RR, Powers TW, Norris-Caneda K, Mehta AS, Angel PM (2018) In situ imaging of N-Glycans by MALDI imaging mass spectrometry of fresh or formalin-fixed paraffin-embedded tissue. Curr Protoc Protein Sci 94 (1):e68
7. Holst S, Heijs B, de Haan N, van Zeijl RJ, Briaire-de Bruijn IH, van Pelt GW, Mehta AS, Angel PM, Mesker WE, Tollenaar RA, Drake RR, Bovée JV, McDonnell LA, Wuhrer M (2016) Linkage-specific in situ sialic acid derivatization for N-glycan mass spectrometry imaging of formalin-fixed paraffin-embedded tissues. Anal Chem 88(11):5904–5913
8. Gustafsson OJ, Briggs MT, Condina MR, Winderbaum LJ, Pelzing M, McColl SR (2015) Everest-Dass AV, packer NH, Hoffmann P. MALDI imaging mass spectrometry of N-linked glycans on formalin-fixed paraffin-embedded murine kidney. Anal Bioanal Chem 407(8):2127–2139
9. Everest-Dass AV, Briggs MT, Kaur G, Oehler MK, Hoffmann P, Packer NH (2016) N-glycan MALDI imaging mass spectrometry on formalin-fixed paraffin-embedded tissue enables the delineation of ovarian cancer tissues. Mol Cell Proteomics 15(9):3003–3016
10. Briggs MT, Condina MR, Ho YY, Everest-Dass AV, Mittal P, Kaur G, Oehler MK, Packer NH, Hoffmann P (2019) MALDI mass spectrometry imaging of early- and late-stage serous ovarian cancer tissue reveals stage-specific N-glycans. Proteomics 19(21-22):e1800482
11. Angel PM, Baldwin HS, Gottlieb Sen D, Su YR, Mayer JE, Bichell D, Drake RR (2017) Advances in MALDI imaging mass spectrometry of proteins in cardiac tissue, including the heart valve. Biochim Biophys Acta Proteins Proteom 1865(7):927–935
12. Heijs B, Holst S, Briaire-de Bruijn IH, van Pelt GW, de Ru AH, van Veelen PA, Drake RR, Mehta AS, Mesker WE, Tollenaar RA, Bovée JV, Wuhrer M, McDonnell LA (2016) Multimodal mass spectrometry imaging of N-glycans and proteins from the same tissue section. Anal Chem 88(15):7745–7753
13. Drake RR, McDowell C, West C, David F, Powers TW, Nowling T, Bruner E, Mehta AS, Angel PM, Marlow LA, Tun HW, Copland JA (2019) Defining the human kidney N-glycome in normal and cancer tissues using MALDI imaging mass spectrometry. J Mass Spectrom 55:e4490
14. Scott DA, Casadonte R, Cardinali B, Spruill L, Mehta AS, Carli F, Simone N, Kriegsmann M, Del Mastro L, Kriegsmann J, Drake RR (2019) Increases in tumor N-glycan Polylactosamines associated with advanced HER2-positive and triple-negative breast cancer tissues. Proteomics Clin Appl 13(1):e1800014
15. West CA, Wang M, Herrera H, Liang H, Black A, Angel PM, Drake RR, Mehta AS (2018) N-linked glycan branching and Fucosylation are increased directly in Hcc tissue as determined through in situ glycan imaging. J Proteome Res 17(10):3454–3462
16. Alagesan K, Everest-Dass A, Kolarich D (2018) Isomeric separation and characterisation of glycoconjugates. Adv Exp Med Biol 1104:77–99
17. West CA, Liang H, Drake RR, Mehta ASA (2020) New enzymatic approach to distinguish fucosylation isomers of N-linked glycans in tissues using MALDI imaging mass spectrometry. J Prot Res 19(8):2989–2996

# Chapter 22

# Glycosylation Profiling of Glycoproteins Secreted from Cultured Cells Using Glycan Node Analysis and GC-MS

Jesús S. Aguilar Díaz de león and Chad R. Borges

## Abstract

Glycan "node" analysis is the process by which pooled glycans within complex biological samples are chemically deconstructed in a way that facilitates the analytical quantification of uniquely linked monosaccharide units (glycan "nodes"). It is based on glycan methylation analysis (a.k.a. linkage analysis) that has historically been applied to pre-isolated glycans. Thus, when using glycan node analysis, unique glycan features within whole biospecimens such as "core fucosylation," "α2-6 sialylation," "β1-6 branching," "β1-4 branching," and "bisecting GlcNAc," are captured as single analytical signals by GC-MS. Here we describe the use of this methodology in cell culture supernatant and in the analysis of IgG (alpha-1 antitrypsin) glycans. The effect of IL-6 and IL-1β cytokines on secreted hepatocyte protein glycan features is demonstrated; likewise, the impact of neuraminidase treatment of IgG is illustrated. For the majority of glycan nodes, the assay is consistent and reproducible on a day-to-day basis; because of this, relatively subtle shifts in the relative abundance of glycan features can be captured using this approach.

**Key words** Glycosylation Profiling, Glycan Nodes, Glycan Permethylation, GC-MS, Cell Culture Supernatant, Aberrant Glycosylation, Glycans, Antibody Glycosylation Profiling, Secreted Glycoproteins

## 1 Introduction

Glycosylation is a common posttranslational modification of mammalian proteins. About 50% of all mammalian proteins are glycosylated [1]. Glycoproteins can be secreted by cells or found in cell membranes, and their glycans play pivotal roles in cell recognition [1], immune evasion [2, 3], and cell signaling [4]. There are over 200 known human glycosyltransferase genes that encode for glycosyltransferase enzymes, each of which assembles glycans in a strict donor, acceptor, and linkage-specific manner [5].

Aberrant glycosylation is common in cancer. Glycosylation changes that occur in this disease are known to help facilitate tumor metastasis [6]. In the antibody therapeutic industry, changes in glycosylation can severely impact antibody stability and

Arnaud Delobel (ed.), *Mass Spectrometry of Glycoproteins: Methods and Protocols*, Methods in Molecular Biology, vol. 2271, https://doi.org/10.1007/978-1-0716-1241-5_22, 

biological function [7]. In both cases, cell culture systems are employed to either study glycosylation in cancer cell lines or to produce antibodies with consistent glycosylation so that their pharmacodynamics and pharmacokinetics will remain constant. In most cases, altered expression of glycosyltransferases serves as the immediate upstream cause of structural changes in glycosylation [7].

While abnormal glycosylation within cancer cells has been extensively studied [8–10], relatively little is known about the mechanisms leading to aberrant glycosylation of the numerous highly abundant blood plasma proteins in cancer, the vast majority of which are secreted by either the liver or B-cells/plasma cells (i.e., IgG) and are the major analytical target of serum glycomics studies [11].

Additionally, structural characterization of therapeutic antibody glycans is required under regulatory guidelines because changes in antibody glycosylation tend to be the major source of batch to batch variability during their production [12]. It is thus important to track therapeutic antibody glycosylation changes to ensure batch to batch consistency and the proper function of the antibody.

Glycans are complex sugar polymers containing a variety of different monosaccharides, α-linkages or β-linkages, linear and branched forms, and sometimes repeating units, all of which add high heterogeneity to the glycan structure and makes the analysis of glycans a difficult task [13]. New methods that expedite the quantification of unique glycan features and that can point directly to dysregulated glycosyltransferases in cell culture supernatant would help facilitate a better understanding of the mechanisms regulating the glycan structures of secreted glycoproteins. Additionally, methods for the direct quantification of specific glycan features in antibodies could help facilitate improved batch to batch glycosylation consistency profiling.

Glycan methylation analysis (a.k.a. glycan linkage analysis) that takes place via generation of partially methylated alditol acetates (PMAAs) and subsequent analysis by GC-MS has been widely used in the past to systematically deconstruct preisolated glycans in order to obtain linkage and branching information [14–17]. Over the past few years we have adapted this approach to the analysis of whole biofluids and employed it as a biomarker development tool in blood plasma/serum [18–23]. In short, the technique involves the simultaneous chemical deconstruction of *N*-, *O*-, and lipid linked glycans in a manner that both conserves and facilitates relative quantification of uniquely linked monosaccharides (glycan nodes). In doing so, the approach captures unique glycan features, such as "core fucosylation," "α2-6 sialylation," "β1-6 branching," "β1-4 branching," and "bisecting GlcNAc," as single analytical signals, some of which serve as 1:1 molecular surrogates of glycosyltransferase activity [18] (Figs. 1 and 2). As such, application of

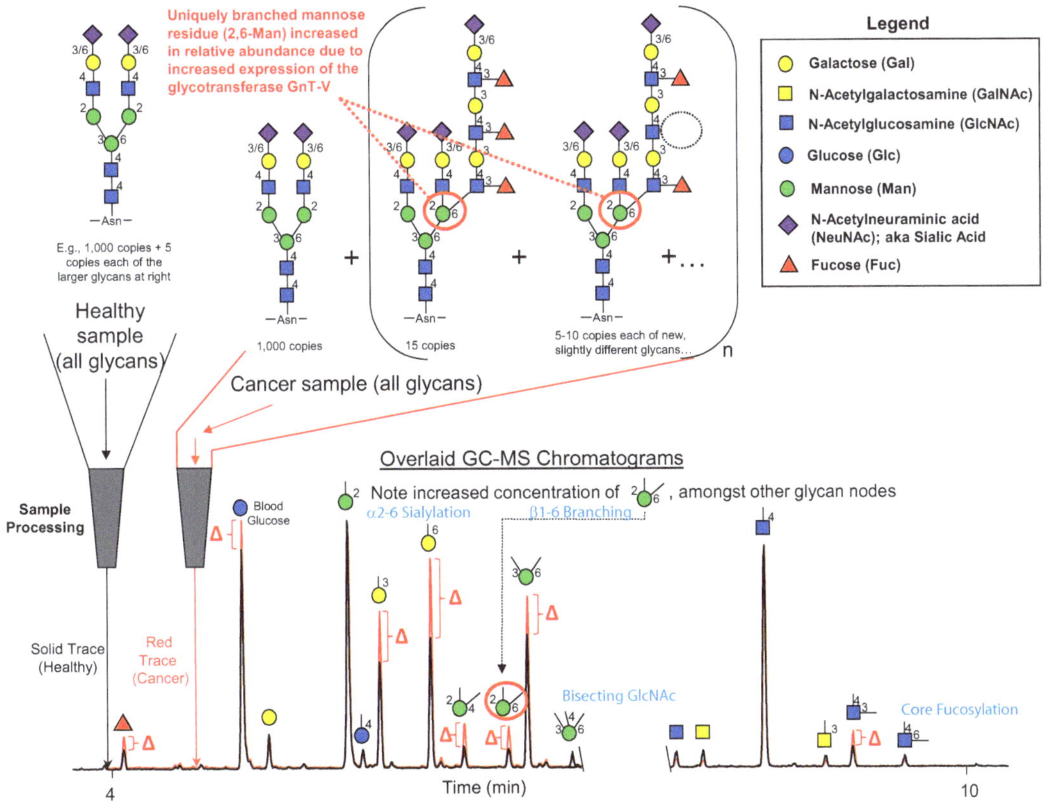


**Fig. 1** Conceptual overview of the glycan "node" analysis concept. The procedure consists of applying glycan linkage (methylation) analysis to intact glycoproteins, whole biofluids or, in this case, unpurified cell culture supernatant. Intact normal and abnormal glycans including *O*-glycans, *N*-glycans, and glycolipids are processed and transformed into partially methylated alditol acetates (PMAAs), each of which corresponds to a particular monosaccharide-and-linkage-specific glycan "node" in the original polymer. As illustrated, analytically pooling together the glycan nodes from among all the aberrant intact glycan structures provides a more direct surrogate measurement of abnormal glycosyltransferase activity than any individual intact glycan while simultaneously converting unique glycan features such as "core fucosylation," "α2-6 sialylation," "bisecting GlcNAc," and "β-1-6 branching" into single analytical signals. Actual extracted ion chromatograms from 9 μl blood plasma samples are shown. Numbers adjacent to monosaccharide residues in glycan structures indicate the position at which the higher residue is linked to the lower residue. (This figure was adapted with permission from ref. 18. Copyright 2013 American Chemical Society)

this approach to profiling the glycosylation of proteins produced by cultured cells should provide a direct means of tracking specific glycan features that reflect important changes in cellular biology and impact the function of secreted glycoproteins and antibodies.

Here we report our detailed methodology for the adaptation of this approach to the analysis of glycoproteins in cell culture supernatants (Table 1 and Figs. 3 and 4) and to antibody glycosylation profiling (Fig. 5). As an example for the application of the assay, the effect of IL-6 and IL-1β cytokines on secreted hepatocyte glycoprotein glycan node profiles is demonstrated using HepG2 cells.

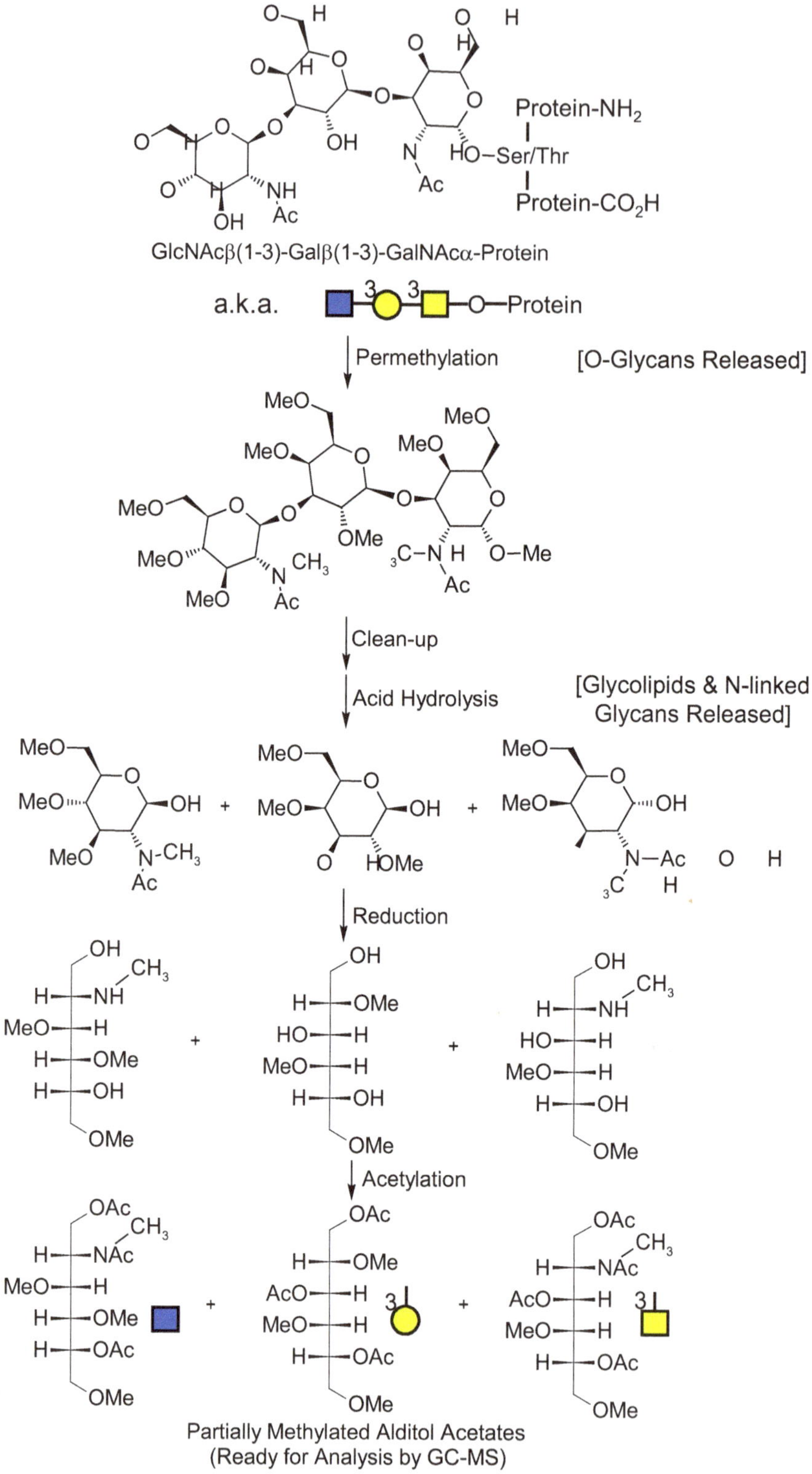


**Fig. 2** Molecular overview of the glycan "node" analysis procedure. *O*-linked glycans are released during permethylation, while *N*-linked glycans and glycolipids are released during acid hydrolysis. The unique pattern of stereochemistry, methylation and acetylation in the final partially methylated alditol acetates (PMAAs) corresponds to the unique glycan "node" in the original glycan polymer and provides the molecular basis for separation and quantification by GC-MS. (Figure adapted with permission from ref. 18. Copyright 2013 American Chemical Society)

**Table 1**
**Analytical reproducibility of glycan nodes in HepG2 cell culture supernatant[a]**

| | | Day 1 ($n = 6$) | | Day 2 ($n = 6$) | | Day 3 ($n = 6$) | | All 3 days ($n = 18$) | |
|---|---|---|---|---|---|---|---|---|---|
| | | Intra-assay % CV[b,c] | | Intra-assay % CV | | Intra-assay % CV | | Inter-assay % CV[b,c] | |
| Specific Glycan feature | Glycan "node" | % of all Hexoses or HexNAcs | % CV | % of all Hexoses or HexNAcs | % CV | % of all Hexoses or HexNAcs | % CV | % of all Hexoses or HexNAcs | % CV |
| *Hexoses* | | | | | | | | | |
| Terminal Fucose | t-Fuc | 19.5 | 14.9 | 17.5 | 18.9 | 19.2 | 9.6 | 18.9 | 14.6 |
| Terminal Galactose | t-Gal | 10.6 | 8.2 | 10.4 | 4.0 | 10.5 | 6.2 | 10.5 | 6.1 |
| | 2-Man | 11.4 | 3.4 | 11.8 | 4.4 | 11.5 | 8.9 | 11.5 | 5.8 |
| | 4-Glc | ND[d] | ND | ND | ND | ND | ND | ND | ND |
| | 3-Man | ND | ND | ND | ND | ND | ND | ND | ND |
| | 2-Gal | ND | ND | ND | ND | ND | ND | ND | ND |
| | 3-Gal | 11.7 | 2.6 | 12.2 | 4.8 | 11.8 | 7.4 | 11.7 | 5.4 |
| α2-6-sialylation | 6-Gal | 6.3 | 3.9 | 6.7 | 5.8 | 6.4 | 9.9 | 6.5 | 7.2 |
| | 3,4-Gal | 1.6 | 10.6 | 1.4 | 7.8 | 1.5 | 11.2 | 1.5 | 11.6 |
| | 2,3-Gal | ND | ND | ND | ND | ND | ND | ND | ND |
| β1-4-branching | 2,4-Man | 15.3 | 4.0 | 16.2 | 6.6 | 14.9 | 5.0 | 15.5 | 6.3 |
| | 4,6-Glc | ND | ND | ND | ND | ND | ND | ND | ND |
| β1-6-branching | 2,6-Man | 3.8 | 7.3 | 4.4 | 4.3 | 4.2 | 6.9 | 4.2 | 7.9 |
| | 3,6-Man | 5.8 | 6.5 | 6.6 | 6.4 | 6.0 | 11.1 | 6.1 | 9.6 |
| | 3,6-Gal | 3.1 | 5.4 | 3.3 | 4.3 | 3.2 | 5.2 | 3.3 | 5.1 |
| Bisecting GlcNAc | 3,4,6-Man | 0.89 | 18.4 | 0.94 | 12.3 | 0.88 | 16.0 | 0.94 | 14.8 |
| *HexNAcs* | | | | | | | | | |
| | t-GlcNAc | 1.9 | 4.0 | 1.9 | 5.4 | 1.9 | 12.7 | 1.9 | 7.6 |
| | t-GalNAc | ND | ND | ND | ND | ND | ND | ND | ND |
| | 4-GlcNAc | 32.0 | 3.7 | 32.4 | 2.4 | 31.0 | 6.9 | 31.8 | 4.8 |
| | 3-GlcNAc | ND | ND | ND | ND | ND | ND | ND | ND |
| | 3-GalNAc | 5.4 | 2.8 | 5.4 | 5.2 | 5.3 | 7.5 | 5.3 | 5.2 |
| | 6-GlcNAc | ND | ND | ND | ND | ND | ND | ND | ND |

(continued)

**Table 1 (continued)**

| | | Day 1 ($n = 6$) | | Day 2 ($n = 6$) | | Day 3 ($n = 6$) | | All 3 days ($n = 18$) | |
|---|---|---|---|---|---|---|---|---|---|
| | | Intra-assay % CV[b,c] | | Intra-assay % CV | | Intra-assay % CV | | Inter-assay % CV[b,c] | |
| Specific Glycan feature | Glycan "node" | % of all Hexoses or HexNAcs | % CV | % of all Hexoses or HexNAcs | % CV | % of all Hexoses or HexNAcs | % CV | % of all Hexoses or HexNAcs | % CV |
| Antennary Fucosylation | 3,4-GlcNAc | 40.7 | 3.0 | 40.1 | 1.1 | 41.1 | 3.5 | 40.8 | 2.8 |
| | 4-GalNAc | ND | ND | ND | ND | ND | ND | ND | ND |
| | 6-GalNAc | ND | ND | ND | ND | ND | ND | ND | ND |
| Core Fucosylation | 4,6-GlcNAc | 10.7 | 5.5 | 10.5 | 2.4 | 10.1 | 4.3 | 10.5 | 4.7 |
| | 3,6-GalNAc | 8.4 | 6.3 | 8.6 | 3.3 | 7.9 | 9.6 | 8.1 | 7.3 |

[a]HepG2 cells were seeded in T75 flasks, and they were grown for 3 days in FBS medium until 70% confluency and then with serum free medium for 48 h. Serum free medium was collected from across all flasks, centrifuged at 1000 × *g* for 5 min, and supernatant was stored in a single container (bulk sample) at −80 °C. Serum free medium was concentrated 30-fold by spin filtration and analyzed by glycan methylation analysis on three different days (6 replicates per day from the same bulk sample)

[b]In general, glycan nodes with the lowest precision (highest % CV values) are those with the lowest relative abundance

[c]These results are consistent with those observed in whole blood plasma/serum [18–23]

[d]ND indicates not detected, but it has been previously detected in other biomatrices [18]

Specific tumor antigens such as β1-6-branching and β1-4-branching are shown to be regulated by IL-6; fewer glycan features appear to be regulated by IL-1β. Additionally, neuraminidase enzyme treatment of alpha-1 antitrypsin IgG demonstrates how the method can be used to detect relative changes in α2-6 sialylation along with corresponding increases in terminal galactose.

## 2 Materials

### *2.1 Cell Culture and Antibody*

1. HepG2 Cells (ATCC, HB-8065).
2. Eagle's Minimum Essential Medium (EMEM) (ATCC, 30-2003).
3. Fetal bovine serum (FBS, US Source, FB-11).
4. 1× PBS, pH 7.4, 1 L.
5. Trypsin 0.25%.

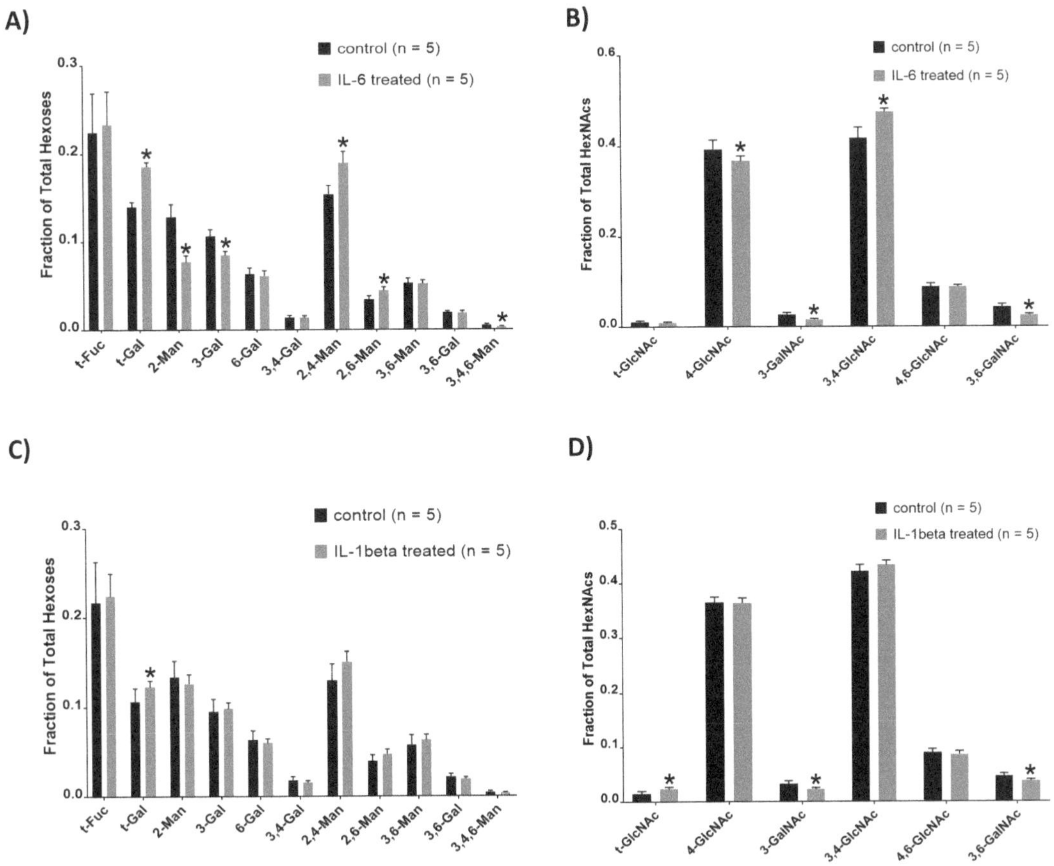


**Fig. 3** Relative abundance of glycan nodes in HepG2 cell culture supernatant with cytokine stimulation. Cells were grown in media containing 10% FBS for 3 days until 70% confluence then washed to remove FBS (Subheading 3.1) and incubated with 3 ml of FBS-free media containing 50 ng/ml IL-6 (**a** and **b**) or IL-1β (**d** and **c**). Media was collected and replaced with fresh cytokine-containing media every 24 h for a total cytokine-exposure period of 48 h, as previously done by Mackievicz et al. [8]. Control cells were cultured in the same way but in the absence of added cytokine. Extracted ion chromatograms corresponding to each glycan node were normalized to the summed area of all hexoses (**a** and **c**) or HexNAcs (**b** and **d**) (as appropriate per that particular glycan node). Error bars represent standard deviation. * Indicates statistically significant differences between control and cytokine-treated cells ($p < 0.001$). Statistical significance determined using multiple *t*-tests with the Holm–Sidak method for multiple comparisons correction (GraphPad v8.2)

6. Human IL-6 (interleukin 6) recombinant protein solution: 200 μg/ml rhIL-6 in 10 mM $NaH_2PO_4$, 150 mM NaCl, pH 7.2.
7. Recombinant human IL-1β: 200 μg/ml rh-IL-1β in 10 mM $NaH_2PO_4$, 150 mM NaCl, pH 7.2.
8. Rabbit anti-human alpha-1-antitrypsin antibody (Dako, A0012).
9. Neuraminidase (Sialidase) from *Clostridium perfringens*; cleaves terminal sialic acids that are α2,3-, α2,6-, or α2,8-linked to Gal, GlcNAc, GalNAc, AcNeu, and GlcNeu, in oligosaccharides, glycolipids, and *N*- or *O*-linked glycoproteins.

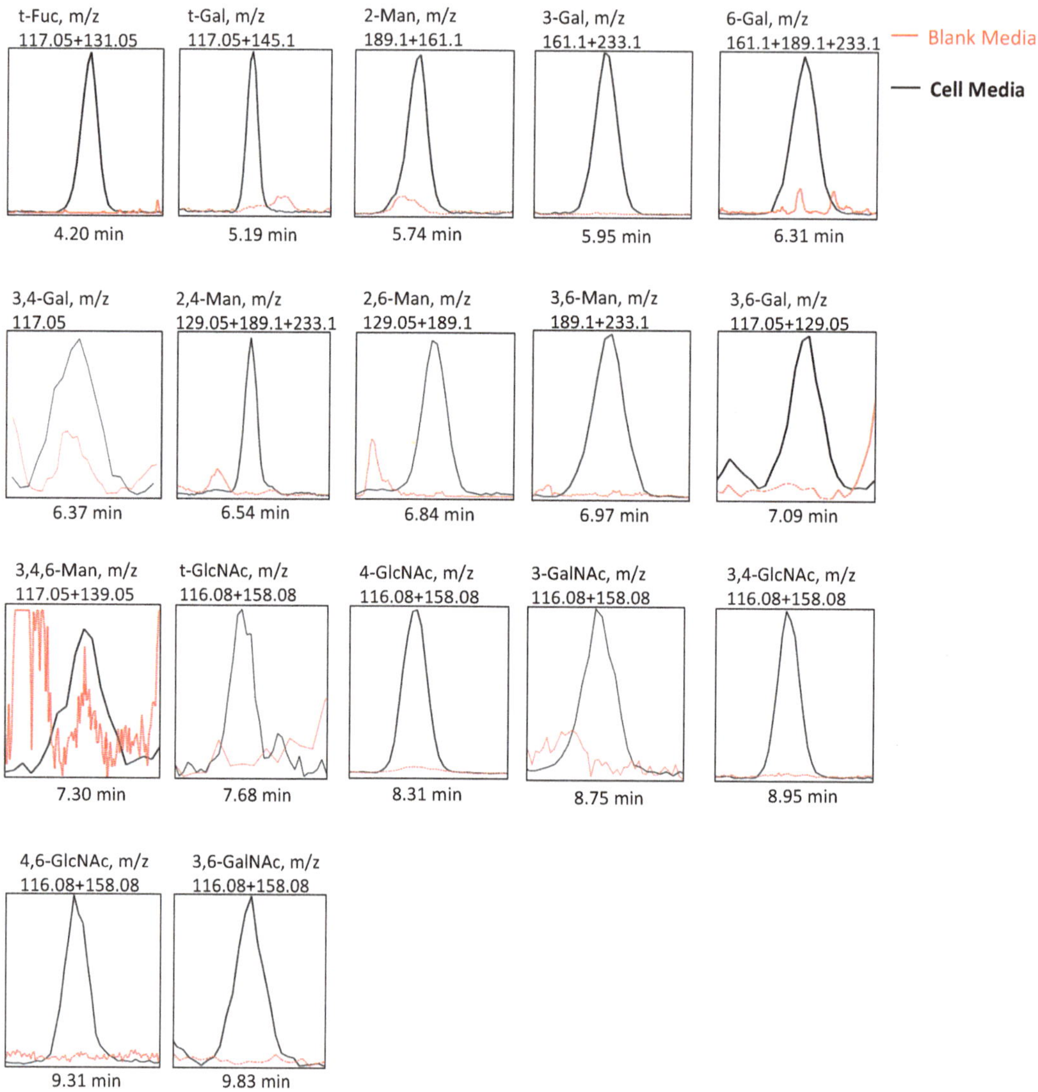


**Fig. 4** Summed extracted ion chromatograms (XICs) for the 17 glycan nodes found in HepG2 cell culture media. Raw XIC traces for EMEM media that was never exposed to cells are shown in red (i.e., "Blank Media"); raw XIC traces for EMEM media exposed to cells then processed by spin filtration (Subheading 3.2) are shown in black (i.e., "Cell-Exposed Media"). Retention times listed correspond to the XIC peak apex

### *2.2 Concentration of Cell Culture Media*

1. 3K MWCO centrifugal spin filters.
2. 10× HBS buffer: make a 1 M HEPES buffer in 1.5 M NaCl solution by dissolving 95.32 g of HEPES and 35.06 g of NaCl in 350 ml of DI water. Bring final pH to 6.8 by adding small solid sodium hydroxide beads (*see* Subheading 2.3) and adjust final volume to 400 ml. Then dilute the HBS buffer solution 10× with distilled water before use. Store at 4 °C.
3. Swing bucket centrifuge.

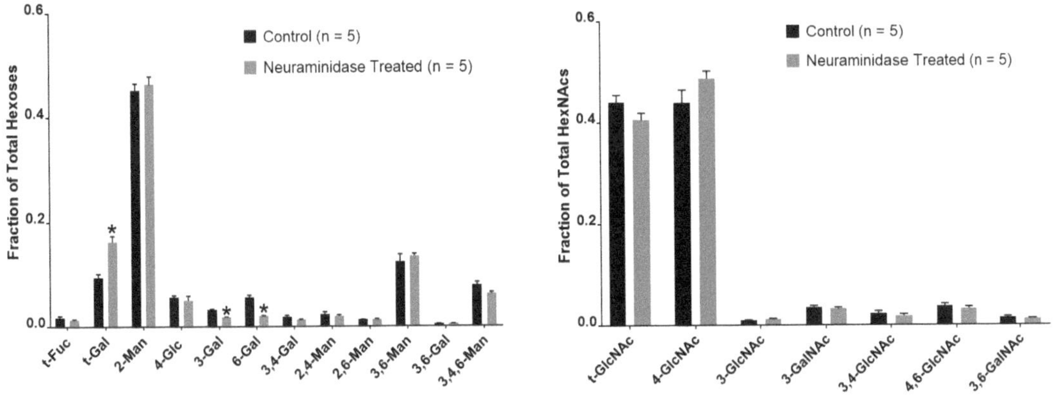


**Fig. 5** Glycan node analysis of purified IgG. Glycan nodes from rabbit anti-alpha-1 antitrypsin IgG before and after treatment of the antibody with neuraminidase. Low amounts of neuraminidase enzyme were intentionally used in order to induce only partial desialylation of IgG. Significant decreases in 6-linked and 3-linked galactose were observed along with an increase in terminal galactose (t-Gal), indicating a partial loss of terminal neuraminic acid residues. Larger quantities of neuraminidase completely eliminate the signal for 6-linked galactose (data not shown). Error bars represent standard deviation. * Indicates statistically significant differences between untreated and neuraminidase-treated antibody ($p < 0.001$). Statistical significance determined using multiple *t*-tests with the Holm–Sidak method for multiple comparisons correction (GraphPad v8.2)

### *2.3 Glycan Node Analysis*

1. Permethylation.
   (a) Spin columns (0.9 ml) with plugs and polyethylene frits.
   (b) DMSO.
   (c) NaOH beads.
   (d) Acetonitrile.
   (e) Iodomethane.
   (f) 0.2 M sodium phosphate buffer, pH 7 containing 0.5 M NaCl: First prepare the dibasic buffer solution by dissolving 29.22 g of NaCl and 28.4 g of $Na_2HPO_4$ in 1 L of water. Second, prepare the monobasic buffer solution by dissolving 23.99 g of $NaH_2PO_4$ and 29.22 g of NaCl in 1 L of water. Then, using both buffer solutions (basic and dibasic) equilibrate buffer to pH 7. Store buffer at room temperature.
   (g) Chloroform.
2. Hydrolysis.
   (a) 2 M trifluoroacetic acid (TFA).
   (b) Heating blocks set at 120 °C.
3. Reduction.
   (a) Freshly prepared 10 mg/ml sodium borohydride solution in 1 M ammonium hydroxide.

   (b) Methanol.
   (c) Glacial acetic acid.
4. Acetylation.
   (a) Acetic anhydride.
   (b) TFA.
   (c) Dichloromethane.
5. Gas chromatography–mass spectrometry (GC-MS).
   (a) Agilent Model A7890 gas chromatograph coupled to a Waters GCT mass spectrometer.
   (b) GC-MS autosampler vials.
   (c) Teflon-lined pierceable caps, 9 mm.
   (d) Acetone.
   (e) GC Column (DB-5 ms, 30 m).

## 3 Methods

### 3.1 Cell Culture

1. Grow HepG2 cells in T75 flasks at 37 °C in 5% $CO_2$ under semiconfluent conditions for 3 days in 10% FBS-supplemented EMEM media.
2. Remove FBS-containing media and wash cells three times with sterile 1× PBS, and once with serum-free EMEM media.
3. Incubate cells with 3 ml of serum free EMEM medium with or without added growth factors or cytokines for 48 h (*see* **Note 1**).
4. Collect and change media every 24 h (replace added growth factors or cytokines if desired).
5. Collect media over the last 24 h, centrifuge it at 1000 × *g* for 5 min to pellet any cell debris, and concentrate supernatant by spin filtration (*see* **Note 2**) (*see* Subheading 3.2).

### 3.2 Concentration of Cell Media by Spin Filtration

1. Place 3 ml of cell culture media into a 4-ml Amicon 3K centrifugal spin filter and centrifuge in a swing bucket rotor at 2950 × *g* for 30 min at room temperature. The final volume is about 500 μl.
2. Resuspend media in 3 ml of 1× HBS buffer and centrifuge again at 2950 × *g* for 30 min.
3. Perform two more washes with 1× HBS buffer, with the third wash centrifuged for 50 min, yielding a final volume of approximately 100 μl (*see* **Notes 3** and **4**).

For the desialylation of alpha 1 antitrypsin antibody:

### 3.3 IgG (Alpha-1 Anti-Trypsin) Antibody Desialylation & Preparation

1. Low amounts of neuraminidase enzyme are used here to facilitate only partial desialylation of IgG, and to show how the assay can capture subtle changes in specific glycan feautures. Larger quantities of neuraminidase completely eliminate the signal for 6-linked galactose (data not shown). To 1.7 μl of antibody (100 μg) add 2 μl of a 0.1 M sodium acetate buffer pH 5 and 1 μl of neuraminidase enzyme (0.1 milliunits). Check the final pH of the sample to be around 5 using Hydrion pH papers.
2. Incubate samples for 4 h at 37 °C.
3. After incubation, bring pH back to 7 with 1 μl of a 0.5 M sodium bicarbonate solution. Then bring final volume to 10 μl with water.
4. Perform glycan node analysis (Subheading 3.4) in at least triplicate.

### 3.4 Glycan Node Analysis

1. If glycan node analysis is a new method in your laboratory, perform the following procedure on three different days (six replicates per day) to verify the reproducibility of the assay in your hands. Reproducibility should be comparable to the results in Table 1.
2. Permethylation of concentrated cell culture medium and IgG: Add 12 μl of whole concentrated cell medium or 10 μl of antibody to a 1.5 ml polypropylene test tube. To this, add 270 μl of dimethyl sulfoxide (DMSO) and 105 μl of iodomethane. Mix this solution and add it to a plugged 1 ml spin column containing sodium hydroxide beads, previously preconditioned with 400 μl of acetonitrile and two rinses with 400 μl of DMSO. Mix samples gently with the sodium hydroxide beads 3–4 times for 10 min using the tip of a 200 μl pipette tip. Then, unplug columns and centrifuge them at 10,000 × *g* for 15 s to collect sample, leaving any NaOH residue behind (*see* **Note 5**). Immediately after, add 300 μl of acetonitrile to the columns to wash off any sample left on the column. Transfer samples and acetonitrile wash to a silanized glass tube containing 3.5 ml of 0.2 M sodium phosphate buffer, pH 7 containing 0.5 M NaCl. To this, add 1.2 ml of chloroform. Perform L/L extractions three times with the sodium phosphate buffer, saving the chloroform layer every time. Then, dry the chloroform under a nitrogen stream in a heating block set at 74 °C.
3. Hydrolysis of permethylated glycans: Prepare a 2 M TFA solution. Add 325 μl of the 2 M TFA solution to each sample, and tightly cap them to prevent evaporation. Then heat samples at 120 °C for 2 h (*see* **Note 6**). Remove TFA by drying sample under nitrogen stream in a heating block set at 74 °C.

4. Reduction of monosaccharides: Add 475 μl of 10 mg/ml sodium borohydride in 1 M ammonium hydroxide to each tube and allow to react for 1 h (*see* **Note 7**). Add 63 μl of methanol to each sample to remove any residual borate, followed by 125 μl of 9:1 methanol: acetic acid. Each time, dry samples under nitrogen in a heating block set at 74 °C. To complete the drying process, place the samples in a vacuum chamber (e.g., vacuum desiccator) at room temperature for at least 20 min.
5. Acetylation: After drying, add 18 μl of DI water to each sample, and mix residues until they are completely dissolved. Then, add 250 μl of acetic anhydride, mix thoroughly, and sonicate in a water batch for 2 min. Incubate samples at 60° C for 10 min, followed by addition of 230 μl of TFA, and incubating again at 60 °C for 10 min. Add 1.8 ml of dichloromethane and 2 ml of DI water to each sample. Then perform L/L extraction twice with water, removing the top layer (containing water) and saving the dichloromethane layer (bottom layer) every time. Add the dichloromethane layer into silanized autosampler vials and dry samples under nitrogen in a heating block at 60 °C. Reconstitute samples with 50 μl of acetone, and place samples onto the GC-MS autosampler rack.
6. Gas Chromatography-Mass Spectrometry (GC-MS): A gas chromatograph coupled to a time-of-flight (TOF) mass spectrometer is used here (*see* **Note 8**). One microliter of each sample is injected in split mode (1:10) onto a silanized glass liner containing a small plug of silanized glass wool held at a temperature of 280 °C. Using helium as the carrier gas, samples are transferred onto the GC column, which is maintained at an initial temperature of 165 °C for 0.5 min, followed by ramping the temperature at 10 °C per minute to 265 °C then immediately ramping at 30 °C per minute to 325 °C and holding for 3 min. Samples eluting from the column are transferred to the mass spectrometer at a transfer line temperature of 250 °C. They are then subjected to electron ionization at 70 eV and 250 °C, and analyzed from $m/z$ 40–800 by TOF-MS in which transients are summed and recorded every 0.2 s.
7. Data Analysis: Identification of each glycan node is made by comparing retention times with those from known partially methylated alditol acetates (PMAAs), for example those obtained by Borges et al., [18] and mass spectra are verified through comparison with the mass spectral library of PMAAs at the University of Georgia's Complex Carbohydrate Research Center website (https://www.ccrc.uga.edu/databases/index.php#). Each glycan node is quantified by the sum of the integrals of a specific set of extracted ion chromatogram peaks

(Fig. 4) [18] using QuanLynx software. Integrated peaks are exported to an Excel spreadsheet to normalize the area of each glycan node by dividing each individual hexose glycan node by the sum of all hexoses, and each individual HexNAc glycan node by the sum of all HexNAcs [18–23].

## 4 Notes

1. Secreted glycoproteins tend to be too dilute for direct analysis in cell culture media. Given the challenges associated with this problem, semiconfluent HepG2 cells are incubated with a minimum volume of serum free media for 48 h, enough to cover the cells in T75 flasks. Reducing the amount of media does not affect cell viability, and the cells sustain well for up to 48 h.
2. Recovered cell culture media must be concentrated and high concentrations of small molecules removed prior to glycan node analysis. Spin filtration accomplishes both of these objectives.
3. The disadvantage of using spin filtration for glycan node analysis is that glycolipids maybe lost during the process.
4. Concentrating whole cell culture media by evaporative concentration using a Speed Vac can help preserve glycolipids. However, due to the presence of salts and other components in the media that decrease protein solubility, a precipitate forms and high chromatographic backgrounds for the blank media are obtained. Hence, this form of protein concentration is not recommended for cell culture media.
5. After spinning down the sodium hydroxide columns to get the permethylated solution, a white precipitate is formed. This white residue should not be added to the 0.2 M sodium phosphate buffer, pH 7 containing 0.5 M NaCl [19]. This can result in poor HexNAc yields.
6. Before incubating samples at 120 °C, securely cap each glass tube and look for any cracks on the glass tubes. A loose cap can cause the sample to evaporate during hydrolysis.
7. The 10 mg/ml sodium borohydride in 1 M ammonium hydroxide should be prepared immediately before use.
8. The GC-MS method is compatible with quadrupole-based mass analyzers operating in selected ion monitoring (SIM) mode. For a list of monitored ions *see* Borges et al. [18].

## References


1. van Kooyk Y, Kalay H, Garcia-Vallejo JJ (2013) Analytical tools for the study of cellular glycosylation in the immune system. Front Immunol 4:451. https://doi.org/10.3389/fimmu.2013.00451
2. Hudak JE, Canham SM, Bertozzi CR (2014) Glycocalyx engineering reveals a Siglec-based mechanism for NK cell immunoevasion. Nat Chem Biol 10(1):69–75
3. Stanczak MA, Siddiqui SS, Trefny MP et al (2018) Self-associated molecular patterns mediate cancer immune evasion by engaging Siglecs on T cells. J Clin Investig 128 (11):4912–4923
4. Qi J, Li N, Fan K et al (2014) β1,6 GlcNAc branches-modified PTPRT attenuates its activity and promotes cell migration by STAT3 pathway. PLoS One 9:e98052. https://doi.org/10.1371/journal.pone.0098052
5. Rini JM, Esko JD (2017) Glycosyltransferases and glycan-processing enzymes. In: Varki A, Cummings RD, Esko JD et al (eds) Essentials of glycobiology, 3rd edn. Cold Spring Harbor, New York
6. Varki A, Kannagi R, Toole B et al (2017) Glycosylation changes in cancer. In: Varki A, Cummings RD, Esko JD et al (eds) Essentials of glycobiology, 3rd edn. Cold Spring Harbor, New York
7. Hossler P, Khattak SF, Li ZJ (2009) Optimal and consistent protein glycosylation in mammalian cell culture. Glycobiology 19:936–949
8. Mackiewicz A, Schultz D, Mathison J, Ganapathi M, Kushner I (1989) Effect of cytokines on glycosylation of acute phase proteins in human hepatoma cell lines. Clin Exp Immunol 75(1):70–75
9. Narisada M, Kawamoto S, Kuwamoto K et al (2008) Identification of an inducible factor secreted by pancreatic cancer cell lines that stimulates the production of fucosylated haptoglobin in hepatoma cells. Biochem Biophys Res Commun 377(3):792–796
10. Klasić M, Krištić J, Korać P et al (2016) DNA hypomethylation upregulates expression of the MGAT3 gene in HepG2 cells and leads to changes in N-glycosylation of secreted glycoproteins. Sci Rep 6:24363. https://doi.org/10.1038/srep24363
11. Drake RR, Jones EE, Powers TW, Nyalwidhe JO (2015) Altered glycosylation in prostate cancer. In: Drake RR, Ball LE (eds) Glycosylation and cancer. Advances in cancer research, vol 126. Elsevier, Amsterdam, pp 345–382
12. Walsh G (2018) Biopharmaceutical benchmarks 2018. Nat Biotechnol 36 (12):1136–1145
13. Mulloy B, Dell A, Stanley P et al (2017) Structural analysis of glycans. In: Varki A, Cummings RD, Esko JD et al (eds) Essentials of glycobiology, 3rd edn. Cold Spring Harbor, New York
14. Ciucanu I, Costello CE (2003) Elimination of oxidative degradation during the per-O-methylation of carbohydrates. J Am Chem Soc 125 (52):16213–16219
15. Kang P, Mechref Y, Klouckova I, Novotny MV (2005) Solid-phase permethylation of glycans for mass spectrometric analysis. Rapid Commun Mass Spectrom 19(23):3421–3428
16. Ciucanu I, Caprita R (2007) Per-O-methylation of neutral carbohydrates directly from aqueous samples for gas chromatography and mass spectrometry analysis. Anal Chim Acta 585(1):81–85
17. Kang P, Mechref Y, Novotny MV (2008) High-throughput solid-phase permethylation of glycans prior to mass spectrometry. Rapid Commun Mass Spectrom 22(5):721–734
18. Borges CR, Rehder DS, Boffetta P (2013) Multiplexed surrogate analysis of glycotransferase activity in whole biospecimens. Anal Chem 85(5):2927–2936
19. Zaare S, Aguilar JS, Hu Y, Ferdosi S, Borges CR (2016) Glycan node analysis: A bottom-up approach to glycomics. J Vis Exp 2016 (111):53961. https://doi.org/10.3791/53961
20. Hu Y, Borges CR (2017) A spin column-free approach to sodium hydroxide-based glycan permethylation. Analyst 142(15):2748–2759
21. Ferdosi S, Ho TH, Castle EP et al (2018) Behavior of blood plasma glycan features in bladder cancer. PLoS One 13(7):e0201208. https://doi.org/10.1371/journal.pone.0201208
22. Ferdosi S, Rehder DS, Maranian P et al (2018) Stage dependence, cell-origin independence, and prognostic capacity of serum glycan fucosylation, β1-4 branching, β1-6 branching, and α2-6 sialylation in cancer. J Proteome Res 17 (1):543–558
23. Hu Y, Ferdosi S, Kapuruge EP et al (2019) Diagnostic and prognostic performance of blood plasma glycan features in the women epidemiology lung cancer (WELCA) study. J Proteome Res 18(11):3985–3998. https://doi.org/10.1021/acs.jproteome.9b00457

# Chapter 23

# Array-Based *N*-Glycan Profiling of Cells in Culture


**Peggi M. Angel, Anand S. Mehta, and Richard R. Drake**



## Abstract

*N*-glycan imaging mass spectrometry (*N*-glycan IMS) enables the detection and characterization of *N*-glycans in thin histological tissue sections. *N*-glycan IMS is used to study *N*-glycan regulation and localization in tissue-specific regions, such as tumor and normal adjacent to tumor, or by cell type within a tissue. Once a specific tissue-localized *N*-glycan signature is found to be associated with by a disease state, it has been challenging to study modulation of the same *N*-glycan signature by conventional molecular biology techniques. Here we describe a protocol that adapts tissue *N*-glycan IMS analysis workflows to cells grown on glass slides in an array format. Cells are grown under normal conditions in a cell culture chamber, fixed to maintain normal morphology, and sprayed with a thin coating of PNGase F to release *N*-glycans for imaging mass spectrometry profiling.



**Key words** *N*-glycan, Glycomics, *N*-glycoform, *N*-glycans in cell culture, *N*-glycan profiling, *N*-glycan cell profiles, Imaging mass spectrometry, MALDI imaging mass spectrometry, Peptide N-glycosidase F, PNGase F


## 1 Introduction

*N*-glycan IMS enables high-throughput detection and localization of hundreds of *N*-glycans within a single thin tissue section, allowing new insights into the role of tissue glycosylation and regulation of disease [1, 2]. *N*-glycan IMS is analogous to data obtained with in-solution *N*-glycomics workflows, but does not generally require any glycan derivatization prior to analysis. *N*-glycan IMS uses robust enzymatic workflows to access the *N*-glycome of archival formalin-fixed, paraffin-embedded (FFPE) tissues, providing information on the localization of the *N*-glycome regulated by cell types and different tissue regions [1, 3, 4]. Using thin histological tissue sections (5-μm) mounted on a standard or conductive microscope slide, an even, thin layer of peptide *N*-glycosidase F (PNGase F) is sprayed onto tissue by an automated sprayer. Brief deglycosylation hydrolyzes *N*-glycans from proteins within the tissue section, and because of the thin coating, the released *N*-glycans remain localized to their tissue features and cell types. For ionization and detection


Arnaud Delobel (ed.), *Mass Spectrometry of Glycoproteins: Methods and Protocols*, Methods in Molecular Biology, vol. 2271,
https://doi.org/10.1007/978-1-0716-1241-5_23,

by matrix-assisted laser desorption ionization (MALDI) mass spectrometry (MS), a chemical matrix is applied as a thin molecular layer across the tissue. *N*-glycans are detected by scanning the tissue using matrix-assisted laser desorption/ionization (MALDI) imaging mass spectrometry. *N*-glycoforms have been reported markers of disease changing with cell status, disease progression, and useful for prediction of survival status [1, 2, 5–11].However, once a target *N*-glycan is identified as regulated by disease state, there are few approaches to investigating the cellular mechanisms of the *N*-glycan. The primary approach to *N*-glycan profiling of cell types uses liquid chromatography coupled to mass spectrometry (LC-MS) [12–14]. These experiments require millions of cells of starting material, significant user expertise, and extensive time consuming sample processing.

Here, we describe a simplified workflow for array-based *N*-glycan profiling of cells in culture adapted from the tissue *N*-glycan IMS workflows [15]. A key step is to culture cells within array wells, followed by cell fixation, delipidation, deglycosylation, and detection by MALDI IMS. This allows quantitative profiles of *N*-glycans to be obtained from cells using standard label free cell culture techniques. Additionally, *N*-glycan turnover may be measured by stable isotope labeling by amino acids (SILAC) in cell culture using the approach isotopic detection of amino sugars with glutamine (IDAWG) [16, 17]. The protocol described here is done after label free or SILAC/IDAWG cell culture and takes around 4 h to prepare five 8-well arrays that have standard dimensions of 75 mm (length) × 25 mm (width) × 1 mm (height). These slides fit into standard imaging mass spectrometry instruments. The procedure can be applied to any cell line that can be cultured on these types of slides.

## 2 Materials

Prepare all solutions in filtered HPLC grade water or ultrapure deionized water with a resistance of 18 MΩ-cm at 25 °C. Follow all safety and waste disposal regulations.

### *2.1 Sample Preparation Solutions*

1. Neutral buffered formalin with 10% formalin, commercially sourced (*see* **Note 1**).
2. Phosphate-buffered saline, commercially sourced and suitable for cell culture experiments.
3. Carnoy's Solution: 10% glacial acetic acid, % HPLC grade chloroform, 60% 200 proof ethanol: Working under a functioning laboratory hood, add 600 mL 200 proof ethanol to a clean 1-L bottle. Carefully add 100 mL glacial acetic acid and

gently mix. Add 300 mL chloroform. Mix thoroughly. Close tightly and store at room temperature in a well ventilated location. Solution is stable for up to 1 month.

4. 25% trifluoroacetic acid (TFA): Add 3 mL HPLC grade water to a clean bottle. Carefully add 1 mL of trifluoroacetic acid to the water and mix. Solution is stable for up to 1 month at room temperature.
5. Matrix solvent (50% Acetonitrile (ACN), 0.1% TFA): Add 25 mL HPLC grade water to a clean 100 mL bottle. Carefully add 100 μL of neat trifluoroacetic acid (≥99%, analytical grade) to the water and mix. Add 50 mL of acetonitrile and mix. Add HPLC grade water to a final volume of 100 mL and mix. Store for up to 1 month at room temperature.
6. MALDI matrix for *N*-glycan profiling (alpha-cyano-4-hydroxycinnamic acid (CHCA), 7 mg/mL in 50% acetonitrile/0.1% TFA): Weigh out 0.0420 ± 0.001 g CHCA. Add the solid CHCA to a clean 50 mL falcon tube. Bring to volume with 6 mL of matrix solvent. Vortex briefly and sonicate for 5 min using a benchtop sonicator. Filter CHCA solution using a 13 mm 0.2 μm PTFE hydrophilic syringe filter graded for use with HPLC solvents (*see* **Note 2**). Prepare immediately before use.
7. Ammonium phosphate, monobasic 5 mM in HPLC grade water. Weigh 0.0287 g of ammonium phosphate monobasic into a clean 50-mL falcon tube. Add 50 mL of HPLC grade water and mix until dissolved. Store for up to 3 months at room temperature.

### 2.2 Enzyme Solutions

1. Preparation of 1 mL of 0.1 μg/μL PNGase F solution. To 100 μg of PNGase F add 1000 μL of HPLC grade water and mix (*see* **Note 3**). Use same day. This preparation is enough enzyme to cover five arrays of dimensions 75 mm × 25 mm × 1 mm.

### 2.3 TM-Sprayer™ Solutions

1. Push solvent (50% methanol/water): Add 500 mL methanol to a clean bottle. Add 500 mL HPLC grade water and mix. Solvent may be kept at room temperature during the duration of use (*see* **Note 4**).

## 3 Methods

*Cell culture*. For the described protocol, cells were cultivated on sterile 8-well Lab-Tek® II Chamber Slide System (Electron microscopy, Hatfield, PA) as these array dimensions (L × W × H 75 mm × 25 mm × 1 mm) are compatible with the majority of imaging mass spectrometry instrumentation. Cells are plated onto the 8-well array followed by desired cell culture experiments. Initial experiments must be done to evaluate cell counts compared to *N*-glycan signal to noise. For example, higher cell counts of human aortic endothelial cells (>10,000) result in signal suppression [15]. For baseline cell profiling, cells should be allowed to adhere prior to analysis. Using a new media aspirator tip on the media aspirator when moving between media blank and cells during media changes prevents cellular *N*-glycan expression from contaminating the media blank. For $^{15}N$ labeling of *N*-glycans, substitute L-glutamine (Amide-15N, 98%+, Cambridge Isotope Laboratories, Inc) for $^{14}N$ glutamine in complete media. Ensure all chemicals used are of ultrapure high-grade quality. It is recommended that small bottles of ultrapure high-grade quality chemicals be reserved and used only for *N*-glycan profiling of cells.

### 3.1 Cell Preparation for MS Profiling

1. After cell culture experiments, keep the removable media chamber walls attached to the arrays. Wash the cells three times with PBS. Ensure that wash meniscus covers the cells completely. Place a new aspirator tip on the aspirator or pipette when moving between media and cells or cells of different experimental conditions. This prevents cross-contamination that will be detected by MS profiling.
2. After final PBS wash, add neutral buffered formalin to the array chambers. Incubate for 20 min at room temperature. Remove neutral buffered formalin and rinse once with PBS.
3. Detach the removable media chamber walls from the arrays. Ensure that all adhesives are removed from the surface of the arrays (*see* **Note 5**).
4. Pour Carnoy's solution into a glass Coplin jar. Solvent should be added to a level that will allow complete immersion of the tissue sections (*see* **Note 6**).
5. In a functioning laboratory hood, delipidate the cells by incubating in Carnoy's solution for 3 min followed by air-drying under the hood for 3 min. It is critical that the array is not agitated during solvent incubation as this may cause loss of cells. During drying, the array should be placed flat with cell side facing upward to minimize streaking and cross-contamination between wells. Repeat the Carnoy's solution wash and air-drying for a total of three cycles.
6. After complete drying, cells may be stored at −20 °C for up to 1 week with minimal degradation prior to further preparation (*see* **Note 7**). It is highly recommended to proceed immediately with sample preparation steps to limit degradation.

### 3.2 Slide Scanning

1. This step produces an optical image needed for selecting tissue regions for imaging analysis. Fiducials, or reference points, are needed to accurately "teach" the instrument where the wells are located. Any recognizable feature on the array that is captured in both the scanned image and the acquisition camera may be used to teach the instrument the well location. For Lab-Tek II slides, the numbers printed near each wells are useful as fiducials.
2. For cell profiling that will be acquired by mass spectrometry at $\geq$100 μm laser step size, scan the whole slide at a minimum of 1200 ppi resolution. For profiling using laser step sizes $\leq$50 μm, scan the slide at a minimum of 4800 ppi resolution. Save the images as JPEG, bitmap (*.bmp) or 8 bit TIFF formats.

### 3.3 PNGase F Application by the M3 TM-Sprayer™

1. For enzyme application, use a syringe pump with a minimum of 0.5% accuracy in pumping 25 μL/min. Enzyme is loaded into the sprayer using a glass or plastic 1-mL syringe with Luer lock.
2. Prepare the PNGase F and load into the glass or plastic 1-mL syringe with Luer lock syringe. Ensure that there are no bubbles in syringe (*see* **Note 8**). Attach the syringe to the clean TM-Sprayer™ line used for enzyme spraying. Ensure that the syringe is fastened securely in the pump.
3. Set the pump to a flow rate of 25 μL/min with an inner diameter matching that of the used syringe. Turn on the pump and allow the spray to stabilize for 3 min. It generally takes about 1–3 min to start emitting solution (*see* **Notes 9** and **10**).
4. Place the arrays with cells facing upward on the TM-Sprayer™ sample area, fastening them with lab tape.
5. Turn on the TM-Sprayer™ and then the controlling computer.
6. Open the nitrogen gas tank valve and set the regulator to 10 psi.
7. In the TM-Sprayer™ software, set the temperature to 45 °C. Temperature will not adjust without the nitrogen gas flowing.
8. Program the TM-Sprayer™ to cover the appropriate number of arrays within a designated sample area. Allow a 5 mm additional edge distance for sprayhead turn round at the edges of the designated sample area (*see* **Note 11**).
9. Program the TM-Sprayer™ method for PNGase F to use 10 passes, crisscross pattern, velocity of 1200, 3.0 mm track spacing, and a dry time of zero. The tip of the sprayhead should be 40 mm distance from the surface of the arrays.

10. Place a blank slide underneath the TM-sprayer ™ nozzle head and start the pump. Once moisture is consistently detected on the blank slide, press "Start" in the TM-Sprayer™ software.
11. PNGase F solution will be applied in a thin layer onto the arrays. During application of the enzyme, moisture will be observed on the slide as the sprayhead passes over an area; this should dry within 30 s (*see* **Note 12**).

### 3.4 Incubation for On-Tissue Digestion

1. Incubation is performed immediately after application of PNGase F.
2. Prepare an incubation chamber for digestion using a plastic 100 × 15 mm cell culture dish. Fit a single layer paper towel (Wypall ×60) on the bottom of the dish. Fold two 4 × 6 Kimwipes and place at opposite sides of the dish. Using a spray bottle of distilled water, add water to saturate the paper towels and Kimwipes. Stop adding water when excess water accumulates in the dish, observed by tilting the dish to one side. For the specified towels and Kimwipes, this is 5–7 mL of water.
3. The incubation chamber should be preheated for 30 min in an oven at 37.5 °C (*see* **Note 13**).
4. Place the slide with the array with cells facing upward into the incubation chamber using the water saturated Kimwipes as a support. Gently push the array down slightly so that when the cover is placed on, the cells do not contact the incubation chamber cover.
5. Incubate 2 h in the oven set at 37.5 ± 1.5 °C (*see* **Note 14**).
6. After incubation, remove the array slowly, holding the surface facing upward and parallel with the countertop. Wipe off the condensation on the back of the array to prevent liquid rolling into wells and causing cross contamination of *N*-glycans.
7. Store the array in a 5 slide mailer to protect the released *N*-glycans. If matrix cannot be sprayed the same day, store briefly in a desiccator (up to 24 h) or at −20 °C for long term (up to 1 week).
8. It is recommended to immediately spray matrix onto the array.

### 3.5 MALDI Matrix Application by the M3 TM-Sprayer™

1. MALDI matrix application is performed after PNGAse F digestion.
2. An isocratic pump with 0.5% accuracy in pumping 70 μL/min is used for matrix application. A glass 5-mL syringe with Luer lock for loading matrix into the sprayer.
3. Ensure that the isocratic pump is set to pump 70 μL/min (*see* **Note 15**). Solvent may be degassed to limit flow variation.
4. Turn on the TM-Sprayer™ and controlling computer.

5. Open the nitrogen gas tank and set the regulator to 10 psi.
6. In the TM-Sprayer™ software, set the temperature to 80 °C. Temperature will not adjust without the nitrogen gas flowing.
7. Program the TM-Sprayer™ method for CHCA matrix application to use 12 passes, crisscross pattern, velocity of 1300, 2.5 mm track spacing and zero dry time. The tip of the nozzle sprayhead should be 40 mm distance from the surface of the slide.
8. Place the arrays on the TM-Sprayer™ sample platform and secure with lab tape.
9. Program the TM-Sprayer™ to cover the appropriate number of arrays, allowing a 5 mm additional edge distance for sprayhead turn round (*see* **Note 11**).
10. Fill a glass 5-mL syringe with the filtered MALDI matrix solution, ensuring that there are no bubbles in the syringe.
11. Fasten the syringe to the TM-Sprayer™ line going to the six-port valve. With the valve switch in "Load" position, inject the MALDI matrix solution into the 5 mL loop.
12. Ensure that the pump is flowing at 70 μL/min and that pump pressure readouts are stable.
13. Move the six-port valve switch to "Spray."
14. Use a blank microscope slide to check the TM-Sprayer™ nozzle for spraying of solution. Once matrix is detected as an opaque film on the dummy slide, press "Start" in the TM-Sprayer™ software.
15. CHCA solution will be applied in a thin layer onto the arrays.
16. Leave the arrays attached to the TM-Sprayer ™ sample platform and proceed with the next step.

### 3.6 Array Application of Ammonium Phosphate Monobasic (AP) Solution

1. This step decreases matrix cluster formation to increase sensitivity of *N*-glycan detection from cell arrays.
2. AP application is performed after MALDI matrix application.
3. An isocratic pump with 0.5% accuracy in pumping 70 μL/min is used for matrix application. A glass 5-mL syringe with Luer lock for loading AP into the sprayer.
4. Ensure that the isocratic pump is set to pump 70 μL/min (*see* **Note 14**). Push solvent may be degassed to limit flow variation.
5. Turn on the TM-Sprayer™ and controlling computer.
6. Open the nitrogen gas tank and set the regulator to 10 psi.
7. In the TM-Sprayer™ software, set the temperature to 60 °C. Temperature will not adjust without the nitrogen gas flowing.

8. Program the TM-Sprayer™ method for AP application to use 2 passes, crisscross pattern, velocity of 1300, 3 mm track spacing and zero dry time. The tip of the nozzle sprayhead should be 40 mm distance from the surface of the slide. Allow a 5 mm additional edge distance for sprayhead turn round (*see* **Note 11**).
9. Fill a glass 5-mL syringe with the filtered AP solution, ensuring that there are no bubbles in the syringe.
10. Fasten the syringe to the TM-Sprayer™ line going to the six-port valve. With the valve switch in "Load" position, inject the AP solution into the 5 mL loop. Repeat at total of three times to completely flush the loop.
11. Load the 5-mL loop with AP solution.
12. Move the six-port valve switch to "Spray."
13. Use a blank microscope slide to check the TM-Sprayer™ nozzle for spraying of solution. Once AP solution is detected as an aqueous build up on the dummy slide, press "Start" in the TM-Sprayer™ software.

### 3.7 Mass Spectrometry Profiling

1. Data from *N*-glycan cell profiles may be collected by imaging mass spectrometry in different ways. For rapid high throughput profiling, a laser stepsize of 300 μm with a random walk raster of diameter 300 μm and 1200 laser shots per data point results in *N*-glycan signal to noise >200, measure on high mannose Man 9-7.
2. For imaging type data, where small clusters of cells to single cells may be resolved, a laser step size of ≤20 μm without random walk raster and 100 laser shots per data point results in signal/noise ratios of >100. To limit lengthy acquisition times in the 8-well (0.8 $cm^2$), this may be used in a target area of the well where cells are confirmed to be present.
3. Evaluation of signal from the preparation is done on media blanks. Peaks at 1663 and 2028 have been observed consistently across many cell types and may be used for evaluation. In the media blank well, these peaks should have the most intense signal.
4. During acquisition, lock masses are used to minimize $m/z$ drift. Lock masses at $m/z$ 1663.5814 and 2028.7136 from media background are useful for minimizing drift. High mannose peaks (e.g., Man9 $m/z$ 1905.6338) may be added to the lock mass list for regions of high cell density.
5. Figure 1 demonstrates signal acquired from a single well using the IDAWG labeling method.

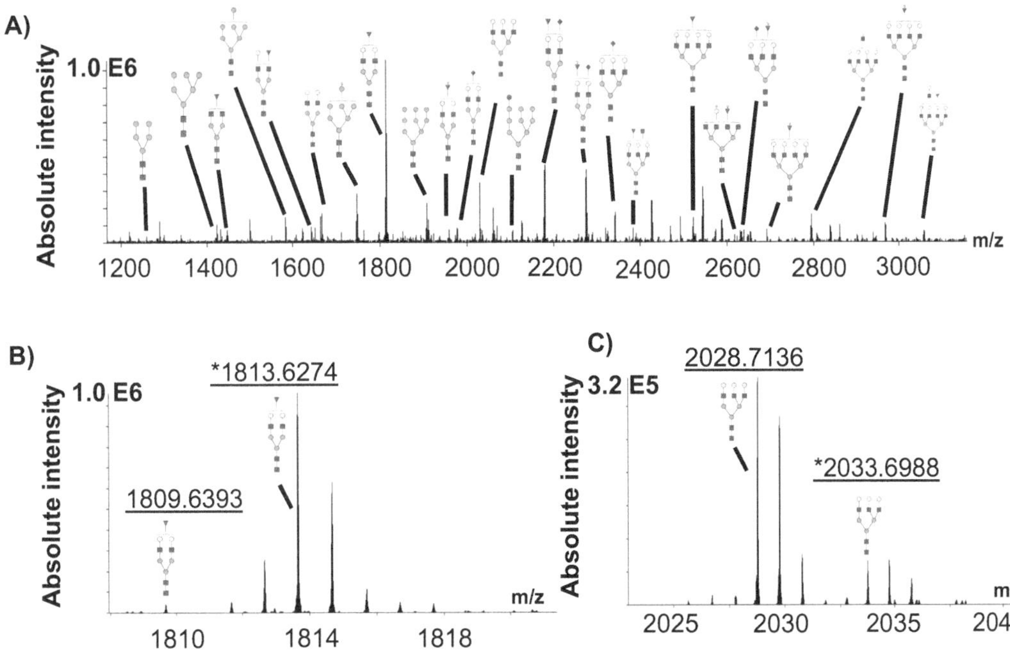


**Fig. 1** Example *N*-glycan profiles from cells using the IDAWG labeling method. Human aortic endothelial cells (HAEC) were grown for 4 days substituting L-glutamine (Amide-$^{15}$N, 98%+, Cambridge Isotope Laboratories, Inc) for $^{14}$N glutamine in complete media. Peaks are detected using MALDI coupled with Fourier Transform Ion Cyclotron Resonance mass spectrometry. (**a**) Typical Overall Average Mass Spectrum with major *N*-glycans annotated. The high mannose series Man5-Man9 present a quality control factor for evaluating cellular derived *N*-glycan signatures. In this profile spectrum, the high mannose series are considered well detected. (**b**) the A2FG peak, *m/z* 1809.6393 with the $^{15}$N isotope detected at 1813.6274. (**c**) A primary peak present in media, *m/z* 2028.7136. Fully labeled 2033.6988 is detected, representing cell turnover of the triantennary peak. Isotopic labeling may be analyzed as described in IDAWG protocols [16, 17]. Turnover rates may be determined by measuring intensity of the $^{15}$N isotopic peak divided by time after accounting for all contributing isotopes

### 3.8 Post-Acquisition Cell Measurements

1. Post-acquisition cell measurements are useful for normalizing wells by cell count.
2. Remove MALDI matrix by incubating arrays for 1 min in 100% ethanol. Repeat to ensure complete removal of all matrix (*see* **Note 15**).
3. Rinse the arrays four times with PBS, gently tapping the arrays to remove excess liquid each time.
4. Rinse the array once in ultrapure HPLC water to remove salts.
5. Incubate the arrays for 1 h at room temperature in a microscope slide mailer filled with Coomassie Blue stain (Simple Blue Stain, Thermo Fisher Scientific).
6. Rinse arrays in fresh changes of PBS until solution is clear.
7. Coverslip each well using xylene or aqueous-based mounting media (*see* **Note 16**).

8. A brightfield microscope or brightfield slide scanner may be used capture images of stained cells per well for cell counting.
9. Cell counts may be done using ImageJ software [18].
10. Higher quality quantitative densitometry may be done using an Odyssey Imaging System (Li-Cor Biosciences) to scan slides at 700 nm with an 84 μm scan line resolution. Measurements per well obtained by Image Studio (Li-Cor Biosciences) as Signal@ 700 nm may be used as a normalization factor to minimize variability between wells (*see* **Note 17**).

## 4 Notes

1. Other fixatives types such as 4% paraformaldehyde have been successfully used.
2. Use of the filter significantly reduces clogging of the solvent lines of the M3 or M5 TM-Sprayer™.
3. Low salt content PNGase F must be used for this assay or excessive MALDI matrix clusters will limit detection of *N*-glycan signal. We use highly concentrated preparations of N-zyme Scientifics PNGase F Prime™ in liquid form (Initial concentration >5 μg/μL; diluted to final concentrations of 0.1 μg/μL).
4. When refilling the push solvent, ensure the bottle is thoroughly cleaned out before adding new solvent to prevent bacterial growth.
5. Fine-tipped tweezers may be used to remove adhesives from between the array wells. A scalpel may be used to gently scrape off adhesive debris.
6. Glass Coplin jars are used as the chloroform in the Carnoy's solution can dissolve many polymer plastics causing spectral interference in *N*-glycan MS profiling.
7. Investigators may ship to sites equipped with imaging mass spectrometers. To limit degradation during shipping, it is essential that cells are maintained at minimum 4 °C using a cooler pack; dry ice is preferred.
8. To eliminate bubbles within the syringe barrel after loading all the PNGase F solution, it is helpful to pull an additional small volume of air into the syringe, creating a single large air bubble. Gently dispense the syringe until the air bubble is gone.
9. The TM-sprayer™ temperature and gas does not need to be turned on at this point.
10. Spray rate should be evaluated prior spraying arrays to ensure accurate deposition of the enzyme. To do this, load a 1-mL syringe with water and start the pump at the required flow rate,

allowing the flow to stabilize for 3 min. Ensure that the TM-sprayer nozzle is at room temperature and has no gas flowing. Place a centrifuge tube under the nozzle head, collecting for 10 min. Measure the 10-min collection. For enzymes in aqueous solution at a rate of 25 μL/min, the maximal volume variance should be ±5.0 μL.

11. Excess solutions will be deposited at the point of the sprayhead turn around location. Ensuring the turnaround point is located off the array well prevents delocalization and excess matrix build up.
12. For higher resolution sampling (≤20 μm laser step size), reduce the flow rate and increase the number of passes to maintain the same amount of enzyme deposition. For instance, we have found that 12 passes at 20 μL/min allows for 20 μm step size sampling, which can visualize certain cell colonies.
13. It is critical that the chamber lid shows a thin layer of condensation, which demonstrates the presence of high humidity, before placing the array inside the incubation chamber. Appropriate humidity conditions are essential to efficient deglycosylation.
14. Frequently, digital readouts on small ovens do not accurately report internal temperature. Ensure that the internal oven temperature is at the correct temperature using a secondary thermometer placed in the oven. This is essential for efficient deglycosylation.
15. Arrays may be inspected under microscope to ensure complete removal of matrix. Cells should be visible without the presence of matrix crystals.
16. Certain xylene-based mounting media may dissolve well edges that are demarked by plastics, adhesives, or markers. Mounting media should be tested before applying to arrays.
17. Signal information per well is obtained from the shape.txt file of the Odyssey Imaging System (Li-Cor Biosciences).

## Acknowledgments

This work was supported by the U01CA242096 (NIH/NCI) to RRD, ASM and PMA. PMA is supported by P20GM103542 (NIH/NIGMS), and in part by pilot research funding, Hollings Cancer Center's Cancer Center Support Grant P30 CA138313 at the Medical University of South Carolina.

## References


1. Powers TW, Jones EE, Betesh LR, Romano PR, Gao P, Copland JA, Mehta AS, Drake RR (2013) Matrix assisted laser desorption ionization imaging mass spectrometry workflow for spatial profiling analysis of N-linked glycan expression in tissues. Anal Chem 85 (20):9799–9806
2. Powers TW, Neely BA, Shao Y, Tang H, Troyer DA, Mehta AS, Haab BB, Drake RR (2014) MALDI imaging mass spectrometry profiling of N-glycans in formalin-fixed paraffin embedded clinical tissue blocks and tissue microarrays. PLoS One 9(9):e106255
3. Groseclose MR, Andersson M, Hardesty WM, Caprioli RM (2007) Identification of proteins directly from tissue: in situ tryptic digestions coupled with imaging mass spectrometry. J Mass Spectrom 42(2):254–262
4. Heijs B, Carreira RJ, Tolner EA, de Ru AH, van den Maagdenberg AMJM, van Veelen PA, McDonnell LA (2015) Comprehensive analysis of the mouse brain proteome sampled in mass spectrometry imaging. Anal Chem 87 (3):1867–1875
5. Powers TW, Holst S, Wuhrer M, Mehta AS, Drake RR (2015) Two-dimensional N-glycan distribution mapping of hepatocellular carcinoma tissues by MALDI-imaging mass spectrometry. Biomol Ther 5(4):2554–2572
6. Drake RR, Powers TW, Jones EE, Bruner E, Mehta AS, Angel PM (2017) MALDI mass spectrometry imaging of N-linked Glycans in cancer tissues. Adv Cancer Res 134:85–116
7. Gustafsson OJR, Briggs MT, Condina MR, Winderbaum LJ, Pelzing M, McColl SR, Everest-Dass AV, Packer NH, Hoffmann P (2015) MALDI imaging mass spectrometry of N-linked glycans on formalin-fixed paraffin-embedded murine kidney. Anal Bioanal Chem 407(8):2127–2139
8. Scott DA, Casadonte R, Cardinali B, Spruill L, Mehta AS, Carli F, Simone N, Kriegsmann M, Mastro LD, Kriegsmann J (2018) Increases in Tumor N-glycan Polylactosamines Associated with Advanced HER2 Positive and Triple Negative Breast Cancer Tissues. Proteomics Clin Appl 13:e1800014
9. Briggs MT, Kuliwaba JS, Muratovic D, Everest-Dass AV, Packer NH, Findlay DM, Hoffmann P (2016) MALDI mass spectrometry imaging of N-glycans on tibial cartilage and subchondral bone proteins in knee osteoarthritis. Proteomics 16(11-12):1736–1741
10. de Haan N, Reiding KR, Haberger M, Reusch D, Falck D, Wuhrer M (2015) Linkage-specific sialic acid derivatization for MALDI-TOF-MS profiling of IgG glycopeptides. Anal Chem 87(16):8284–8291
11. West CA, Wang M, Herrera H, Liang H, Black A, Angel PM, Drake RR, Mehta AS (2018) N-linked glycan branching and fucosylation are increased directly in hepatocellular carcinoma tissue as determined through in situ glycan imaging. J Proteome Res 17 (10):3454–3462. https://doi.org/10.1021/acs.jproteome.8b00323
12. Aich U, Lakbub J, Liu A (2016) State-of-the-art technologies for rapid and high-throughput sample preparation and analysis of N-glycans from antibodies. Electrophoresis 37 (11):1468–1488
13. Shajahan A, Heiss C, Ishihara M, Azadi P (2017) Glycomic and glycoproteomic analysis of glycoproteins—a tutorial. Anal Bioanal Chem 409(19):4483–4505
14. Domann PJ, Pardos-Pardos AC, Fernandes DL, Spencer DIR, Radcliffe CM, Royle L, Dwek RA, Rudd PM (2007) Separation-based glycoprofiling approaches using fluorescent labels. Proteomics 7(S1):70–76
15. Angel PM, Saunders J, Clift CL, White-Gilbertson S, Voelkel-Johnson C, Yeh E, Mehta A, Drake RR (2019) A rapid Array-based approach to N-glycan profiling of cultured cells. J Proteome Res 18 (10):3630–3639
16. Orlando R, Lim J-M, Atwood Iii JA, Angel PM, Fang M, Aoki K, Alvarez-Manilla G, Moremen KW, York WS, Tiemeyer M (2009) IDAWG: Metabolic incorporation of stable isotope labels for quantitative glycomics of cultured cells. J Proteome Res 8 (8):3816–3823
17. Fang M, Lim JM, Wells L (2010) Quantitative glycomics of cultured cells using isotopic detection of aminosugars with glutamine (IDAWG). Curr Protoc Chem Biol 2(2):55–69
18. Collins TJ (2007) ImageJ for microscopy. BioTechniques 43(1 Suppl):25–30

# Chapter 24

# Analysis of Oligomeric and Glycosylated Proteins by Size-Exclusion Chromatography Coupled with Multiangle Light Scattering

**Kathryn Hastie, Vamseedhar Rayaprolu, and Erica Ollmann Saphire**

## Abstract

Analytical size-exclusion chromatography (SEC) is a powerful technique that separates proteins based on their hydrodynamic radii. This approach can provide some rudimentary information about the molecular weight of proteins, but results are also influenced by the in-solution protein conformation and hydrophobicity. SEC also can be affected by nonspecific interactions with the column matrix that influence protein separation. Light scattering (LS) is an absolute and highly accurate measurement of protein molecular weight. Coupling analytical size-exclusion chromatography with multiangle light scattering (SEC-MALS) yields a more robust and accurate method for determining multiple biophysical parameters of proteins while avoiding SEC artifacts. This union of two techniques can help determine the absolute molecular stoichiometry, homo- and heteroassociation of sample components, the nature of protein conjugates, and the molar mass of single molecules and multisubunit complexes. In this chapter, we provide several examples of analysis of glycosylated protein conjugates to showcase the power of SEC-MALS.

**Key words** Size-exclusion chromatography (SEC), Light scattering, Glycosylation, Glycoprotein, Molar mass, Oligomer, Multiangle light scattering (MALS)

## 1 Introduction

Successful biophysical characterization of proteins in solution involves analysis of the molecular weight, degree of any conjugation (e.g., glycosylation), and oligomeric state. These analyses indicate whether the correct protein has been isolated as well as whether the protein forms the expected oligomer or if it exists as high-molecular weight aggregates. Several methods can be used to determine the molecular weight of a protein, including, but not limited to, SDS-PAGE, native PAGE, analytical ultracentrifugation (AUC), mass spectrometry (MS), and analytical size-exclusion chromatography (SEC). SDS-PAGE can facilitate estimation of purity and protein molecular weight based on comparison to a set of standard proteins. However, SDS-PAGE typically involves

Arnaud Delobel (ed.), *Mass Spectrometry of Glycoproteins: Methods and Protocols*, Methods in Molecular Biology, vol. 2271, https://doi.org/10.1007/978-1-0716-1241-5_24, 

dissociation of native oligomers or high-molecular weight aggregates and is unsuitable for analysis of native biomolecules and quaternary associations. Although native-PAGE can, in theory, maintain the correct oligomeric state of a protein, this approach often requires optimization on a per-protein basis and the results can be difficult to interpret or reproduce effectively, particularly for glycosylated or other conjugated proteins. AUC is a robust and accurate method for quantitative mass determination and can accurately assess native oligomeric state. However, the experimental timeframe for this technique can be long, and significant expertise is required for data interpretation. Mass spectrometry can provide precise information about the mass and charge of proteins in solution, but it too requires specialized expertise and training for data collection and analysis. SEC separates proteins in solution state through a packed column and provides quantitative data on the distinct species present in a given solution. For analytical SEC, a set of reference proteins is used to construct a calibration curve that relates elution volume to molecular weight. This curve can then be used to estimate the molecular weight of the protein of interest. Importantly, there are two challenges to interpretation. First, SEC separates molecules not by mass, but rather by hydrodynamic radius. Second, the standards used to create the curve are generally globular proteins that elute as single species and do not interact with the column matrix. Hence, use of analytical SEC to accurately determine mass (and therefore also oligomeric state) relies on the protein of interest behaving like a reference standard, with globular shape, hydrodynamic radius similar to the standards, and without propensity to form hydrophobic or electrostatic interactions with the column matrix that might delay elution. As such, mass calculation using SEC standards fails for many proteins, including those that have hydrophobic or charged surfaces; those with modifications such as glycosylation or pegylation; those that are detergent-solubilized, like membrane proteins; and those proteins that form nonspherical, higher order assemblies.

In contrast, light scattering (LS) is a simple, rigorous, and highly reproducible spectroscopic technique that is widely used to determine the absolute mass of both simple and complex proteins and assemblies in solution [1]. When coupled with SEC, light-scattering can provide molar mass calculations of samples, independent of the elution volume. Many comprehensive review articles have been published on the theory of light scattering and its use in mass determination for macromolecules in solution (*see* for examples [2, 3]). Briefly, the average mass of an analyte can be calculated using the proportionality:

$$M \propto \frac{I}{\left(\frac{dn}{dc}\right)^2 \times c}$$

where $M$ is the molecular weight of the analyte, $I$ is the intensity of scattered light as measured by a LS detector, $c$ is the concentration of the analyte, and $dn/dc$ is the refractive index increment of the analyte that indicates the change in the refractive index ($dn$) with change in solute concentration ($dc$). In SEC coupled to multi-angle light scattering (SEC-MALS), the sample is first fractionated over a column, with LS used to analyze each separated species individually. Importantly, the mass determination by LS is independent of the elution volume of the sample. This is an important distinction from standard SEC in which the elution volume is assumed to directly reflect the mass.

For simple unmodified polypeptides, either an ultraviolet (UV) or refractive index (RI) detector can be used to calculate the concentration of the eluting sample. Use of RI as the concentration source is preferable as: (1) it can adequately analyze molecules with very low extinction coefficients, minimizing quantity of sample required to achieve a suitable UV signal, as well measure the concentration of samples that do not absorb UV; and (2) concentration determination by RI requires only the value for $dn/dc$, which for nearly all proteins is 0.185 mL/g ± 1–2% [4, 5]. If a UV detector is used as the concentration source, a priori knowledge of the extinction coefficient of each species present is essential and requires that the exact sequence of the protein of interest be known.

For accurate mass determination of complex proteins, such as glycoproteins or membrane proteins, both UV and RI detectors, in addition to the LS detector are required. In these cases, the UV detector is sensitive to only the protein concentration (as glycans and detergents do not absorb UV light) while the RI detector is sensitive to the concentration of the entire complex. Software that can integrate all three signals allows complete and simple mass calculation of such samples. As one example, the ASTRA® software package that comes standard with the Wyatt detectors regularly used in our lab provides a module to perform analysis of conjugated proteins. As another, OmniSEC software provided by Malvern Instruments for Viscotek systems can also be used to analyze conjugated proteins. Regardless of the manufacturer, the theory behind use of the UV-LS-RI detector triad is similar [6–9]. In brief, the protein conjugate can be analyzed based on the principle of mass conservation. The difference in the $dn/dc$ and $UV_{280nm}$ of the protein, versus that of the conjugate, can be exploited to determine how the individual per unit concentration of each component contributes to the total concentration of the complex. Since the amount of light scattered is proportional to the molar mass and concentration of the sample, the three-detector system can be used to determine not only the total molar mass of the complex, but also the individual mass contributions of the protein and conjugate within the complex. For example, we determined that a glycoprotein eluted as a 190 kDa species, with 156 kDa contributed by polypeptide and 34 kDa contributed by carbohydrate [10]. Based on an expected protein monomer mass of ~50 kDa, SEC-MALS analysis revealed the protein exists as a trimer in solution.

This chapter outlines the instrumentation, workflow, and data analysis frequently used for SEC-MALS and provides several examples of successful mass determination for glycosylated, oligomeric proteins. SEC-MALS is a workhorse for solution-based molecular weight analyses of proteins, and other techniques such as ion exchange chromatography can be coupled with MALS to perform additional high-resolution molecular weight analyses [11].

## 2 Materials

### 2.1 Instruments and LC Column (See Note 1)

1. High-performance liquid chromatography (HPLC) or fast protein liquid chromatography (FPLC) pump that can produce 0.3–1.0 mL/min flow rates with minimal pressure fluctuation (e.g., ÄKTA PURE, GE Healthcare).
2. In-line mobile phase filter with 0.1 μm pore size, installed between the pumps and the SEC column (*see* **Note 2**).
3. UV detector capable of signal export to LS instrumentation.
4. Multiangle static light scattering detector (e.g., miniDawn, Wyatt Technology).
5. Differential refractometer (e.g., dRI, Optilab, Wyatt Technology).
6. Computer and software for data collection (e.g., Unicorn software for ÄKTA FPLCs and ASTRA® for Wyatt instruments, Windows 7 Pro or 10 Pro 32-bit/64-bit).
7. Size-exclusion chromatography column suitable for fractionation of the protein of interest (*see* **Note 3**).

### 2.2 Reagents and Supplies

1. Bottle top filters with 0.1 μm pore size (Millipore).
2. Buffer: Typical running buffers for proteins are phosphate-buffered saline or Tris buffers with 50–300 mM NaCl. Variations in pH, salt concentration, and inclusion of reducing agents are protein-specific and must be determined empirically by the user. The mobile phase must be compatible with the SEC column used for protein separation. At least 1 L of buffer should be freshly prepared using de-ionized, ultrapure water, filtered to 0.1 μM and extensively degassed (*see* **Note 4**).
3. Protein standard: ~200 μL 1–2 mg/mL BSA prepared in SEC buffer (*see* **Note 5**).
4. Sample of interest: 50–500 μg in a volume corresponding to no more than 5% of the column volume and ideally less than 3%. The amount of sample needed for accurate mass calculation

depends primarily on the size of the molecule, as large molecules scatter more light than small molecules. Ideally, samples should be prepurified on a preparative scale and the particular peak of interest used for subsequent SEC-MALS analysis (*see* **Note 6**). Required information includes the $UV_{280nm}$ (in mL/mg cm) of both the protein of interest and, if analyzing a modified protein, the UV extinction coefficient ($UV_{280nm}$, mL/(mg cm)) and d*n*/d*c* (mL/g) of the modifier (*see* **Note 7**).

## 3 Methods

### 3.1 System Setup

Prior to the first run, all instruments and respective software packages for control and analysis should have been installed and requisite calibration constants and other settings entered according to the manufacturer's instructions. Detectors should be installed downstream from the column in the following order: UV, LS, then RI. The RI detector is last in the series due to the fragility of its flow cell and lower pressure tolerance. Export any outputs from the FPLC to the MALS instrument, such as the UV or autoinject signal, via the I/O box of the FPLC. Use of tubing with a narrow inner diameter of 0.25 mm throughout the system will minimize delay volumes between detectors and limit band-broadening. However, tubing with an inner diameter of 0.75 mm should be used for final output from the RI detector to the fraction collector or waste container (*see* **Note 8**).

### 3.2 System and Column Equilibration

1. Fill the system with the size-exclusion running buffer. Set the flowpath to bypass the column and set the RI detector flow cell to purge. Start a flow of 0.5–1 mL/min (or a rate appropriate for the selected column) such that the buffer flows through the injection loop (if using a manual injection system) and all detectors. Clean buffers, tubing and instruments should produce minimal noise, not exceeding 50–100 μV for the 90-degree LS detector.
2. Switch from bypass to the column position of choice. Equilibrate the column for at least 2–3 column volumes, with all detectors in-line and RI on purge mode, or until the baseline has acceptable noise and drift levels (*see* **Note 9**). Prior to sample application, the RI purge should be turned off and the RI signal assessed for minimal drift. For some columns, overnight equilibration may be necessary to ensure that the system is particle free. If overnight equilibration is needed, users can either prepare sufficient buffer to run the system at the desired flow rate for the required amount of time, or equilibrate at a lower flow rate and gradually increase the flow rate to prevent additional column shedding caused by abrupt changes in

column pressure. Use the FPLC in continuous flow mode and ensure that the flow does not stop until all SEC-MALS experiments are complete.

### 3.3 System Configuration and Validation

Prior to analysis of an actual sample, several configuration parameters must be established for the MALS system, including peak alignment, band broadening, and, for LS detectors with multiple angles, angular normalization (*see* **Note 10**). The same experimental data used for system configuration can also be used for system validation, which must be performed prior to running the conjugated/glycoprotein sample by checking that the same molar mass is obtained for a nonconjugated standard regardless of the concentration source. The validation procedures are outlined below and assume that the user has created a standard method for online UV-LS-RI analysis. Example data are shown in Fig. 1 and Table 1.

1. Inject >200 μg BSA standard and allow collection to proceed for 1.25 column volumes.
2. Set appropriate baselines to encompass the level of pure mobile phase on either side of the peak. Ensure that any automatically detected baselines are flat and not set to end on a noise peak.
3. Choose peaks for analysis. Select the central ~50% of the peak. BSA elutes primarily as a monomeric species, although dimers and trimers can be detected. The width of the peak selected is a user-defined parameter and largely depends on the experimental goals. Only the area chosen in the peak will contribute to mass calculation. If using BSA, input values for $dn/dc = 0.185$ and $UV_{280nm} = 0.667$ mL/(mg cm) under each peak.
4. Perform peak alignment, band-broadening correction, and normalization procedures according to software recommendations. Save these values as a new method and create a default method containing the saved configuration parameters.
5. ASTRA® software uses the RI signal for mass calculation by default. Under the Results section, the mass calculation for the BSA monomer should agree with the sequence-based mass of 66.4 kDa to within 5% (the usual and expected accuracy of SEC-MALS mass calculation).
6. Under the configuration setting, change the concentration source from RI to UV. The mass calculation based on the known $UV_{280nm}$ of 0.667 mL/(mg cm) should be within 5% of the mass calculated by RI (*see* **Note 11**).

Here, we describe the analysis of complex proteins using the "protein conjugate" method specific to the ASTRA® software package and use a glycoprotein as an example. In principle, this analysis can

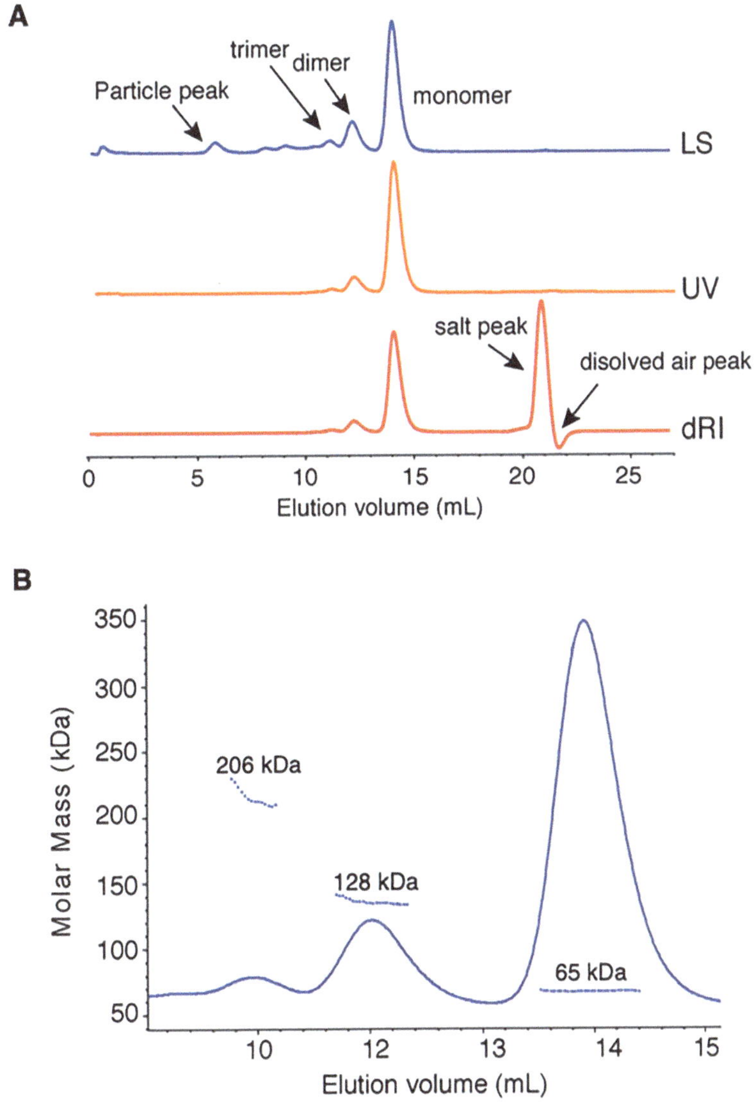


**Fig. 1** SEC-MALS analysis of bovine serum albumin (BSA). (**a**) Chromatogram traces, corresponding to the light-scattered (LS, 90° detector), ultraviolet (UV) and refractive index (RI) of BSA, separated on a Superdex 200increase 10/300 size-exclusion column. Traces were normalized to the monomer peak and are offset for clarity. Features such as the particle peak in the LS trace and the salt and dissolved air peak in the RI trace are common artifacts. (**b**) Excellent separation of the monomer, dimer and trimer species of BSA allow for accurate mass determination. The calculated mass, using the RI detector as the concentration source, is shown for each species and is highly homogeneous across the dimer and monomer peaks

be performed on any conjugated protein, such as pegylated proteins or membrane proteins around which detergents form a micelle (*see* **Note 12**). Examples of SEC-MALS using three detectors and conjugate analysis are shown in Fig. 2 and Table 2.

**Table 1**
**BSA mass calculation via SEC-MALS. The predicted molar mass for BSA monomers, dimers, and trimer is displayed, along with the experimentally determined molar mass of each species using either the RI or UV as a concentration source**

| | Predicted MM (kDa) | MM calculated by concentration source (kDa)[a] RI | UV | Discrepancy (RI/UV, %) |
|---|---|---|---|---|
| Monomer | 66.4 | 65 ± 1.2 | 67.1 ± 1.4 | 1.2 |
| Dimer | 132.8 | 127.5 ± 1.2 | 131 ± 2.1 | 1.6 |
| Trimer | 199.2 | 205.7 ± 5.2 | 189.6 ± 6.1 | 8.5 |

*SEC* size-exclusion chromatography, *UV* ultraviolet, *LS* light scattering, *RI* refractive index, *MM* molar mass
[a]Average mass calculated from three measurements; errors represent the standard deviation

### 3.4 Determination of the Molar Mass of a Glycoprotein Using Protein Conjugate Analysis in ASTRA®

1. Open a new default method containing the configuration created in Subheading 3.3.
2. Inject sample onto column and allow collection to proceed for 1.25 column volumes.
3. Set appropriate baselines.
4. Choose peaks for analysis.
5. Input $dn/dc$ and $UV_{280nm}$ values specific for the protein component of the sample.
6. Under the "conjugate analysis" input the $dn/dc$ and $UV_{280nm}$ values corresponding to the conjugate of interest. For glycosylated proteins produced in insect cells, we use a $dn/dc$ of 0.14; glycans do not absorb at $UV_{280nm}$ and therefore, this field is left blank.
7. In the ASTRA® software, hit "apply" and then right click on the file name and navigate to the "protein conjugate" method. This method contains additional processing parameters and algorithms that consider both the protein and conjugate $dn/dc$ and $UV_{280nm}$ to calculate the molar mass. The Results section will no longer list a concentration source, as both UV and RI are used. Instead, the total molar mass of the protein conjugate, as well as the individual mass contributions of the protein and conjugate are reported.

### 3.5 Determination of the Molar Mass of a Glycoprotein–Fab Complex Protein Conjugate Analysis in ASTRA®

The protein conjugate method in ASTRA® provides a straightforward way to characterize conjugated proteins and allows the individual mass contributions to be calculated. However, this method can also be used for more complex analyses, such as antigen–antibody interactions. In this section, we will outline the procedure we use to calculate the stoichiometry of binding between the antigen-binding fragment (Fab) of an antibody and its target, a viral glycoprotein. This procedure requires two steps: (a) determination of the

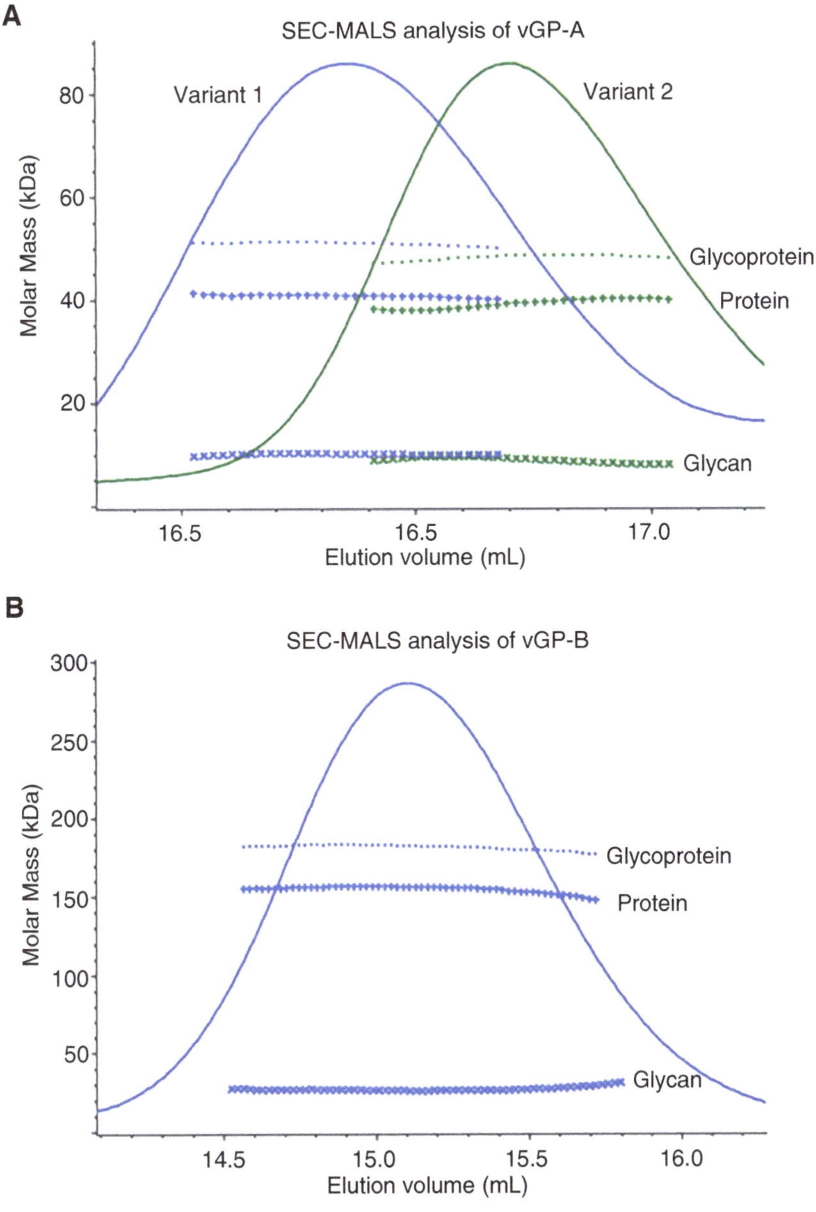


**Fig. 2** Mass calculation of glycoproteins separated on a Superose 6increase 10/300 size-exclusion column coupled with UV-LS-RI measurements. (**a**) SEC-MALS analysis of two variants of the same viral glycoprotein demonstrates that the proteins have equal molar mass and glycan content, yet different hydrodynamic radii. (**b**) SEC-MALS analysis of a self-associating viral glycoprotein shows the presence of a single, trimeric species. For each panel, the *X*-axis corresponds to the elution volume for each species and the *Y*-axis plots the molar mass distribution. The glycoprotein mass distributions are shown as squares, and the mass contributions for the protein and glycan components are shown as crosses and x's, respectively

glycoprotein molar mass and the glycoprotein conjugate $dn/dc$ and $UV_{280nm}$ values; and (b) antibody–glycoprotein complex analysis. This method, like the more simplistic single-species conjugate analysis, also requires that the user acquire all three signals (i.e., UV, LS,

**Table 2**
**Theoretical and calculated molar mass for each viral glycoprotein (vGP) is displayed, along with the predicted and experimentally determined stoichiometry for each protein**

| | | MM determined from SEC-UV-LS-RI analysis (kDa) | | | | Stoichiometry | |
|---|---|---|---|---|---|---|---|
| | Expected mass of monomer (kDa) | Glycoprotein | Protein | Glycan | Fractional weight, protein | Calculated | Expected |
| vGP-A1 | 38.75 | 50.8 | 40.9 | 10.0 | 0.80 | 1.06 | 1 |
| VGP-A2 | 38.75 | 48.1 | 39.3 | 8.9 | 0.82 | 1.01 | 1 |
| vGP-B | 52.30 | 181.4 | 154.2 | 27.2 | 0.85 | 2.95 | 3 |

*SEC* size-exclusion chromatography, *UV* ultraviolet, *LS* light scattering, *RI* refractive index, *MM* molar mass

RI) and that the d*n*/d*c* and extinction coefficient of each component of the complex are known or can be obtained experimentally. In some situations, three detectors are not available, the d*n*/d*c* and/or extinction coefficients are not known for all components, or, in other cases, these values are not sufficiently different to allow the protein conjugate method to work effectively. Hence, we will also describe an alternative method to obtain the molar mass of multiprotein complexes. Example chromatograms and analysis of a glycoprotein alone and in complex with Fab are shown in Fig. 3 and Tables 3 and 4.

*Analysis A:*

1. First determine the molar mass and oligomeric state of the antigen and, if necessary, use the protein conjugate method outlined in Subheading 3.4 to determine the d*n*/d*c* and $UV_{280nm}$ of the conjugate.

*Analysis B*:

2. Incubate an appropriate ratio of pre-SEC purified components and allow complexes to form (*see* **Note 13**).
3. Inject sample onto the column and collect data for 1.25 column volumes.
4. For each eluting peak:
   (a) Input the d*n*/d*c* and $UV_{280nm}$ values specific for one component of the complex in the "protein" processing parameters section and the d*n*/d*c* and $UV_{280nm}$ values specific for the other component of the complex in the "conjugate" parameters section. In the example result outlined in Fig. 3, the glycoprotein antigen is considered the "protein" component, while the Fab is treated as the "conjugate."

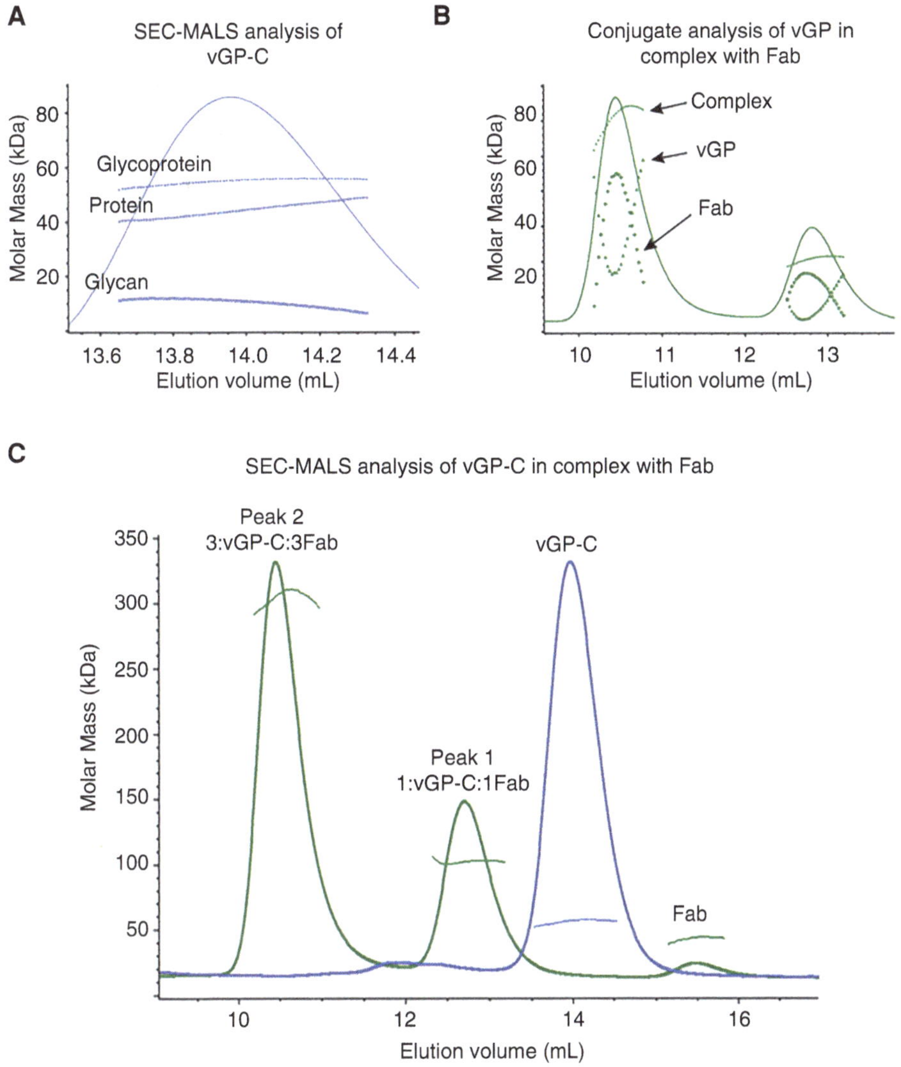


**Fig. 3** SEC-MALS analysis of a viral glycoprotein alone and in complex with an antibody fragment that induces trimerization. (**a**) SEC-MALS conjugate analysis of the vGP alone to determine the $UV_{280nm}$ and d*n*/d*c* values for the glycoprotein for use in downstream analyses. (**b**) SEC-MALS conjugate analysis of vGP bound to an antibody fragment (Fab). The complex elutes as two peaks, which represent a trimer of vGP in complex with three Fab molecules and one vGP in complex with one Fab [14]. Note that the total mass of the complex remains relatively constant throughout the peak, but the calculated mass contributions of vGP and Fab are heterogeneous due to the inability to deconvolute the mass contributions of each species. (**c**) Overlay of SEC-MALS analysis of the vGP-Fab complex and vGP alone. Mass calculations were performed using only the RI detector as a concentration source and a d*n*/d*c* of 0.177, which corresponds to the weight-averaged d*n*/d*c* for a 1:1 complex of vGP and Fab. For each panel, the *X*-axis corresponds to the elution volume for each species while the *Y*-axis corresponds to molar mass distribution. The mass distributions for the glycoprotein (panel **a**) or glycoprotein–Fab complex (panel **b** and **c**) are shown as squares, the mass contribution for individual components (protein and glycan for panel **a**, glycoprotein and Fab for panel **b**) are shown as crosses and triangles, respectively

**Table 3**
**Mass determination and stoichiometry of a glycoprotein in complex with an antibody fragment, using the protein conjugate method. The molar mass of each eluting peak derived from the "protein conjugate" method is displayed. The total molar mass of the glycoprotein–Fab complex in each peak corresponds to a monomer and trimer of vGP bound to one and three Fabs, respectively**

| | MM each component (kDa) | MM complex (kDa) | MM of vGP (kDa) | MM of Fab (kDa) | Stoichiometry based on 1:1 MM |
|---|---|---|---|---|---|
| vGP[a] | 55.9 | | | | |
| Fab[b] | 44.7 | | | | |
| MM, 1:1 complex | 100.6 | | | | |
| vGP-Fab conjugate, Peak 1 | | 103.6 | 62.3 | 41.3 | 1vGP: 1Fab |
| vGP-Fab conjugate, peak 2 | | 303.5 | 147.8 | 155.6 | 3vGP: 3Fab |

[a]Mass of the glycoprotein was calculated using SEC-UV-LS-RI conjugate analysis. $UV_{280nm}$ (1.21) and $dn/dc$ (0.174) values for the glycoprotein obtained from the analysis in Part A were used in the analysis in Part B

[b]The molar mass of the Fab was calculated from the excess Fab peak. A $dn/dc$ value of 0.185 and a $UV_{280nm}$ value of 1.46, derived from the known sequence of the Fab, was used in the conjugate analysis

**Table 4**
**Mass determination and stoichiometry of a glycoprotein in complex with Fab, using the RI detector as a concentration source. Theoretical $dn/dc$ values were calculated for a 1:1 complex between vGP and Fab, and for the vGP and Fab alone, based on conservation of mass. The $dn/dc$ values derived from these three stoichiometries encompass the range of possible $dn/dc$ values for all available stoichiometries (1:2, 2:1, 3:2, etc.). Note: errors in $dn/dc$ will result in a twofold error in the mass calculation. In this test case, the difference between a $dn/dc$ value corresponding to the extremes of no Fab bound to vGP (0.174) and no vGP bound to Fab (0.185) is ~6%. This difference translates to ~12% error in the calculation of the molar mass for each eluting peak. This error in mass calculation is near the expected error for SEC-MALS overall and is less than a monomer of either species**

| Stoichiometry for $dn/dc$ estimation | $dn/dc$ | MM complex, Peak 1 (kDa) | Stoichiometry based on 1:1 MM | MM complex, Peak 2 (kDa) | Stoichiometry based on 1:1 MM |
|---|---|---|---|---|---|
| vGP only | 0.174[a] | 106.2 | 1.1 | 313.2 | 3.1 |
| 1vGP: 1Fab | 0.177 | 104.4 | 1.0 | 307.9 | 3.1 |
| Fab only | 0.185 | 99.9 | 1.0 | 294.6 | 2.9 |

[a]Derived from analysis of the glycoprotein alone

(b) Apply the input parameters and navigate to the "protein conjugate" method. In the Results section, the total molar mass of the complex as well as the individual mass contributions of each component of the complex are reported.

*Analysis C*

We often encounter issues with multiprotein "protein conjugate" analyses in which the overall mass calculation is accurate, but the mass contribution of each individual component is not. This situation can likely be due to errors in the precision of calculating the $dn/dc$ and $UV_{280nm}$ of the glycoprotein conjugate, and/or the relative similarity between the glycoprotein $dn/dc$ and $UV_{280nm}$ values and those of the Fab. In such situations, accurate deconstruction of the mass contributions of each component is difficult. Thus, for these complex systems, the use of only the RI detector to calculate concentration and a $dn/dc$ value calculated off-line, based on the properties of conservation of mass for both the glycoprotein and the bound Fab, is sufficient to estimate the stoichiometry of binding.

1. Perform Analysis A to obtain the glycoprotein $dn/dc$. In the example outlined in Fig. 3, we obtain a conjugate (glycoprotein) $dn/dc$ of 0.174 and a $UV_{280nm}$ of 1.21 and determined that our glycoprotein exists as an ~55 kDa monomer in solution.
2. Perform Analysis B. In our example, we can use the RI detector as the concentration source and a $dn/dc$ of 0.185 to calculate the molar mass of the excess Fab as ~46 kDa. In lieu of applying the conjugate analysis method, we can instead calculate the possible range of $dn/dc$ values (no Fab present to no GP present) for the glycoprotein complex using the equation:

$$\frac{dn}{dc}_{\text{complex}} = Fw_{\text{gp}} \times \frac{dn}{dc}_{\text{GP}} + Fw_{\text{Fab}} \times \frac{dn}{dc}_{\text{Fab}}$$

where $F_w$ indicates the fractional weight contributed by each species to the complex as a whole. The $dn/dc$ values obtained from this "offline" calculation can assist in the estimation of the mass of such types of complex associations. From our experience, the values obtained are remarkably similar to the total mass calculated for the complex using the conjugate method (*see* Table 4). Note that although errors in the $dn/dc$ value will translate to a twofold error in mass calculation, the discrepancy between the estimated and true $dn/dc$ of a complex sample is usually less than ~5%.

## 4 Notes

1. Analysis of glycoproteins requires three detectors: UV, MALS and RI. Prior to running a SEC-MALS experiment, all instruments and software should be installed per the manufacturer's instructions and stable communication between the instruments must be confirmed. Here, we describe procedures

specific to an ÄKTA Pure FPLC (GE Healthcare) with a 0.2–0.5 cm UV flow cell, connected in-line through a variable valve to a miniDawn (Wyatt Technology) MALS detector and Optilab dRI (Wyatt Technology). The Unicorn and ASTRA® software packages used here for data collection and analysis are specific for these instruments. However, SEC-MALS may be performed using any standard HPLC, UHPLC or FPLC, MALS, and RI equipment. A single-angle LS detector is sufficient for analysis of proteins up to ~500 kDa; additional angles increase accuracy of mass determination and are required for proteins with molecular masses >500 kDa.

2. A 0.1 μm filter placed between the pumps (after the mixer in ÄKTA systems) and column is essential to obtain a suitable LS baseline and will retain any large particles present in the mobile phase or shed from the pump heads. A second in-line filter (e.g., 2 μm PEEK frit) with a low dead volume can also be installed between the injection loop and the column to trap protein aggregates present in the sample or formed during injection. Both filters should be replaced when the system operating pressure increases by more than ~5%.
3. Column selection is protein dependent. SEC columns are rated for separation of specific ranges of molecular mass (or more appropriately by specific radii). Many columns on the market are suitable for SEC-MALS analysis. Primary considerations when choosing a column are matrix type, column stability in the mobile phase, separation resolution and quality of the light-scattering baseline after the column has been preequilibrated. Ideally, the protein of interest will elute at the mid- to lower end of the fractionation range and will be well-separated from both the column exclusion volume and any peaks associated with dissolved air and the mobile phase (*see* Fig. 1a). After initial column conditioning and equilibration, the light-scattering baseline noise should not exceed 50–100 μV. Our laboratory regularly uses Superdex 75increase, 200increase, and Superose 6increase columns (GE Healthcare), which provide both good separation for the variety of proteins we analyze and compatibility with a wide range of buffer compositions. These columns also do not excessively shed particles.
4. Buffer preparation is a central factor for obtaining high-quality SEC-MALS data. We have found that buffers prepared fresh from powder stocks produce quieter baselines than those prepared from 10× solutions. All bottles should be washed thoroughly and rinsed with de-ionized water. Bottle-top filters (0.1 μm) should be precleaned either with de-ionized water or 50–100 mL buffer to remove particulates and preservatives from the dry filter. The remaining buffer can then be filtered into a clean bottle. Buffers should be extensively degassed

using a stir bar for 2–3 h or overnight and then capped to prevent dust from entering. Buffers should be equilibrated to the running temperature of the instruments prior to the start of the experiment to prevent off-gassing. Buffers that have been stored for longer than 1 week should be filtered and degassed prior to reuse.

5. Any isotropic, monodisperse, globular protein with well-characterized molecular mass, d*n*/d*c*, and $UV_{280nm}$ can be used for system calibration. We regularly use BSA due to the extensive biochemical information available for this protein. Other suitable proteins include any of those routinely used for SEC calibration. A proper standard should fractionate in the mid-range of the column, well away from both the exclusion volume and solvent peak of the column.
6. Samples should be either centrifuged for 10 min at 10,000 × *g* or filtered using a 0.2 μm low-protein binding syringe or spin filter to remove any large particles that might coelute with the peak of interest. If using a syringe or spin filter, preequilibrate with buffer to remove particles and preservatives from the dry filters.
7. We have found inaccurate $UV_{280nm}$ values to be one of the greatest sources of divergence from the expected molar mass calculation for glycoproteins. Ensure any additions or modifications to the protein (e.g., affinity tags, point mutations) are included in the $UV_{280nm}$ calculation. For most proteins, a standard d*n*/d*c* value of 0.185 can be used, although this value may be more variable for proteins <10 kDa [4]. Glycans do not absorb UV and hence a $UV_{280nm}$ value is not required. We commonly use a d*n*/d*c* of 0.14 for glycans [12]. The "Refractive Increment Data-Book for Polymer and Biomolecular Scientists" [13] is an additional source for d*n*/d*c* values of nonprotein moieties.
8. SEC-MALS is nondestructive technique and as such, fractions from SEC separation can be collected using the FPLC fraction collector. Many systems have precalibrated delay volumes to account for the tubing between the UV detector and fractionation valve. With additional detectors in-line, it is important to consider these additional delay volumes. Hence, the peak elution volumes displayed on the FPLC and LS software will not correspond to a particular fraction. Our users are encouraged to collect all fractions between the exclusion volume (~8 mL on the 24 mL SEC columns regularly used in our lab) and ~5 mL after the protein has eluted and examine each fraction for protein concentration to assess which should be pooled. NB: due to the sensitivity of the RI flow cell to high back pressure, do not use smaller than 0.7 mm I.D. tubing at the outlet to the fraction collector or waste.

9. Proper system equilibration is critical for obtaining high-quality SEC-MALS data. The LS baseline should be stable and free of substantial noise and the RI detector should demonstrate minimal drift. Low baseline noise is particularly important when working with low sample amounts of smaller proteins. For 50 kDa proteins, we find 200 μg is sufficient sample to obtain peaks with excellent signal-to-noise ratios, provided the guideline of <100 μVolts of noise is followed. Noise levels greater than this value will require an increase in total sample and degrade the overall quality of the UV-LS-RI traces and data.
10. Alignment and band broadening, which account for the tubing length and dilution of a sample as it travels from one detector to another, using an isotropic and monodisperse sample, such as BSA, should be performed at the time of instrument installation. These parameters should be redetermined if tubing (length or diameter) between the instruments is altered. Normalization needs to be performed once for a given mobile phase but should be redetermined if the mobile phase is appreciably different between runs (e.g., inclusion of glycerol, high salt, or markedly different pH). Our lab reassesses alignment, band broadening, and normalization regularly (once per month, or each time an SEC-MALS experiment is performed if an experiment is performed fewer times than once per month).
11. The molar mass across the monomer peak should be uniform within 2–5%. We find freshly made BSA stocks give cleaner, single peaks as compared to frozen stocks, which may contain a higher proportion of fragments. If the molar mass obtained with the RI concentration source is incorrect, first ensure the correct $dn/dc$ value parameters have been entered. If the $dn/dc$ value is correct, the RI detector may need to be cleaned or recalibrated according to the manufacturer's instructions. If the mass calculated with the UV concentration source is incorrect, ensure that the appropriate UV value, correct cell length, and AU/V response factor have been entered.
12. The mass fraction of the glycans or other conjugates must be >3–5% of the protein mass for the analysis to distinguish the mass of each component. Protein conjugate analysis also requires that the two components of the conjugate have different $dn/dc$ and extinction coefficients.
13. We regularly use a 1.5–3-fold molar excess of Fab to ensure all antigen is bound and find that 1 h incubation is sufficient to achieve antigen saturation. However, binding ratios and kinetics should be determined empirically for each complex sample.

## Acknowledgments

This work was supported by the National Institutes of Health grants U19AI10972, R01AI13244, and R21AI116112.

### References


1. Wyatt PJ (1993) Light scattering and the absolute characterization of macromolecules. Anal Chim Acta 272(1):1–40. https://doi.org/10.1016/0003-2670(93)80373-S
2. Folta-Stogniew E, Williams KR (1999) Determination of molecular masses of proteins in solution: implementation of an HPLC size exclusion chromatography and laser light scattering service in a core laboratory. J Biomol Tech 10(2):51–63
3. Wen J, Arakawa T, Philo JS (1996) Size-exclusion chromatography with on-line light-scattering, absorbance, and refractive index detectors for studying proteins and their interactions. Anal Biochem 240(2):155–166. https://doi.org/10.1006/abio.1996.0345
4. Zhao H, Brown PH, Schuck P (2011) On the distribution of protein refractive index increments. Biophys J 100(9):2309–2317. https://doi.org/10.1016/j.bpj.2011.03.004
5. McMeekin T, Groves ML, Hipp NJ (1964) Refractive indices of amino acids, proteins and related substances. Amino acids and serum proteins. American Chemical Society, Washington, DC
6. Arakawa T, Wen J (2001) Determination of carbohydrate contents from excess light scattering. Anal Biochem 299(2):158–161. https://doi.org/10.1006/abio.2001.5432
7. Kendrick BS, Kerwin BA, Chang BS, Philo JS (2001) Online size-exclusion high-performance liquid chromatography light scattering and differential refractometry methods to determine degree of polymer conjugation to proteins and protein-protein or protein-ligand association states. Anal Biochem 299 (2):136–146. https://doi.org/10.1006/abio.2001.5411
8. Pollock JF, Ashton RS, Rode NA, Schaffer DV, Healy KE (2012) Molecular characterization of multivalent bioconjugates by size-exclusion chromatography with multiangle laser light scattering. Bioconjug Chem 23 (9):1794–1801. https://doi.org/10.1021/bc3000595
9. Folta-Stogniew E (2006) Oligomeric states of proteins determined by size-exclusion chromatography coupled with light scattering, absorbance, and refractive index detectors. Methods Mol Biol 328:97–112. https://doi.org/10.1385/1-59745-026-X:97
10. Hastie KM, Igonet S, Sullivan BM, Legrand P, Zandonatti MA, Robinson JE, Garry RF, Rey FA, Oldstone MB, Saphire EO (2016) Crystal structure of the prefusion surface glycoprotein of the prototypic arenavirus LCMV. Nat Struct Mol Biol 23(6):513–521. https://doi.org/10.1038/nsmb.3210
11. Amartely H, Avraham O, Friedler A, Livnah O, Lebendiker M (2018) Coupling multi angle light scattering to ion exchange chromatography (IEX-MALS) for protein characterization. Sci Rep 8(1):6907. https://doi.org/10.1038/s41598-018-25246-6
12. Pletcher Carol HR, Robert M, Wei GJ, Bloomfield VA, Nelsestuen GL (1980) Deglycosylated prothrombin fragment 1. J Biol Chem 255:7433–7438
13. Theisen A, Johann C, Deacon MP, Harding SE (2000) Refractive increment data-book for polymer and biomolecular scientists. Nottingham University Press, Nottingham UK
14. Hastie KM, Zandonatti MA, Kleinfelter LM, Heinrich ML, Rowland MM, Chandran K, Branco LM, Robinson JE, Garry RF, Saphire EO (2017) Structural basis for antibody-mediated neutralization of Lassa virus. Science 356(6341):923–928. https://doi.org/10.1126/science.aam7260

# Chapter 25

# Analysis of Glycoproteins by ATR-FTIR Spectroscopy: Comparative Assessment

**Allison Derenne, Kheiro-Mouna Derfoufi, Ben Cowper, Cédric Delporte, Claire I. Butré, and Erik Goormaghtigh**

## Abstract

FTIR spectroscopy has been widely used to characterize biopharmaceuticals for many years, in particular to analyze protein structure. More recently, it was demonstrated to be a useful tool to study and compare protein samples in terms of glycosylation. Based on a spectral region specific to carbohydrate absorption, we present here a detailed protocol to compare the FTIR spectra of glycoproteins in terms of global glycosylation level and in terms of glycan composition. This FTIR information is compared to MS information. Both approaches yield consistent results but it appears FTIR analysis is easier and more rapid to perform comparisons.

**Key words** FTIR spectroscopy, Infrared, Glycosylation, Monoclonal antibodies, Glycoproteins, biosimilar

## Abbreviations

| | |
|---|---|
| ATR | attenuated total reflection |
| FTIR | Fourier-transform infrared |
| mAb | monoclonal antibody |
| MCT | mercury–cadmium–telluride |
| PC | principal component |
| PCA | principal component analysis |

## 1 Introduction

A set of efficient and complementary analytical tools are required to characterize glycoproteins and ensure the quality and safety of biopharmaceuticals. In the context of monitoring batch-to-batch consistency and biosimilar development, Fourier Transform

Arnaud Delobel (ed.), *Mass Spectrometry of Glycoproteins: Methods and Protocols*, Methods in Molecular Biology, vol. 2271, https://doi.org/10.1007/978-1-0716-1241-5_25, 

Infrared (FTIR) spectroscopy is becoming a powerful, reliable, and simple tool for comparability assessment.

FTIR spectroscopy relies on the fact that most molecules absorb light in the infrared region of the electromagnetic spectrum, converting it to molecular vibrations. In fact, IR light interacts with a chemical bond only if the energy of the radiation matches the vibrational transition energy of this bond. The absorption wavenumber of a chemical bond depends on the exact structure of the molecules. As such, the IR spectrum provides a precise and usually unique fingerprint of the glycans [1].

Infrared spectroscopy uses several modes of recording, among which attenuated total reflection (ATR) is particularly convenient. In ATR mode, the infrared radiation is directed into a high refractive index crystal. The reflection at the surface of the crystal, called the internal reflection element (IRE), creates an evanescent wave perpendicular and outside the surface of the IRE. For biological and macromolecular samples, ATR-FTIR has multiple advantages which are described elsewhere [2–4].

FTIR spectroscopy has been widely used to characterize biopharmaceuticals for many years, in particular to analyze protein structure. These analyses are mainly based on the region of the FTIR spectrum present between 1700 and 1500 $cm^{-1}$. This region contains the two most intense bands found in the FTIR spectrum of proteins, the amide I and II bands resulting respectively from the absorption of the carbonyl and N-H groups present in the peptide bond [5, 6].

Carbohydrates per se have also been studied using FTIR spectroscopy. One recent example is the quantitative analysis of sugar types in honey. Honey is a mix of simple sugars (monosaccharides or disaccharides). It is a complex product coming from different floral sources with a broad diversity of components. Therefore, analysis of honey is challenging but the recording of FTIR spectra combined with statistical tools allows to quantify all major sugars [7, 8] demonstrating the sensitivity of the FTIR spectrum to subtle variations in saccharide structure. Another paper has demonstrated the possibility to use FTIR spectroscopy to monitor sugars in biological culture, microalgae for instance [9]. Research has also been carried out on mono- and polysaccharides alone. This study [10] has evidenced four spectral regions to analyze in the FTIR spectra of saccharides:

1. The region between 3600 and 3050 $cm^{-1}$ characteristics of O–H vibrations;
2. The region between 3050 and 2800 $cm^{-1}$, specific to $CH/CH_2$ stretching vibrations;
3. The region between 1500 and 1200 $cm^{-1}$ dominated by deformational mode of $CH/CH_2$.

4. The region between 1200 and 800 $cm^{-1}$ assigned to the stretching vibrations of C–O and C–C groups.

For the study of glycans in glycoproteins, the region between 1200 and 800 $cm^{-1}$ is particularly interesting as proteins have very low signal in this region. The contribution of glycosylation to the FTIR spectra of proteins was first addressed by Khajehpour et al. in 2006, who recorded the spectra of five glycoproteins (and one protein without glycosylation) and analyzed the region between 1200 and 1000 $cm^{-1}$. The glycosylation profiles of these five proteins are very different as each of them contains specific distinct monosaccharides. This study showed that the intensity of these spectral bands was related to the amount of carbohydrates contained in the glycoprotein, and also that the shape of these bands varies with the type of monosaccharides present in the protein [11].

In addition, a recent study has shown that FTIR spectra of glycoproteins provide a global but accurate fingerprint of the glycosylation profile. This fingerprint is not only sensitive to large differences such as the presence or absence of several monosaccharides but also to smaller modifications of the glycan and monosaccharide content [12].

The use of FTIR methods to analyze glycosylation offers three key advantages. Firstly, in comparison with all existing methods for glycan analysis, it requires very limited sample preparation. Analysis is performed on intact proteins, which represents a major advantage. Indeed, most existing methods for glycans and monosaccharides analysis involve several preparation steps: glycan release, labeling, separation, and hydrolysis for monosaccharide analysis. Secondly, the processing time is extremely short. The measurement of a FTIR spectrum takes a maximum of 5 min; analysis in 96 or 384-well plates is established and the use of microarrays of proteins is under development [13], drastically increasing sample throughput. In addition, data analysis can be fully automated, dramatically reducing spectral interpretation time. Finally, FTIR spectroscopy can be used as a Multi-Attribute Methodology (MAM) [6]. Analysis of other critical parameters for therapeutic proteins (such as the protein structure or protein concentration) can be performed simultaneously.

The main limitation of this approach is that FTIR spectroscopy is a global method. Information on specific glycosidic residues or on specific glycosylation site cannot be extracted.

In this protocol, we detail two procedures: one to compare FTIR spectra of glycoproteins in terms of global glycosylation level, and the other to compare FTIR spectra of glycoproteins in terms of glycan composition.

## 2 Materials

### 2.1 Chemicals

1. Liquid nitrogen.
2. Sodium chloride, NaCl (Sigma-Aldrich).
3. Ultrapure water.
4. Micro-Biospin™ P-6 Gel Columns (Bio-Rad #7326221, Tris buffer, sample volume 10–75 μL, 6000 Daltons MW limit).
5. Nitrogen flow.

### 2.2 Equipment

1. Fourier-transform infrared (FTIR) spectrophotometer with mid-infrared capabilities and equipped, either directly or through a special accessory, to record measurement in attenuated total reflection (ATR) mode.
2. A liquid nitrogen–cooled MCT (mercury–cadmium–telluride) detector.
3. An internal reflection element (IRE), such as diamond (*see* **Note 1**).

### 2.3 Samples

1. Prepare the protein or mAb at a concentration between 1 and 50 mg/mL (*see* **Note 2**).
2. The original buffer is replaced by NaCl 0.9% aqueous solution using Micro-Biospin™ (*see* **Note 3**).
3. The following samples have been used in the examples shown in this protocol.
   (a) Therapeutic mAbs obtained from Hospital Saint-Pierre (Brussels).
   (b) Alpha1-acid glycoprotein (G9885-10 mg), alpha-crystalline (C4163-5 mg), egg white avidin (A9275-10 mg), and fetuin from fetal bovine serum(F2379-100 mg) from Merck.

## 3 Methods

### 3.1 Preparation of the FTIR Spectrometer and the IRE

1. If required for the instrument used, set up the ATR accessory.
2. Check that all the optical parts of the spectrometer are purged with dried air. Alternatively, desiccant air dryer such as silica gel can be used but should be replaced frequently.
3. Fill the MCT detector with liquid nitrogen. Always follow the security rules for liquid nitrogen as it can be extremely harmful.
4. Wait for at least 30 min to reach a stable signal.
5. Load your recording parameters (*see* **Note 4**).

6. Clean the IRE with ultrapure water before each measurement. Gently rub the diamond surface with tissue soaked with water, rinse with pure water and dry immediately using a piece of tissue. Usually glycoproteins do not require a specific cleaning procedure. If a problem is encountered with a particular protein, refer to the manufacturer's notice to check solvents and detergents compatible with the crystal.
7. Check if enough signal reaches the detector. It is expected that ca 50% of the signal intensity measured in the absence of the ATR setup reaches the detector in its presence (without sample). If this value falls below 20%, signal will be noisy and a realignment of the ATR setup is required.
8. Record a background (*see* **Note 5**) with the parameters used for the sample spectrum. The background should be recorded just before the start of the measurements and before every new sample. Make sure to clean the crystal before recording a background.

### 3.2 Sample Spreading and Spectrum Recording

1. Load the sample using a micropipette tip. Do not hesitate to touch the IRE with the tip. The volume of sample to spread depends strongly on the IRE used. Typically, we spread 0.5 μL of protein with a concentration above 1 mg/mL on the diamond crystal.
2. Under a gentle nitrogen flow (i.e., 140 mL/min), evaporate the liquid water (*see* **Note 6**). If the previsualization mode is available (continuous scan and display), display the full spectrum and watch the intensity of the major water band (3000–3500 $cm^{-1}$) to control the evaporation process. Wait until this band reaches a stable level.
3. Start recording the measurement (*see* **Note 7**).
4. Record at least three spectra for each sample.

### 3.3 Spectral Processing

All processing steps described here can be performed on any spectral analysis software including programs provided with recent spectrometers. We perform these steps in Matlab, using an interface developed by Prof. Erik Goormaghtigh called "Kinetics."

1. Subtract water vapor contribution from your sample spectrum. First, record a water vapor spectrum. To do so, record a new background and then, after opening slightly the sample compartment for a few seconds, record a new spectrum. Second, determine the subtraction coefficient. This can be conveniently achieved by using a specific water vapor peak (i.e., the 1559 $cm^{-1}$ peak). Integrate the peak between 1562 and 1555 $cm^{-1}$ on both the sample and the water vapor spectra. The ratio of the two areas corresponds to the subtraction coefficient which should be applied to remove the water vapor contribution.

2. Apply a baseline correction on the entire spectrum. Select the points at the edge of the bands or at the lowest intensity between bands. Straight lines should be interpolated between these points and should be subtracted from the spectra. The same baseline points must be used for all the spectra to be compared. Typically, for the figures presented in this chapter, the following wavenumbers have been used as baseline points: 3700, 3000, 2800, 1718, 1476, 1179, 965, and 865 $cm^{-1}$.
3. Normalize the spectra for equal area, either between 1718 and 1476 $cm^{-1}$, or between 1179 and 965 $cm^{-1}$, after subtraction of a straight baseline passing through these two points (*see* **Note 8**).
4. Smooth the spectrum to 4 $cm^{-1}$ by convolution by a 4 $cm^{-1}$ (full width at half height) Gaussian line.

### 3.4 Spectral Analysis: Comparison of the Global Glycosylation Level

The global glycosylation level is defined as the weight ratio between sugars and proteins.

1. To reveal variations in the global glycosylation level, apply the normalization step between 1718 and 1476 $cm^{-1}$. This allows for a comparison of the glycan content for the same, normalized, quantity of proteins. The spectral differences observed in the region related to glycan absorption will thus be mainly due to differences in the weight ratio between glycans and proteins. An example on six proteins with distinct global levels of glycosylation is presented in Fig. 1.

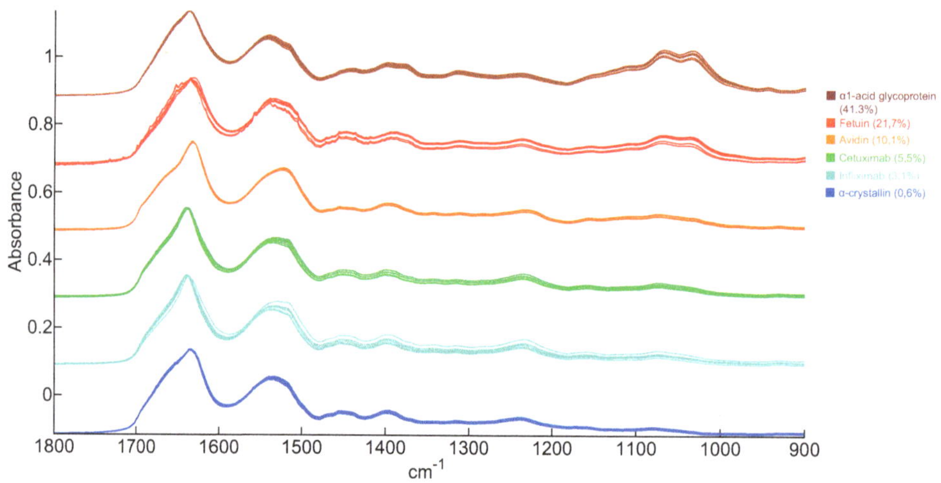


**Fig. 1** Example FTIR spectra recorded for six glycoproteins with various global glycosylation levels. The glycosylation levels are indicated in the legend and have been determined using MALDI-TOF (*see* **Notes 9** and **10**). The raw data (before the processing steps) are presented between 1800 and 900 $cm^{-1}$ in this figure. Six spectra of each sample have been recorded. Each protein is depicted in a unique color. Spectra of each protein have been offset for better readability

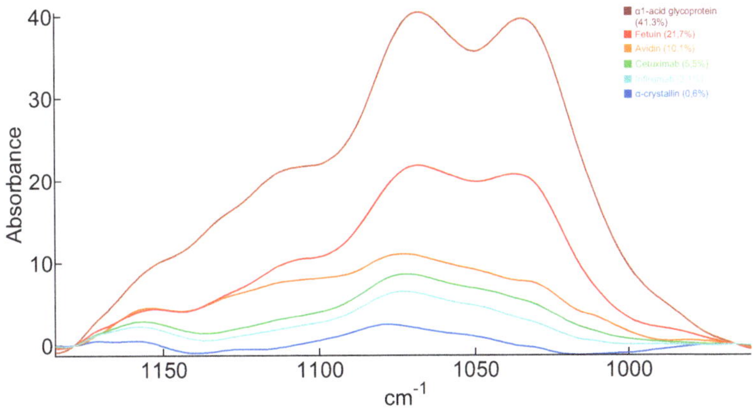


**Fig. 2** Average of the six FTIR spectra recorded for each protein (presented in Fig. 1) and depicted between 1185 and 960 cm$^{-1}$. The spectra have been processed as explained above. Each color corresponds to one specific protein. The glycosylation levels indicated in the legend and have been determined using MALDI-TOF (*see* **Notes 9** and **10**). The intensity of the spectral band shown on this figure clearly increases according to the global glycosylation level

2. Associate a color with each specific sample.
3. After all the processing steps, average the spectra of each sample in order to provide a visual comparison (*see* Fig. 2).
4. Enlarge the spectral region associated with glycosylation (1179 and 965 cm$^{-1}$) for a comparison by visual inspection.
5. Integrate the band between 1179 and 965 cm$^{-1}$ to obtain an index related to the global level of glycosylation. Perform the integration on individual spectra in order to use average and standard deviation for statistical comparison. A graph comparing the peak area for the six proteins is shown on Fig. 3.

### 3.5 Spectral Analysis: Comparison of the Glycan Composition

1. To observe variations in the glycan composition, apply the normalization step between 1179 and 965 cm$^{-1}$. This enables comparison of samples considering the same quantity of carbohydrates and removes spectral variations due to the mass ratio between glycosylation and protein. The spectral differences observed in the region related to glycan absorption will thus be mainly due to differences in glycan composition. An example with 8 batches of 2 mAbs (originators and biosimilars) is presented on Fig. 4.
2. Associate a color with each specific sample.
3. Enlarge the spectral region associated with glycosylation (1179 and 965 cm$^{-1}$) (*see* Fig. 5).
4. Apply principal component analysis (PCA) using any spectral analysis software. The covariance matrix should be diagonalized to calculate new variables, the so-called principal

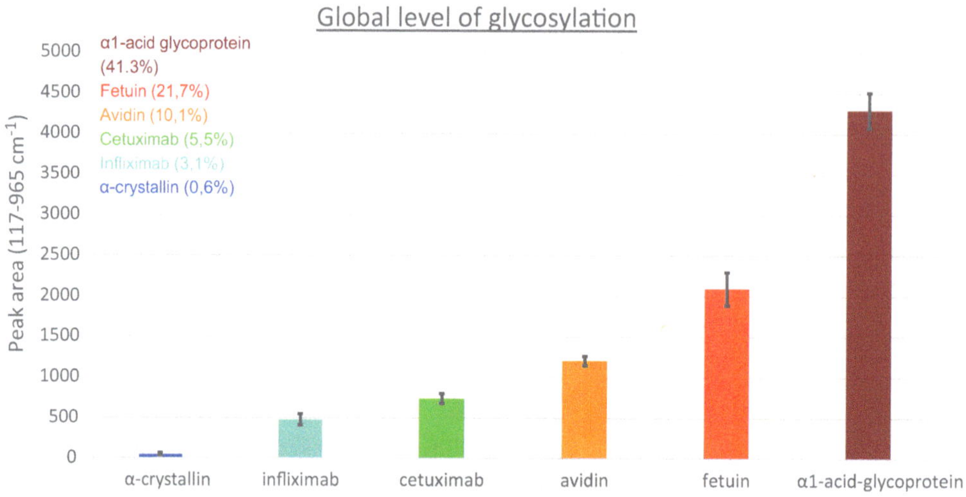


**Fig. 3** Comparison of the area of the spectral band corresponding to glycosylation for each protein. The area has been calculated by integrating the spectra between 1179 and 965 $cm^{-1}$. Vertical error bars correspond to the standard deviation obtained from the six FTIR spectra recorded for each sample. The peak area is clearly related to the global glycosylation level indicated in the legend, which has been determined by MALDI-TOF MS (*see* **Notes 9** and **10**)

components (PC). Original spectra should be projected on the space of these new variables to obtain score plots. The score plot for the spectra shown on Fig. 5 is presented on Fig. 6 (*see* **Note 11**).

5. Compare the projection of the spectra on the first four principal components to highlight similarities and discrepancies among the data.
6. Apply multivariate analysis of variance (MANOVA) considering the scores of the first four principal components to test whether a group is significantly different from others. *P*-values resulting of the MANOVA for the spectra shown on Fig. 5 are presented on Table 1.

## 4 Notes

1. Many materials might be used for ATR in infrared spectroscopy (e.g., Germanium, ZnSe, KRS-5, Silicon, Diamond) and all will give good results. For practical reasons, we will recommend to use either Ge or diamond for their higher refractive index (germanium) and especially hardness (diamond) allowing easier handling and cleaning procedures.
2. If possible, the protein concentration should be above 5 mg/mL to obtain a good signal to noise ratio.

A

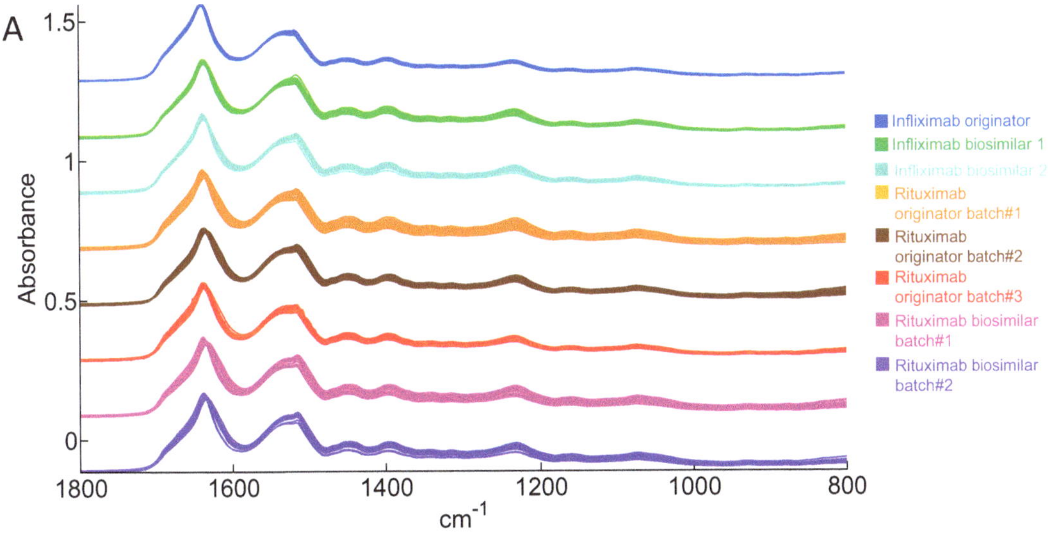


B

| Sample name | Mass ratio (%) | | | | | | |
|---|---|---|---|---|---|---|---|
| | FA2 | FA2G1 | FA2G2 | M5 | High mannose (M6, M7, M8) | Sialylated glycans | α-Gal- Gal-containing glycans |
| Infliximab originator | 44.30 | 27.67 | 4.09 | 2.26 | 0.00 | 6.07 | 2.37 |
| Infliximab biosimilar 1 | 39.82 | 40.40 | 6.26 | 2.78 | 0.00 | 8.40 | 0.59 |
| Infliximab biosimilar 2 | 36.63 | 42.00 | 6.89 | 2.44 | 0.00 | 10.04 | 0.75 |
| Rituximab originator batch#1 | 33.65 | 47.11 | 12.16 | 0.78 | 0.55 | 3.64 | 0.00 |
| Rituximab originator batch#2 | 39.88 | 41.81 | 9.43 | 1.31 | 1.70 | 3.16 | 0.00 |
| Rituximab originator batch#3 | 40.29 | 41.13 | 9.36 | 1.28 | 1.32 | 3.65 | 0.00 |
| Rituximab biosimilar batch#1 | 42.42 | 41.09 | 8.11 | 2.29 | 0.97 | 2.60 | 0.00 |
| Rituximab biosimilar batch#2 | 41.15 | 41.81 | 8.54 | 2.27 | 0.99 | 2.51 | 0.00 |

**Fig. 4** Examples of FTIR spectra recorded for eight batches (originators and biosimilars) of two mAbs (rituximab and infliximab) with similar but not identical glycan composition. (**a**) The raw data (before the processing steps) are presented between 1800 and 900 $cm^{-1}$ in this figure. Twelve to eighteen spectra of each sample have been recorded. Each protein is identified by a unique color. Spectra have been offset for better readability. (**b**) Mass percentages (calculated using relative peak areas) of major *N*-glycans or *N*-glycan types for each mAb. Relative peak areas were derived from UPLC-FLR-MS analysis of the *N*-glycans released, labeled, and purified using the GlycoWorks RapiFluor-MS *N*-glycan kit from Waters. MS data were used to identify *N*-glycans and FLR data was used for the relative quantification

3. The sample preparation is very limited: only a desalting step is required. Nevertheless, this step is crucial as many salts, detergent and buffers interfere with the measurements. Most phosphate and sugar-containing excipients have strong absorption in the spectral region related to glycosylation. This can be realized with desalting columns. NaCl 0.9% aqueous solution is used to maintain ionic strength. NaCl has no absorption band in FTIR spectra. The standard procedure of Micro-Biospin™ should be followed and repeated three times to ensure complete elimination of potential interfering components.
4. The recording parameters should be appropriate for proteins or biological measurements. Resolution and number of scans impact the signal-to-noise ratio. To obtain the spectra

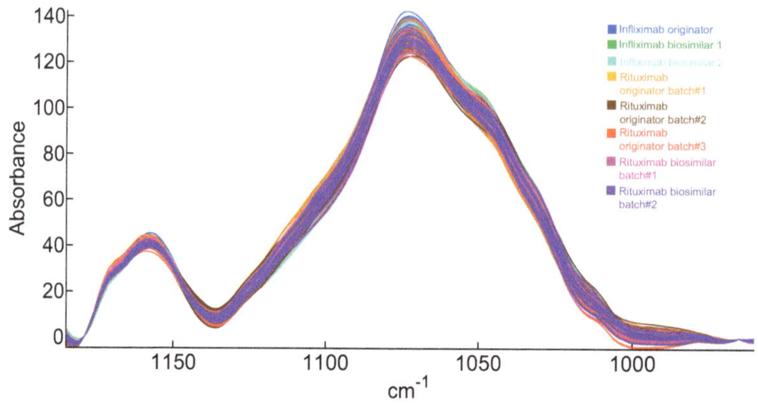


**Fig. 5** Processed spectra as explained above and rescaled between 1185 and 960 cm$^{-1}$. Each color corresponds to one specific protein. The shape of the spectra are very similar and multivariate and statistical tools are required to highlight differences among antibody glycosylation in this set of data

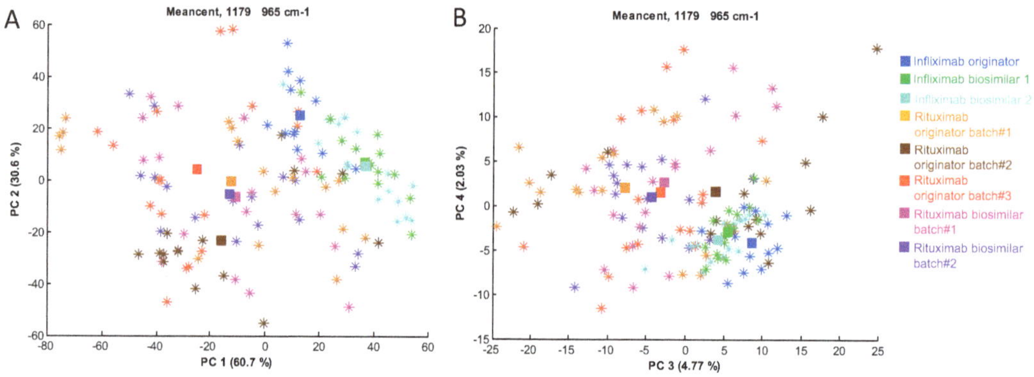


**Fig. 6** PCA score plots of the 128 individual preprocessed FTIR spectra (for the eight batches of the two mAbs mentioned above) in the PC1-PC2 space (**a**) and in the PC3-PC4 space (**b**) performed on the 1179–965 cm$^{-1}$ spectral region. Each star represents one spectrum. For the sake of clarity, a color is associated with each sample. Percentages on the axis labels indicate the variance described by PC1 (60.7%), PC2 (30.6%), PC3 (4.77%), and PC4 (2.03%). A mean centering was applied prior to computing the PCA. The square is the position of the mean of each sample. Even if the separation is not perfectly clear, all the spectra of rituximab (originator and biosimilar) are on one side of each plot and all infliximab batches (originator and biosimilar) are on the other side of the plot. While the representation provided in figure reveals significant degrees of separation between the antibodies along the first 4 PCs, considering them all together requires use of MANOVA (*see* Table 1)

presented in the figures of this protocol, the following parameters have been employed using a Tensor 27 spectrometer (Bruker).

(a) Recording:
   - Resolution: 2 cm$^{-1}$.
   - Number of scans: 128.
   - Spectral range: 4000–800 cm$^{-1}$.

**Table 1**
***p*-values calculated by multivariate analysis of variance (MANOVA) of the first 4 principal components between each pair of monoclonal antibodies. Principal components are computed between 1179 and 965 $cm^{-1}$. The scores of the first 4 components of all the 12 to 18 spectra for each antibody were submitted to MANOVA. When comparing the two mAbs (rituximab and infliximab), the differences are particularly clear and all the *p*-values are lower than 0.001 (cf. orange cells in the table). In contrast, when comparing rituximab batches, *p*-values are generally higher than 0.001 (cf. blue cells in the table). Finally, when comparing the infliximab biosimilar with the originator, the *p*-values are lower than 0.001 (cf. green cells). It is, however, not the case when the two infliximab biosimilars are compared (yellow cell)**

| **P-values MANOVA (4 first PCs)** | Infliximab originator | Infliximab biosimilar 1 | Infliximab biosimilar 2 | Rituximab originator batch#1 | Rituximab originator batch#2 | Rituximab originator batch#3 | Rituximab biosimilar batch#1 | Rituximab biosimilar batch#2 |
|---|---|---|---|---|---|---|---|---|
| Infliximab originator | 1.00E+00 | | | | | | | |
| Infliximab biosimilar 1 | 3.10E-05 | 1.00E+00 | | | | | | |
| Infliximab biosimilar 2 | 5.08E-05 | 8.64E-01 | 1.00E+00 | | | | | |
| Rituximab originator batch#1 | 8.90E-14 | 3.09E-17 | 4.04E-19 | 1.00E+00 | | | | |
| Rituximab originator batch#2 | 2.61E-14 | 2.51E-18 | 1.19E-17 | 1.78E-02 | 1.00E+00 | | | |
| Rituximab originator batch#3 | 9.44E-16 | 2.87E-19 | 8.40E-20 | 6.13E-01 | 6.18E-04 | 1.00E+00 | | |
| Rituximab biosimilar batch#1 | 1.45E-13 | 1.45E-16 | 4.52E-17 | 3.91E-01 | 2.33E-01 | 3.99E-01 | 1.00E+00 | |
| Rituximab biosimilar batch#2 | 3.97E-14 | 8.88E-18 | 1.63E-18 | 4.87E-01 | 1.21E-01 | 5.39E-01 | 9.55E-01 | 1.00E+00 |

(b) Optic:
- Source setting: MIR-source.
- Beamsplitter: KBr.
- Optical filter setting: OPEN.
- Aperture setting: 6 mm.
- Detector setting: MCT.
- Scanner velocity: 20.0 kHz.
- Sample signal gain: Automatic.
- Background signal gain: Automatic.

(c) Acquisition:
- Wanted high frequency limit: 8000 $cm^{-1}$.
- Wanted low frequency limit: 0 $cm^{-1}$.
- Low pass filter: Automatic.
- Acquisition mode: Double sided; Forward-backward.
- Correlation mode: ON.

(d) FT:
- Phase resolution: 16.
- Phase correction mode: Mertz.
- Apodization function: Norton-Beer, Medium.
- Zerofilling factor: 2.

5. A background is a FTIR spectrum with nothing on the crystal. It is automatically subtracted from any subsequent FTIR measurement.
6. Elimination of the water molecules prevents overlapping of the large water absorption peaks with the sample absorption spectrum.
7. Due to buffer composition, we often notice that a significant increase in hydration water band (3000–3500 $cm^{-1}$) upon nitrogen flow removal. We thus suggest to leave a soft nitrogen flow on the sample during the acquisition of the spectrum.
8. A normalization step is required as the sample thickness on the ATR crystal is not identical for all measurements. This is critical to compare sample in terms of chemical composition.
9. For each protein, the intact mass was determined using MALDI-TOF MS. The mass related to glycan is calculated by subtracting the theoretical mass of the amino acid sequence from the intact mass obtained by MALDI-TOF MS. We assume that no other post-translational modification is present. The global level of glycosylation is the ratio between the mass of glycans and the intact mass.

10. MALDI-TOF MS or other reference measurements are not required. They are presented here in order to confirm the observations made on FTIR spectra.
11. In FTIR spectra, each wavenumber is a variable. With more than 200 wavenumbers associated with the absorption of glycosylation and with 16–18 spectra for each sample, the number of variables submitted to statistical analysis quickly becomes extremely large. Data are best handled after Principal Component Analysis (PCA), which is an unsupervised multivariate method that enables a reduction of variables by building linear combinations of wavenumbers that vary together [14]. Diagonalization of the covariance matrix of the data provides new variables, the so-called principal components (PC) holding all the correlated original variables on which original spectra are finally projected. The first principal component accounts for most of the variance present in the data set; the second is built with the residual variance and is uncorrelated to the first one. The subsequent components are constructed in the same way and account for the residual variance. In practice, almost all the variance of the original data can be explained with the four to six first PC, reducing the description of each spectrum to four to six numbers. Simultaneously, the weight of each PC in a spectrum characterizes a spectrum and allows an unsupervised classification of the spectra as such an observation does not suppose any a priori information on these groups [14]. In the analyses reported here, the collection of spectra was mean-centered (the mean was removed from the individual spectra). Whether a group is significantly different from the others was tested by multivariate analysis of variance (MANOVA) where each spectrum is identified by its projection on the first 4 principal components [15].

## References


1. Stuart B (2004) Introduction. In: Infrared spectroscopy: fundamentals and applications, pp 1–13
2. Martin I, Goormaghtigh E, Ruysschaert J-M (2003) Attenuated total reflection IR spectroscopy as a tool to investigate the orientation and tertiary structure changes in fusion proteins. Biochim Biophys Acta 1614:97–103
3. Vigano C, Manciu L, Buyse F, Goormaghtigh E, Ruysschaert J-M (2000) Attenuated total reflection IR spectroscopy as a tool to investigate the structure, orientation and tertiary structure changes in peptides and membrane proteins. Biopolymers 55:373–380
4. Ruysschaert J-M, Raussens V (2018) ATR-FTIR analysis of amyloid proteins. Methods Mol Biol 1777:69–81
5. Goormaghtigh E, Cabiaux V, Ruysschaert JM (1994) Determination of soluble and membrane protein structure by Fourier transform infrared spectroscopy. I. Assignments and model compounds. Subcell Biochem 23:329–362
6. Groß PC, Zeppezauer M (2010) Infrared spectroscopy for biopharmaceutical protein analysis. J Pharm Biomed Anal 53:29–36
7. Lichtenberg-Kraag B, Hedtke C, Bienefeld K (2002) Infrared spectroscopy in routine quality analysis of honey. Apidologie 33:327–337
8. Anjos O, Campos MG, Ruiz PC, Antunes P (2015) Application of FTIR-ATR spectroscopy to the quantification of sugar in honey. Food Chem 169:218–223

9. Girard J-M, Deschênes J-S, Tremblay R, Gagnon J (2013) FT-IR/ATR univariate and multivariate calibration models for in situ monitoring of sugars in complex microalgal culture media. Bioresour Technol 144:664–668
10. Wiercigroch E, Szafraniec E, Czamara K, Pacia MZ, Majzner K, Kochan K, Kaczor A, Baranska M, Malek K (2017) Raman and infrared spectroscopy of carbohydrates: a review. Spectrochim Acta A Mol Biomol Spectrosc 185:317–335
11. Khajehpour M, Dashnau JL, Vanderkooi JM (2006) Infrared spectroscopy used to evaluate glycosylation of proteins. Anal Biochem 348:40–48
12. Derenne A, Derfoufi K-M, Cowper B, Delporte C, Goormaghtigh E (2020) FTIR spectroscopy as an analytical tool to compare glycosylation in therapeutic monoclonal antibodies. Anal Chim Acta 1112:62–71
13. De Meutter J, Vandenameele J, Matagne A, Goormaghtigh E (2017) Infrared imaging of high density protein arrays. Analyst 142:1371–1380
14. Johnson RA, Wichern DW (1998) Principal Components. In: Applied multivariate statistical analysis. Prentice Hall, Upper Saddle River, pp 458–513
15. Johnson HE, Lloyd AJ, Mur LAJ, Smith AR, Causton DR (2007) The application of MANOVA to analyse *Arabidopsis thaliana* metabolomic data from factorially designed experiments. Metabolomics 3:517–530

# Chapter 26

# Deploying Mass Spectrometric Data Analysis in the Amazon AWS Cloud Computing Environment

Jonathan E. Katz

## Abstract

There are many advantages for deploying a mass spectrometry workflow to the cloud. While "cloud computing" can have many meanings, in this case, I am simply referring to a virtual computer that is remotely accessible over the Internet. This "computer" can have as many or few resources (CPU, RAM, disk space, etc.) as your demands require and those resources can be changed as you need without requiring complete reinstalls. Systems can be easily "checkpointed" and restored. I will describe how to deploy virtualized, remotely accessible computers on which you can perform your basic mass spectrometry data analysis. This use is a quite restricted microcosm of what is available under the umbrella of "cloud computing" but it is also the (useful!) niche use for which straightforward how-to documentation is lacking.

This chapter is intended for people with little or no experience in creating cloud computing instances. Executing the steps in this chapter, will empower you to instantiate a computer with the performance of your choosing with preconfigured software already installed using the Amazon Web Service (AWS) suite of tools. You can use this for use cases that span when you need limited access to high end computing thru when you give your collaborators access to preconfigured computers to look at their data.

**Key words** Cloud computing, Virtual computers, Mass spectrometry, AWS, EC2, Systems management

## 1 Introduction

For the academic or other smaller scale mass spectrometry laboratory, it is quite typical that the control and analysis computer are one in the same. This is a suboptimal workflow. Data analysis during acquisition runs the increased risk of poorly timed computer crashes. User access to control instrumentation invites unintended configuration management issues. Similarly, analysis tasks run the range of demands for computational resources——a simple view of the TIC and manual interrogation of some spectra requires considerably less than what is required when performing proteomic searches or metabolomic feature extraction. Further, which workflows are being utilized from day to day can vary tremendously——

Arnaud Delobel (ed.), *Mass Spectrometry of Glycoproteins: Methods and Protocols*, Methods in Molecular Biology, vol. 2271, https://doi.org/10.1007/978-1-0716-1241-5_26, 

for those that facilitate users this can create a burdensome administration load making sure the right computers with the right resources installed with the right software are available to the myriad of users and collaborators that want to see their data.

Currently, there are several computation resource vendors that provide Internet based access to user-defined remote-accessible virtual computers. In this example, I will focus on solutions provided by Amazon thru their Amazon Web Services (AWS) offering, however similar solutions exist from Google, Oracle, and so on. As alluded above, the range of resources offered by these providers runs from specialized storage and database applications, thru load balancing distributed processing and massive data analytics. My goals for this document naturally limit my discussions to those most commonly used to augment the mass spectrometry data analysis workflow.

In this chapter, I describe how to create a virtual computer using Amazon's EC2 service. I describe virtual hard drives that you can easily attach/detach from your new virtual computer allowing for easy backups/recovery and cloning. I provide a limited discussion on security and, finally, I provide some sample use cases.

In this case, the virtual computer will be physical hardware in Amazon's computational centers that is partitioned and configured, at your guidance, to appear as any other network attached computer. Typically, usage charges will be quoted on an hourly basis but prorated to the nearest second—Amazon does have a free tier which you can use to explore the features of cloud computing. This tier is only available for 1 year on an account and it does have a number of limitations (machine configurations, etc.). The charges will vary based on the resources that you are using. More memory, more powerful CPUs, more disk space—these will all increase your hourly charges. Depending on your workflow, there can be a significant cost difference between when the computer is "turned-on" and when it is off. While you will continue to be billed for hard drives (storage) you are using, you will no longer be billed for hourly CPU costs when the computer is shutdown. As an example, as of this writing, using Amazon's Cost Calculator [1], a t3.medium instance (Intel Xeon Platinum 8000 series processor, 2 cores, 4 GB RAM) with a 30 GB SSD "primary" drive and a 500 GB magnetic "data" drive will cost ~$70/month. Of this cost, ~$3/month are for the 30 GB SSD Drive, ~$22.50/month is for the 500 GB magnetic drive, and ~$44 is for the CPU ($30) and Microsoft software licenses ($14). Thus, when the computer is "off" the CPU and licensing costs stop and you are only being billed $25.50 for the persistent data storage.

While I, for the most part, will not write about configuration optimization, it is worth noting that the above described a la carte approach to configuration means that you can choose an expensive CPU with lots of memory when doing your proteomic search but then shutting down your computer and relaunching it with less resources when you are browsing your results. Additionally, I will not be talking about licensing enforcement methodologies. For commercial software that uses network based digital restriction management this is often transparent. For commercial software that uses USB dongles or network card "MAC address" based restrictions, these can be less trivial to overcome.

Here is the general outline we will follow.

1. Get an AWS account.
2. Select and instantiate your first cloud computer.
3. Install software you will need to perform analysis.
4. Copy over your data.
5. Enjoy your new cloud computer as we explore advanced workflows such as backups and data sharing.

## 2 Materials

1. *Payment Method*. By far the easiest method to pay for AWS services is via major credit card. Additional payment methods are listed on Amazon's AWS website [2].
2. *Computer/Network Connection*. Your computer will be acting as a terminal into a remote computer. The computational "heavy lifting" is being performed by your AWS resources and the RDP protocol is very efficient (*see* **Note 1**). Unfortunately, there is no way to virtualize screen resolution—I recommend that any computer you use have at least a true screen resolution of 1920 × 1080 or as your mass spectrometry software suggests.
3. *Web Browser*. To access and configure AWS resources.
4. *RDP Client*. RDP is the protocol that you will need to support in order to connect to your AWS Windows instances. There are many options available depending on your operating system, here are the ones I most commonly use:
   (a) Mac/Windows: **Microsoft Remote Desktop** client.
   (b) Chromebook/Android: Microsoft's **Remote Desktop (RD Client)** client.
   (c) Linux: **remmina**.

## 3 Methods

### 3.1 Creation of an Amazon AWS Account

Create an Amazon AWS account (or log in if you have one). You may have an amazon.com account for retail purchases, this is a different account that is created to manage AWS services.

1. Navigate your browser to https://console.aws.amazon.com/. If you already have an account, proceed to Subheading 3.2, otherwise continue to create an account.
2. After you enter your **email** and **password** (twice to prove you can), there is an option to add an **AWS Account Name**. This optional account name will be an alias for your 12-digit account number. When you (or a sub-account you create) log in, either the account number or the **AWS Account Name** will need to be provided.
3. Next you are asked to provide contact information. In this dialog, there is a question if this is a **Personal** or **Professional** account. While the choice you make will change some of the requested information (most notably, the request for a *Company name*) there is no difference in features between the two accounts.
4. You will next be asked for payment information. There are alternatives beyond the scope of this document, for this example I am assuming that you will be entering credit card information.
5. Validate your contact telephone number (by SMS or call) and choose a support plan (I choose the free **Basic** plan).
6. You will now be directed back to the dialogue you originally were presented with.

### 3.2 Log into Your Account

1. At this point, (assuming you have just followed the above instructions) you do not have any named subaccounts so you will need to log in with the email and password you previously provided. Amazon refers to this as **root account credentials**.
2. Please *see* Fig. 1 for an example of the interface after you have logged in.

### 3.3 Optional Additional Account Configurations

The following two steps describe mechanisms that allow more granular control over billing and account access from individual entities. This can be done at any time, so, if unimportant for now, proceed to Subheading 3.4.

1. An **organization** allows you to have several administratively distinct AWS accounts that are tied together with shared billing and subject to organization wide policy controls. The organization managers can see the breakdown of the charges by account number which can be used for accounting or chargeback purposes. An example use case would be a chemistry department that wants individual labs to be able to manage their own access and resource deployment.
2. In the upper right menu, click the tab that matches your account name (Fig. 1, **red dashed circle**). In the pull down, select "My Organization." There will be a large dialog displaced that is asking you to "create organization." Click this.

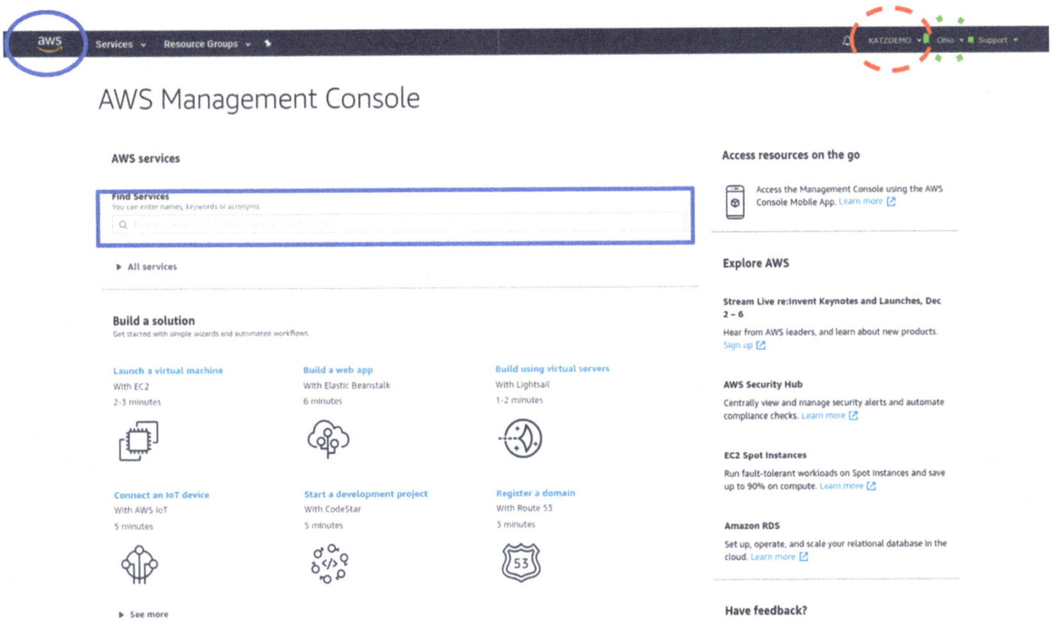


**Fig. 1** The AWS Primary Console Screen. The AWS primary console screen has several dynamic portions including the main body and the side bars. Consistent will be the title bar along the top. You can always return to this view by clicking the AWS logo in the top left indicated by the **blue solid circle**. From this screen you can quickly navigate to other functional views by searching for the appropriate subsystem in the search box indicated by the **blue solid rectangle**. Other features of interest include your account management pulldown indicated by the **red dashed circle** and your current region indicated by the **green dotted circle**

3. After you have validated your email and your billing information you will then be able to either invite existing AWS members into your organization or create new accounts within your organization.
4. These are distinct administrative groups each of which can control its own access controls, set their own policies, etc. They are subject to the global organization policies.
5. The Amazon **Identity and Access Management (IAM)** service allows you to create additional accounts (subject to differential security policies) under a single AWS account. The login credentials used to create the AWS account are the "root" credentials for that account. It is a recommended best practice to only use the root credentials for specific administrative tasks—for general use, it is recommended to use IAM to create an alternate login to your AWS account. These IAM logins can be governed by defined policies on a group or individual basis. An example use case would be a chemistry laboratory that wants individual research scientists to be able to instantiate their own compute resources subject to the policies of the laboratories AWS account manager.

6. If you are not already logged in, navigate to https://console.aws.amazon.com/ and log in with your AWS root credentials (the email/password you used to create the AWS account).
7. If you have not already done so, you will need to enable IAM access to billing data. This will be required for the creation of IAM accounts with administrative access.
8. In the top right of the menu bar is your account name (Fig. 1, **red dashed circle**), click on that and in the pull-down menu, choose **My Account**.
9. In the plethoric catalog of choices, about half-way down, will be a section titled **IAM User and Role Access to Billing Information**. Click **edit**.
10. Click the checkbox to enable access and then click **Update**.
11. Now navigate to https://console.aws.amazon.com/iam/.
12. In the center dialog box, click the inverted chevron associated with "**Create individual IAM users**" then click **manage users** (*see* Fig. 2).
13. Click on the "**add user**" button. You will be asked to provide account authentication information. Commonly, and especially if this is your first **IAM** user on this account, you will probably want to select the option for **Access Type** as AWS **Management Console Access**.

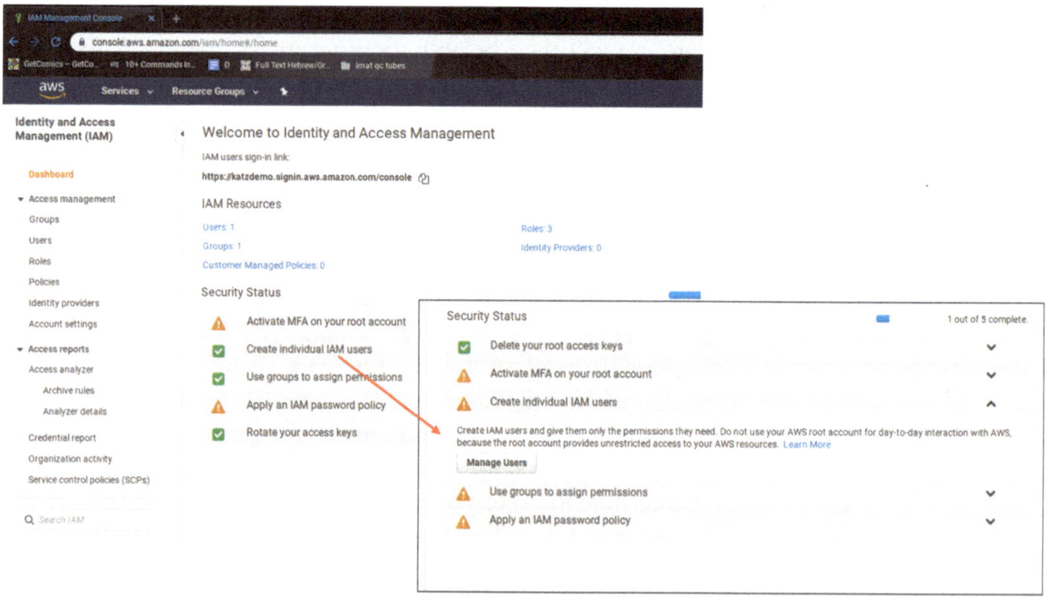


**Fig. 2** Adding IAM Users. From the main IAM console, clicking on the chevron next to the header Create individual IAM users will prompt you with a dialogue asking if you want to manage users

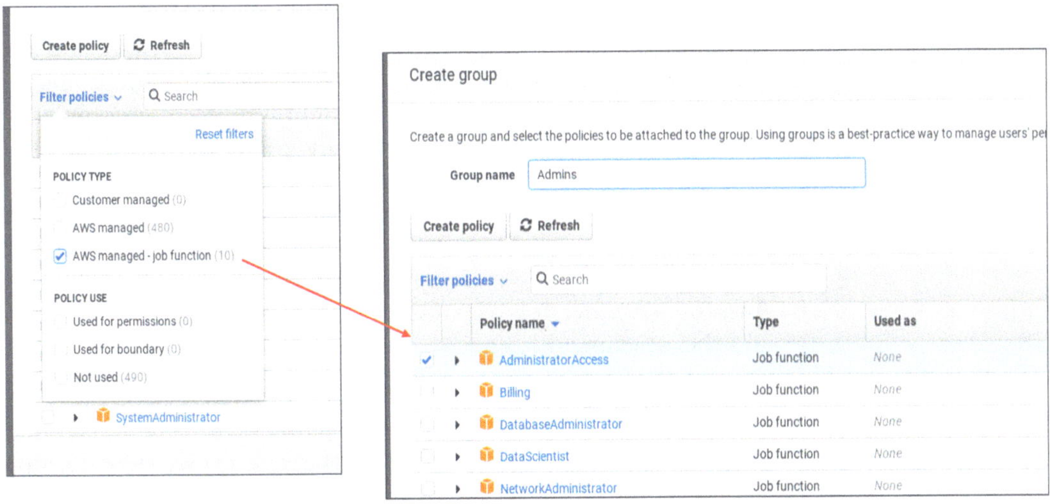


**Fig. 3** Configuring permission credential groups for IAM users. When you create a permission group for IAM users, there is an exhaustively detailed collection of individual permissions that can be allowed or denied. To simplify administrative choices, common collections of permissions are group together. Show is the selection of AWS managed policies by job function and then they selection of the AdministratorAccess job. This particular selection will create a group with full permissions

14. You will now be asked to set the permissions for the user. The default case creates a user with no permissions at all which can be stifling. We will create our first account with full control. Start by clicking **Create Group**.
15. If you click on the chevron next to **Filter Polices** a small dialog will open, from there select **AWS managed - job function**. *See* Fig. 3.
16. There will now be about 10 listings. Select the checkbox next to **AdministratorAccess**, add a name in the text box next to **Group name** and then click on the **Create group** button. You will want to sure the group you have created is selected before proceeding. As you learn EC2 you will want to explore restrictions you may want to put in place.
17. Next step asks if you want to assign **tags** to the account. Tags are for your benefit and can be used for internal auditing or clustering of your accounts. It is ok to leave these blank.
18. Finally confirm all your settings and your new **IAM** account has been created.
19. Before leaving the **IAM** console note that near the top there is a direct link to log into your account. There is an option to the right of that link that allows you to **customize** that link. It is an alias for the IAM login.
20. Try logging out of your **root credential** account and try logging into your **IAM** account. Navigate to https://console.aws.amazon.com/ and enter your account ID or account alias to continue to the IAM login.

### 3.4 Create Your First Data Drive

With hard drives, there is an obvious logical divide between the drive one uses for operating system and installed software and the drive one uses for actual "data" files on which you are going to operate. For a lot of workflows, you might want to have completely different operating system and software configurations, but, typically, your data will remain a constant. In keeping with this mindset, while, perhaps, less intuitive, I will first describe the creation of a virtual hard drive (*see* **Note 2**) to hold your data rather than on the OS drive. Note that creating additional storage is not part of Amazon's free tier, it is ok to skip this step and perform the creation and the attachment later.

1. An easy way to navigate the AWS console is to click on the **AWS** logo in the upper left of the title bar (Fig. 1, **blue circle**) and then type the desired service in the search box (Fig. 1, **blue rectangle**). In this case, we are going to type **EC2** and press enter.
2. In the upper right of the title bar, a location ("availability zone," Fig. 1, **dotted green circle**) will be listed. When you provision services they will be instantiated in a data center in that locale. There are some optimization choices that can be made here (e.g., some services are not available in all locales and the same service can be priced *modestly* differently in different locales), but a good rule of thumb is to choose a location that is closest to your physical location—Amazon has not quite solved the "speed of light" issue and so there will be an imperceptible timing difference if you choose a more distal data center. What *is* important is that your data drives and your virtual computers are created in the same availability zones (i.e., the same functional portion of the data center—I will be walking through this).
3. There is a menu on the left-hand side of your screen. In that menu, there is a section **Elastic Block Storage**. Under that heading, select **Volumes** (think "virtual hard drive").
4. Click on the button **Create Volume.**
5. You will be presented with several options; let us walk through a few (which I have also reflected in Fig. 4).
6. For a data drive, a spinning physical disk will be cheaper for the same amount of storage but have less performance than an SSD drive. I typically select **Throughput optimized HDD**.
7. Now set the size at **500** GiB (feel free to go higher or lower as your needs require.
8. **Important:** note your availability zone .... I choose **us-east-1a** for this example, but that does not matter .... What *does* matter is making sure you choose the same zone when creating your virtual computer.

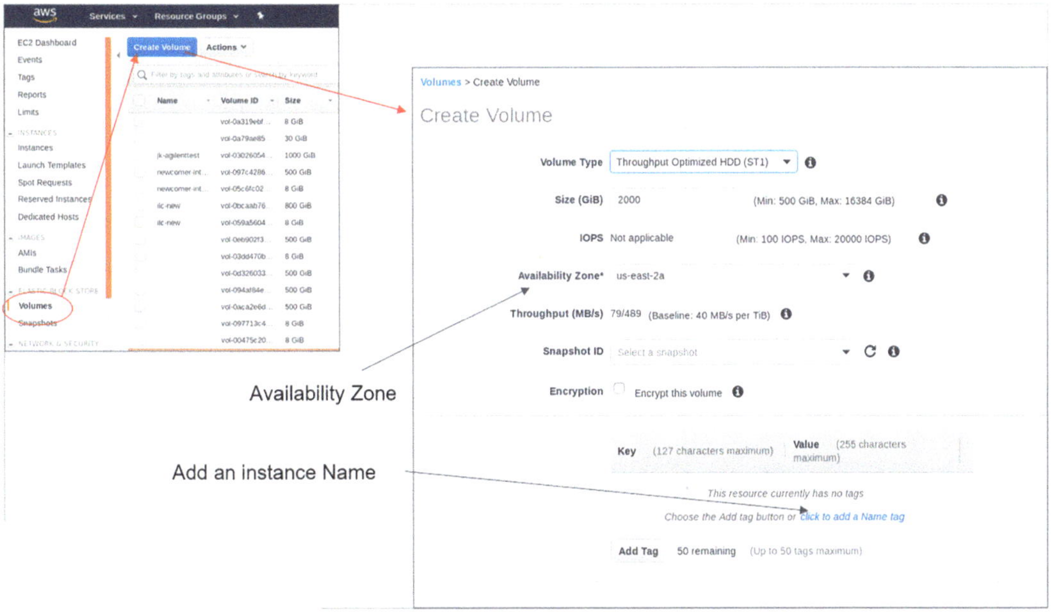


**Fig. 4** Creating an EBS Data Volume. This figure shows key features of the AWS user interface with which you will interact when creating an EBS data volume. Please note the locations where you select the availability zone as well as where you can provide a name to your virtual hard drive

9. **Snapshot ID** is useful, but not for your first hard drive. In the future, you might create a data drive filled with lots of good resources. If you take a snapshot of that drive, you can then create new "hard drives" based on these snapshots. I will be going through this workflow a bit later.
10. **Key/Value** pairs are arbitrary tags that you can use to organize your volumes. The same mechanism is used for your virtual computers. I *strongly* urge you to name all of your virtual hard drives and computers. If you create a **Name** key with a value descriptive to you, that name will appear in the listings of resources. It is so useful to add this tag that you will note, below the list of entries, a link that says **"click here to add a Name tag"** prompting you to do so if you have not already. For this example, I added the tag "MS Data Disk 001" and then clicked **Create Volume** for which I was rewarded with a "Volume created successfully" notice and a volume ID which I did not bother recording since I remembered to add a name tag. The drive you created will appear as unformatted when you attach it to your virtual computer. This is easy to remedy and we will walk through this later, but if you do not recognize that the drive starts as unformatted, it can be difficult to troubleshoot why your new drive is not working.
11. You can now click the **Close** button. You will now be back at the "volumes" screen and should see your new volume. Note

that the name tag content appears in the name column for the volume. If you had not named the volume this can get quite confusing when you have more than one of them.

12. To delete your volume (volumes cost money!), make sure you are on the volumes management screen (if not: click **AWS** in the title bar (Fig. 1, **blue circle**), search **EC2** (Fig. 1, **blue rectangle**), select **Volumes** under the **Elastic Block Storage** of the left menu). Now select your volume and select **Actions** → **Detach Volume** (assume it is attached to an instance), then select **Actions** → **Delete Volume**. Please note that this is an irrevocable action. Note that any data on the volume reflected in a snapshot will be retained, so this will not reduce your storage costs unless you delete snapshots you created as well.

### 3.5 Creation of Computer Credentials

Beyond your AWS login credentials, you need to create credentials to log into the computer instances you create. AWS credentials allow you to log into the AWS control panel. Computer credentials allow you to log into the computer you created. AWS delivers computer login credentials to you via a public/private key mechanism. This workflow is shown in Fig. 5.

1. If not already on the EC2 page, click the **AWS** logo in the title bar (Fig. 1, **blue circle**), then search for **EC2** (Fig. 1, **blue rectangle**), then on the left hand side menu, scroll down until you hit the **Network and Security** section, in there, click **Key Pairs**.
2. Click **Create Key Pair**. Name your pair informatively like "keypair1" then click **Create.** Your browser will automatically download a file that matches the key pair you just created—in this case, "keypair1.pem" should have been downloaded.
3. Now, when you create your compute instances, or other resources that require authentication you can reference this keypair (*see* **Note 3**).

### 3.6 "Provision" Your First Compute Instance

You will now "provision" your first compute instance. This workflow is shown in Fig. 6.

Click on the **Launch Instance** button, if you do not see the **Launch Instance** button, click on the **AWS** in the left side of the title bar (Fig. 1, **blue circle**), search for **EC2** (Fig. 1, **blue rectangle**), press enter. You will either see the **Launch Instance** button, or you can click the **Instances** item in the menu that is on along the left side of your dashboard (Fig. 6).

1. After you have started the launch dialog, your first choice will now be which operating system you want to have "installed" on your new instance. I have had success with Microsoft **Windows Server 2008 R2 Base** (64bit). All the Microsoft

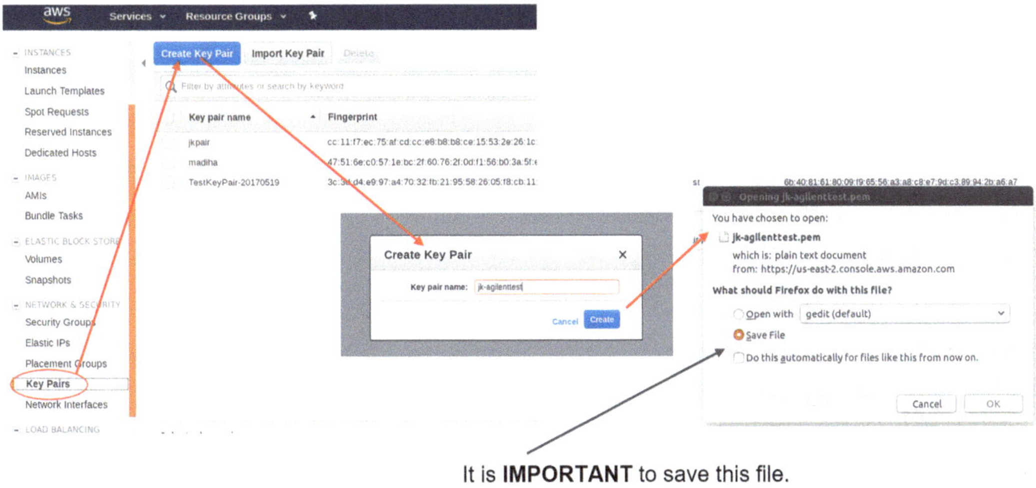


**Fig. 5** Creating a Key Pair. This figure shows key features of the AWS user interface with which you will interact when creating a Key Pair. It is important that you save the produced .pem file (your file save dialogue may be different than that shown)

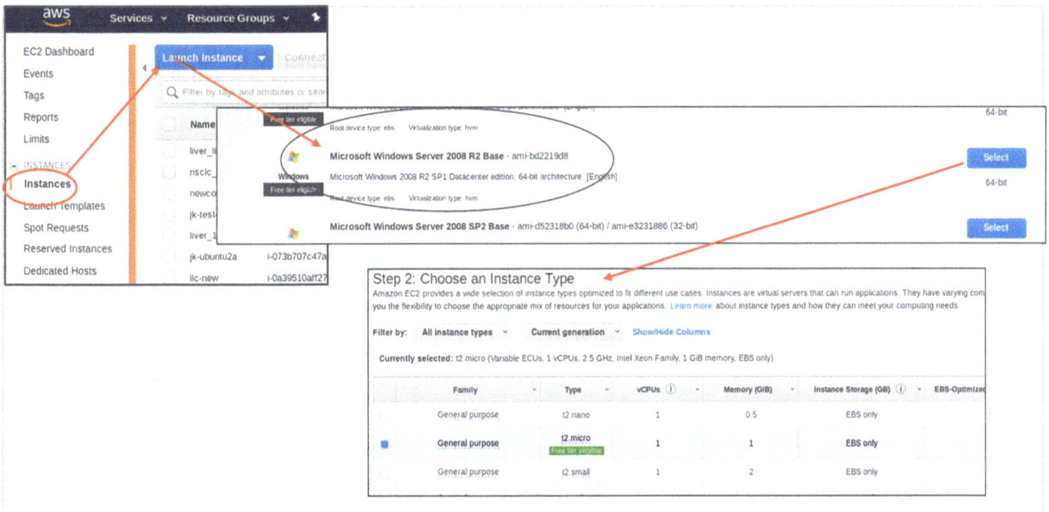


**Fig. 6** Instantiating your Compute Instance - Feature Selection. This figure shows key features of the AWS user interface with which you will interact when selecting the operating system and compute hardware for your compute instance

instances will be of the server variety, this one does have the feel of Windows 7 which works well with many of the mass spectrometry software needs (which tend to be a generation behind in OS). If you need Windows 10, you can select an appropriate alternative.

2. Your next choice will be the hardware resources available to your instance. For this example, I will choose **t3.large** with its 2 vCPUs and 8 GB of ram; you should choose a configuration

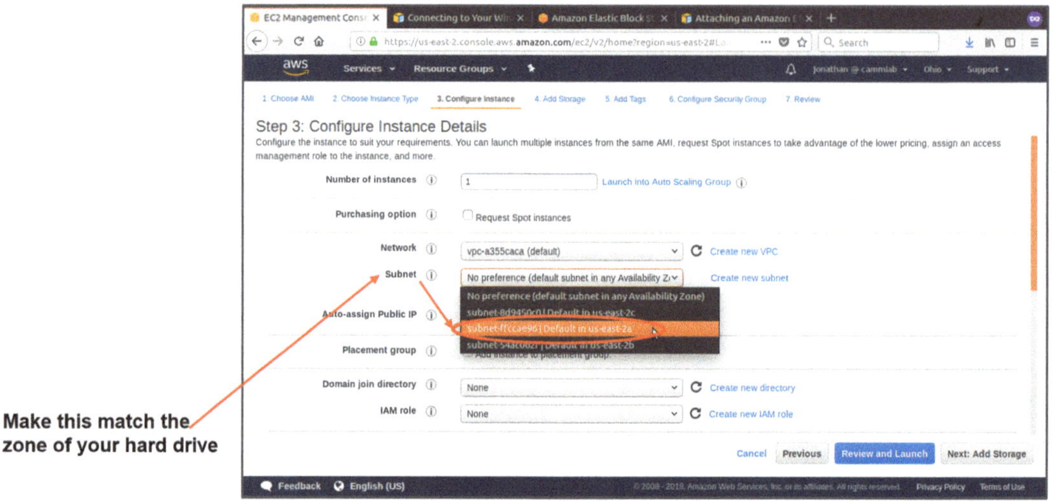


**Fig. 7** Instantiating your Compute Instance—Availability Zone Selection. In step 3, make sure that you choose the subnet that matches the Availability Zone of any storage resources that this instance will need to connect to

that matches your needs. Please *see* **Note 4**. After selecting, click **Next: Configure Instance Details**.

3. On the instance details page, for your purposes, the only crucial parameter is the **Subnet**, click on that and choose the availability zone that matches the data drive you created above (*see* Fig. 7). I will choose **us-east-1a** to match the choice I made for the hard drive storage. Do not feel compelled to choose this same zone—choose the one near you—just make sure it matches the zone in which your data drive resides. Now click **Next: add storage**.
4. You are now asked to create the initial drives for your instance. I tend to think of these as the operating system volumes. You are creating the C: drive which will hold the Windows operating system and probably and software you install. I will be choosing the default **30 GB SSD** as my option and then clicking **Next: Add Tags**. By default, this drive will have the same name as the “instance name” which means if you do not name your instance this will be blank. Regarding size, later on I will show you how to expand the size of your OS volume, however, if you know ahead of time that you will need more storage, feel free to select that now. Realistically, I typically run my root volume as **100 GB**.
5. As with the hard drive storage, Key/Value pairs are arbitrary tags that you can use to organize your compute instances. Click on “**click here to add a Name tag**” prompting you to do so if you have not already. For this example, I added the tag “**MS_Mach_001**” and then clicked **Next: Configure Security Group**.

6. Feel free to just click **Review and Launch**. For **step 6**, you are asked to create the security group. Imagine that you are configuring a virtual firewall. The default is all outgoing connections and incoming RDP connections are allowed. If you wish to explore this, for example, to restrict access to specific machines or to allow your computer to act as a web server, please *see* **Note 5**.
7. Finally, you will be asked to review your choices and click **Launch**. At this point, you will be prompted for they keypair you want to use. *Make sure you have a copy of the .pem file for the keypair you select.* You can also choose to create a new key pair. Acknowledge you have access to the .pem file and click **Launch Instances**.
8. You will be told the instance is being created and invited to click **View Instances**.
9. To delete your instance (instances cost money!), make sure you are on the instances management screen (if not: click **AWS** in the title bar, search **EC2**, select **Instances** under the **Instances** section of the left menu). Now select your instance and select **Actions** → **Instance State** → **Terminate**. Please note that this is an irrevocable action. Also note that hard drives that were created as part of creating this instance (e.g., the operating system hard drive) will also be destroyed. So if you want to preserve that for some reason, you should make a snapshot first.

### 3.7 Connecting to Your New Cloud Computer

Based on the default configuration of your new computer and the default security rules, you can only connect to your new Windows computer using the RDP protocol. You will also need the .pem file associated with the keypair used to create the instance.

1. If you are not on the page showing your instances, click **AWS** in the title bar, then search for **EC2** then, on the left menu, click **Instances** in the **Instances** section.
2. Click the checkbox next to your instance and then click the box labeled **Connect**.
3. You will now be presented with two options (Fig. 8). You are prompted to get an easy to use RDP file, and you are prompted to get the administrator password for your instance. Also, you are informed of the IP name/address of your new compute instance. If your RDP client can use the RDP file, that is great. If not, just write down the IP name of your computer. Now click **Get Password**.
4. You will now be asked for the .pem file that is associated with your key pair. **Choose File** allows you to select that file, then you can **Decrypt Password** to get the login password for your

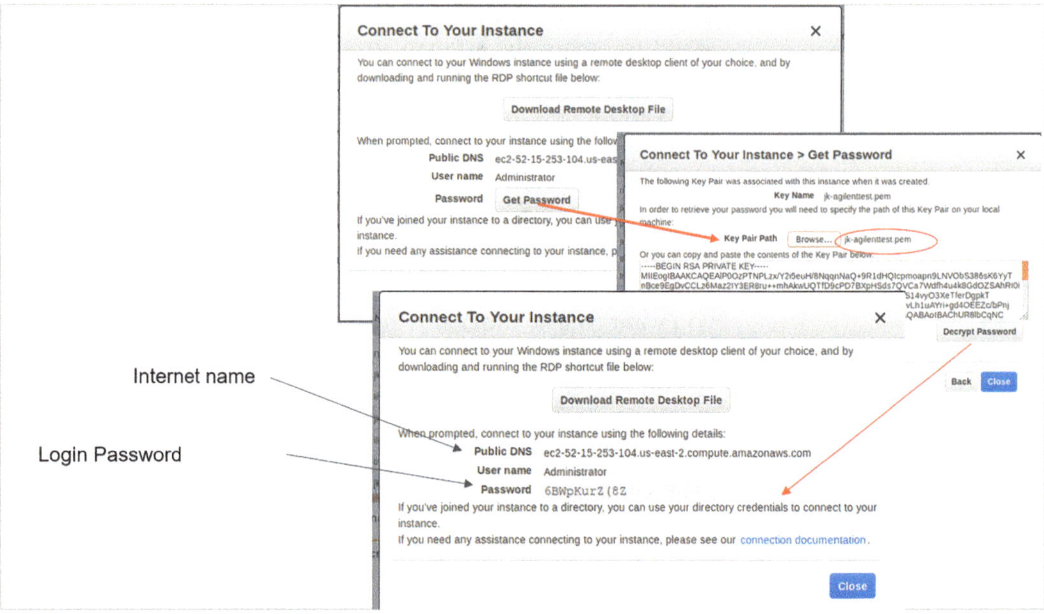


**Fig. 8** Connecting to your Instance. You will be using an RDP client to connect to your instance. In order to do this, you will need the IP hostname and the administrator password. This exemplar shows what that dialogue looks like and where that information may be obtained

new compute instance (Fig. 8). If your RDP client supports this, not that you can copy the password to the clipboard by clicking the icon to the right of the displayed password.

5. Open your RDP client and choose to connect to the IP address/name associated with instance. Log in as administrator and use the password you decrypted above. Woo Hoo!
6. I tend to do things at this point for my convenience and taste, like, download Firefox. Note that in your initial install, the security is set such that the Microsoft web browser will prompt you to go to any website that is not part of its trusted domains. Either lower the security setting, or just keep adding Internet addresses as prompted. Also, if you are not used to using Windows Server, you will encounter minor differences, such as the security mentioned above and be prompted for reasons when you do shutdowns.

### 3.8 Attach and Configure Your Data Drive

Note that you can attach volumes to compute instances that are not running.

1. Click on **Volumes** under the **Elastic Block Store** in the left hand menu—if it is not there, click the **AWS** logo in the title bar, search for **EC2** and then click enter. You should be presented with a menu of all your cloud based hard drives called *volumes*. If you have done nothing other than these

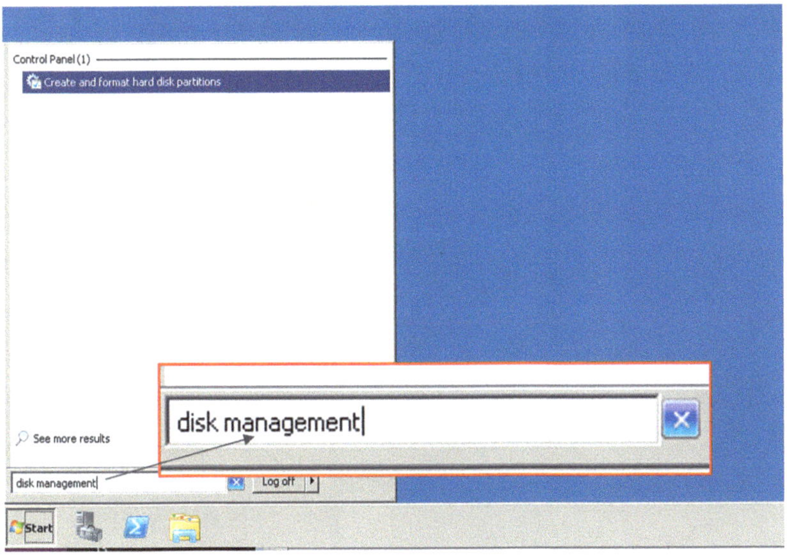


**Fig. 9** Starting the Windows Disk Management Tool. Click the Windows button in the lower left, then type "disk management" and press enter

instructions, you should have one volume named after your compute instance that is shown as "attached" and you should have the MS-Data-01 drive you created marked as "available."

2. Check the **MS-Data-01** drive, click on **Actions** and then choose to **Attach Volume**. You now see the joy of naming everything you do descriptively! If you click on the instance, you can type the name of the machine you created or select it from the dialog it prompts you with. Then click **Attach**. You will be sent back to the volume page with a prompt showing that efforts are being undertook to attach your volume to your instance. Remember, if this device has not been configured, your compute instance will treat it as an attached, unformatted hard drive. **Important:** A volume can only be attached to one instance at a time and you will have to detach a volume from an instance before you can delete it.
3. Format your virtual hard drive. **Important:** this only needs to be done *once* when the drive is first attached and before you use it. Switch over to your windows instance, click on the windows start button in the lower left corner. Type **disk management** and press enter (*see* Fig. 9). You should be presented with the windows disk management dialog which will show your functional "C" drive (probably as Disk 0) and the "Unallocated" "D" drive (probably as Disk 1).
4. We now need to initialize (format) and partition the hard drive so Window can use it. When you start disk management, it may present you with the initialization dialog immediately. If not, right-click on the "Unknown" volume and select **Initialize Disk.**

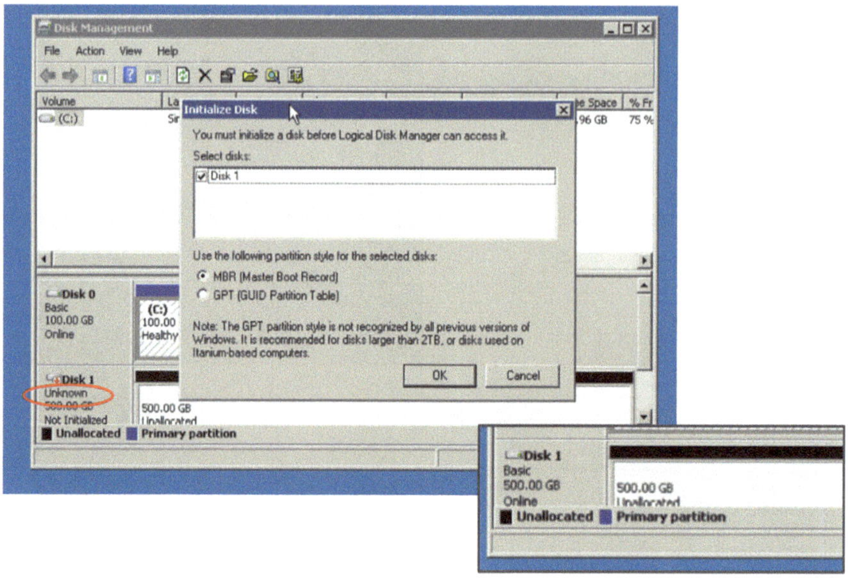


**Fig. 10** Initializing Your Storage Device. Typically, you will be presented with this dialog when you start "disk management" and there is an uninitialized storage device attached—these will be shown as "unknown" devices as shown by the red circle. If you cancel the "Initialize" dialog or otherwise want to open it, you can right-click on the "unknown" disk and select initialize

5. Select **MBR** and select ok (*see* Fig. 10); there may be cases when you should choose **GPT**, *see* **Note 6** for a discussion. The drive will switch status from "unknown" to "basic."
6. Now we will partition the drive and add filesystems. Right-click on the "Unallocated" drive and choose "**New Simple Volume. . . .**".
7. Follow the wizard choosing all default steps (unless you are wiser). You will end up with a single partition hard drive, formatted as NTFS and allocated as the D: drive (assuming that is what you assigned).
8. It is strongly recommended that you store all data and result files on this data disk. Unless you take active action, the C drive is ephemeral and will vanish if you ever terminate (vs. stop) your compute instance.

### 3.9 Transferring Data to Your New Computer (Including Vendor Software)

The general workflow that has worked for me is to use another vendor that has addressed the "data sharing" problem. Upload your data or install files from your control to desktop computer and then access them on your compute instance.

1. **Google File Stream** allows you to present your Google based storage as a separate drive device (e.g., "g:"). You can get the download here https://support.google.com/a/answer/7491144?hl=en. While this works incredibly well, I tend to

recommend that you change your cache drive from the default C: drive to your D: data drive as the C: drive will probably be too small (*see* **Note 7** for instructions).

2. **Dropbox** also works quite well. It presents your files as part of your home folder within Windows. Dropbox presents files as part of your file tree rather than as a separate device. I find this to be less to my liking with regards to keeping data/volumes and files cleanly managed.
3. At this point I will also install my mass spectrometry analysis software. Online resources (e.g., "R") I will get online. If I have install files from a vendor, I will use Dropbox or Google File-Stream to import the vendor install files.

### 3.10 Creating Snapshots

Think of a snapshot as a complete backup/clone of one of your hard drives in its current state. These can, of course, be used for data recovery, but they can also be used to build new cloned hard drives (possibly with different sizing). If you created an operating system volume when you created your instance, be aware it will be destroyed when your instance is destroyed. A snapshot of your operating system volume will allow you to clone your installation in the future as well as access data that may have been stored on that volume.

1. If you are cloning the hard drive that contains your operating system you will want to stop the instance first. In general, you only want to snapshot a volume that is not actively getting written to. For example, do not snapshot your data volume while you are doing a Mascot search.
   (a) Log into the instance, click the **windows button** and select **shutdown**.
   (b) -OR- From the AWS console, click **AWS** in the title bar, search for **EC2**, select **Instances** on the left menu in the instances section. Find your instance in the main panel (you did remember to use good names, yes?), select it, then click **Actions** → **Instance State** → **Stop**.
2. I tend to perform snapshots on the volume screen. If you are not there already, click **AWS** in the title bar, search **EC2**, select **Volumes** under the **Elastic Block Storage** of the left menu.
3. Now select the volume for which you want to create a snapshot and then click **Actions** → **Create Snapshot** (*see* **Note 8** regarding snapshot costs).
4. To Delete a snapshot, navigate to the snapshots management screen by clicking **AWS** in the title bar, search **EC2**, select **Snapshots** under the **Elastic Block Storage** of the left menu. Select the snapshot(s) you wish to delete, then click **Actions** → **Delete**.

### 3.11 Workflow Examples

In the following examples, I will describe some common tasks I have done that leverage the advantages of a workflow in the cloud putting together the processes I described above. These workflows take advantage of the fact that cloning hard drives, creating snapshots of hard drives and moving hard drives between computers of different configurations (beefy vs. not) are all *trivial* on the cloud and executed via simple mouse clicks.

#### 3.11.1 Clone Your OS Drive and Create a New Compute Instance from it

1. Shutdown your compute instance (Subheading 3.10). **Important:** Note the password required to log into this instance – in these instructions, I will tell you how you can swap out operating system "hard drives." The password administration configuration files will remain associated with the snapshot of the operating system hard drive.
2. Create a snapshot of your OS drive (Subheading 3.10). **Important:** Write down or copy the snapshot ID of the snapshot you wish to use as the basis for your cloned OS drive (and you did save the login password, from above, yes?). Note that the administration password used during the creation of this instance will be the same password you use for any instance you create in the future for which you make this the operating system drive.
3. Create a new volume (Subheading 3.4—**Important**: when you get to Subheading 3.4, **step 9**, you will choose the snapshot you created above—This is where having written down the ID or copying it to your clipboard is really useful). If your volume is larger than your snapshot, we will have to address this in Subheading 3.11.2 below.
4. Create a new instance (Subheading 3.6). Often, I will create an instance and do software installs with less CPU and RAM than is required when doing analysis. When creating the new instance, you can scale the CPU or RAM as you see fit. When you create an instance, it comes with an attached new storage device we will have to remove.
5. Now shut the instance down (Subheading 3.10, **step 1**).
6. Detach (and delete) the volumes that were created when you created your new instance (Subheading 3.4, **step 12**).
7. Attach the volume you created in Subheading 3.11.1, **step 3** above. To do this, click **AWS** in the title bar, search **EC2**, select **Volumes** under the **Elastic Block Storage**. Select the drive you want to attach as the operating system drive and select **Actions** → **Attach Volume**. Choose your stopped instance by name or id. For device.... **IMPORTANT :** Type in **/dev/sda1**. Yes, this is documented, no not in an obvious places.

8. Now restart your instance. From the instances main panel (if not there, click **AWS** in the title bar, search for **EC2**, select **Instances** on the left menu in the instances section) select the instance, then choose **Actions** → **Instance State** → **Start** and connect as normal.

*3.11.2 Increase the Size of Your OS Volume*

In following the steps of Subheading 3.11.1 above, at **step 3**, it is possible you may have created your new instance with a larger volume size than that from which you created your snapshot. If this is the case, Windows will still think your volume is the size when the snapshot was originally created (as referenced in Subheading 3.11.1, **step 3**). Here is one way to resolve this.

1. Connect to your instance (Subheading 3.7).
2. Press the windows key, type **cmd**, right-click on the "cmd" app and select "run as administrator" (*see* Fig. 11).
3. Type **diskpart** in the command shell.
4. At the "diskpart" prompt type list volume, note the volume number of the volume you want to expand. If this is the system volume, it will most likely be volume 0.
5. Type select volume 0 and, if needed, replace 0 with the actual volume you want to expand.

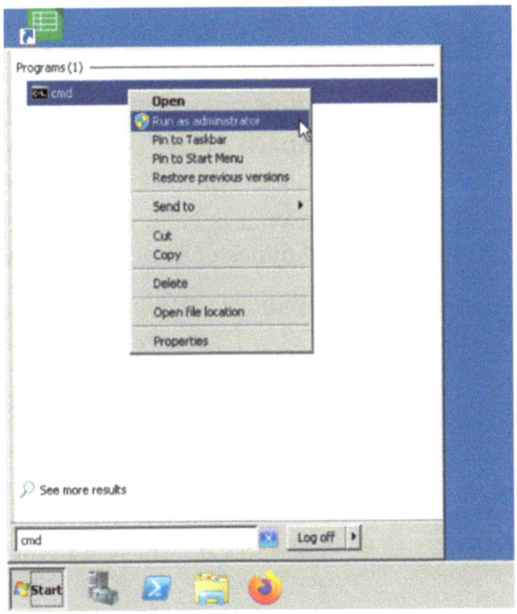


**Fig. 11** Starting an "Administrator" Command Shell. Click the Windows button in the lower left, then type "cmd", **do not** press enter, but rather right-click on the "cmd" entry and choose "Run as Administrator"

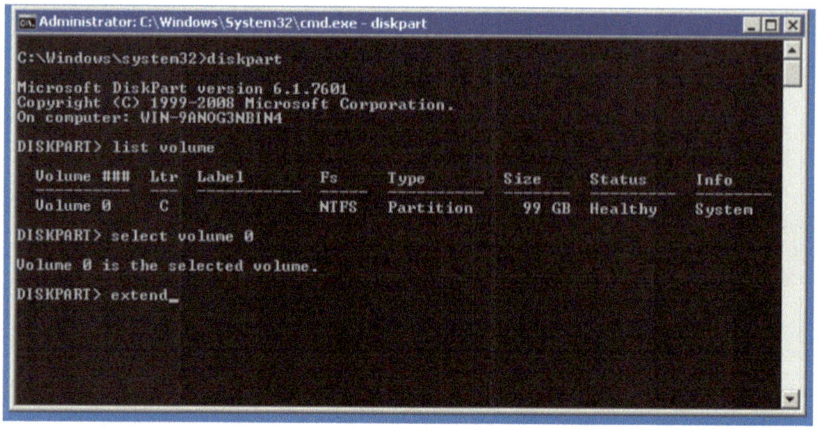


**Fig. 12** Sample DiskPart Dialogue Expanding a Volume. This exemplar shows a sample interaction with the "diskpart" command to extend a volume

6. Type **extend**. You should be rewarded with a status message that diskpart was able to successfully extend the partition size (*see* Fig. 12).
7. You can now leave diskpart by typing **exit**.

### *3.11.3 Create a Data Analysis System for a Collaborator*

If a collaborator wishes to perform some basic analysis on their data on their own this can easily be accomplished by creating a computing instance loaded with just their software and data.

1. Create an instance and install it with all the software your collaborator will need for their analysis.
2. Follow Subheading 3.11.1 to create a new instance based on a clone of your master image from the above step.
3. Attach a clone of their data drive to their instance or otherwise copy their data so that it is available on their instance.
4. If your collaborator has troubles with RDP, consider installing **Anydesk** [3] or **Teamviewer** [4] for them to use instead.
5. Important considerations.
   (a) Make sure that this "computer" you created for your collaborator has no personal information on it (e.g., account logins, browser caches, etc.).
   (b) Make sure that the "computer" does not have write access to any important resources. For example, the primary data repository. Assume that anything you give to your collaborator will unintentionally be loaded with every piece of malware you can think of—not every collaborator is this bad, but enough are.
   (c) Resources cost money! You will want to shut down the instances when you are not using them and free up and hard drive space you have allocated that you no longer are using.

*3.11.4 Cheap Laptops and Cloud Resources Make for Safer and More Robust Bench-Side Computing*

Within my laboratory, it is becoming more common for students and researchers to desire to use their laptops at the bench. This is very convenient for real time documentation as well as access to protocols and Internet based resources. However, to use the same laptop inside and outside of the laboratory presents safety concerns that need to be addressed. This can be addressed with cloud-based resources.

1. Acquire an inexpensive laptop. This laptop will be dedicated to use within the laboratory. We currently use Chromebooks for this task; if this works with your policies and workflows, it allows for people to easily replicate their environment and access resources across several laptops (e.g., one inside and one outside the lab). Further Chromebooks support Microsoft RDP.
2. As necessary, use the laptop to connect to an AWS instance to record or analyze data and protocols. Since the AWS instance is accessible from any other computer, it allows the user to have access to the same environment and data inside and outside the lab without the risk of carrying a contaminated device.

## 4 Notes

1. For perspective, I often perform remote administration via a low-end Chromebook (Intel Celeron, 4 GB Ram) connected over a 4G hotspot (1.5 Mbps up/10 Mbps down). While sufficient, improved network performance certainly does not hurt.
2. Throughout this chapter, for simplicity, I will often refer to "hard drives" when referring to the storage available to your cloud instance. Technology-wise, storage can either be SSD or magnetic based. For large bulk storage (e.g., data files) the magnetic drives will typically be cheaper. The computer "boot drives" in Amazon need to be SSD based. When you request a storage volume and attach it to your instance, it functionally behaves very much like a physical "hard drive"; thus, aside from performance choices (such as SSD vs magnetic) the underlying implementation details are not important. The suite of storage solutions is referred to as **EBS**, the virtual hard drives are referred to as **volumes**.
3. Superficially, a "keypair" is a cryptographic concept that references a mathematical system in which one number of the pair can encrypt messages to the holder of the second number of the pair and vice versa. When you create the keypair, Amazon keeps one and gives you the other. Now, when Amazon wants to send you information (like a password) it uses this system.

Since Amazon servers do not keep the half of the keypair they sent you, it can always be used as a way to authenticate yourself back to Amazon.

4. It is worth checking the AWS cost calculator [1] when configuring instances. You can also check https://docs.aws.amazon.com/AWSEC2/latest/UserGuide/instance-types.html for a complete description of all the instances and what their nuances of difference are. As an example, costs can be different depending on what region you instantiate your images; the free tier eligible configurations are not always the cheapest options when you have to pay—for example, t2.micro is "Free tier" but t3 are often cheaper than t2 even though next generation.
5. The default configuration allows all outgoing connections and only allows incoming connections using a computer remote control protocol, RDP, from all addresses (denoted by the 0.0.0.0/0). If all connections to your resource will come from a specific computer or collection of computers, you might want to limit the incoming RDP connections to be from a specific machine or subnet on the Internet. As an example, if I am coming from a machine with a known IP address of 128.125.1.15, I could add a rule that restricts incoming access to 128.125.1.15/32. If I am coming from any one of the machines on the 128.125.0.0 subnet, I could specify that as 128.125.0.0/16. There are many subnet mask calculators online that will help you format these correctly for other use cases [5]. You can additionally add rules here to allow other types of incoming connections (e.g., if you were running a MaxQuant server).
6. Hard drives need to be initialized and partitioned. Partitioning allows you to make a single hard drive appear as multiple virtual hard drives. Initialization takes a raw device and adds the minimal data structure required for the operating system to start using it—this includes the definition of the partition table. You have two options when you initialize your hard drive—MBR and GPT. MBR is the older, and traditionally more universally recognized format. However, there are some limitations—MBR can only have 4 physical partitions, but, even worse, MBR can only operate, assuming some common assumptions, with "hard drives" that are 2 TB or smaller. I tend to select MBR unless my operational needs require GPT (e.g., larger volumes).
7. To change the Google FileStream cache location in Windows, you are looking to change the registry value of ContentCachePath the process of which is described here: https://support.google.com/a/answer/7644837?hl=en. I recommend creating **D:\GoogleFSCache** and setting ContentCachePath to

point to that directory. To edit the registry values, click on the **Windows Button**, type **regedit** then navigate through **HKEY_LOCAL_MACHINE**, **SOFTWARE**, **Google**, **DriveFS**. When you get to **DriveFS**, you will click on the name rather than expanding it. Then do **Edit** → **New** → **String Value** and create **ContentCachePath**. If you double click on this, you can set it to your new cache directory (in my case, D:\GoogleFSCache). You will now need to reboot your instance or restart Google File Stream. Reboots are easier to execute so I will assume you did that; in either case, when FileStream next comes online, you should see a confirmatory notice that your cache has moved.

8. With storage, you are billed for the disk space you consume—when you originally create a snapshot, no additional space is required (it is all part of the original volume). Only if you make changes to the volume, then the snapshot has to store the differences and your storage costs go up. If you delete the underlying volume, then the snapshot will consume as much space as that volume did.

## Acknowledgments

I thank David B. Agus and the USC Center for Applied Molecular Medicine for generous institutional support and Naim Matasci for insightful discussions.

### References

1. https://calculator.s3.amazonaws.com/index.html
2. https://aws.amazon.com/premiumsupport/knowledge-center/accepted-payment-methods/
3. https://anydesk.com
4. https://www.teamviewer.com
5. https://www.calculator.net/ip-subnet-calculator.html

# Index



Arnaud Delobel (ed.), *Mass Spectrometry of Glycoproteins: Methods and Protocols*, Methods in Molecular Biology, vol. 2271, https://doi.org/10.1007/978-1-0716-1241-5, © Springer Science+Business Media, LLC, part of Springer Nature 2021